LONDON MATHEMATICAL SOCIETY LECTURE NOTI

Managing Editor: Professor Endre Süli, Mathematical Institute, Univ
Woodstock Road, Oxford OX2 6GG, United Kingdom

The titles below are available from booksellers, or from Cambridge U
www.cambridge.org/mathematics

London Mathematical Society Lecture Note Series: 491

Groups and Graphs, Designs and Dynamics

Edited by

R. A. BAILEY
University of St Andrews, Scotland

PETER J. CAMERON
University of St Andrews, Scotland

YAOKUN WU
Shanghai Jiao Tong University, China

CAMBRIDGE
UNIVERSITY PRESS

Shaftesbury Road, Cambridge CB2 8EA, United Kingdom

One Liberty Plaza, 20th Floor, New York, NY 10006, USA

477 Williamstown Road, Port Melbourne, VIC 3207, Australia

314–321, 3rd Floor, Plot 3, Splendor Forum, Jasola District Centre, New Delhi – 110025, India

103 Penang Road, #05–06/07, Visioncrest Commercial, Singapore 238467

Cambridge University Press is part of Cambridge University Press & Assessment, a department of the University of Cambridge.

We share the University's mission to contribute to society through the pursuit of education, learning and research at the highest international levels of excellence.

www.cambridge.org
Information on this title: www.cambridge.org/9781009465953

DOI: 10.1017/9781009465939

When citing this work, please include a reference to the DOI 10.1017/9781009465939

First published 2024

Printed in the United Kingdom by TJ Books Limited, Padstow Cornwall

A catalogue record for this publication is available from the British Library

A Cataloging-in-Publication data record for this book is available from the Library of Congress

ISBN 978-1-009-46595-3 Paperback

Contents

Authors

R. A. Bailey
University of St Andrews, St Andrews, Fife, U.K.
rab24@st-andrews.ac.uk

Mike Boyle
University of Maryland, College Park, MD, USA
mboyle@umd.edu

Peter J. Cameron
University of St Andrews, St Andrews, Fife, U.K.
pjc20@st-andrews.ac.uk

Tullio Ceccherini-Silberstein
Università del Sannio, Benevento, Italy
tullio.cs@sbai.uniroma1.it

Nobuaki Obata
Tohoku University, Sendai, Japan
obata@tohoku.ac.jp

Fabio Scarabotti
Università degli Studi di Roma "La Sapienza", Roma, Italy
fabio.scarabotti@sbai.uniroma1.it

Scott Schmieding
Pennsylvania State University, State College, PA, USA
sks7247@psu.edu

Filippo Tolli
Università Roma Tre, Roma, Italy
filippo.tolli@uniroma3.it

Preface

The conference and PhD–Masters summer school G2D2 (Groups and Graphs, Designs and Dynamics), part of the G2 series of conferences[1], was held in the Three Gorges Mathematical Research Center in Yichang, China, in 2019.

The material presented in the conference provided participants with a Chinese banquet of beautiful and important mathematics. The main courses were short lecture courses on the four topics in the title of the conference, records of which make up the present volume. In addition there were introductory lectures on each of the four topics, together with many invited and contributed talks. We would like to invite you warmly to participate in this banquet.

The base of a Chinese banquet is rice, and this role is played here by linear algebra. Dating from the nineteenth century, linear algebra is now a central part of mathematics and is fundamental to all the areas described here; indeed, it forms a unifying thread running through four chapters describing very different parts of mathematics.

The basic interpretation of a square matrix in linear algebra is that it represents a linear transformation from a vector space to itself; we are concerned with properties of the transformation independent of the means of representation. A natural extension of this is to consider a set of such matrices, and there are good reasons for taking the set to be a group; this leads us to the important topic of group representation. Its importance is indicated by comparing the two editions of William Burnside's book on finite groups, the first such book in English. In the first edition, he explains why he has omitted all mention of representation theory, since "it would be difficult to find a result that could be most directly obtained by the consideration of groups of linear transformations." However, by the second edition, he explains how recent re-

---

[1] The history of this series of conferences, designed to encourage international cooperation, can be found at https://ekonsta.github.io/Slides/G2-series.pdf

sults (by Frobenius, Schur and Burnside himself) have led him to change his opinion, and devote more than half the book to representation theory.

The chapter by Tullio Ceccherini-Silberstein, Fabio Scarabotti and Filippo Tolli is an exposition of representation theory, strongly influenced by its applications to harmonic analysis, random walks, Gelfand pairs, and representations of general linear groups.

In general, the concept of positivity links algebra and analysis. Equipped with a ∗-operation (involution), an algebra allows us to define a positive linear functional, which plays the role of expectation (mean value) in probability theory. In this aspect, the idea of algebraic probability emerges and provides a quantum probabilistic approach to the study of discrete structures. The adjacency matrix of a simple graph is a real symmetric nonnegative matrix, so its spectrum is an important isomorphism invariant, which carries important information about the graph. The chapter by Nobuaki Obata shows how the spectra of graphs in various families which have a limit object can be calculated, using the techniques of quantum probability theory.

The purpose of statistics is to derive useful information from data. If the data come from an experiment, the scientist has the option of using a design that will more efficiently yield the information, because the random variables representing estimated values of the parameters have smallest possible variance. Since this is a multidimensional optimization problem, there is no unique solution, and several optimality measures have been proposed. The chapter by R. A. Bailey and Peter Cameron shows that, in the case where a block design is used, all common optimality criteria can be expressed in terms of the spectrum of the Laplacian matrix of a certain graph, the concurrence graph of the design. The Laplacian has applications in many other fields such as electrical networks and random walks, and these applications are also described.

Symbolic dynamics grew from the work of Poincaré in the early twentieth century; for some questions, complicated continuous dynamical systems can be replaced by subsystems of the shift operation on the set of infinite sequences over a finite alphabet. In this case, unlike the others, the linear algebra required is not "off-the-shelf", since it involves matrices over a semiring such as the natural numbers. This leads naturally to the concepts of stability and K-theory, all of which are explained in detail in the chapter by Mike Boyle and Scott Schmieding, after an introduction to topological dynamics.

The authors have all been aware of connections between their topics and other parts of mathematics, whether these be representation theory and Markov chains, graph spectra and quantum probability, optimal designs and electrical circuits, or symbolic dynamics and algebraic K-theory.

We thank our authors for their work on the project, and we hope that bringing these surveys on different topics into a single book will help highlight the similarities between them as well as their individual aesthetic attraction. On this, let us quote four mathematicians:

I am just a student all my life. From the very beginning of my life I was trying to learn. And for example now, when listening to the talks and reading notes of this conference, I discover how much I still do not know and have to learn. Therefore, I am always learning. In this sense I am a student. Never a "Führer".

Israel M. Gelfand, a dinner talk in the conference "The Unity of Mathematics"

The cross-fertilization of ideas is crucial for the health of the science and mathematics.

Mikhael Gromov, Possible Trends in Mathematics in the Coming Decades, an open letter addressed to the French mathematical community in 1998

Mathematics has three purposes. It delivers an instrument for the study of nature, but this is not all. It has a philosophical purpose and, I dare say, an aesthetic purpose. These purposes can not be separated and the best way to achieve one purpose is to aim at the other ones, or at least not to lose sight of them.

Henri Poincaré

My own mathematical works are always quite unsystematic, without mode or connection. Expression and shape are almost more to me than knowledge itself. But I believe that, leaving aside my own peculiar nature, there is in mathematics itself, in contrast to the experimental disciplines, a character which is nearer to that of free creative art.

Hermann Weyl

We hope that our readers will experience some of what these mathematicians describe.

The chapters of the book are in the order of the subjects in the title: Groups, Graphs, Designs, Dynamics. Although all have their roots in linear algebra, and there are a number of cross-references, we stress that the chapters are independent, and can be read in any order. Also, many topics, ranging from major fields like probability theory to niche subjects like distance-regular graphs, recur in several chapters. We invite readers who are expert in one of our four topics to explore some of the others.

According to Donald Knuth in *The Art of Computer Programming*,

The book *Dynamic Programming* by Richard Bellman is an important, pioneering work in which a group of problems is collected together at the end of some chapters under the heading "Exercises and Research Problems," with extremely trivial questions appearing in the midst of deep, unsolved problems. It is rumored that someone once asked

Dr. Bellman how to tell the exercises apart from the research problems, and he replied: "If you can solve it, it is an exercise; otherwise it's a research problem."

We have not inflicted this on our readers: exercises and problems are clearly labelled. But some authors have scattered exercises through the text, while others have gathered them at the end of sections. In particular, in the first chapter, the exercises form an integral part of the development of the theory. In general, we urge readers to try the exercises for themselves, as students at the original courses were encouraged to do.

We want to express our gratitude to many people who have made this book possible, especially the authors for the care and attention with which they have produced their chapters. We thank Xiaoli Wu at Higher Education Press and David Liu and Anna Scriven at Cambridge University Press for help in turning four individual chapters into a coherent book. For the G2D2 event, we thank Xuemei Deng, Zongzhu Lin, Shuang Yi from Three Gorges Mathematical Research Center and Sergey Goryainov, Da Zhao, Yinfeng Zhu from Shanghai Jiao Tong University; we also acknowledge financial support from Three Gorges Mathematical Research Center, National Natural Science Foundation of China and International Linear Algebra Society.

R. A. Bailey, Peter J. Cameron, Yaokun Wu

1

Topics in representation theory of finite groups

Tullio Ceccherini-Silberstein, Fabio Scarabotti and Filippo Tolli

Abstract

This is an introduction to representation theory and harmonic analysis on finite groups. This includes, in particular, Gelfand pairs (with applications to diffusion processes *à la* Diaconis) and induced representations (focusing on the little group method of Mackey and Wigner). We also discuss Laplace operators and spectral theory of finite regular graphs. In the last part, we present the representation theory of $GL(2, \mathbb{F}_q)$, the general linear group of invertible 2×2 matrices with coefficients in a finite field with q elements. More precisely, we revisit the classical Gelfand–Graev representation of $GL(2, \mathbb{F}_q)$ in terms of the so-called multiplicity-free triples and their associated Hecke algebras. The presentation is not fully self-contained: most of the basic and elementary facts are proved in detail, some others are left as exercises, while, for more advanced results with no proof, precise references are provided.

Keywords: finite group, group representation, character, Gelfand pair, spherical function, spherical Fourier transform, Mackey–Wigner little group method, Markov chain, random walk, Ehrenfest diffusion process, ergodic theorem, finite graph, spectral graph theory, Laplace operator, distance-regular graph, strongly regular graph, association scheme, finite field, affine group over a finite field, general linear group over a finite field, Gelfand–Graev representation, multiplicity-free triple, Hecke algebra

Mathematics Subject Classification: 20C15, 20C08, 20C30, 20C35, 20G05, 05C50, 43A35, 43A65, 43A30, 43A90

1.1 Introduction

The present text constitutes an expanded and more detailed exposition of the lecture notes of a course on Representation Theory delivered by the first named author at the International Conference and PhD-Master Summer School on Groups and Graphs, Designs and Dynamics (G2D2) held in Yichang (China) in August 2019.

One of the main features of Harmonic Analysis is the study of linear operators that are invariant with respect to the action of a group. In the classical abelian setting, for instance, this is used to express the solutions of a constant coefficients differential equation (such as the heat equation) in terms of infinite sums of exponentials (Fourier series).

Here, we consider a finite (possibly non-abelian) counterpart. Let G be a finite group, let $K \leq G$ be a subgroup, and consider the G-module $L(G/K)$ of all complex valued functions on the (finite) homogenous space G/K of left cosets of K in G. The corresponding space of linear G-invariant operators we alluded to above, the so-called commutant $\text{End}_G(L(G/K))$, bears a natural structure of an involutive unital algebra that turns out to be isomorphic to the algebra $^K L(G)^K$ of all bi-K-invariant complex valued functions on G. When these algebras are commutative, we say that (G,K) is a Gelfand pair: the terminology originates from the seminal paper by I. M. Gelfand [40] in the setting of Lie groups. Finite Gelfand pairs, when G is a Weyl group or a Chevalley group over a finite field, or the symmetric group $S_n = \text{Sym}(\{1,2,\ldots,n\})$, were studied by Ph. Delsarte [25], motivated by applications to association schemes of coding theory, Ch F. Dunkl [30, 31, 32, 33] and D. Stanton [67] with relevant contributions to the theory of special functions, E. Bannai and T. Ito [3] who initiated Algebraic Combinatorics, J. Saxl [59] in the study of Finite Geometries and Designs, and A. Terras [69] with applications to number theory. A special mention deserves the work in Probability Theory by P. Diaconis and collaborators [26] with remarkable applications to the study of diffusion processes and asymptotic behaviour of finite Markov chains. A. Okounkov and A. M. Vershik [55] (see also [16]) used methods from the theory of finite Gelfand pairs in order to give a new approach to the representation theory of the symmetric groups. Further expositions of the theory of finite Gelfand pairs and association schemes can be found in the monographs by R. A. Bailey [2], P.-H. Zieschang [73], as well as in the survey paper [14] and in our first monograph [15]. We conclude this bibliographical overview by mentioning the work of R. I. Grigorchuk [43] (see also [5, 23, 24]) in connection with the theory of the so-called self-similar groups.

Given a Gelfand pair (G,K), the simultaneous diagonalization of all G-

invariant operators can be achieved by means of a particular basis of $^K L(G)^K$. The elements of this basis, called spherical functions, are the analogues of the exponentials in the classical case and can be defined both intrinsically and as matrix coefficients of particular representations (the spherical representations). Besides the trivial though interesting case when the group G is abelian, an important example of a Gelfand pair is given by $(G \times G, \widetilde{G})$, with $\widetilde{G}$ the diagonal subgroup: in this case, the spherical functions are nothing but the normalized characters of G, showing that the theory of central functions on a group can be treated in the setting of the Gelfand pairs, as a particular case.

By virtue of the Ergodic Theorem, the rate of convergence to the stationary distribution of the n-step distributions μ_n of a finite (ergodic and symmetric) Markov chain can be estimated in terms of the second largest eigenvalue modulus of the corresponding transition matrix. An example of a Gelfand pair is $(S_n, S_k \times S_{n-k})$, where $S_n = \mathrm{Sym}(\{1, 2, \ldots, n\})$ is the symmetric group of degree n, and, for $1 \leq k \leq n/2$, we regard $S_k = \mathrm{Sym}(\{1, 2, \ldots, k\})$ and $S_{n-k} = \mathrm{Sym}(\{k+1, k+2, \ldots, n\})$ as subgroups of S_n. In the 80s Diaconis and Shahshahani [28] (see also [14, 15]), were able to use this Gelfand pair to find very precise asymptotics of $(\mu_n)_{n \in \mathbb{N}}$ for the Bernoulli–Laplace model of diffusion. In particular, they showed that an interesting phenomenon occurs: the transition from order to chaos is concentrated in a relatively small interval of time: this is the *cut-off phenomenon*. Other important examples, where the theory of spherical functions plays a central role, are the Ehrenfest model of diffusion (see Section 1.5.2) and the random transpositions model [26, 27, 14, 15].

The G-module $L(G/K)$ can be seen as the representation space of the induced representation $\mathrm{Ind}_K^G \iota_K$ of the trivial representation ι_K of K, and we have that (G, K) is a Gelfand pair if and only if $\mathrm{Ind}_K^G \iota_K$ decomposes without multiplicity. More generally, if θ is an irreducible K-representation, the algebra $\mathrm{End}_G(\mathrm{Ind}_K^G \theta)$ of intertwiners is isomorphic to a suitable convolution algebra $\mathscr{H}(G, K, \theta)$ of complex valued functions on G, and we say that (G, K, θ) is a multiplicity-free triple if these algebras are commutative; equivalently, if $\mathrm{Ind}_K^G \theta$ decomposes without multiplicity. Multiplicity-free triples were partially studied by I. G. Macdonald [50], by D. Bump and D. Ginzburg [9], and in [19, Chapter 13] when $\dim \theta = 1$; a generalization to higher dimensions, with a complete analysis of the spherical functions, is treated in our papers [61, 62, 63, 64] and the recent monograph [20]. An earlier application, where a problem of Diaconis on the Bernoulli-Laplace diffusion model with many urns was solved, was presented in the second named author's PhD thesis and published in [60]. As pointed out in [19, Chapter 14], our theory of multiplicity-free triples shed light on the representation theory of $\mathrm{GL}(2, \mathbb{F}_q)$, the general linear group of 2×2 matrices with coefficients in the field with q elements, as developed by I. I. Piatetski-Shapiro in [57].

These lecture notes are organized as follows. In Section 1.2, we briefly recall the basics of the representation theory of finite groups: this includes Schur's lemma, some character theory, and the Peter–Weyl theorem. In Sections 1.2.2, 1.2.3, and 1.2.4 we study Gelfand pairs in detail, focusing on spherical functions, the spherical Fourier transform, and the harmonic analysis of invariant operators. Then, in Sections 1.5.1 and 1.5.2 we present the applications of Gelfand pairs to Markov chains, culminating in the celebrated Diaconis–Shahshahani upper-bound lemma, and describe the asymptotics for the Ehrenfest model of diffusion. In Sections 1.6.1 and 1.6.2 we study induced representations, Frobenius reciprocity, and Mackey theory, and then, in Section 1.6.3, we apply this machinery to obtain the Mackey–Wigner little group method. In Section 1.6.4 we introduce the Hecke algebras $\widetilde{\mathscr{H}}(G, K, \theta)$ and $\mathscr{H}(G, K, \theta)$ and show that they are both isomorphic to the commutant $\mathrm{End}_G(\mathrm{Ind}_K^G \theta)$. In Section 1.6.5 we then define multiplicity-free triples and present their general theory. After a short overview of the basics of finite fields and their characters (Section 1.7.1), as an application of the little group method of Mackey and Wigner we describe all irreducible representations of $\mathrm{Aff}(\mathbb{F}_q)$, the affine group over the field with q elements. The last two sections are devoted to the general linear group $\mathrm{GL}(2, \mathbb{F}_q)$ and its representations: in relation with the latter, we limit ourselves to the description of the decomposition of the Gelfand–Graev representation.

Our presentation is mostly self-contained. However, for the sake of brevity, some of the proofs are either omitted (but with clear references for a complete exposition), or sketched, or left as an exercise to the reader. Several other exercises are proposed as complements and further developments.

Acknowledgments. We express our deep gratitude to Yaokun Wu and Da Zhao for many valuable comments and remarks. We also thank Rosemary Bailey and Peter Cameron for their most precious help and concern in the editing process.

1.2 Representation theory and harmonic analysis on finite groups

In this section, we present the basics of the representation theory of finite groups and we introduce and study the notion of a finite Gelfand pair, thus providing a setting for a suitable extension of the classical Fourier analysis.

Our exposition is inspired by Diaconis' book [26] and to Figà–Talamanca's

lecture notes [37] and our monographs [15, 19]. We also took a particular bene-
fit from the monographs by Alperin and Bell [1], Fulton and Harris [39], Isaacs
[46], Naimark and Stern [52], Serre [65], Simon [66], and Sternberg [68]. Expositions of the theory of Gelfand pairs are also presented in the monographs
by J. Dieudonné [29], H. Dym and H. P. McKean [34] , J. Faraut [36], A. Figà-
Talamanca and C. Nebbia [38], S. Helgason [44] and J. Wolf [71] for the general case of locally compact groups.

1.2.1 Representations

Let G be a finite group.

Definition 1.2.1 (Representation)   A *representation* of G (also called a G-
representation) is a pair (ρ, V), where V is a finite dimensional complex vector
space and $\rho\colon G \to \mathrm{GL}(V)$ is a group homomorphism from G into the group
$\mathrm{GL}(V)$ of all *invertible linear transformations* of V.

If (ρ, V) is a representation of G, then one has:

- $\rho(1_G) = I_V$
- $\rho(g_1 g_2) = \rho(g_1)\rho(g_2)$
- $\rho(g^{-1}) = \rho(g)^{-1}$
- $\rho(g)(av + bw) = a\rho(g)v + b\rho(g)w$

for all $g, g_1, g_2 \in G$, $v, w \in V$, and $a, b \in \mathbb{C}$, where $1_G \in G$ is the identity element
and $I_V\colon V \to V$ is the identity transformation.

Equivalently, a representation can be viewed as an *action* $\alpha\colon G \times V \to V$ of
G on V by linear transformations by setting $\alpha(g, v) := \rho(g)v$ for all $g \in G$ and
$v \in V$.

In the following, for the sake of brevity, when a given representation (ρ, V)
is clear from the context, we shall denote it simply by either ρ or V.

The dimension $d_\rho := \dim V$ of the vector space V is called the *dimension*
of ρ.

Definition 1.2.2 (Sub-representation)   Let (ρ, V) be a G-representation. A
subspace $W \leq V$ is G-*invariant* if $\rho(g)w \in W$ for all $g \in G$ and $w \in W$. The
pair (ρ_W, W), where $\rho_W(g) := \rho(g)|_W$ for all $g \in G$, is a G-representation,
called a *sub-representation* of (ρ, V). We shall then write $(\rho_W, W) \leq (\rho, V)$.

Clearly, $d_{\rho_W} \leq d_\rho$.

Definition 1.2.3 (Irreducible representation)   A G-representation (ρ, V) is *irreducible* if V admits no nontrivial G-invariant subspaces, that is, the only G-
invariant subspaces $W \leq V$ are $W = \{0\}$ and $W = V$.

We denote by $\mathrm{Irr}(G)$ the set of all irreducible representations of G.

The representation of dimension zero is considered to be neither reducible nor irreducible, just as the number 1 is considered to be neither composite nor prime.

It is obvious that every one-dimensional representation is irreducible.

Definition 1.2.4 (Equivalent representations)   Two G-representations (ρ_1, V_1) and (ρ_2, V_2) are *equivalent* if there exists a linear isomorphism $T: V_1 \to V_2$ such that

$$T \circ \rho_1(g) = \rho_2(g) \circ T$$

for all $g \in G$. We then write $\rho_1 \sim \rho_2$. We shall refer to T as to an *intertwining isomorphism*.

If (ρ_1, V_1) is equivalent to a sub-representation of (ρ_2, V_2) we write $\rho_1 \preceq \rho_2$, and we say that ρ_1 is *contained* in ρ_2.

Note that $\sim$ is an equivalence relation in the set of all G-representations, which preserves irreducibility and dimension (exercise).

Definition 1.2.5 (Unitary representation)   Suppose that a complex vector space V is equipped with an inner product $\langle \cdot, \cdot \rangle_V$. A G-representation (ρ, V) is *unitary* if, for every $g \in G$, the linear operator $\rho(g)$ is unitary, that is,

$$\langle \rho(g)v_1, \rho(g)v_2 \rangle_V = \langle v_1, v_2 \rangle_V$$

for all $v_1, v_2 \in V$.

Note that, if (ρ, V) is a unitary representation, then

• $\rho(g^{-1}) = \rho(g)^*$

for all $g \in G$, where * denotes the *adjoint* operation.

Exercise 1.2.6 (Unitarizability of representations)   Suppose that a complex vector space V is equipped with an inner product $\langle \cdot, \cdot \rangle_V$. Let (ρ, V) be a G-representation. Then when equipping V with the new inner product $(\cdot, \cdot)_V$ defined by

$$(v_1, v_2)_V := \frac{1}{|G|} \sum_{g \in G} \langle \rho(g)v_1, \rho(g)v_2 \rangle_V$$

for all $v_1, v_2 \in V$, the representation (ρ, V) becomes unitary.

By virtue of the previous exercise, from now on, we shall consider only unitary representations. This will not affect equivalence as the next exercise shows.

Exercise 1.2.7 Let (ρ_1, V_1) and (ρ_2, V_2) be two unitary G-representations. Suppose that $\rho_1 \sim \rho_2$. Then there exists a unitary operator $U : V_1 \to V_2$ such that

$$U \circ \rho_1(g) = \rho_2(g) \circ U$$

for all $g \in G$.

Hint: Use the *polar decomposition* $T = U|T|$ for an intertwining isomorphism $T : V_1 \to V_2$ (for more details, see [19, Lemma 10.1.4]).

We can rephrase the result in the above exercise by saying that two equivalent unitary representations are *unitarily equivalent*.

Definition 1.2.8 (Dual of a group) The *dual* of the group G is the quotient $\widehat{G} := \mathrm{Irr}(G)/\sim$. In the following we shall also refer to $\widehat{G}$ as to a complete set of irreducible pairwise non-equivalent G-representations.

We shall see later (cf. Theorem 1.2.36) that $|\widehat{G}| < \infty$.

Definition 1.2.9 (Direct sum) Let (ρ_1, V_1) and (ρ_2, V_2) be two G-representations. We equip $V := V_1 \oplus V_2$ with the inner product $\langle \cdot, \cdot \rangle_V$ defined by setting

$$\langle v_1 + v_2, v_1' + v_2' \rangle_V := \langle v_1, v_1' \rangle_{V_1} + \langle v_2, v_2' \rangle_{V_2}$$

for all $v_1, v_1' \in V_1$ and $v_2, v_2' \in V_2$. The (unitary) G-representation (ρ, V) defined by setting

$$\rho(g)(v_1 + v_2) := \rho_1(g)v_1 + \rho_2(g)v_2$$

for all $g \in G$ and $v_1 \in V_1$, $v_2 \in V_2$, is called the *direct sum* of (ρ_1, V_1) and (ρ_2, V_2) and is denoted by $(\rho_1 \oplus \rho_2, V_1 \oplus V_2)$.

Note that $d_{\rho_1 \oplus \rho_2} = d_{\rho_1} + d_{\rho_2}$ and that $\rho_i \preceq \rho_1 \oplus \rho_2$ for $i = 1, 2$.

Definition 1.2.10 (Conjugate representation) Let (ρ, V) be a G-representation and let V' denote the dual vector space. The *conjugate representation* of ρ is the unitary representation (ρ', V') defined by setting

$$[\rho'(g)f](v) := f(\rho(g^{-1})v)$$

for all $g \in G$, $f \in V'$, and $v \in V$.

It is an exercise to check that ρ' is unitary (resp. irreducible) if and only if ρ is unitary (resp. irreducible).

Exercise 1.2.11 (Orthogonal complement)   Suppose that (ρ, V) is a (unitary) G-representation and let $W \leq V$ be a nontrivial G-invariant subspace. Show that

$$W^{\perp} = \{v \in V : \langle v, w \rangle_V = 0 \text{ for all } w \in W\}$$

is also G-invariant. Deduce that $\rho = \rho_W \oplus \rho_{W^{\perp}}$.

From the above exercise and an obvious inductive argument, one immediately deduces the following:

Theorem 1.2.12   *Every G-representation is the direct sum of finitely many irreducible G-representations.* $\square$

The above theorem may be rephrased as follows. Suppose that (ρ, V) is a G-representation. Then there exist a positive integer n and (not necessarily distinct) $\rho_1, \rho_2, \ldots, \rho_n \in \widehat{G}$ such that $\rho \sim \rho_1 \oplus \rho_2 \oplus \cdots \oplus \rho_n$.

Example 1.2.13 (Trivial representation)   The *trivial representation* of a group G, denoted $(\iota_G, \mathbb{C})$, is the one-dimensional representation defined by setting $\iota_G(g) = \text{Id}_{\mathbb{C}}$ for all $g \in G$.

Given a finite group G, we denote by $L(G)$ the complex vector space of all functions $f: G \to \mathbb{C}$. We equip $L(G)$ with the *convolution product* $*$ defined by setting, for $f_1, f_2 \in L(G)$,

$$(f_1 * f_2)(g) = \sum_{h \in G} f_1(gh^{-1}) f_2(h) \quad \text{for all } g \in G. \tag{1.1}$$

With the product $*$, the space $L(G)$ becomes an algebra, called the $\mathbb{C}$-*group algebra* of G. Note that $L(G)$ is unital, with unity element δ_{1_G}. Moreover, the map $f \mapsto f^*$, where $f^*(g) := \overline{f(g^{-1})}$ for all $g \in G$, is an involution.

Example 1.2.14 (Regular representations)   Let G be a finite group. Then the *left* (resp. *right*) *regular representation* of G is the (unitary) representation $(\lambda_G, L(G))$ (resp. $(\rho_G, L(G))$) defined by setting

$$[\lambda_G(g)f](h) = f(g^{-1}h) \quad (\text{resp. } [\rho_G(g)f](h) = f(hg))$$

for all $g, h \in G$ and $f \in L(G)$.

Exercise 1.2.15   Show that the left (resp. right) regular representation is unitary when $L(G)$ is endowed with the scalar product $\langle \cdot, \cdot \rangle_{L(G)}$ defined by setting

$$\langle f_1, f_2 \rangle_{L(G)} := \sum_{g \in G} f_1(g) \overline{f_2(g)}$$

for all $f_1, f_2 \in L(G)$.

Example 1.2.16 (Representations of a cyclic group)   Let

$$G = C_n = \{1, a, a^2, \ldots, a^{n-1}\} \cong \mathbb{Z}/n\mathbb{Z}$$

denote the *cyclic group* of order n. Consider the primitive nth root of unity $\omega :=
e^{2\pi i/n}$ and, for $k \in \mathbb{Z}$, let $(\rho_k, \mathbb{C})$ denote the (unitary) representation defined by

$$\rho_k(a^h) = \omega^{kh} \mathrm{Id}_{\mathbb{C}}$$

for all $h = 0, 1, \ldots, n-1$. Note that $\rho_k = \rho_{k'}$ if $k \equiv k' \bmod n$ and that $\rho_k \not\sim \rho_{k'}$
if $k \not\equiv k' \bmod n$. In fact, $\widehat{C_n} = \{\rho_k : k = 0, 1, \ldots, n-1\}$.

Example 1.2.17 (Two particular representations of the symmetric group)   Let
$G = S_n = \mathrm{Sym}(\{1, 2, \ldots, n\})$ denote the *symmetric group* of degree n, that is
the group of all bijective maps (*permutations*) $g \colon \{1, 2, \ldots, n\} \to \{1, 2, \ldots, n\}$.

The *sign representation* of S_n is the one-dimensional representation $(\mathrm{sign}, \mathbb{C})$
defined by

$$\mathrm{sign}(g) = \begin{cases} \mathrm{Id}_{\mathbb{C}} & \text{if } g \in A_n \\ -\mathrm{Id}_{\mathbb{C}} & \text{if } g \in S_n \setminus A_n \end{cases}$$

for all $g \in S_n$, where $A_n \leq S_n$ is the *alternating subgroup* (consisting of all
permutations which can be expressed as a product of an even number of transpositions).

Let V be an n-dimensional vector space equipped with a scalar product.
Fix an orthonormal basis $\{e_1, e_2, \ldots, e_n\} \subset V$. The *permutation representation*
of S_n (cf. Definition 1.2.50) is the (unitary) representation (ρ, V) defined by
setting

$$\rho(g)e_i = e_{g(i)}$$

for all $g \in S_n$ and $i = 1, 2, \ldots, n$.

Exercise 1.2.18   Let $G = S_n$ be the symmetric group of degree n.

Show that the sign representation $(\mathrm{sign}, \mathbb{C})$ is indeed a unitary representation.

Show that the permutation representation (ρ, V) is indeed a unitary representation. Let $W \leq V$ denote the one-dimensional subspace spanned by the
vector $e_1 + e_2 + \cdots + e_n$. Show that W is G-invariant. Show that

$$W^\perp = \{\sum_{i=1}^n \alpha_i e_i : \alpha_i \in \mathbb{C} \text{ and } \alpha_1 + \alpha_2 + \cdots + \alpha_n = 0\}$$

is equal to the linear span of $\{e_i - e_{i-1} : i = 2, 3, \ldots, n\}$, and is irreducible. Deduce that $V = W \oplus W^\perp$ is the decomposition of V into irreducible components.

Definition 1.2.19 (Commutant)   The *commutant* of two G-representations (ρ_1, V_1) and (ρ_2, V_2) is the vector space

$$\mathrm{Hom}_G(V_1, V_2) := \{T : V_1 \to V_2 : T \text{ is linear and } T\rho_1(g) = \rho_2(g)T \text{ for all } g \in G\}.$$

We refer to its elements as to the *intertwiners* of ρ_1 and ρ_2. When $V_1 = V_2 = V$ we denote the *commutant* $\mathrm{Hom}_G(V, V)$ by $\mathrm{End}_G(V)$. It has a natural structure of an algebra.

Exercise 1.2.20   Let (ρ_1, V_1), (ρ_2, V_2), and (ρ, V) be unitary G-representations. Given $T \in \mathrm{Hom}_G(V_1, V_2)$, let $T^* : V_2 \to V_1$ denote the adjoint operator. Show that $T^* \in \mathrm{Hom}_G(V_2, V_1)$. Show that the commutant $\mathrm{End}_G(V)$ has a natural structure of a $*$-algebra.

The following is a celebrated, elementary but extremely useful result of Schur.

Lemma 1.2.21 (Schur's lemma)   *Let* (ρ_1, V_1) *and* (ρ_2, V_2) *be two irreducible G-representations. If* $T \in \mathrm{Hom}_G(V_1, V_2)$, *then either* $T = 0$ *or* T *is an isomorphism (and* $\rho_1 \sim \rho_2$*).*

Proof   The kernel $\ker T \leq V_1$ and the image $\mathrm{ran}\, T \leq V_2$ are G-invariant subspaces, and by the irreducibility of ρ_1 and ρ_2 they must be trivial. If $\ker T = \{0\}$, then $\mathrm{ran}\, T = V_2$ and therefore T is an isomorphism; and if $\ker T = V_1$, then $T \equiv 0$.   □

Corollary 1.2.22   *Let* (ρ, V) *be an irreducible G-representation and consider* $T \in \mathrm{End}_G(V)$. *Then* $T \in \mathbb{C}I_V$.

Proof   Let $\lambda \in \mathbb{C}$ be an eigenvalue of T, so that $T - \lambda I_V$ cannot be an isomorphism. As $T - \lambda I_V \in \mathrm{End}_G(V)$, Schur's lemma (Lemma 1.2.21) ensures that $T - \lambda I_V \equiv 0$, that is, $T = \lambda I_V$.   □

Exercise 1.2.23   Let G be a group. Show that if G is abelian and (ρ, V) is a G-representation, then ρ is irreducible if and only if $d_\rho = 1$. Show that, vice versa, if every irreducible G-representation is one-dimensional, then G is abelian. *Hint.* For the converse implication, use the following steps:

- A representation (ρ, V) of G is *faithful* provided that $\rho(g) \neq I_V$ for all $g \in G \setminus \{1_G\}$. Show that the regular representations (cf. Example 1.2.14) of G are faithful.
- Apply Theorem 1.2.12 to the left regular representation of G and deduce that for every $g \in G \setminus \{1_G\}$, there exists an irreducible representation (ρ_g, V_g) of G such that $\rho_g(g) \neq I_V$.

- If G is nonabelian, there exist $g_1, g_2 \in G$ such that $g_1 g_2 \neq g_2 g_1$. Use the previous step, with $g = g_1 g_2 g_1^{-1} g_2^{-1} \neq 1_G$ to show that (ρ_g, V_g) cannot be one-dimensional.

For an alternative solution, see Remark 1.2.43.

Definition 1.2.24 (Matrix coefficient)   Let (ρ, V) be a G-representation and $v, w \in V$. The *matrix coefficient* associated with the pair (v, w) is the function $u_{v,w}^{\rho} : G \to \mathbb{C}$ defined by setting

$$u_{v,w}^{\rho}(g) = \langle \rho(g)w, v \rangle_V \quad \text{for all } g \in G.$$

If $\{v_1, v_2, \ldots, v_{d_\rho}\}$ is a basis of V, the matrix coefficient u_{v_i, v_j}^{ρ} will be simply denoted by $u_{i,j}^{\rho}$. Observe that the matrix $(u_{i,j}^{\rho})_{i,j}$ is the matrix representing the operator $\rho(g) \in \text{End}(V)$ with respect to the basis $\{v_1, v_2, \ldots, v_{d_\rho}\}$.

Lemma 1.2.25 (Orthogonality relations)   *Let (ρ_1, V_1) and (ρ_2, V_2) be two irreducible G-representations and suppose that $\rho_1 \nsim \rho_2$. Then every matrix coefficient of ρ_1 is orthogonal to every matrix coefficient of ρ_2.*

Proof   Let $v_1, w_1 \in V_1$ and $v_2, w_2 \in V_2$ and define

$$
\begin{array}{cccc}
L: & V_1 & \longrightarrow & V_2 \\
 & v & \mapsto & \langle v, w_1 \rangle_{V_1} w_2
\end{array}
$$

and

$$\tilde{L} = \sum_{g \in G} \rho_2(g^{-1}) L \rho_1(g).$$

It is easy to check that $\tilde{L}$ belongs to $\text{Hom}_G(V_1, V_2)$ so that, by Schur's lemma, $\tilde{L} = 0$. Thus

$$
\begin{aligned}
0 = \langle \tilde{L} v_1, v_2 \rangle_{V_2} &= \sum_{g \in G} \langle L \rho_1(g) v_1, \rho_2(g) v_2 \rangle_{V_2} \\
&= \sum_{g \in G} \langle \rho_1(g) v_1, w_1 \rangle_{V_1} \langle w_2, \rho_2(g) v_2 \rangle_{V_2} \\
&= \sum_{g \in G} u_{w_1, v_1}^{\rho_1}(g) \overline{u_{w_2, v_2}^{\rho_2}(g)} \\
&= \langle u_{w_1, v_1}^{\rho_1}, u_{w_2, v_2}^{\rho_2} \rangle_{L(G)}.
\end{aligned}
$$

This shows that the matrix coefficients $u_{w_1, v_1}^{\rho_1}$ and $u_{w_2, v_2}^{\rho_2}$ are orthogonal. $\square$

Lemma 1.2.26   *Let (ρ, V) be an irreducible G-representation. If $\{v_1, \ldots, v_{d_\rho}\}$ is an orthonormal basis of V, then one has*

$$\langle u_{i,j}^{\rho}, u_{k,\ell}^{\rho} \rangle_{L(G)} = \frac{|G|}{d_\rho} \delta_{i,k} \delta_{j,\ell}$$

for all $1 \leq i,j,k,\ell \leq d_\rho$.

Proof We leave the proof as an exercise as a slight modification of the previous one. Note that in the present setting we have $\widetilde{L} \in \mathbb{C}I_V$. □

Exercise 1.2.27 Let (ρ,V) be a (not necessarily irreducible) G-representation and fix an orthonormal basis $\{v_1,\ldots,v_{d_\rho}\}$ of V. Show that:

- $u^\rho_{i,j}(g^{-1}) = \overline{u^\rho_{j,i}(g)}$;
- $u^\rho_{i,j}(g_1 g_2) = \sum_{k=1}^{d_\rho} u^\rho_{i,k}(g_1) u^\rho_{k,j}(g_2)$;
- $\sum_{j=1}^{d_\rho} u^\rho_{j,i}(g)\overline{u^\rho_{j,k}(g)} = \delta_{i,k}$ (*dual orthogonality relations*);

for all $g,g_1,g_2 \in G$ and $1 \leq i,j,k \leq d_\rho$.

Let V be a finite dimensional vector space. We recall that the *trace* is the linear map tr: $\mathrm{End}(V) \to \mathbb{C}$ that satisfies the following two properties:

(T1) $\mathrm{tr}(xy) = \mathrm{tr}(yx)$ for all $x,y \in \mathrm{End}(V)$
(T2) $\mathrm{tr}(I_V) = \dim V$.

Note that if $\{v_1,v_2,\ldots,v_d\}$ is an orthogonal basis of V, then $\mathrm{tr}(x) = \sum_{i=1}^{d}\langle xv_i,v_i\rangle_V$ for all $x \in \mathrm{End}(V)$.

Definition 1.2.28 The *character* of a G-representation (ρ,V) is the map $\chi^\rho: G \to \mathbb{C}$ defined by setting

$$\chi^\rho(g) = \mathrm{tr}(\rho(g)) = \sum_{i=1}^{d_\rho} u^\rho_{i,i}(g)$$

for all $g \in G$, where, for the last term, the diagonal matrix coefficients are relative to an (= any) orthonormal basis of V.

Remark 1.2.29 Let ρ,σ be G-representations. We denote the unitary group of complex numbers by $\mathbb{T} = \{z \in \mathbb{C}: |z| = 1\} \subset \mathbb{C}$. Then:

- If $d_\rho = 1$, then $\chi^\rho \equiv \rho: G \to \mathbb{T}$;
- if $\rho \sim \sigma$, then $\chi^\rho = \chi^\sigma$ (cf. Corollary 1.2.35);
- $\chi^\rho(1_G) = d_\rho$;
- $\chi^\rho(ghg^{-1}) = \chi^\rho(h)$;
- $\chi^\rho(g^{-1}) = \overline{\chi^\rho(g)}$,

for all $g,h \in G$.

Exercise 1.2.30 Show that $|\chi^\rho(g)| \leq d_\rho$ for all $g \in G$.

Corollary 1.2.31 *Let ρ_1 and ρ_2 be two irreducible G-representations. Then*

$$\langle \chi^{\rho_1}, \chi^{\rho_2} \rangle_{L(G)} = \begin{cases} |G| & \text{if } \rho_1 \sim \rho_2 \\ 0 & \text{otherwise.} \end{cases}$$

In other words, the characters constitute an orthogonal system in $L(G)$. □

Corollary 1.2.32 *Let ρ and σ be two G-representations. Suppose that $\rho = \rho_1 \oplus \rho_2 \oplus \cdots \oplus \rho_k$ is a decomposition of ρ into irreducible representations and that σ is irreducible. Then, setting $m_\sigma^\rho := |\{i : \sigma \sim \rho_i\}|$, we have that*

$$m_\sigma^\rho = \frac{1}{|G|} \langle \chi^\rho, \chi^\sigma \rangle_{L(G)}.$$ □

Definition 1.2.33 The (nonnegative) integer m_σ^ρ is called the *multiplicity of σ in ρ*.

Corollary 1.2.34 *Let ρ and σ be two G representations. Suppose that $\rho \sim \oplus_{\theta \in \widehat{G}} m_\theta^\rho \theta$ and $\sigma \sim \oplus_{\theta \in \widehat{G}} m_\theta^\sigma \theta$. Then*

$$\frac{1}{|G|} \langle \chi^\rho, \chi^\sigma \rangle_{L(G)} = \sum_{\theta \in \widehat{G}} m_\theta^\rho m_\theta^\sigma.$$ □

Corollary 1.2.35 *Let ρ and σ be two G-representations. Then*

• *ρ is irreducible if and only if $\frac{1}{|G|} \langle \chi^\rho, \chi^\rho \rangle_{L(G)} = 1$;*
• *$\rho \sim \sigma$ if and only if $\chi^\rho = \chi^\sigma$.* □

The following is a fundamental result on the representation theory of finite groups: it provides a complete description of the decomposition of the regular representation. It was proved, in the more general setting of compact groups, by Hermann Weyl and his student Fritz Peter [56].

Theorem 1.2.36 (Peter–Weyl) (1) *Every irreducible representation $(\rho, V_\rho) \in \widehat{G}$ appears in the left regular representation $(\lambda_G, L(G))$ with multiplicity equal to its dimension:*

$$L(G) \sim \bigoplus_{\rho \in \widehat{G}} d_\rho V_\rho.$$

(2) *The set $\mathfrak{U} = \{u_{i,j}^\rho : 1 \le i, j \le d_\rho, \rho \in \widehat{G}\}$ of matrix coefficients is a complete orthogonal system in $L(G)$.*

(3) *$|G| = \sum_{\rho \in \widehat{G}} d_\rho^2$.*

Proof   (1) Let $g, h \in G$. Since $\lambda_G(g)\delta_h = \delta_{gh}$ we have

$$\chi^{\lambda_G}(g) = \begin{cases} |G| & \text{if } g = 1_G \\ 0 & \text{otherwise.} \end{cases}$$

Therefore, if $\rho \in \widehat{G}$ the multiplicity of ρ in λ_G is given by

$$m_\rho^{\lambda_G} = \frac{1}{|G|} \langle \chi^\rho, \chi^{\lambda_G} \rangle_{L(G)} = \frac{1}{|G|} \sum_{g \in G} \overline{\chi^\rho(g)} \chi^{\lambda_G}(g) = \overline{\chi^\rho(1_G)} = d_\rho.$$

(2) and (3) follow easily by observing that $|G| = \dim L(G) = \sum_{\rho \in \widehat{G}} d_\rho^2$ and $|\mathfrak{U}| = \sum_{\rho \in \widehat{G}} d_\rho^2$. $\square$

Definition 1.2.37   Let G be a finite group. A function $f \in L(G)$ is said to be *central* if the following equivalent conditions hold:

(1) f is constant on each *conjugacy class* $\mathscr{C}(g) := \{h^{-1}gh : h \in G\}$, $g \in G$, of G;
(2) $f(gh) = f(hg)$ for all $g, h \in G$;
(3) $f * f' = f' * f$ for all $f' \in L(G)$.

Exercise 1.2.38   Let G be a finite group.

(1) Show that the conditions (1), (2), and (3) in Definition 1.2.37 are equivalent.
(2) Show that the set $\mathscr{A}$ of all central functions in $L(G)$ forms a $*$-subalgebra.
(3) Show that $f * \phi * f^* \in \mathscr{A}$ for all $f \in L(G)$ and $\phi \in \mathscr{A}$.
(4) Show that $\chi^\rho \in \mathscr{A}$ for all G-representations ρ.

Theorem 1.2.39   *The characters constitute an orthogonal basis of the vector space of central functions of $L(G)$. In particular, $|\widehat{G}|$ equals the number of conjugacy classes of G.*

Proof   See [15, Theorem 3.9.10] and/or [19, Theorem 10.3.13.(ii)]. $\square$

Definition 1.2.40   Let G be a finite group. A function $\phi\colon G \to \mathbb{C}$ is said to be *positive-definite* if the following equivalent conditions hold:

(1) $\sum_{g,h \in G} \phi(h^{-1}g) f(g) \overline{f(h)} \geq 0$ for all $f \in L(G)$;
(2) $\sum_{i,j=1}^n c_i \overline{c_j} \phi(g_j^{-1} g_i) \geq 0$ for all $c_1, c_2, \ldots, c_n \in \mathbb{C}$, $g_1, g_2, \ldots, g_n \in G$, and $n \geq 1$;
(3) there exists a (unitary) representation (σ_ϕ, V_ϕ) of G and a cyclic vector $v_\phi \in V_\phi$ such that $\phi(g) = \langle \sigma_\phi(g) v_\phi, v_\phi \rangle_{V_\phi}$ for all $g \in G$.

In condition (3) above, the vector $v_\phi \in V_\phi$ being *cyclic* means that the vectors $\sigma_\phi(g) v_\phi$, $g \in G$, span V_ϕ.

Exercise 1.2.41   Let G be a finite group.

(1) Show that the conditions (1), (2), and (3) in Definition 1.2.40 are equivalent.
(2) Show that a linear combination with positive coefficients of positive-definite functions is positive-definite as well.
(3) Show that characters of G-representations are positive-definite functions.

Hint. For the implication (1) $\Longrightarrow$ (3), define $\widetilde{\phi} \colon L(G) \to \mathbb{C}$ by setting

$$\widetilde{\phi}(f) := \sum_{g \in G} \phi(g) f(g)$$

for all $f \in L(G)$, and define $\ll \cdot, \cdot \gg \colon L(G) \times L(G) \to \mathbb{C}$ by setting

$$\ll f_1, f_2 \gg \; := \widetilde{\phi}(f_2^* * f_1) \equiv \sum_{g,h \in G} \phi(h^{-1}g) f_1(g) \overline{f_2(h)}$$

for all $f_1, f_2 \in L(G)$. Show that $\ll \cdot, \cdot \gg$ defines a semi-definite sesquilinear form on $L(G)$. Show that the degenerate elements $f \in L(G)$ which satisfy $\ll f, f \gg \; = 0$ form a left ideal $\mathscr{I}$ of $L(G)$. The quotient space $V_\phi := L(G)/\mathscr{I}$ is a complex vector space with an inner product defined by setting

$$\langle f_1 + \mathscr{I}, f_2 + \mathscr{I} \rangle_{V_\phi} := \; \ll f_1, f_2 \gg$$

for all $f_1, f_2 \in L(G)$: check that the above is well defined.

Finally, define the G-representation (σ_ϕ, V_ϕ) by setting

$$\sigma_\phi(g)(f + \mathscr{I}) := \lambda_G(g) f + \mathscr{I}$$

for all $g \in G$ and $f \in L(G)$, where λ_G is the left-regular representation of G, and set

$$v_\phi := \delta_{1_G} + \mathscr{I} \in V_\phi,$$

where $\delta_{1_G} \in L(G)$ is the Dirac function at the identity element 1_G of G.

The triple $(V_\phi, \sigma_\phi, v_\phi)$ above is called the *GNS-construction*, after Israel M. Gelfand, Mark A. Naimark, and Irving E. Segal.

The following is a finite group version of a celebrated theorem of Salomon Bochner stating that the finite positive Borel probability measures on a locally compact abelian group G (e.g., $G = \mathbb{R}$) are the Fourier transform of continuous positive-definite functions on the Pontryagin dual $\widehat{G}$ of G (note that $\widehat{\mathbb{R}} \cong \mathbb{R}$) which take value 1 at $1_{\widehat{G}}$ (cf. [58, Theorem IX.9]).

Proposition 1.2.42   *Let G be a finite group and let $\phi \in L(G)$. Then the following conditions are equivalent:*

(1) *ϕ is central, positive-definite, and $\phi(1_G) = 1$;*

(2) *there exists* $(\alpha_\sigma)_{\sigma \in \widehat{G}}$, *where* $\alpha_\sigma \in [0,1]$ *and* $\sum_{\sigma \in \widehat{G}} \alpha_\sigma = 1$, *such that*

$$\phi = \sum_{\sigma \in \widehat{G}} \frac{\alpha_\sigma}{d_\sigma} \chi^\sigma.$$

Proof   This is left as an exercise (see also [7, Proposition 1.6]).   □

Condition (2) above may be rephrased by saying that ϕ is a convex combination of normalized characters.

Remark 1.2.43   Let G be a group such that $d_\rho = 1$ for all $(\rho, V) \in \widehat{G}$. It follows from Theorem 1.2.36.(2) and Theorem 1.2.39 that $L(G)$ is a commutative algebra. The latter is easily seen to be equivalent to G being abelian. This constitutes an alternative (more advanced) solution to Exercise 1.2.23.

Definition 1.2.44   Let V_1 and V_2 be two complex vector spaces endowed with scalar products. Their *tensor product* $V_1 \otimes V_2$ is the linear span of $\{v_1 \otimes v_2 : v_1 \in V_1, v_2 \in V_2\}$, where $v_1 \otimes v_2$ denotes the anti-bilinear form on $V_1 \times V_2$ defined by setting

$$(v_1 \otimes v_2)(u_1, u_2) = \langle v_1, u_1 \rangle_{V_1} \langle v_2, u_2 \rangle_{V_2}$$

for all $(u_1, u_2) \in V_1 \times V_2$. We equip $V_1 \otimes V_2$ with the scalar product defined by

$$\langle v_1 \otimes v_2, w_1 \otimes w_2 \rangle := \langle v_1, w_1 \rangle_{V_1} \langle v_2, w_2 \rangle_{V_2} \quad \text{for all } v_1, w_1 \in V_1, v_2, w_2 \in V_2.$$

If $A_i \in \text{End}(V_i)$, $i = 1, 2$, we define their tensor product $A_1 \otimes A_2 \in \text{End}(V_1 \otimes V_2)$ by setting $(A_1 \otimes A_2)(v_1 \otimes v_2) = (A_1 v_1) \otimes (A_2 v_2)$ for all $v_1 \in V_1$ and $v_2 \in V_2$.

Let (ρ_1, V_1) (resp. (ρ_2, V_2)) be a representation of a group G_1 (resp. G_2). Their *outer tensor product* is the $(G_1 \times G_2)$-representation $(\rho_1 \boxtimes \rho_2, V_1 \otimes V_2)$ defined by setting

$$(\rho_1 \boxtimes \rho_2)(g_1, g_2) = \rho_1(g_1) \otimes \rho_2(g_2) \quad \text{for all } (g_1, g_2) \in G_1 \times G_2.$$

Similarly, if (ρ_1, V_1) and (ρ_2, V_2) are two representations of the same group G, their *internal tensor product* is the G-representation $(\rho_1 \otimes \rho_2, V_1 \otimes V_2)$ defined by setting

$$(\rho_1 \otimes \rho_2)(g) = \rho_1(g) \otimes \rho_2(g) \quad \text{for all } g \in G.$$

After identifying G with the diagonal subgroup $\widetilde{G} = \{(g,g) : g \in G\}$ of $G \times G$, we observe that $\rho_1 \otimes \rho_2 = \text{Res}_{\widetilde{G}}^{G \times G}(\rho_1 \boxtimes \rho_2)$.

Exercise 1.2.45   Let ρ_1, ρ_1' (resp. ρ_2, ρ_2') be two G_1 (resp. G_2)-representations. Show that

(1) $\chi^{\rho_1 \boxtimes \rho_2}(g_1, g_2) = \chi^{\rho_1}(g_1) \chi^{\rho_2}(g_2)$ for all $(g_1, g_2) \in G_1 \times G_2$;

(2) $\rho_1 \boxtimes \rho_2$ is irreducible if and only if ρ_1 and ρ_2 are irreducible;

(3) $\rho_1 \boxtimes \rho_2 \sim \rho_1' \boxtimes \rho_2'$ if and only if $\rho_1 \sim \rho_1'$ and $\rho_2 \sim \rho_2'$.

From the above exercise one deduces the following:

Theorem 1.2.46 *Let G_1 and G_2 be two groups. Then $\widehat{G_1 \times G_2} = \widehat{G_1} \boxtimes \widehat{G_2}$.* $\square$

Exercise 1.2.47 Let (ρ, V) be a unitary representation of a group G and let (ρ', V') be its conjugate representation (cf. Definition 1.2.10). Show that the trivial representation of G (cf. Definition 1.2.13) satisfies $\iota_G \preceq \rho \otimes \rho'$.

If A is a (finite) abelian group, then there exist $d_1, d_2, \ldots, d_n \in \mathbb{N}$ with $d_i | d_{i+1}$, $i = 1, 2, \ldots, n-1$ such that $A \cong C_{d_1} \times C_{d_2} \times \cdots \times C_{d_n}$ (recall that $C_d \cong \mathbb{Z}/d\mathbb{Z}$ is the cyclic group of order d). The dual of an abelian group has a natural structure of an abelian group and we have

$$\widehat{A} \cong \widehat{C_{d_1}} \times \widehat{C_{d_2}} \times \cdots \times \widehat{C_{d_n}} \cong C_{d_1} \times C_{d_2} \times \cdots \times C_{d_n} \cong A.$$

The group isomorphism $A \cong \widehat{A}$ is not canonical, but we have the canonical *Pontryagin duality* between A and its *bidual* (the dual of the dual) $\widehat{\widehat{A}}$:

$$A \ni g \mapsto \psi_g \in \widehat{\widehat{A}}, \quad \text{with } \psi_g(\chi) = \chi(g) \text{ for all } \chi \in \widehat{A}.$$

1.2.2 Finite Gelfand pairs

Let G be a finite group and let $K \leq G$ be a subgroup.

A function $f \in L(G)$ is *K-invariant* on the *right* (resp. on the *left*) if $f(gk) = f(g)$ (resp. $f(kg) = f(g)$) for all $g \in G$ and $k \in K$. Then f is *bi-K-invariant* if it is *K*-invariant both on the left and the right. We denote by $L(G)^K$ (resp. ${}^K L(G)$) the subspace of $L(G)$ of *K*-invariant functions on the right (resp. on the left).

Let $X = G/K = \{gK : g \in G\}$ be the *homogeneous space* of *left cosets* of K in G and observe that we can identify $L(G)^K$ with $L(X) = \{f : X \to \mathbb{C}\}$: indeed, the map $L(X) \ni f \mapsto \tilde{f} \in L(G)^K$, defined by

$$\tilde{f}(g) := f(gK) \quad \text{for all } g \in G$$

yields a linear isomorphism from $L(X)$ onto $L(G)^K$.

More generally, suppose that G acts transitively on a set X (that is, for all $x_1, x_2 \in X$ there exists $g \in G$ such that $gx_1 = x_2$; equivalently, for all $x \in X$ the *G*-orbit $Gx := \{gx : g \in G\}$ of x is all of X). Fix $x_0 \in X$ and denote by $K = \text{Stab}_G(x_0) := \{g \in G : gx_0 = x_0\}$ the *stabilizer* of x_0 in G. Then the map $X \ni gx_0 \mapsto gK \in G/K$ is a bijection.

Similarly, if $K\backslash G/K = \{KgK : g \in G\}$ denotes the set of *double cosets* of K in G, then the space of bi-K-invariant functions

$$^K L(G)^K := \{f \in L(G) : f(k_1 g k_2) = f(g), \forall k_1, k_2 \in K, g \in G\}$$

is isomorphic to both $L(K\backslash G/K)$ and

$$^K L(X) := \{f \in L(X) : f(kx) = f(x), \text{ for all } x \in X \text{ and } k \in K\};$$

the second isomorphism is given, again, by the map $f \mapsto \tilde{f}$ restricted to $^K L(X)$.

Note that if $f_1, f_2 \in L(X)$, then

$$\langle f_1, f_2 \rangle_{L(X)} = \frac{1}{|K|} \langle \tilde{f}_1, \tilde{f}_2 \rangle_{L(G)}.$$

Exercise 1.2.48 Let G act transitively on a set X. Consider the diagonal action of G on $X \times X$ given by $g(x_1, x_2) := (gx_1, gx_2)$ for all $g \in G$ and $x_1, x_2 \in X$. Fix $x_0 \in X$ and let $K = \text{Stab}_G(x_0)$. Show that the following quantities are all equal:

(1) the number of G-orbits on $X \times X$;
(2) the number of K-orbits on X;
(3) $|K\backslash G/K|$.

Hint. Show that the map that associates with a G-orbit Θ on $X \times X$ the subset $\Omega := \{x \in X : (x, x_0) \in \Theta\}$ yields a bijection between the set of all G-orbits on $X \times X$ and the set of all K-orbits on X. Also show that the map $KgK \mapsto Kgx_0$ yields a bijection between the set of all double cosets of K in G and the set of all K-orbits on X.

Exercise 1.2.49 (1) Show that for $f_1, f_2 \in L(G)$ we have that $f_1 * f_2$ is K-invariant on the left (resp. right) if f_1 (resp. f_2) is K-invariant on the left (resp. right). Deduce that $^K L(G)^K$ is a two-sided ideal of $L(G)$.

(2) Check that the map $L(G) \ni f \mapsto f^K \in L(G)^K$, with

$$f^K(g) := \frac{1}{|K|} \sum_{k \in K} f(gk), \forall g \in G$$

is well defined and it is the orthogonal projection onto the subspace of right K-invariant functions.

(3) Check that the map $L(G) \ni f \mapsto {}^K f^K \in {}^K L(G)^K$, with

$$^K f^K(g) := \frac{1}{|K|^2} \sum_{k_1, k_2 \in K} f(k_1 g k_2), \forall g \in G$$

is well defined and it is a *conditional expectation*, that is, $^K(f_1 * f * f_2)^K = f_1 * {}^K f^K * f_2$ for all $f_1, f_2 \in {}^K L(G)^K$ and $f \in L(G)$.

(4) Show that $(f_1 * f_2)^* = f_2^* * f_1^*$ for all $f_1, f_2 \in L(G)$.

Definition 1.2.50 Suppose that G acts transitively on a set X. The *permutation representation* $(\lambda, L(X))$ is the G-representation defined by setting

$$[\lambda(g)f](x) = f(g^{-1}x) \quad \text{for all } f \in L(X), g \in G, x \in X.$$

Exercise 1.2.51 Show that the permutation representation $(\lambda, L(X))$ is unitary.

Proposition 1.2.52 *Suppose that G acts transitively on a set X. Let $x_0 \in X$ and set $K = \mathrm{Stab}_G(x_0)$. Then $\mathrm{End}_G(L(X))$ and $^K L(G)^K$ are isomorphic as algebras.*

Proof Given a linear map $T \colon L(X) \to L(X)$, there exists a matrix $(r(x,y))_{x,y \in X}$ such that

$$[Tf](x) = \sum_{y \in X} r(x,y)f(y) \quad \text{for all } f \in L(X) \text{ and } x \in X. \tag{1.2}$$

We have that $T \in \mathrm{End}_G(L(X))$ if and only if $r(gx, gy) = r(x,y)$ for all $g \in G$ and $x, y \in X$, and this is in turn equivalent to saying that r is constant on the G-orbits on $X \times X$. Define $\psi \colon X \to \mathbb{C}$ by setting

$$\psi(x) = r(x, x_0) \quad \text{for all } x \in X. \tag{1.3}$$

Note that ψ is K-invariant: $\psi(kx) = r(kx, x_0) = r(kx, kx_0) = r(x, x_0) = \psi(x)$, so that $\widetilde{\psi} \in {}^K L(G)^K$. Moreover (1.2) becomes

$$[\widetilde{Tf}](g) = [Tf](gx_0) = \frac{1}{|K|} \sum_{h \in G} r(gx_0, hx_0) f(hx_0)$$

$$= \frac{1}{|K|} \sum_{h \in G} f(hx_0) r(h^{-1}gx_0, x_0) = \frac{1}{|K|} [\widetilde{f} * \widetilde{\psi}](g), \tag{1.4}$$

and we say that $\frac{1}{|K|} \widetilde{\psi} \in {}^K L(G)^K$ is the *kernel* of T.

However, if $T_1, T_2 \in \mathrm{End}_G(L(X))$ and $\widetilde{\psi}_1, \widetilde{\psi}_2$ are the associated kernels, we have that the kernel of $T_1 \circ T_2$ is $\frac{1}{|K|^2} \widetilde{\psi}_2 * \widetilde{\psi}_1$. Thus, if we set $f^{\sharp}(g) = f(g^{-1})$ for all $f \in L(G)$ and $g \in G$, we deduce that the desired isomorphism is given by $T \mapsto \frac{1}{|K|} (\widetilde{\psi})^{\sharp}$. $\qquad\square$

Definition 1.2.53 (Gelfand pair) (G, K) is a *Gelfand pair* if the algebra $^K L(G)^K$ is commutative.

Exercise 1.2.54 (Symmetric Gelfand pairs) Let G be a finite group and let $K \leq G$ be a subgroup. Suppose that

$$g^{-1} \in KgK \quad \text{for all } g \in G. \tag{1.5}$$

Show that (G,K) is a Gelfand pair. We then say that (G,K) is a *symmetric* Gelfand pair.

More generally, we have:

Exercise 1.2.55   Suppose that there exists $\tau \in \text{Aut}(G)$ such that

$$g^{-1} \in K\tau(g)K \quad \text{for all } g \in G. \tag{1.6}$$

Show that (G,K) is a Gelfand pair. We then say that (G,K) is a *weakly symmetric* Gelfand pair.

Exercise 1.2.56   Let $\widetilde{G} = \{(g,g) : g \in G\}$ denote the diagonal subgroup of $G \times G$.

(1) Show that $(G \times G, \widetilde{G})$ is a Gelfand pair. (See [62] for a more general construction.)
(2) Show that the Gelfand pair $(G \times G, \widetilde{G})$ is symmetric if and only if G is *ambivalent*, that is, every element $g \in G$ is conjugate in G to its inverse g^{-1}.

Exercise 1.2.57   Suppose that G acts transitively on a set X; let $x_0 \in X$, set $K = \text{Stab}_G(x_0)$, and consider the diagonal action of G on $X \times X$. Show that (G,K) is a symmetric Gelfand pair if and only if the G-orbits on $X \times X$ are symmetric, i.e. $G(x,y) = G(y,x)$ for all $x,y \in X$.

Exercise 1.2.58   Let G act on a metric space (X,d) and suppose that the action is *two-point homogeneous* (or *distance transitive*) i.e., $G(x_1,y_1) = G(x_2,y_2)$ if $d(x_1,y_1) = d(x_2,y_2)$. Show that (G,K) is a symmetric Gelfand pair.

Definition 1.2.59   A G-representation (ρ,V) is *multiplicity-free* if it does not contain two equivalent irreducible representations, in formulæ,

$$\rho = \oplus_{\theta \in \widehat{G}} m_\theta^\rho \theta \Rightarrow m_\theta^\rho \leq 1 \text{ for all } \theta \in \widehat{G}.$$

Theorem 1.2.60   *The following conditions are equivalent:*

(1) (G,K) *is a Gelfand pair;*
(2) $\text{End}_G(L(X))$ *is commutative;*
(3) $(\lambda,L(X))$ *is multiplicity-free.*

Proof   The equivalence between (1) and (2) follows from Proposition 1.2.52. Suppose that (3) holds, i.e.

$$L(X) = \oplus_{i=0}^N V_i$$

with V_i irreducible and $V_i \nsim V_j$ if $i \neq j$.
  If $T \in \text{End}_G(L(X))$, then $T_i = T|_{V_i}$ is either trivial ($T_i = 0$) or injective (since

$\ker T_i$ is a G-invariant subspace of V_i). In the latter case $\operatorname{ran} T_i \leq L(X)$ is G-invariant and isomorphic to V_i. Thus $\operatorname{ran}(T_i) \cap V_j \leq V_j$ is either 0 or V_j. This holds only if $j = i$. Therefore, by Schur's lemma there exists $\lambda_i \in \mathbb{C}$ such that $T_i = \lambda_i P_{V_i}$ (with P_{V_i} the orthogonal projection on V_i) and $T = \sum_{i=0}^{N} \lambda_i P_{V_i}$. If $S \in \operatorname{End}_G(L(X))$ is another intertwiner, then $S = \sum_{i=0}^{N} \mu_i P_{V_i}$ and therefore we have $ST = \sum_{i=0}^{N} \mu_i \lambda_i P_{V_i} = TS$, showing the commutativity of $\operatorname{End}_G(L(X))$.

Vice versa, suppose that (2) holds. If $L(X)$ is not multiplicity-free, then $L(X) = V \oplus W \oplus U$ with $V \sim W$ irreducible. Let $R \in \operatorname{Hom}_G(V, W)$ be an isomorphism. Consider the linear operators $S, T : L(X) \to L(X)$ defined by setting

$$S(v + w + u) = R^{-1}w$$
$$T(v + w + u) = Rv$$

for all $v \in V$, $w \in W$, $u \in U$. We have that $T, S \in \operatorname{End}_G(L(X))$: indeed, for $v \in V$, $w \in W$, $u \in U$, and $g \in G$,

$$
\begin{aligned}
T\lambda(g)(v + w + u) &= T(\lambda(g)v + \lambda(g)w + \lambda(g)u) \\
&= R\lambda(g)v \\
&= \lambda(g)Rv \\
&= \lambda(g)T(v + w + u).
\end{aligned}
$$

The proof for S is completely analogous. Observe that $STv = v$, while $TSv = 0$, thus showing that $\operatorname{End}_G(L(X))$ is not commutative. $\qquad\square$

With the notation of the above proof, we have the following:

Corollary 1.2.61 *The map* $\operatorname{End}_G(L(X)) \ni T \mapsto (\lambda_0, \lambda_1, \ldots, \lambda_N) \in \mathbb{C}^{N+1}$ *is an algebra isomorphism.* $\qquad\square$

From the above results we deduce :

$$
\begin{aligned}
N + 1 &= |\{\text{irreducible sub-representations in } L(X)\}| \\
&= \dim \operatorname{End}_G(L(X)) \\
&= \dim {}^K L(G)^K \\
&= |K \backslash G / K| \\
&= |\{K\text{-orbits on } X\}|.
\end{aligned}
$$

1.2.3 Spherical functions

In this section we suppose that (G, K) is a Gelfand pair.

Definition 1.2.62 (Spherical function)   A function $\phi \in {}^K L(G)^K$ is *spherical* if

- for every $f \in^K L(G)^K$ there exists $\lambda_f \in \mathbb{C}$ such that $f * \phi = \lambda_f \phi$;
- $\phi(1_G) = 1$.

Note that if ϕ is a spherical function and $f \in {}^K L(G)^K$, then $\lambda_f = [f * \phi](1_G)$.

Lemma 1.2.63 *Let ϕ be a spherical function. Define $\Phi: L(G) \to \mathbb{C}$ by setting*

$$\Phi(f) = \sum_{g \in G} f(g)\phi(g^{-1}) \tag{1.7}$$

for all $f \in L(G)$. Then Φ is a linear multiplicative functional on ${}^K L(G)^K$, that is, $\Phi(f_1 * f_2) = \Phi(f_1)\Phi(f_2)$ for all $f_1, f_2 \in {}^K L(G)^K$. Vice versa, every nontrivial multiplicative linear functional on ${}^K L(G)^K$ is determined by a unique spherical function.

Proof   We leave it to the reader to check that Φ is a multiplicative linear functional.

Vice versa, suppose that Φ is a multiplicative linear functional on ${}^K L(G)^K$. Then we can extend Φ to a linear functional $\widetilde{\Phi}$ on $L(G)$ by setting $\widetilde{\Phi}(f) = \Phi({}^K f^K)$ for all $f \in L(G)$. By Riesz' representation theorem there exists $\psi \in L(G)$ such that $\widetilde{\Phi}(f) = \sum_{g \in G} f(g)\psi(g^{-1})$. We leave it to the reader to check that the function $\phi := {}^K \psi^K \in {}^K L(G)^K$ is spherical and satisfies (1.7) for all $f \in {}^K L(G)^K$.   □

Proposition 1.2.64 (Basic properties of spherical functions)   *Let ϕ and ψ be two distinct spherical functions. Then:*

- $\phi(g^{-1}) = \overline{\phi(g)}$ *for all $g \in G$;*
- $\phi * \psi = 0$;
- $\langle \lambda(g_1)\phi, \lambda(g_2)\psi \rangle_{L(G)} = 0$ *for all $g_1, g_2 \in G$ (in particular $\phi \perp \psi$).*

Proof   We leave it to the reader.   □

Theorem 1.2.65

$$|\{spherical\ functions\}| = |K \backslash G / K| = \dim{}^K L(G)^K.$$

In particular, the spherical functions constitute an orthogonal basis for the space of all bi-K-invariant functions on G.

Proof   By Proposition 1.2.52 and Corollary 1.2.61, the algebras ${}^K L(G)^K$ and $\mathbb{C}^{N+1}$ are isomorphic. The statement follows by observing that the only multiplicative linear functionals on $\mathbb{C}^{N+1}$ are the maps

$$\Phi_j: \quad \begin{array}{ccc} \mathbb{C}^{N+1} & \to & \mathbb{C} \\ (\alpha_0, \alpha_1, \dots, \alpha_N) & \mapsto & \alpha_j \end{array}$$

$j = 0, 1, \dots, N$.   □

For $f \in L(G)^K$ define $\check{f} \in L(X)$ by setting $\check{f}(gx_0) = f(g)$ for all $g \in G$ (as usual, $x_0 = K \in G/K$; equivalently, $x_0 \in X$ and $K = \text{Stab}_G(x_0)$).

Theorem 1.2.66 *Let* $\phi_0, \phi_1, \ldots, \phi_N \in {}^K L(G)^K$ *be the spherical functions. Set*

$$V_i = \text{span}\{\lambda(g)\check{\phi}_i : g \in G\} \leq L(X)$$

for $i = 0, 1, \ldots, N$. *Then*

$$L(X) = \bigoplus_{i=0}^{N} V_i$$

is the decomposition of the permutation representation into irreducible subrepresentations.

Proof Each V_i is G-invariant and, being cyclic (that is, G-generated by a single vector), is irreducible. Moreover $V_i \perp V_j$ if $i \neq j$ (cf. Proposition 1.2.64). Since there are exactly $N + 1$ irreducible components of $L(X)$, we conclude that the V_i's exhaust all of $L(X)$. $\qquad\square$

Definition 1.2.67 $(\lambda|_{V_i}, V_i)$ is called the *spherical representation* associated with ϕ_i.

We always choose $\phi_0 \equiv 1$ so that V_0 is the trival representation.

Exercise 1.2.68 The spherical functions of the Gelfand pair $(G \times G, \widetilde{G})$ (cf. Exercise 1.2.56) are the normalized characters of G, namely, the bi-$\widetilde{G}$-invariant functions φ_σ, $\sigma \in \widehat{G}$, defined by

$$\varphi_\sigma(g, h) = \frac{1}{d_\sigma} \chi^\sigma(g^{-1}h)$$

for all $g, h \in G$.

Let (ρ, V) be a G-representation. We denote by

$$V^{\rho, K} = \{v \in V : \rho(k)v = v, \text{ for all } k \in K\}$$

the subspace of *K-invariant* vectors. If the representation ρ is clear from the context we will simply write V^K for $V^{\rho, K}$. However, note that $L(G)^K = L(G)^{\rho_G, K}$ while ${}^K L(G) = L(G)^{\lambda_G, K}$ (cf. Example 1.2.14) and we write ${}^K L(X)$ for $L(X)^{\lambda, K}$ (cf. Definition 1.2.50).

For the proof of Theorem 1.2.71 we need a couple of classical results from the theory of group actions, namely the so-called *Burnside lemma* and the *Wielandt lemma*. The first result is not due to Burnside himself, who merely quotes it in his book [10], attributing it instead to Frobenius, although it was already known to Cauchy (cf. [54, 72]). For a proof we refer to [6] (see also [15, Lemma 3.11.1]).

Exercise 1.2.69 (Burnside's lemma)   Let G be a finite group acting (not necessarily transitively) on a finite set Ω. Denote by $(\lambda, L(\Omega))$ the *permutation representation*, defined by setting $[\lambda(g)f](\omega) = f(g^{-1}\omega)$ for all $g \in G$, $f \in L(\Omega)$, and $\omega \in \Omega$. Denote by $\chi = \chi^\lambda$ the associated character. Show that

$$\frac{1}{|G|} \sum_{g \in G} \chi(g) = \frac{1}{|G|} \sum_{\omega \in \Omega} |\mathrm{Stab}_G(\omega)| = \text{number of } G\text{-orbits in } \Omega.$$

The result in the next exercise was surely known to Schur and, possibly, even to Frobenius. A standard reference is the book by Helmut Wielandt [70] (see also [15, Theorem 3.13.3]).

Exercise 1.2.70 (Wielandt's lemma)   Let $K \leq G$ be finite groups and set $X := G/K$. Let $L(X) = \bigoplus_{i=0}^N m_i V_i$ be a decomposition into irreducible G-subrepresentations of the associated permutation representation, where m_i denotes the multiplicity of V_i. Then

$$\sum_{i=0}^N m_i^2 = \text{number of } G\text{-orbits on } X \times X = \text{number of } K\text{-orbits on } X. \quad (1.8)$$

Theorem 1.2.71   *(G, K) is a Gelfand pair if and only if $\dim V^{\rho, K} \leq 1$ for all $(\rho, V) \in \widehat{G}$. If this is the case, then $\dim V^{\rho, K} = 1$ if and only if (ρ, V) is equivalent to a spherical representation.*

Proof   Let $(\rho, V) \in \widehat{G}$ with $\dim V^{\rho, K} \geq 1$. Pick $u_0 \in V^{\rho, K}$ and define $T : V \to L(X) = L(G/K)$ by setting $[Tv](gK) = (\langle v, \rho(g)u_0 \rangle_V)$. Now $T \in \mathrm{Hom}_G(V, L(X))$, and by Schur's lemma we deduce that $V \sim V_{\bar{i}}$ for some $0 \leq \bar{i} \leq N$. Since $L(X) = \bigoplus_{i=0}^N V_i$,

$$N + 1 = \dim L(X)^{\lambda, K} = (\dim^K L(G)^K),$$

and $L(X)^{\lambda, K} = \bigoplus_{i=0}^N V_i^{\lambda, K}$, we deduce that $\dim V_i^{\lambda, K} \leq 1$ for all $i = 0, 1, \ldots, N$. This in turn implies that $\dim V^{\rho, K} = \dim V_{\bar{i}}^{\lambda, K} \leq 1$.

Vice versa, suppose that $\dim V^{\rho, K} \leq 1$ for all $(\rho, V) \in \widehat{G}$. Let $L(X) = \bigoplus_{i=0}^H m_i W_i$ be the decomposition of the permutation representation into irreducible components. If $N + 1$ is the number of K-orbits on X we have (keeping in mind (1.8))

$$\sum_{i=0}^H m_i^2 = N + 1 = \sum_{i=0}^H m_i \dim W_i^{\lambda, K} \leq \sum_{i=0}^H m_i. \quad (1.9)$$

This forces $m_i = 1$ for all $i = 0, 1, \ldots, H$ and $H = N$. $\square$

1.2.4 Harmonic analysis of finite Gelfand pairs

Let (G,K) be a finite Gelfand pair and denote by $\phi_0 = 1, \phi_1, \ldots, \phi_N \in {}^K L(G)^K$ the associated spherical functions.

Definition 1.2.72 The linear map $\mathscr{F} \colon {}^K L(X) \to \mathbb{C}^{N+1}$ defined by setting

$$[\mathscr{F}f](i) = \langle f, \check{\phi}_i \rangle_{L(X)} = \sum_{x \in X} f(x) \overline{\check{\phi}_i(x)}$$

for all $f \in {}^K L(X)$ and $i = 0, 1, \ldots, N$, is called the *spherical Fourier transform* associated with the Gelfand pair (G,K).

Exercise 1.2.73 (Inversion formula) Let $f \in {}^K L(X)$. Show that

$$f(x) = \frac{1}{|X|} \sum_{i=0}^{N} d_i [\mathscr{F}f](i) \check{\phi}_i(x), \tag{1.10}$$

where, as usual, $d_i = \dim(V_i)$ is the dimension of the ith spherical representation.

Proposition 1.2.74 *Let $T \in \mathrm{End}_G(L(X))$. Then, for all $i = 0, 1, \ldots, N$,*

$$T|_{V_i} = \lambda_i I_{V_i},$$

where $\lambda_i = [\mathscr{F}\psi](i)$ and $\psi \in {}^K L(X)$ is as in (1.2) and (1.3).

Proof Let $x_0 \in X$ be the point stabilized by K. Then, for all $g \in G$ we have

$$
\begin{aligned}
[T\check{\phi}_i](gx_0) &= \frac{1}{|K|}[\phi_i * \tilde{\psi}](g) \\
&= \frac{1}{|K|}[\phi_i * \tilde{\psi}](1_G)\phi_i(g) \\
&= \frac{1}{|K|}\left(\sum_{h \in G} \phi_i(h^{-1})\tilde{\psi}(h)\right)\phi_i(g) \\
&= \frac{1}{|K|}\langle \tilde{\psi}, \phi_i \rangle_{L(G)} \phi_i(g) \\
&= \langle \psi, \check{\phi}_i \rangle_{L(X)} \check{\phi}_i(gx_0) \\
&= [\mathscr{F}\psi](i)\check{\phi}_i(gx_0).
\end{aligned}
$$

As T is an intertwiner, we deduce that $Tv = \lambda_i v$ for all $v \in V_i$, where $\lambda_i = [\mathscr{F}\psi](i)$. $\qquad\square$

Remark 1.2.75 Let $\Omega \subseteq X$ be a K-orbit and denote by $1_\Omega \in L(X)$ its characteristic function. Then

$$[\mathscr{F}1_\Omega](i) = |\Omega|\overline{\check{\phi}_i(x)}$$

where $x \in \Omega$ is arbitrary (spherical functions are constant on K-orbits).

Definition 1.2.76 (Convolution in $L(X)$)   Let G be a finite group acting transitively on a finite set X. Let $x_0 \in X$ and set $K := \text{Stab}_G(x_0)$. The *convolution* of two functions $f_1, f_2 \in L(X)$ is the function $f_1 * f_2 \in L(X)$ defined by setting

$$f_1 * f_2 := \frac{1}{|K|} (\tilde{f}_1 * \tilde{f}_2)^{\vee}.$$

Given $f \in L(X)$ we write $f^{*1} := f$ and, for $n \geq 2$, we recursively set $f^{*n} := f * (f^{*(n-1)})$.

Exercise 1.2.77   Let $f_1, f_2 \in {}^K L(X)$. Show that $f_1 * f_2 \in {}^K L(X)$ and

$$\mathscr{F}(f_1 * f_2) = \mathscr{F}(f_1)\mathscr{F}(f_2). \qquad (1.11)$$

Exercise 1.2.78   Show that the orthogonal projection $P_i \colon L(X) \to V_i$ is given by

$$[P_i f](gx_0) = \frac{d_i}{|X|} \langle f, \lambda(g)\check{\phi}_i \rangle_{L(X)}$$

for all $f \in L(X)$.

1.3 Laplace operators and spectra of random walks on finite graphs

In this section we present some elementary theory of finite regular simple graphs and the spectral theory of their associated adjacency (resp. Markov, resp. Laplace) matrices. A particular emphasis is given for a particular, yet significant, subclass of such graphs, namely that of distance-regular graphs. We refer to our monographs [15, 19] for other related aspects of finite graph theory.

1.3.1 Finite graphs and their spectra

Definition 1.3.1 (Finite graph)   A *finite graph* is a pair $\mathcal{G} = (X, E)$, where X is a finite set of *vertices* and E is a subset of $\{\{x, y\} : x, y \in X\}$, called the set of *edges*. An edge of the form $e = \{x\} \in E$ is called a *loop* at x.

Let $\mathcal{G} = (X, E)$ be a finite graph. Given $e = \{x, y\} \in E$, we say that the vertices x and y are *adjacent* (or *neighbours*) and we write $x \sim y$. Given $x \in X$, we denote by $\deg(x) := |\{y \in X : x \sim y\}|$ the number of adjacent vertices (including x itself if there is a loop at x). If $\deg(x) = \deg(y) =: k$ for all $x, y \in X$, one says that $\mathcal{G}$ is *regular* of *degree* k.

Given a subset $Y \subset X$ of vertices, the *subgraph induced* by Y is the graph $\mathcal{G}_Y = (Y, E_Y)$ where $E_Y := \{e = \{x, y\} \in E : x, y \in Y\}$.

A *path* in $\mathcal{G}$ is a sequence $\pi = (x_0, x_1, \ldots, x_n)$ of vertices $x_i \in X$ such that $x_i \sim x_{i+1}$ for all $i = 0, 1, \ldots, n-1$. The vertices x_0 and x_n are termed the *initial* and *terminal* vertices of π, and one says that π connects them. The integer n is called the *length* of the path π, denoted $\ell(\pi)$.

We introduce an equivalence relation $\approx$ on X by declaring that $x \approx y$ if there exists a path $\pi = \pi(x, y)$ connecting them. The subgraph induced by an $\approx$ equivalence class is called a *connected component* of $\mathcal{G}$. If there exists a unique such connected component, one says that $\mathcal{G}$ is *connected*.

If $\mathcal{G}$ is connected, given two vertices x and y, the nonnegative integer $d(x, y)$ $:= \min_\pi \ell(\pi)$, where π ranges among all paths $\pi = \pi(x, y)$, is called the *distance* of x and y. The nonnegative integer $\operatorname{diam}(X) := \max\{d(x, y) : x, y \in X\}$ is called the *diameter* of the connected graph $\mathcal{G} = (X, E)$.

This way, our finite graphs are simple, i.e., with no multiple edges, and undirected.

Exercise 1.3.2   Let $\mathcal{G} = (X, E)$ be a finite graph. Show that the map $d \colon X \times X \to [0, +\infty)$ is a distance function.

Definition 1.3.3 (Adjacency and Markov matrices and Laplacian)   Let $\mathcal{G} = (X, E)$ be a finite graph. The matrix $A = (A(x, y))_{x, y \in X}$ where

$$A(x, y) := \begin{cases} 1 & \text{if } x \sim y, \\ 0 & \text{otherwise,} \end{cases}$$

for all $x, y \in X$, is called the *adjacency matrix* of $\mathcal{G}$.

If $\mathcal{G}$ is regular of degree k, the matrices

$$M := \frac{1}{k}A \ \text{ and } L := I - M \equiv I - \frac{1}{k}A, \tag{1.12}$$

where $I = (\delta_{x,y})_{x,y \in X}$ is the identity matrix, are called the *Markov matrix* and the *discrete Laplacian* on $\mathscr{G}$, respectively.

Note that the Laplacian can be defined for arbitrary graphs, not necessarily regular: see Definition 3.2.11.

In the following we shall limit ourselves to the case where $\mathscr{G}$ is k-regular.

Recall that the spectrum $\sigma(T)$ of $T \in \text{End}(L(X))$ is the set of all eigenvalues of T:

$$\sigma(T) := \{\lambda \in \mathbb{C} : T - \lambda I \text{ is not invertible in } \text{End}(L(X))\}.$$

If T is symmetric, then the spectrum is real: $\sigma(T) =: \{\lambda_0, \lambda_1, \ldots, \lambda_m\} \subseteq \mathbb{R}$, and, denoting by V_i the T-eigenspace associated with λ_i, $i = 0, 1, \ldots, m$, we have the decomposition

$$L(X) = \bigoplus_{i=0}^{m} V_i. \tag{1.13}$$

In our setting, we have $\sigma(A) \subseteq [-k, k]$, $\sigma(M) \subseteq [-1, 1]$, and $\sigma(L) \subseteq [0, 2]$. Moreover, given the simple expressions (1.12) relating A, M, and L, the corresponding spectra are set-theoretically related by the expressions $\sigma(M) = \frac{1}{k}\sigma(A)$ and $\sigma(L) = 1 - \sigma(M) = 1 - \frac{1}{k}\sigma(A)$, and the corresponding eigenspaces coincide (we leave it as an exercise to check the details). For this reason, we limit ourselves to the analysis of the Markov matrix M.

We first note that $\lambda_0 := 1$ is an eigenvalue of M: indeed, any constant function $f \in L(X)$ (or, more generally, any function $f \in L(X)$ which is constant on each connected component of $\mathscr{G}$) is an M-eigenvector corresponding to the eigenvalue 1, that is, $Mf = f$. More precisely, we have the following:

Proposition 1.3.4 *Let $\mathscr{G} = (X, E)$ be a k-regular finite graph. Let $V_0 \leq L(X)$ denote the M-eigenspace corresponding to the eigenvalue $\lambda_0 = 1$. Then $\dim(V_0)$ equals the number of connected components of $\mathscr{G}$.*

Proof Let $\mathscr{G}_i = (X_i, E_i)$, $i = 1, 2, \ldots, n$, be the connected components of $\mathscr{G}$. It is obvious that if f is constant on each $\mathscr{G}_i$, then $Mf = f$. As the characteristic functions $\chi_{X_i} \in L(X)$ are linearly independent, this shows that $\dim(V_0) \geq n$. Conversely, suppose that $Mf = f$ with $f \in L(X)$ non-identically zero and real valued. Fix $i \in \{1, 2, \ldots, n\}$ and denote by $x_i \in X_i$ a maximum point for $|f|$ in X_i, i.e. $|f(x_i)| \geq |f(y)|$ for all $y \in X_i$; we may suppose, up to passing to $-f$, that $f(x_i) \geq 0$. Then $f(x_i) = \sum_{y \in X_i} m(x_i, y) f(y)$ and as $\sum_{y \in X_i} m(x_i, y) = 1$ we have $\sum_{y \in X_i} m(x_i, y)[f(x_i) - f(y)] = 0$. Since $m(x_i, y) \geq 0$ and $f(x_i) \geq f(y)$ for all $y \in X_i$, we deduce that $f(y) = f(x_i)$ for all $y \sim x_i$. Let now $z \in X_i$; then, by definition, there exists a path $p = (x_i, x_i', \ldots, x_i'' = z)$ connecting x_0 to z. In the previous step we have established that $f(x_i') = f(x_i) \geq f(y)$ for all $y \in X_i$ so that we can iterate the same argument to show that $f(x_i) = f(x_i') = \cdots = f(x_i'') = f(z)$, i.e., f is constant in X_i. This shows that V_0 is spanned by the χ_{X_i}s. We deduce that $\dim(V_0) = n$. $\qquad\qquad\qquad\qquad\qquad\qquad\qquad\square$

Definition 1.3.5 A graph $\mathcal{G} = (X, E)$ is *bipartite* if there exists a nontrivial partition $X = X_0 \sqcup X_1$ of the set of vertices such that every edge joins a vertex in X_0 with a vertex in X_1; that is, $E \subseteq \{\{x_0, x_1\} : x_0 \in X_0, x_1 \in X_1\}$.

Note that a bipartite graph has no loops and that, if $\mathcal{G}$ is connected, then the partition of the set of vertices is unique.

Example 1.3.6 Figure 1.1 shows the bipartite graph $\mathcal{G} = (X, E)$ with vertex set $X = X_0 \sqcup X_1$, where $X_0 = \{x, y\}$ and $X_1 = \{u, v, z\}$, and edge set $E = \{\{x, u\}, \{x, v\}, \{y, v\}, \{y, z\}\}$.

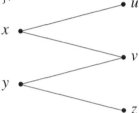

Figure 1.1 A bipartite graph

Exercise 1.3.7 Let $\mathcal{G} = (X, E)$ be a connected graph. Show that the following conditions are equivalent:

(1) $\mathcal{G}$ is bipartite;
(2) $\mathcal{G}$ is *bicolorable*, i.e., there exists a map $\phi : X \to \{0, 1\}$ such that $x \sim y$ implies $\phi(x) \neq \phi(y)$ for all $x, y \in X$;
(3) every closed path in $\mathcal{G}$ has even length;
(4) there exists $x_0 \in X$ such that every closed path containing x_0 has even length;
(5) given $x, y \in X$, then for all paths p connecting x and y one has $|p| \equiv d(x, y)$ mod 2, that is $|p| - d(x, y)$ is even.

Proposition 1.3.8 *Let $\mathcal{G} = (X, E)$ be a k-regular connected graph and denote by M the associated Markov matrix. Then the following conditions are equivalent:*

(1) *$\mathcal{G}$ is bipartite;*
(2) *the spectrum $\sigma(M)$ is symmetric: $\lambda \in \sigma(M)$ if and only if $-\lambda \in \sigma(M)$;*
(3) *$-1 \in \sigma(M)$.*

Proof   Suppose that $\mathscr{G}$ is bipartite with $X = X_0 \sqcup X_1$ and that $Mf = \lambda f$. Define $\widetilde{f} \in L(X)$ by setting $\widetilde{f}(x) := (-1)^j f(x)$ for all $x \in X_j$, $j = 0,1$. Then, for $x \in X_j$ we have:

$$[M\widetilde{f}](x) = \sum_{y:y\sim x} m(x,y)\widetilde{f}(y)$$
$$= (-1)^{j+1} \sum_{y:y\sim x} m(x,y)f(y)$$
$$= (-1)^{j+1}\lambda f(x)$$
$$= -\lambda\widetilde{f}(x).$$

We have shown that if λ is an eigenvalue for f, then $-\lambda$ is an eigenvalue for $\widetilde{f}$; this gives the implication (1) $\implies$ (2).

Since we always have $1 \in \sigma(M)$ (cf. Proposition 1.3.4), the implication (2) $\implies$ (3) is obvious.

Finally suppose that $Mf = -f$ with $f \in L(X)$ nontrivial and real valued. Denote by $x_0 \in X$ a point of maximum for $|f|$; then, up to switching f to $-f$, we may suppose that $f(x_0) > 0$. We then have that $-f(x_0) = \sum_{y:y\sim x_0} m(x_0,y)f(y)$ implies $\sum_{y:y\sim x_0} m(x_0,y)[f(x_0) + f(y)] = 0$. Since $f(x_0) + f(y) \geq 0$ we deduce $f(y) = -f(x_0)$ for all $y \sim x_0$. Set $X_j := \{y \in X : f(y) = (-1)^j f(x_0)\}$ for $j = 0,1$. We claim that $X = X_0 \sqcup X_1$: indeed $\mathscr{G}$ is connected and if $p = (x_0, x_1, \ldots, x_m)$ is a path, then $f(x_j) = (-1)^j f(x_0)$. Finally, if $y \sim z$ we clearly have $f(y) = -f(z)$ so that $\mathscr{G}$ is bicolorable, that is, it is bipartite. $\square$

Definition 1.3.9 (Distance-regular graphs)   (See also Definition 2.5.4.) A finite graph $\mathscr{G} = (X,E)$ with no loops is called *distance-regular* if there exist two sequences of constants, called the $\mathscr{G}$-*parameters*, $b_0, b_1, \ldots, b_N$ and $c_0, c_1, \ldots, c_N$, where $N = \mathrm{diam}(\mathscr{G})$, such that, for any pair of vertices $x,y \in X$ with $d(x,y) = i$ one has

$$|\{z \in X : d(x,z) = 1, d(y,z) = i+1\}| = b_i$$
$$|\{z \in X : d(x,z) = 1, d(y,z) = i-1\}| = c_i$$

for all $i = 0,1,\ldots,N$. In other words, if $d(x,y) = i$, then x has b_i neighbors at distance $i+1$ from y and c_i neighbors at distance $i-1$ from y. In particular, taking $x = y$ we get $b_0 = |\{z \in X : d(x,z) = 1\}|$, for all $x \in X$, that is, $\mathscr{G}$ is regular of degree b_0.

Exercise 1.3.10   Let $\mathscr{G}$ be a distance-regular graph. Show that the following hold:

(1) $b_N = 0 = c_0$;
(2) $c_1 = 1$;

(3) for $x,y \in X$ with $d(x,y) = i$ one has $|\{z \in X : d(x,z) = 1, d(y,z) = i\}| = b_0 - b_i - c_i$;

(4) for any $x \in X$, the cardinality $k_i := |\{y \in X : d(x,y) = i\}|$ of the sphere of radius i centered at x is given by $k_i = b_0 b_1 \cdots b_{i-1}/c_2 c_3 \cdots c_i$, for $i = 2, 3, \ldots, N$.

Let $\mathcal{G} = (X, E)$ be a distance-regular graph. For $j = 0, 1, \ldots, N$, we define the matrix $A_j = (A_j(x,y))_{x,y \in X}$ by setting

$$A_j(x,y) := \begin{cases} 1 & \text{if } d(x,y) = j \\ 0 & \text{otherwise.} \end{cases} \tag{1.14}$$

Note that $A_0 = I$ and A_1 is the adjacency matrix of $\mathcal{G}$. We denote by $\mathscr{A}$ the subalgebra of $\text{End}(L(X))$ generated by $A_0, A_1, \ldots, A_N$. It is called the *Bose–Mesner algebra* associated with $\mathcal{G}$ (see [3, 4] and [2]).

Proposition 1.3.11 *Let $\mathcal{G} = (X, E)$ be a distance-regular graph as in Definition 1.3.9. Then the following hold.*

(1) *For $j = 0, 1, \ldots, N$,*

$$A_j A_1 = b_{j-1} A_{j-1} + (b_0 - b_j - c_j)A_j + c_{j+1} A_{j+1}, \tag{1.15}$$

where $A_{N+1} = 0$.

(2) *For $j = 0, 1, \ldots, N$ there exists a real polynomial p_j of degree j such that*

$$A_j = p_j(A_1). \tag{1.16}$$

In particular, $\mathscr{A} = \{p(A_1) : p \text{ polynomial over } \mathbb{C}\}$ is commutative, and its dimension is $N + 1$. In fact, $A_0, A_1, \ldots, A_N$ constitute a vector space basis for $\mathscr{A}$.

(3) *Let*

$$L(X) = \oplus_{i=0}^n V_i \tag{1.17}$$

denote the decomposition into distinct eigenspaces of A_1, with V_0 the one-dimensional space of constant functions. Then $n = N$ and each V_i is invariant for all operators $A \in \mathscr{A}$. Moreover, if V_0 is the subspace of constant functions, the eigenvalue λ_0 of A_1 corresponding to V_0 is equal to the degree of X, that is, $\lambda_0 = b_0$.

(4) *Denote by E_i the orthogonal projection onto V_i and let λ_i denote the eigenvalue of A_1 corresponding to V_i. Then,*

$$A_j = \sum_{i=0}^N p_j(\lambda_i) E_i, \tag{1.18}$$

where p_j is the polynomial in (1.16). Similarly, the projection $E_i := q_i(A_1)$ for some polynomial q_i.

Proof   (1) For $f \in L(X)$ and $y \in X$ one clearly has

$$(A_j A_1 f)(y) = \sum_{\substack{z \in X: \\ d(z,y)=j}} (A_1 f)(z) =$$

$$= \sum_{\substack{z \in X: \\ d(z,y)=j}} \sum_{\substack{x \in X: \\ d(x,z)=1}} f(x) =$$

$$= \sum_{\substack{z \in X: \\ d(z,y)=j}} \left(\sum_{\substack{x \in X: \\ d(x,z)=1 \\ d(x,y)=j-1}} f(x) + \sum_{\substack{x \in X: \\ d(x,z)=1 \\ d(x,y)=j}} f(x) + \sum_{\substack{x \in X: \\ d(x,z)=1 \\ d(x,y)=j+1}} f(x) \right) =$$

$$= b_{j-1} \sum_{\substack{x \in X: \\ d(x,y)=j-1}} f(x) +$$

$$+ (b_0 - b_j - c_j) \sum_{\substack{x \in X: \\ d(x,y)=j}} f(x) +$$

$$+ c_{j+1} \sum_{\substack{x \in X: \\ d(x,y)=j+1}} f(x) =$$

$$= b_{j-1}(A_{j-1} f)(y) + (b_0 - b_j - c_j)(A_j f)(y) + c_{j+1}(A_{j+1} f)(y)$$

because for any x with $d(x,y) = j - 1$ there exist b_{j-1} elements $z \in X$ such that $d(x,z) = 1$ and $d(z,y) = j$, and therefore $f(x)$ appears b_{j-1} times in the above sums. A similar argument holds for $d(x,y) = j$ or $j+1$ (also recall (3) in Exercise 1.3.10). This shows (1.15).

(2) From (1) we get

$$A_1^2 = b_0 A_0 + (b_0 - b_1 - c_1)A_1 + c_2 A_2 \qquad (1.19)$$

that is

$$A_2 = \frac{1}{c_2} A_1^2 - \frac{b_0 - b_1 - c_1}{c_2} A_1 - \frac{b_0}{c_2} I =: p_2(A_1),$$

and the general case follows by induction (note that as X is connected, one always has $c_2, c_3, \ldots, c_N > 0$). In particular, $\mathscr{A}$ is commutative. Moreover $\{A_0 = I, A_1, \ldots, A_N\}$ is a vector space basis for $\mathscr{A}$. Indeed, for any polynomial p one has that $p(A_1)$ is a linear combination of $A_0 = I, A_1, \ldots, A_N$ (this is a converse to (1): as in (1.19), it follows from a repeated application of (1.15)). Moreover, if $\alpha_0, \alpha_1, \ldots, \alpha_N \in \mathbb{C}$ and $x, y \in X$, one has

$$\left(\sum_{j=0}^{N} \alpha_j A_j \delta_y \right)(x) = \alpha_{d(x,y)}$$

thus showing that $A_0, A_1, \ldots, A_N$ are also independent. We deduce that $\dim(\mathscr{A}) = N + 1$.

(3) Since $A_j = p_j(A_1)$ we have that V_i is also an eigenspace of the self-adjoint operator A_j with corresponding eigenvalue $p_j(\lambda_i)$ (below we shall prove that $n = N$). The fact that the eigenvalue λ_0 corresponding to the eigenspace V_0 equals the degree of X is nothing but a reformulation of the fact that, a graph X is connected (if and) only if 1 is an eigenvalue of multiplicity 1 of the Markov operator $M = \frac{1}{b_0} A_1$ (see Proposition 1.3.4).

(4) Denote by E_i the orthogonal projection onto V_i. From the preceding facts we deduce that $A_j = \sum_{i=0}^{n} p_j(\lambda_i) E_i$ for all $j = 0, 1, \ldots, N$. As the spaces V_i's are orthogonal, the corresponding projections E_i's are independent. Moreover, they belong to $\mathscr{A}$ as they are expressed as polynomials in A_1:

$$E_i = \frac{\prod_{j \neq i}(A_1 - \lambda_j I)}{\prod_{j \neq i}(\lambda_i - \lambda_j)}. \tag{1.20}$$

As a consequence, the operators $E_0, E_1, \ldots, E_n$ constitute another vector space basis for $\mathscr{A}$, and therefore $n = N$. □

Let $\mathscr{G} = (X, E)$ be a distance-regular graph and set $d_i := \dim V_i$ (cf. (1.17)). From the above theorem it follows that there exist *real* coefficients $\phi_i(j)$, $i, j = 0, 1, \ldots, N$ such that

$$E_i = \frac{d_i}{|X|} \sum_{j=0}^{N} \phi_i(j) A_j \tag{1.21}$$

for $i = 0, 1, \ldots, N$.

Definition 1.3.12 (Spherical function on a distance-regular graph) The function $\phi_i \in L(\{0, 1, \ldots, N\})$ is called the *spherical function* of X associated with V_i.

The factor $\frac{d_i}{|X|}$ in (1.21) is just a normalization constant.
The matrices

$$P = (p_j(\lambda_i))_{j,i=0,1,\ldots,N}$$

and

$$Q = \left(\frac{d_i}{|X|} \phi_i(j) \right)_{i,j=0,1,\ldots,N}$$

are called the *first* and the *second eigenvalue matrix* of X, respectively.

Lemma 1.3.13 (1) $P^{-1} = Q$ (that is $\frac{d_i}{|X|} \sum_{j=0}^{N} \phi_i(j) p_j(\lambda_h) = \delta_{i,h}$);
(2) $\phi_i(j) = \frac{1}{k_j} p_j(\lambda_i)$, where k_j is as in Exercise 1.3.10.(4), for all $i, j = 0, 1, \ldots, N$;
(3) $\phi_0(j) = 1$ for all $j = 0, 1, \ldots, N$;

(4) $\phi_i(0) = 1$ for all $i = 0, 1, \ldots, N$;

(5) $\lambda_i = b_0 \phi_i(1)$.

Proof We leave the proof as an exercise (see [15, proof of Lemma 5.1.8]). □

Theorem 1.3.14 (1) *The spherical functions satisfy the following orthogonality relations:*

$$\sum_{j=0}^{N} k_j \phi_i(j) \phi_h(j) = \frac{|X|}{d_i} \delta_{i,h} \qquad (1.22)$$

for all $i, h = 0, 1, \ldots, N$.

(2) *We have the following finite difference equations:*

$$c_j \phi_i(j-1) + (b_0 - b_j - c_j)\phi_i(j) + b_j \phi_i(j+1) = \lambda_i \phi_i(j) \qquad (1.23)$$

for all $i, j = 0, 1, \ldots, N$ (we use the convention that $\phi_i(-1) = \phi_i(N+1) = 0$).

Proof (1) This is easily established by explicitly writing the coefficients in $QP = I$ and then using Lemma 1.3.13 in order to express $p_j(\lambda_i) = k_j \phi_i(j)$.

(2) From (1.18) and Lemma 1.3.13 we deduce

$$A_j = \sum_{i=0}^{N} k_j \phi_i(j) E_i. \qquad (1.24)$$

From Proposition 1.3.11 and (1.24) we deduce

$$A_1 A_j = b_{j-1} A_{j-1} + (b_0 - b_j - c_j) A_j + c_{j+1} A_{j+1} =$$

$$= \sum_{i=0}^{N} \left[b_{j-1} k_{j-1} \phi_i(j-1) + (b_0 - b_j - c_j) k_j \phi_i(j) + \right. \qquad (1.25)$$

$$\left. + c_{j+1} k_{j+1} \phi_i(j+1) \right] E_i.$$

On the other hand, as $A_1 E_i = E_i A_1 = \lambda_i E_i$ (recall that $A_1 = \sum_{i=0}^{N} \lambda_i E_i$), multiplying both sides of (1.24) by A_1 we obtain

$$A_1 A_j = \sum_{i=0}^{N} k_j \phi_i(j) \lambda_i E_i. \qquad (1.26)$$

Equating the two expressions of $A_1 A_j$ in (1.25) and (1.26) we obtain

$$b_{j-1} \frac{k_{j-1}}{k_j} \phi_i(j-1) + (b_0 - b_j - c_j)\phi_i(j) + c_{j+1} \frac{k_{j+1}}{k_j} \phi_i(j+1) = \lambda_i \phi_i(j).$$

Then (1.23) follows from Exercise 1.3.10.(4). □

In the monograph [53] one may find several examples of orthogonal polynomials satisfying systems of equations such as (1.23).

Example 1.3.15 (The discrete circle)   As a first example of distance-regular graph, we examine the discrete circle $\mathscr{C}_n$ on $n \geq 3$ vertices. The vertex set $X := \{0, 1, \ldots, n-1\}$ and the edges are $e = \{x, x+1\}$ with $x \in X$ with summation modulo n. Clearly its diameter is given by $\mathrm{diam}(\mathscr{C}_n) = [n/2]$, the integer part of $n/2$. We leave it as an exercise to check that $\mathscr{C}_n$ is distance-regular with $N = [n/2]$ and parameters $b_0 = 2, b_1 = b_2 = \ldots = b_{N-1} = 1, c_1 = c_2 = \ldots = c_{N-1} = 1$ and, finally, $c_N = 1$ if n is odd and $c_N = 2$ if n is even.

In the present setting, the difference equations (1.23) become

$$\begin{cases} \phi_i(j-1) + \phi_i(j+1) = 2\phi_i(1)\phi_i(j) & \text{for } 1 \leq j \leq N-1 \\ \phi_i(N-1) + \phi_i(N) = 2\phi_i(1)\phi_i(N) & \text{if } n \text{ is odd} \\ 2\phi_i(N-1) = 2\phi_i(1)\phi_i(N) & \text{if } n \text{ is even} \end{cases}$$

for all $i = 0, 1, \ldots, N$. Recalling the prosthaphæresis formula

$$\cos \alpha + \cos \beta = 2\cos((\alpha + \beta)/2)\cos((\alpha - \beta)/2)$$

we deduce that $\phi_i(j) = \cos(2\pi i j / n)$ for all $0 \leq i, j \leq N$.

Keeping in mind the decomposition (1.13) (where now "m" is replaced by "N"), we compute the dimension d_i of the subspaces V_i, for $i = 0, 1, \ldots, N$. Suppose first that n is even, so that $N = n/2$. We have $k_0 = 1$ and, from the orthogonality relations (1.22), the parameters yield (cf. Exercise 1.3.10.(4))

$$k_i = b_0 b_1 \cdots b_{i-1} / c_2 c_3 \cdots c_i = 2$$

for all $1 \leq i \leq N-1$, and

$$k_N = b_0 b_1 \cdots b_{N-1} / c_2 c_3 \cdots c_N = 1.$$

We have $\dim(V_0) = 1$ (this is the dimension of the constant valued functions). Moreover, for $1 \leq i \leq N-1$ we have

$$\sum_{j=0}^{n/2} k_j \phi_i^2(j) = \phi_i^2(0) + 2 \sum_{j=1}^{n/2-1} \cos^2(2\pi i j / n) + \phi_i^2(n/2)$$

$$= 2 + 2 \sum_{j=1}^{n/2-1} \frac{1 + \cos(4\pi i j / n)}{2}$$

$$= 2 + (n/2 - 1) + \sum_{j=1}^{n/2-1} \cos(4\pi i j / n)$$

$$= n/2 \equiv N,$$

where, denoting by ω a primitive Nth root of unity, we used the equality

$$\sum_{j=1}^{n/2-1} \cos(4\pi i j/n) = \sum_{j=1}^{N-1} \cos(2\pi i j/N) = \Re(\omega + \omega^2 + \cdots + \omega^{N-1}) = -1.$$

We deduce that $d_i = n/(\sum_{j=0}^{n/2} k_j \phi_i^2(j)) = n/(n/2) = 2$. Finally, as $\phi_{n/2}^2(j) = 1$ for all $j = 0, 1, \ldots, n/2$, we deduce that $d_N = 1$.

We leave it as an exercise to check that, for n odd, one has $d_0 = 1$ and $d_i = 2$ for all $i = 1, 2, \ldots, N = (n-1)/2$.

Example 1.3.16 (Complete graph)   The *complete graph* $K_n = (X, E)$ on n vertices is defined by setting $X := \{1, 2, \ldots, n\}$ and

$$E := \{\{x, y\} : x, y \in X, \ x \neq y\}$$

(Figure 1.2). It is clear that $N := \mathrm{diam}(K_n) = 1$ and that K_n is a distance regular graph with parameters $b_0 = n - 1$, $b_1 = b_N = 0$, $c_0 = 0$, and $c_1 = 1$.

  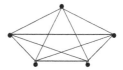

Figure 1.2   The complete graphs K_3, K_4, and K_5.

The decomposition (1.13) becomes $L(X) = \sum_{i=0}^{N} V_i = V_0 \oplus V_1$, where, as usual V_0 is the one-dimensional subspace of constant valued functions on X, and $V_1 = \{f \in L(X) : \sum_{x \in X} f(x) = 0\}$ is the (orthogonal) $(n-1)$-dimensional subspace of 0-mean-valued functions on X. The spherical functions $\phi_0, \phi_1 \in L(\{0, 1\})$ are given by $\phi_0(j) = 1$ for $j = 0, 1$, and $\phi_1(0) = 1$ and $\phi_1(1) = -1/(n-1)$, as one easily deduces (exercise) from the orthogonality relations (1.22).

Example 1.3.17 (Hamming scheme and hypercube)   Set $X_{n,m+1} := \{0, 1, 2, \ldots, m\}^n$. The map $d : X_{n,m+1} \times X_{n,m+1} \to \mathbb{N}$ defined by setting

$$d(x, y) := |\{k : x_k \neq y_k\}|$$

for all $x = (x_1, x_2, \ldots, x_n), y = (y_1, y_2, \ldots, y_n) \in X_{n,m+1}$, is easily seen to be a metric on $X_{n,m+1}$, called the *Hamming distance* on $X_{n,m+1}$.

We define a graph $\mathscr{G}_{n,m+1}^{\mathrm{H}} = (X_{n,m+1}, E_{n,m+1})$ by setting

$$E_{n,m+1} := \{\{x, y\} : x, y \in X_{n,m+1} \text{ such that } d(x, y) = 1\}.$$

Then $\mathscr{G}_{n,m+1}^{\mathrm{H}}$ is a distance regular graph with diameter n and parameters

$$c_i = i, \qquad i = 1, 2, \ldots, n$$
$$b_i = (n-i)m, \qquad i = 0, 1, \ldots, n-1.$$

In particular, its degree is $b_0 = nm$. We leave it as an exercise to check for the details. See also page 134.

Note that for $n = 1$, the graph $\mathscr{G}_{1,m+1}^{\mathrm{H}}$ coincides with the complete graph K_{m+1} on $m+1$ vertices. Moreover, in this case, we always have $b_i + c_i = b_0$ (cf. Exercise 1.3.10).

For $m = 1$, the graph $\mathscr{G}_{n,2}^{\mathrm{H}}$ is called the *n-hypercube*, denoted Q_n (Figure 1.3).

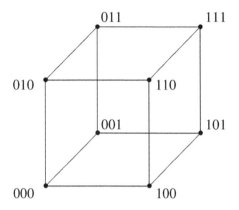

Figure 1.3 The *3-hypercube Q_3*.

We refer to [15, Section 2.6 and Section 5.3] for the expressions of the spherical functions and the computation of the dimensions of the corresponding eigenspaces for Q_n and $\mathscr{G}_{n,m+1}^{\mathrm{H}}$, respectively. Note that the spherical functions constitute an important family of orthogonal polynomials, called the *Krawtchouk polynomials*.

Example 1.3.18 (Johnson scheme)   Let n be a positive integer. For $0 \le m \le n$ denote by $\Omega_{m,n}$ the set of all m-subsets of $\{1, 2, \ldots, n\}$. The map $d \colon \Omega_{m,n} \times \Omega_{m,n} \to \mathbb{N}$ defined by setting

$$d(A,B) := m - |A \cap B|$$

for all $A, B \in \Omega_{m,n}$ is easily seen to be a metric on $\Omega_{m,n}$, called the *Johnson distance* on $\Omega_{m,n}$.

We define a graph $\mathscr{G}_{m,n}^{\mathrm{J}} = (\Omega_{m,n}, E_{m,n})$ by setting

$$E_{m,n} := \{\{A,B\} : A, B \in \Omega_{m,n} \text{ such that } d(A,B) = 1\}.$$

We leave it as an exercise to check that $\mathscr{G}^J_{m,n}$ is a distance regular graph with diameter $\min\{m, n-m\}$ and parameters $c_i = i^2$ and $b_i = (n-m-i)(m-i)$ for $i = 0, 1, \ldots, \min\{m, n-m\}$.

We refer to [15, Section 6.1] for the expression of the spherical functions and the computation of the dimensions of the corresponding eigenspaces.

The book by Brouwer, Cohen, and Neumaier [8] is an encyclopedic treatment of distance-regular graphs.

1.3.2 Strongly regular graphs

This section is devoted to an interesting subclass of distance regular graphs (see also Section 2.5 for asymptotic aspects as well as Sections 3.2 and 3.4 for more combinatorial aspects of distance regular graphs).

Definition 1.3.19 A finite simple graph $\mathscr{G} = (X, E)$ without loops is called *strongly regular* with *parameters* (v, k, λ, μ) if

(1) it is regular of degree k and $|X| = v$;

(2) for all $\{x, y\} \in E$ there exist exactly λ vertices adjacent to both x and y;

(3) for all $x, y \in X$ with $x \neq y$ and $\{x, y\} \notin E$ there exist exactly μ vertices adjacent to both x and y.

It is customary to exclude graphs which satisfy the definition trivially, namely those graphs which are the disjoint union of one or more equal-sized complete graphs, and their complements (cf. Exercise 1.3.23 below). But we warn the reader that this convention is not adopted in Section 3.1.8, where these examples play an important role.

Note that, in the above definition, $0 \leq \lambda \leq k-1$ and $0 \leq \mu \leq k$. Moreover, if $\mu > 0$ then $\mathscr{G}$ is connected. In the following we shall always assume that $\mu > 0$.

Remark 1.3.20 Let $\mathscr{G} = (X, E)$ be a strongly regular graph with parameters (v, k, λ, μ) such that $\mu > 0$. By our assumptions on μ, given any two non-adjacent vertices there exists $z \in X$ such that $x \sim z$ and $z \sim y$, so that $d(x, y) = 2$. It follows that $N := \operatorname{diam}(\mathscr{G}) = 2$. Then, it is easy to check that $\mathscr{G}$ is a distance regular graph with parameters $(b_0, b_1, b_2) = (k, k-1-\lambda, 0)$ and $(c_0, c_1, c_2) = (0, 1, \mu)$.

Proposition 1.3.21 *Let* $\mathscr{G} = (X, E)$ *be a connected strongly regular graph with parameters* (v, k, λ, μ) *and denote by* A *its adjacency matrix. Let* $L(X) =$

$V_0 \oplus V_1 \oplus V_2$ *denote the decomposition (1.13), where, as usual V_0 is the one-dimensional subspace of constant-valued functions on X. The associated spherical function ϕ_0, ϕ_1, and ϕ_2 are given by*

- $\phi_0(j) = 1$ *for all* $j = 0,1,2$;
- $\phi_1(0) = 1$, $\phi_1(1) = \frac{\lambda - \mu + \sqrt{\Delta}}{2k}$, *and* $\phi_1(2) = -\frac{\mu(2 + \lambda - \mu + \sqrt{\Delta})}{2k(k-1)}$;
- $\phi_2(0) = 1$, $\phi_2(1) = \frac{\lambda - \mu - \sqrt{\Delta}}{2k}$, *and* $\phi_2(2) = -\frac{\mu(2 + \lambda - \mu - \sqrt{\Delta})}{2k(k-1)}$,

where $\Delta = (\lambda - \mu)^2 + 4(k - \mu)$. The dimensions $d_i = \dim(V_i)$, $i = 0,1,2$ are given by

- $d_0 = 1$;
- $d_1 = \frac{1}{2}\left[(v-1) - \frac{2k + (v-1)(\lambda - \mu)}{\sqrt{\Delta}}\right]$;
- $d_2 = \frac{1}{2}\left[(v-1) + \frac{2k + (v-1)(\lambda - \mu)}{\sqrt{\Delta}}\right]$.

Proof Formula (1.15) for $j = 1$ becomes (recall that $A = A_1$):

$$A^2 = kI + \lambda A + \mu(J - I - A)$$

which is equivalent to

$$A^2 + (\mu - \lambda)A + (\mu - k)I = \mu J,$$

where J is the $X \times X$ matrix consisting only of ones ($J_{x,y} = 1$ for all $x, y \in X$). Since the operator J, restricted to the subspace

$$L(X) \ominus V_0 := \{f \in L(x) : \sum_{x \in X} f(x) = 0\}$$

is the 0 operator, we deduce that the eigenvalues $t_1, t_2 \in \sigma(A) \setminus \{k\}$ satisfy the equation

$$t^2 + (\mu - \lambda)t + (\mu - k) = 0.$$

Therefore, up to a transposition of the indices, we have

$$t_1 := \frac{\lambda - \mu + \sqrt{\Delta}}{2} \quad \text{and} \quad t_2 := \frac{\lambda - \mu - \sqrt{\Delta}}{2}.$$

From Lemma 1.3.13(5), we obtain the above values $\phi_i(1)$ for $i = 1, 2$. Finally, the values of $\phi_i(2)$ for $i = 1, 2$ are easily deduced from the orthogonality relations (1.22).

For the dimensions of the eigenspaces, as usual we have $d_0 = \dim(V_0) = 1$. Moreover, from the identities $d_1 + d_2 = \dim(L(X)) - 1 = |X| - 1 = v - 1$ and $0 = \text{tr}(A) = k + t_1 d_1 + t_2 d_2$ one deduces the corresponding expressions for d_1 and d_2. $\square$

Exercise 1.3.22   Let $m \geq 4$ and denote by X the set of all 2-element subsets of $\{1, 2, \ldots, m\}$. The *triangular graph* $T(m)$ is the finite graph with vertex set X and such that two distinct vertices are adjacent if they are not disjoint.

Show that $T(m)$ is strongly regular with parameters $v = \binom{m}{2}$, $k = 2(m-2)$, $\lambda = m - 2$, and $\mu = 4$.

Exercise 1.3.23 (Complement of a graph)   Let $\mathscr{G} = (X, E)$ be a finite simple graph without loops. The *complement* of $\mathscr{G}$ is the graph $\overline{\mathscr{G}}$ with vertex set X and edge set $\overline{E} = \{\{x, y\} : x, y \in X, x \neq y, \{x, y\} \notin E\}$.

(1) Show that if $\mathscr{G}$ is strongly regular with parameters (v, k, λ, μ), then $\overline{\mathscr{G}}$ is strongly regular with parameters $(v, v - k - 1, v - 2k + \mu - 2, v - 2k + \lambda)$.
(2) From (1) deduce that the parameters of a strongly regular graph satisfy the inequality $v - 2k + \mu - 2 \geq 0$.
(3) Suppose that $\mathscr{G}$ is strongly regular. Show that $\mathscr{G}$ and $\overline{\mathscr{G}}$ are both connected if and only if $0 < \mu < k < v - 1$. If this is the case, one says that $\mathscr{G}$ is *primitive*.
   Hint: show that $\mu = 0$ implies $\lambda = k - 1$ and write $\mu < k$ in the form $v - 2k + \mu - 2 < (v - k - 1) - 1$.

Example 1.3.24 (Petersen graph)   The complement of the triangular graph $T(5)$ (see Exercise 1.3.22) is the celebrated *Petersen graph* (see Figure 1.4). It is a connected strongly regular graph with parameters $(10, 3, 0, 1)$. The monograph [45] is entirely devoted to this graph which turned out to serve as a counterexample to several important conjectures.

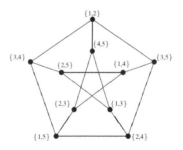

Figure 1.4 The Petersen graph

Example 1.3.25 (Clebsch graph)   The *Clebsch graph* (see Figure 1.5) is defined as follows. The vertex set X consists of all subsets of even cardinality of the set $\{1, 2, 3, 4, 5\}$. Moreover, two vertices $A, B \in X$ are adjacent if $|A \triangle B| = 4$ (here $\triangle$ denotes the symmetric difference of two sets). We leave it as an exercise to show that it is a strongly regular graph with parameters $(16, 5, 0, 2)$.

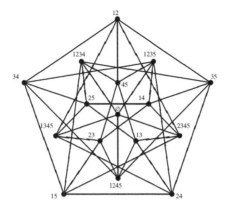

Figure 1.5 The Clebsch graph

For more on strongly regular graphs we refer to the monographs by van Lint and Wilson [49] and Godsil and Royle [42].

1.4 Association schemes

In this section we give the definition of an association scheme and discuss some examples. Association schemes constitute a central notion in Algebraic Combinatorics, which is "the approach to combinatorics – formulated in Ph. Delsarte's monumental and epochal thesis [25] in 1973 – enabling us to look at a wide range of combinatorial problems from a unified viewpoint" [3]. There are several beautiful books devoted to this subject: we mention, among others, those by Eiichi Bannai and Ito [3] and the new edition, written in collaboration with Etsuko Bannai and Rie Tanaka [4], Bailey [2], Godsil [41], van Lint and Wilson [49], Cameron [11], Cameron and van Lint [12], MacWilliams and Sloane [51], and by P.-H. Zieschang [73].

We finally present a generalization expressed in terms of hypergroups. We refer to the monograph [22] by Corsini and Leoreanu for a comprehensive treatment of the theory of hypergroups.

Definition 1.4.1 Let X be a finite set. An *association scheme* on X is a partition

$$X \times X = \mathscr{C}_0 \sqcup \mathscr{C}_1 \sqcup \ldots \sqcup \mathscr{C}_N,$$

where the sets $\mathscr{C}_i$ (called the *associate classes*) satisfy the following properties:

(1) $\mathscr{C}_0 = \{(x,x) : x \in X\}$ is the diagonal;

(2) for each $i = 1, 2, \ldots, N$, there exists i' with $1 \leq i' \leq N$ such that $\mathscr{C}_{i'} = \mathscr{C}_i^*$, where $\mathscr{C}_i^* := \{(y,x) \in X \times X : (x,y) \in \mathscr{C}_i\}$;

(3) there exist nonnegative integers (called the *parameters* of the scheme) $p_{i,j}^k$, $i, j, k = 0, 1, \ldots, N$, such that

$$|\{z \in X : (x,z) \in \mathscr{C}_i, (z,y) \in \mathscr{C}_j\}| = p_{i,j}^k$$

for all $(x,y) \in \mathscr{C}_k$.

Moreover, the association scheme is called *commutative* (resp. *symmetric*) provided $p_{i,j}^k = p_{j,i}^k$ (resp. $\mathscr{C}_i = \mathscr{C}_i^*$; equivalently, $i' = i$) for all $1 \leq i, j, k \leq N$.

Note that symmetry implies commutativity.

Let X be a finite set and let $(\mathscr{C}_j)_{j=0}^N$ be an association scheme on X. For $j = 0, 1, \ldots, N$, we define the matrix $A_j = (A_j(x,y))_{x,y \in X}$ by setting

$$A_j(x,y) := \begin{cases} 1 & \text{if } (x,y) \in \mathscr{C}_j \\ 0 & \text{otherwise.} \end{cases} \tag{1.27}$$

Note that $A_0 = I$. The subalgebra $\mathscr{A} \subseteq \text{End}(L(X))$ generated by $A_0, A_1, \ldots, A_N$ is called the *adjacency algebra* (or, when it is commutative, the *Bose–Mesner algebra*) associated with the association scheme $(\mathscr{C}_j)_{j=0}^N$ on X (see [3, 4, 2]). We remark that condition (3) in Definition 1.4.1 is equivalent to the following condition on $\mathscr{A}$:

$$A_i A_j = \sum_{k=0}^N p_{i,j}^k A_k \tag{1.28}$$

for all $0 \leq i, j \leq N$.

Example 1.4.2 (Groups as association schemes)   Every finite group naturally gives rise to an association scheme over its underlying set. Indeed, given a finite group G, for $g \in G$ set

$$\mathscr{C}_g := \{(h,k) \in G \times G : h^{-1}k = g\}.$$

We then have, $\mathscr{C}_{1_G} = \{(g,g) : g \in G\}$ is the diagonal. Moreover, $\mathscr{C}_g^* = \mathscr{C}_{g^{-1}}$, in other words, $g' = g^{-1}$, for all $g \in G$. Finally, the parameters

$$p_{g,h}^k := \begin{cases} 1 & \text{if } k = gh \\ 0 & \text{otherwise,} \end{cases}$$

for all $g, h, k \in G$, trivially satisfy (3).

Note that G is commutative if and only if the corresponding association scheme is commutative. Also the association scheme is symmetric exactly if every nontrivial element in G has period 2 (that is, G is an elementary abelian 2-group).

We leave it as an easy exercise to check that the associated adjacency algebra $\mathscr{A}$ is isomorphic to the group algebra $L(G) = \{f : G \to \mathbb{C}\}$ of G (equipped with the convolution product (1.1)).

Example 1.4.3 (Association scheme associated with a group action)   Let G be a finite group acting transitively on a set X. Consider the diagonal action of G on $X \times X$ and denote by $\mathscr{C}_0, \mathscr{C}_1, \ldots, \mathscr{C}_N$ (with $\mathscr{C}_0 = \{(x,x) : x \in X\}$) the corresponding orbits. Let us show that $\mathscr{C}_0, \mathscr{C}_1, \ldots, \mathscr{C}_N$ form an association scheme over X. The fact that $\mathscr{C}_0, \mathscr{C}_1, \ldots, \mathscr{C}_N$ form a partition of $X \times X$ and that $\mathscr{C}_0$ is the diagonal (cf. Definition 1.4.1(1)) immediately follows from the definitions. Let $1 \leq i \leq N$ and let $(x,y) \in \mathscr{C}_i$. Then denoting by $\mathscr{C}_{i'}$ the G-orbit of (y,x), we clearly have $\mathscr{C}_i^* = \mathscr{C}_{i'}$. This shows (2). Finally, let $1 \leq i, j, k \leq N$ and suppose that $(x,y), (x',y') \in \mathscr{C}_k$. Let $X_{x,y} := \{z \in X : (x,z) \in \mathscr{C}_i \text{ and } (z,y) \in \mathscr{C}_j\}$ and $X_{x',y'} := \{z' \in X : (x',z') \in \mathscr{C}_i \text{ and } (z',y') \in \mathscr{C}_j\}$. Let $g \in G$ such that $(gx, gy) = g(x,y) = (x',y')$. Then the map $\varphi \colon X_{x,y} \to X_{x',y'}$ defined by setting $\varphi(z) := gz$ for all $z \in X_{x,y}$ is well defined and bijective. Indeed, we have $(x,z) \in \mathscr{C}_i$ (resp. $(z,y) \in \mathscr{C}_j$) if and only if $(x', \varphi(z)) = g(x,z) \in \mathscr{C}_i$ (resp. $(\varphi(z), y') = g(z,y) \in \mathscr{C}_j$), and the inverse map is $\varphi^{-1}(z') := g^{-1}z'$ for all $z' \in X_{x',y'}$. This shows that the parameter $p_{i,j}^k$ is well defined, and (3) follows as well.

Let $x_0 \in X$ and denote by $K = \mathrm{Stab}_G(x_0)$ its stabilizer in G. We leave to the reader the following exercise:

(1) Show that the associated Bose–Mesner algebra $\mathscr{A}$ is isomorphic to the algebras $\mathrm{End}_G(L(X))$ and $^K L(G)^K$ (cf. Proposition 1.2.52 and its proof).
(2) Show that the association scheme in (1) is commutative if and only if (G, K) is a Gelfand pair.
(3) Show that the association scheme in (1) is symmetric if and only if (G, K) is a symmetric Gelfand pair (cf. Exercise 1.2.57).

Example 1.4.4 (Association scheme associated with conjugacy classes on a finite group)   Let G be a finite group. Given $g \in G$, denote by $C(g) := \{h^{-1}gh : h \in G\}$ its conjugacy class. Let $C := \{C(g) : g \in G\}$ be the set of all conjugacy classes of G and denote by $c_0 := C(1_G) = \{1_G\}$ the conjugacy class of the identity element of G. Note that $C(g^{-1}) = \{h^{-1}g^{-1}h : h \in G\} = \{(h^{-1}gh)^{-1} : h \in G\} = C(g)^{-1}$ for all $g \in G$. For $c \in C$ we then set

$$\mathscr{C}_c := \{(x,y) \in G \times G : x^{-1}y \in c\}.$$

We have $\mathscr{C}_0 := \mathscr{C}_{c_0} = \{(g,g) : g \in G\}$ is the diagonal. Moreover, $\mathscr{C}_c^* = \mathscr{C}_{c^{-1}}$, in other words, $c' = c^{-1}$, for all $c \in C$. Let now $c_1, c_2, c_3 \in C$ and suppose that

$(x_1, y_1), (x_2, y_2) \in \mathcal{C}_{c_1}$, that is, $x_i^{-1} y_i \in c_1$. Let also $z_1 \in G$. Then $(x_1, z_1) \in \mathcal{C}_{c_2}$ and $(z_1, y_1) \in \mathcal{C}_{c_3}$ if and only if $x_1^{-1} z_1 \in c_2$ and $z_1^{-1} y_1 \in c_3$; equivalently,

$$z_1 \in x_1 c_2 \cap y_1 (c_3)^{-1} = x_1(c_2 \cap x_1^{-1} y_1 (c_3)^{-1}).$$

Analogously, for $z_2 \in G$ one has $(x_2, z_2) \in \mathcal{C}_{c_2}$ and $(z_2, y_2) \in \mathcal{C}_{c_3}$ if and only if $z_2 \in x_2(c_2 \cap (x_2)^{-1} y_2 (c_3)^{-1})$. As $x_1^{-1} y_1, (x_2)^{-1} y_2 \in c_1$, there exists $t \in G$ such that $(x_2)^{-1} y_2 = t^{-1}(x_1^{-1} y_1) t$. We deduce that

$$\begin{aligned} |x_1(c_2 \cap x_1^{-1} y_1 (c_3)^{-1})| &= |c_2 \cap x_1^{-1} y_1 (c_3)^{-1}| \\ &= |t^{-1}(c_2 \cap x_1^{-1} y_1 (c_3)^{-1}) t| \\ &= |c_2 \cap (x_2)^{-1} y_2 (c_3)^{-1}| \\ &= |x_2(c_2 \cap (x_2)^{-1} y_2 (c_3)^{-1})|. \end{aligned}$$

This shows that the parameter $p_{c_2, c_3}^{c_1} = |c_2 \cap x^{-1} y(c_3)^{-1}|$ is well defined, that is, it does not depend on the choice of $(x, y) \in c_1$. This completes the proof that the $\mathcal{C}_c$s form an association scheme.

It is easy to see that the association scheme is commutative. Clearly, it is symmetric if and only if every $g \in G$ is conjugate to its inverse g^{-1}, a condition which is usually expressed by saying that the group G is *ambivalent*.

We leave it as an easy exercise to check that the associated Bose–Mesner algebra $\mathscr{A}$ is isomorphic to the subalgebra

$$L_c(G) := \{f \in L(G) : f(g) = f(h^{-1} gh) \text{ for all } g, h \in G\}$$

of *conjugacy-invariant functions* on G (equipped with the convolution product (1.1)).

Another interesting class of association schemes is provided by distance-regular graphs:

Proposition 1.4.5 (Distance-regular graphs are association schemes) *Let $\mathscr{G} = (X, E)$ be a distance-regular graph with diameter N, and set*

$$\mathscr{C}_i := \{(x, y) \in X \times X : d(x, y) = i\},$$

for $i = 0, 1, \ldots, N$. Then $\mathscr{C}_0, \mathscr{C}_1, \ldots, \mathscr{C}_N$ form a symmetric association scheme over X.

Proof We clearly have (1) $\mathscr{C}_0 = \{(x, x) : x \in X\}$ is the diagonal and (2) $\mathscr{C}_i$ is *symmetric* for $i = 1, 2, \ldots, N$. Consider the matrices $A_0, A_1, A_2, \ldots, A_N$ defined in (1.14). Recall that these constitute a vector space basis for the corresponding Bose–Mesner algebra $\mathscr{A} \subset \mathrm{End}(L(X))$ (cf. Proposition 1.3.11(2)). These are exactly the matrices defined in (1.14). In this setting, (1.28) (which is equivalent to condition (3) in Definition 1.4.1) follows from Proposition 1.3.11(2).   □

As a consequence, the Hamming scheme (cf. Example 1.3.17) and the Johnson scheme (cf. Example 1.3.18) are symmetric (and therefore commutative) association schemes.

A peculiarity of a distance-regular graph is that, as remarked above (cf. Proposition 1.3.11(2)), its Bose–Mesner algebra is singly generated, namely by A_1. This is no longer true for general symmetric association schemes: see, for instance, [15, Chapter 7].

The following definition yields a generalization of the notion of an association scheme.

Definition 1.4.6 (Hypergroups)   A finite *(algebraic) hypergroup* is a pair $(X, *)$, where X is a nonempty finite set equipped with a multi-valued map, called *hyperoperation* and denoted $*$, from $X \times X$ to $\mathscr{P}^*(X)$, the set of all nonempty subsets of X, satisfying the following properties:

(1) $(x * y) * z = x * (y * z)$ for all $x, y, z \in X$ (*associative property*);
(2) $x * X = X * x = X$ for all $x \in X$ (*reproduction property*),

where, for subsets $Y, Z \subset X$, one defines $Y * Z = \bigcup_{y \in Y, z \in Z} y * z \subset X$.
  If, in addition one has

(3) $x * y = y * x$ for all $x, y \in X$ (*commutative property*)

one says that $(X, *)$ is commutative.
  Also, an element $e \in X$ is called a *unit* provided that

(4) $x \in (e * x) \cap (x * e)$ for all $x \in X$.

Finally, given $x \in X$, an element $y \in X$ such that there is a unit e with

(5) $e \in (x * y) \cap (y * x)$

is called an *inverse* of x.

For another equivalent definition, under the name of *functional hypergroup*, we refer to [30, 31, 48] (see also [20, Appendix 3.3 and Example A.1]).
  We remark that conditions (1), (4), and (5) imply condition (2). Suppose indeed that the hyperoperation $*$ is associative, that a unit $e \in X$ exists, and every element $x \in X$ has an inverse. Given $x, z \in X$, let $y \in X$ be an inverse of x. Then $x * X \supset x * (y * z) = (x * y) * z \supset (e * z) \ni z$, and, similarly, $X * x \ni z$. As z was arbitrary, this proves (2).

Example 1.4.7 (Association schemes are hypergroups)   Let $\overline{X}$ be a finite set and let $\mathscr{C}_0, \mathscr{C}_1, \ldots, \mathscr{C}_N$ be an association scheme on $\overline{X}$. Set $X := \{\mathscr{C}_0, \mathscr{C}_1, \ldots, \mathscr{C}_N\}$. For $0 \leq i, j \leq N$ we set

$$\mathscr{C}_i * \mathscr{C}_j := \{\mathscr{C}_k : p_{i,j}^k \neq 0\}.$$

Let us show that $*$ is a hyperoperation turning $(X, *)$ into a hypergroup. Let A_j, $j = 0, 1, \ldots, N$ denote the matrices as in (1.27). Keeping in mind (1.28), the associative property of $*$ is easily deduced from the associative property of the product of matrices. Moreover, the element $e := \mathscr{C}_0$ is, clearly, a unit. Finally, it is straightforward that, for every $0 \leq j \leq N$, the class $\mathscr{C}_{j'} = \mathscr{C}_j^*$ is an inverse of $\mathscr{C}_j$. It follows from the above remark that $(X, *)$ is a hypergroup.

Example 1.4.8 (Dual of a finite group as hypergroup)   Let G be a finite group. Then $X = \widehat{G}$, the dual of the group G, is an algebraic hypergroup after setting $x * y = \{z \in X : z \preceq x \otimes y\}$ for all $x, y \in X$. Moreover, the trivial representation $\iota_G \in X$ serves as a unit for the hypergroup and, given any $x \in X$, the conjugate representation $x' \in X$ (cf. Definition 1.2.10) serves as an inverse (cf. Exercise 1.2.47).

Finally in this section, we note that symmetric association schemes reappear in the definition of partially balanced designs in Section 3.1.8.

1.5 Applications of Gelfand pairs to probability

In this section, we illustrate the methods developed by Persi Diaconis and his collaborators which use representation theory of finite groups to determine the asymptotic behaviour of several mixing processes (typically finite Markov chains, e.g., random walks, invariant under the action of a finite group of symmetries). We illustrate this focusing on the Ehrenfest diffusion model which presents the so-called *cut-off phenomenon*. The standard reference is Diaconis' book [26]. We also based our exposition on our own monograph [15].

1.5.1 Markov chains

Definition 1.5.1   A *finite Markov chain* is a triple (X, P, ν_0), where X is a finite set, called the *state space*, $P = (p(x,y))_{x,y \in X}$, called the *transition matrix*, is a *stochastic matrix*, i.e.,

$$\begin{cases} p(x,y) \geq 0 & \text{for all } x, y \in X \\ \sum_{y \in X} p(x,y) = 1 & \text{for all } x \in X, \end{cases}$$

and $v_0 \colon X \to [0,1]$, called the *initial distribution*, is a *probability distribution* on X, i.e.,

$$\begin{cases} v_0(x) \geq 0 & \text{for all } x \in X \\ \sum_{x \in X} v_0(x) = 1. \end{cases}$$

In the standard definition, a (discrete-time) Markov chain is a sequence of random variables $X_1, X_2, X_3, \ldots$ with the so-called *Markov property*, namely that the probability of moving to the next state depends only on the present state and not on the previous states, but, for our purposes, Definition 1.5.1 suffices. We can interpret a finite Markov chain (X, P, v_0) as a *random walk* on X: at time $t = 0$ the random walker is in state $x = x(0)$ with probability $v_0(x)$. If at time t he or she is in state $x = x(t) \in X$, then at time $t+1$ he or she moves to state $y \in X$ with probability $p(x,y)$. Then, for $m \in \mathbb{N}$, the mth power matrix $P^m = \left(p^{(m)}(x,y)\right)$ is still stochastic and $p^{(m)}(x,y)$ is the probability of reaching state y at time $t + m$ given that at time t the random walker is in state x.

Definition 1.5.2 (*m*th iterate and uniform distributions) Let (X, P, v_0) be a finite Markov chain. For $m \in \mathbb{N}$, the probability distribution $v^{(m)} := v_0 P^m$, that is,

$$v^{(m)}(x) := \sum_{y \in X} v_0(y) p^{(m)}(y,x)$$

for all $x \in X$, is called the *m*th *iterate distribution*.

The probability distribution $u \colon X \to [0,1]$ defined by setting $u(x) := \frac{1}{|X|}$ for all $x \in X$ is called the *uniform distribution* on X.

Remark 1.5.3 Note that the 0th iterate distribution satisfies $v^{(0)} \equiv v_0$. Moreover, $\mu * u = u$ for all probability distributions μ on X, where $*$ denotes the convolution product (cf. Example 1.2.13).

Definition 1.5.4 A stochastic matrix $P = (p(x,y))_{x,y \in X}$ is called *ergodic* (or *primitive*) if there exists m_0 such that $p^{(m_0)}(x,y) > 0$ for all $x, y \in X$.

Note that if a stochastic matrix P is ergodic and $p^{(m_0)}(x,y) > 0$ for all $x, y \in X$, then for all $m \geq m_0$ one has $p^{(m)}(x,y) > 0$ for all $x, y \in X$.

Given a finite Markov chain (X, P, v_0), a probability distribution π is called a *stationary distribution* for P provided that $\pi P = \pi$, that is, $\sum_{y \in X} \pi(y) p(y,x) = \pi(x)$ for all $x \in X$.

Theorem 1.5.5 (Ergodic theorem) *Let (X, P, v_0) be an ergodic Markov chain. Then there exists a unique, strictly positive, stationary distribution π for P and it is given as the limit of the mth iterate distributions, in formulæ,*

$$\lim_{m \to \infty} v^{(m)} = \pi.$$

Proof   We shall not prove this theorem in its full generality. The interested reader may find a complete proof in [15, Theorem 1.4.1]. However, we shall present a proof for two particular, yet significant, cases where equivalent conditions for ergodicity of the Markov chain are exploited (cf. Theorem 1.5.7 and Theorem 1.5.16).   □

Definition 1.5.6 (Simple random walk on a finite regular graph)   Let $\mathscr{G} = (X,E)$ be a k-regular finite graph. Given an initial distribution ν_0 on X, the associated *simple random walk* (*SRW*, for short) on $\mathscr{G}$ is the Markov chain (X,P,ν_0) where

$$p(x,y) := \begin{cases} 1/k & \text{if } x \sim y \\ 0 & \text{otherwise} \end{cases}$$

for all $x,y \in X$.

We remark that the simple random walk can be defined on any graph, not necessarily regular; the more general case has applications such as Jerrum's Markov chain for choosing a random orbit in a finite group action, or a random conjugacy class in a finite group [47].

Theorem 1.5.7 (Ergodic theorem for SRW on a regular graph)   *Let $\mathscr{G} = (X,E)$ be a k-regular finite graph. Suppose that $\mathscr{G}$ is connected and not bipartite. Let (X,P,ν_0) denote the Markov chain associated with the simple random walk on $\mathscr{G}$ and initial distribution ν_0 on X. Then the mth iterate distributions converge to the uniform distribution u on X:*

$$\lim_{m \to \infty} \nu^{(m)} = u.$$

Proof   Let $\lambda_0 \geq \lambda_1 \geq \lambda_2 \geq \cdots \geq \lambda_N$, where $N = |X| - 1$, denote the eigenvalues of $P = (p(x,y))_{x,y \in X}$. Recall that $\lambda_0 = 1 > \lambda_1$ by Proposition 1.3.4, since $\mathscr{G}$ is connected, and $\lambda_N > -1$ by Proposition 1.3.8, since $\mathscr{G}$ is not bipartite. Since P is symmetric, we can find an orthogonal matrix O and a diagonal matrix D (whose diagonal entries are the eigenvalues) such that $P = ODO^t$. As a consequence, $P^n = OD^nO^t$, where the diagonal matrix D^n has, as diagonal entries, the nth powers of the eigenvalues $\lambda_0, \lambda_1, \ldots, \lambda_N$. Also recall that the columns of the orthogonal matrix O are exactly the normalized eigenvectors $v_0, v_1, \ldots, v_N \in \mathbb{R}^{N+1}$ corresponding to the eigenvalues $\lambda_0, \lambda_1, \ldots, \lambda_N$. In particular, $v_0 = (1/\sqrt{|X|}, 1/\sqrt{|X|}, \ldots, 1/\sqrt{|X|})$. As $\lambda_i^m \to 0$ as $m \to \infty$ for all $i = 1, 2, \ldots, N$, then denoting by Q the diagonal matrix with 1 at position $(0,0)$

and vanishing elsewhere, we deduce that

$$\lim_{m \to \infty} P^m = \lim_{m \to \infty} OD^m O^t$$
$$= O(\lim_{m \to \infty} D^m)O^t$$
$$= OQO^t$$
$$= \Pi$$

where $\Pi_{x,y} = 1/|X|$ for all $x,y \in X$.   □

Note that, since $u(x) > 0$ for all $x \in X$, from the sign-permanence theorem we deduce that the Markov chain P in the theorem above is ergodic.

Definition 1.5.8   Let X be a finite set and let G be a finite group acting on X. A stochastic matrix $P = (p(x,y))_{x,y \in X}$ is *G-invariant* provided that $p(gx,gy) = p(x,y)$ for all $x,y \in X$ and $g \in G$.

Exercise 1.5.9   Let (X,d) be a metric space. Let G be a finite group acting isometrically on (X,d). Show that a stochastic matrix $P = (p(x,y))_{x,y \in X}$ is *G*-invariant if and only if $p(x,y)$ depends only on $d(x,y)$ for all $x,y \in X$.

Proposition 1.5.10   *Let X be a finite set and let G be a finite group acting on X. Let $P = (p(x,y))_{x,y \in X}$ be a G-invariant stochastic matrix. Let $x_0 \in X$ and set $K := \mathrm{Stab}_G(x_0)$. Then the map $v \colon X \to [0,1]$ defined by $v(x) = p(x_0,x)$ for all $x \in X$ is a K-invariant probability distribution on X and (cf. the notation in Definition 1.2.76)*

$$p^{(m)}(x_0,x) = v^{*m}(x)$$

for all $x \in X$ and $m \in \mathbb{N}$.

Proof   We limit ourselves to show the equality for the case $m = 2$. Let $x \in X$ and let $g \in G$ such that $x = gx_0$. Then, using *G*-invariance of P in lines 4 and 7, and $y = hx_0$ in line 5,

$$v^{*2}(x) = [v * v](x)$$

$$= \frac{1}{|K|} \sum_{h \in G} \widetilde{v}(gh)\widetilde{v}(h^{-1})$$

$$= \frac{1}{|K|} \sum_{h \in G} p(x_0, ghx_0)p(x_0, h^{-1}x_0)$$

$$= \frac{1}{|K|} \sum_{h \in G} p(g^{-1}x_0, hx_0)p(hx_0, x_0)$$

$$= \sum_{y \in X} p(g^{-1}x_0, y)p(y, x_0)$$

$$= p^{(2)}(g^{-1}x_0, x_0)$$

$$= p^{(2)}(x_0, gx_0)$$

$$= p^{(2)}(x_0, x).$$

We leave it to the reader to prove the general case. $\qquad\square$

The following is immediate.

Corollary 1.5.11 *Let (X, P, v_0) be a finite Markov chain and let G be a finite group acting on X. Suppose that P is G-invariant. Fix $x_0 \in X$ and set $K = \mathrm{Stab}_G(x_0)$. Then the map $v \colon X \to [0,1]$ defined by $v(x) = p(x_0, x)$ for all $x \in X$ is a K-invariant probability distribution on X and*

$$v^{(m)} = v_0 * v^{*m} \tag{1.29}$$

for all $m \geq 1$ (cf. Definitions 1.5.2 and 1.2.76). $\qquad\square$

Proposition 1.5.12 *Let (X, P, v_0) be a Markov chain. Let G be a finite group acting on X. Let $x_0 \in X$ and set $K = \mathrm{Stab}_G(x_0)$. Suppose that (G, K) is a Gelfand pair and denote by $\phi_0 = 1, \phi_1, \ldots, \phi_N$ and by $d_0 = 1, d_1, \ldots, d_N$ the spherical functions and the dimensions of the corresponding spherical representations. Then*

$$v^{*m} = \frac{1}{|X|} \sum_{i=0}^{N} d_i [(\mathscr{F}v)(i)]^m \check{\phi}_i \tag{1.30}$$

and

$$\|v^{*m} - u\|^2_{L(X)} = \frac{1}{|X|} \sum_{i=1}^{N} d_i [(\mathscr{F}v)(i)]^{2m}, \tag{1.31}$$

where u denotes the uniform distribution on X.

Proof   Formula (1.30) follows immediately from the inversion formula (1.10) and the property (1.11).

We now observe that $u = \frac{\phi_0}{|X|}$ so that, by (1.30), we have

$$v^{*m} - u = \frac{1}{|X|} \sum_{i=1}^{N} d_i \left[(\mathscr{F} v)(i) \right]^m \check{\phi}_i.$$

Formula (1.31) then follows from the orthogonality relations for the spherical functions. $\qquad\Box$

Definition 1.5.13   Let X be a finite set. The *total variation distance* of two probability measures μ and v on X is

$$\|\mu - v\|_{TV} := \max_{A \subseteq X} \left| \sum_{x \in A} (\mu(x) - v(x)) \right| = \max_{A \subseteq X} |\mu(A) - v(A)|.$$

Exercise 1.5.14   Given $f \in L(X)$ we denote by $\|f\|_{L^1(X)} := \sum_{x \in X} |f(x)|$ its L^1-norm. Show that $\|\mu - v\|_{TV} = \frac{1}{2} \|\mu - v\|_{L^1(X)}$.

The following is the celebrated upper bound lemma of Diaconis and Shahshahani.

Corollary 1.5.15 (Upper bound lemma)   *With the same notation as in Proposition 1.5.12,*

$$\|v^{*m} - u\|_{TV}^2 \leq \frac{1}{4} \sum_{i=1}^{N} d_i |\mathscr{F} v(i)|^{2m},$$

where u is the uniform distribution on X.

Proof   Using the Cauchy–Schwarz inequality in the second line and (1.31) in the third,

$$
\begin{aligned}
\|v^{*k} - u\|_{TV}^2 &= \frac{1}{4} \|v^{*m} - u\|_{L^1(X)}^2 \\
&\leq \frac{1}{4} \|v^{*m} - u\|^2 \cdot |X| \\
&= \frac{1}{4} \sum_{i=1}^{N} d_i |\mathscr{F} v(i)|^{2m}. \qquad\Box
\end{aligned}
$$

We define the *support* of $f \in L(X)$ as the subset of X given by $\operatorname{supp}(f) := \{x \in X : f(x) \neq 0\} \subseteq X$.

Theorem 1.5.16   *Let (X, P, v_0) be a Markov chain. Let G be a finite group acting on X and suppose that P is G-invariant. Let $x_0 \in X$ and set $K = \operatorname{Stab}_G(x_0)$. Suppose that (G, K) is a Gelfand pair and denote by $\phi_0 = 1, \phi_1, \ldots, \phi_N \in {}^K L(X)$*

the associated spherical functions. Suppose that there exists $m_0 \geq 1$ such that $\text{supp}(v^{*m_0}) = X$. *Then the mth iterates $v^{(m)}$ converge to the uniform distribution u on X.*

Proof   By virtue of the upper bound lemma (cf. Corollary 1.5.15) and (1.29) combined with Remark 1.5.3, it suffices to show that $|\mathscr{F}v(i)| < 1$ for all $i = 1, 2, \ldots, N$. Since $\mathscr{F}v^{*m} = (\mathscr{F}v)^m$ for all $m \in \mathbb{N}$ (cf. Exercise 1.2.77), the above condition is clearly equivalent to $|\mathscr{F}v^{*m_0}(i)| < 1$ for all $i = 1, 2, \ldots, N$. Since $\text{supp}(v^{*m_0}) = X$, this follows from the expression of $\mathscr{F}v^{*m_0}$ after observing that the spherical functions ϕ_i with $i \geq 1$ satisfy (i) $|\phi_i(x)| \leq 1$, (ii) $\phi_i(x_0) = 1$, and (iii) there exists $y \in X$ such that $\Re\phi_i(y) < 0$ (the latter follows from (ii) and $\sum_{x \in X} \phi_i(x) = 0$). $\square$

1.5.2 The Ehrenfest diffusion model

The Ehrenfest model of diffusion was proposed by Paul and Tatiana Ehrenfest in 1907 [35] to explain the second law of thermodynamics. We are given two urns numbered 0 and 1 and n balls numbered $1, 2, \ldots, n$. A *configuration* is a placement of the balls into the urns: there are 2^n configurations (2 choices for each ball).

The *configuration space* is $X = \mathscr{P}(\{1, 2, \ldots, n\})$, the set of all subsets of $\{1, 2, \ldots, n\}$: a subset $A \subseteq \{1, 2, \ldots, n\}$ corresponds to the balls contained in urn 0 (the remaining balls, namely those in $\{1, 2, \ldots, n\} \setminus A$, are in urn 1).

The *initial configuration* is $A_0 = \{1, 2, \ldots, n\}$: at time $t = 0$ all balls are in urn 0, while urn 1 is empty (Figure 1.6).

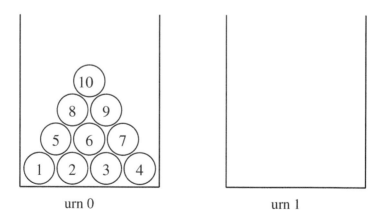

urn 0   urn 1

Figure 1.6 The *initial configuration* for the Ehrenfest diffusion model

Then, at each time t, a ball is randomly chosen (each ball might be chosen with probability $1/n$) and it is moved to the other urn (Figures 1.7 and 1.8).

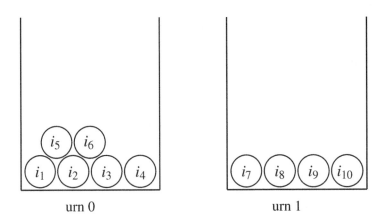

Figure 1.7 A configuration at time t in the Ehrenfest diffusion model

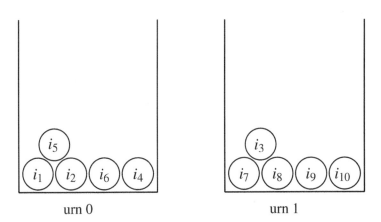

Figure 1.8 The configuration at time $t+1$ if the chosen ball is i_3

This process can be seen as a Markov chain on X with initial probability distribution $v_0 = \delta_{A_0}$, the Dirac delta at A_0, and transition matrix P' given by

$$p'(A,B) = \begin{cases} \frac{1}{n} & \text{if } |A\triangle B| = 1 \\ 0 & \text{otherwise} \end{cases}$$

for all $A, B \in X$, where $A\triangle B = (A \setminus B) \cup (B \setminus A)$ is the symmetric difference of A and B. Since the above stochastic matrix is not ergodic $((p')^{(2n+1)}(A,A) = 0$ for all $A \in X)$, we will consider a slight variation, namely the stochastic matrix P defined by

$$p(A,B) = \begin{cases} \frac{1}{n+1} & \text{if } |A\triangle B| = 1 \\ \frac{1}{n+1} & \text{if } A = B \\ 0 & \text{otherwise}; \end{cases}$$

in other words at each time t we allow the possibility (with probability $\frac{1}{n+1}$) to remain in the same state (i.e., to not change the configuration at time t).

Define the *Hamming distance* d_H on $\{0,1\}^n$ by setting

$$d_H((a_1, a_2, \ldots, a_n), (b_1, b_2, \ldots, b_n)) = |\{k \in \{1, 2, \ldots, n\} : a_k \neq b_k\}|$$

for all $(a_1, a_2, \ldots, a_n), (b_1, b_2, \ldots, b_n) \in \{0,1\}^n$ (cf. Example 1.3.17). The metric space $Q_n = (\{0,1\}^n, d_H)$ is called the n-dimensional *hypercube*. We then regard Q_n as an undirected graph with loops, with vertex set $\{0,1\}^n$ and edges the pairs of vertices with Hamming distance equal to either 0 or 1. (This is the usual hypercube graph with a loop at each vertex.) We then identify X and Q_n via the bijection $\Phi \colon X \to Q_n$ given by

$$\Phi(A) = (a_1, a_2, \ldots, a_n), \quad \text{where } a_k = \begin{cases} 1 & \text{if } k \in A \\ 0 & \text{if } k \notin A. \end{cases}$$

Note that $|A\triangle B| = d_H(\Phi(A), \Phi(B))$ for all $A, B \in X$. This way, the Ehrenfest diffusion model (with n balls) can be seen as the *simple random walk* on the hypercube Q_n.

The wreath product $G = S_2 \wr S_n = (S_2 \times S_2 \times \cdots \times S_2) \rtimes S_n$ acts on Q_n by setting

$$(\sigma_1, \sigma_2, \ldots, \sigma_n; \theta)(a_1, a_2, \ldots, a_n) = (\sigma_1 a_{\theta^{-1}(1)}, \sigma_2 a_{\theta^{-1}(2)}, \ldots, \sigma_n a_{\theta^{-1}(n)})$$

for all $\sigma_i \in S_2$, $\theta \in S_n$, $a_i \in \{0,1\}$, and $i = 1, 2, \ldots, n$.

Exercise 1.5.17 Show that the above action is isometric and two-point homogeneous.

Let $x_0 = (0, 0, \ldots, 0) \in Q_n$ and set $K = \text{Stab}_G(x_0) \cong S_n$. From the above exercise we deduce that $(G, K) = (S_2 \wr S_n, S_n)$ is a symmetric Gelfand pair.

Set

$$V_0 = \{f \in L(\{0,1\}) : f \text{ is constant}\} \text{ and } V_1 = \{f \in L(\{0,1\}) : f(0) + f(1) = 0\}.$$

Then we have the orthogonal decomposition $L(\{0,1\}) = V_0 \oplus V_1$ into S_2-irreducible representations and, in turn,

$$L(Q_n) = L(\{0,1\}^n) \equiv (L(\{0,1\}))^{\otimes n} = (V_0 \oplus V_1)^{\otimes n}.$$

Setting $W_j = \text{span}\{f_1 \otimes f_2 \otimes \cdots \otimes f_n : f_i \in V_0 \cup V_1, |\{i : f_i \in V_1\}| = j\}$, for $j = 0, 1, \ldots, n$, we have:

Theorem 1.5.18 $L(Q_n) = \bigoplus_{j=0}^{n} W_j$ *is the decomposition into* $(S_2 \wr S_n)$-*irreducible pairwise inequivalent sub-representations.*

The jth spherical function $\phi_j \in W_j$ *is*

$$\phi_j(g) = \frac{1}{\binom{n}{j}} \sum_{t=\max\{0,j-n+\ell\}}^{\min\{\ell,j\}} (-1)^t \binom{\ell}{t}\binom{n-\ell}{j-t},$$

where $\ell = d_h(gx_0, x_0)$, *for all* $g \in G$.

Proof   See [15, Proposition 5.4.3].   □

We refer to [13, 18] for a far-reaching generalization of the decomposition of $L(Q_n)$ in the above theorem.

Remark 1.5.19   The functions ϕ_j are the so-called *Krawtchouk polynomials*.

The following theorem describes a very interesting feature for the asymptotics of the Ehrenfest diffusion process.

Theorem 1.5.20 (Diaconis–Shahshahani)   *With the above notation we have the following.*

(1) *For* $k = \frac{1}{4}(n+1)(\log n + c)$ *with* $c > 0$

$$\|v^{*k} - u\|_{TV}^2 \le \frac{1}{2}(e^{e^{-c}} - 1).$$

(2) *For* $k = \frac{1}{4}(n+1)(\log n - c)$ *with* $c \in (0, \log n)$ *and* n *large*

$$\|v^{*k} - u\|_{TV} \ge 1 - 20e^{-c}.$$

Proof   See [15, Theorem 2.4.3]. Note that the Upper bound lemma (Corollary 1.5.15) plays a crucial role.   □

The above theorem shows that $k^* = \frac{1}{4}(n+1)\log(n)$ steps are necessary and sufficient to reach the uniform distribution in the Ehrenfest model of diffusion. Moreover, the so-called *cut-off phenomenon* occurs (see Figure 1.9): the transition from order to chaos is concentrated in a small neighborhood of time $t = k^*$.

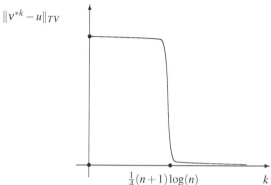

Figure 1.9 The cut-off phenomenon.

1.6 Induced representations and Mackey theory

This section is devoted to the basic theory of induced representations. We reformulate the classical Mackey theory, developed by Mackey in the setting of locally compact groups, in the finite group case. In particular, we present the *Little Group Method* due to Mackey and Wigner. Standard references for induced representations are the monographs by: Alperin and Bell [1], Fulton and Harris [39], Isaacs [46], Naimark and Stern [52], Serre [65], Simon [66], and Sternberg [68]. See also our monographs [16, 18, 19, 20] as well as the research-expository paper [17]. Finally, we generalize the notion of a finite Gelfand pair by introducing the reader to the theory of *multiplicity-free triples* developed in [20], where the role of the classical commutant $\mathrm{End}_G(L(G/K))$ is now played by a so-called *Hecke algebra*.

1.6.1 Induced representations

Definition 1.6.1 Let G be a finite group and let $K \leq G$ be a subgroup. The *induced representation* of a K-representation (σ, V) is the G-representation

$(\text{Ind}_K^G \sigma, \text{Ind}_K^G V)$ defined by setting

$$\text{Ind}_K^G V = \{f \in V[G] : f(gk) = \sigma(k^{-1})f(g), \text{ for all } g \in G, k \in K\},$$

where $V[G]$ is the complex vector space of all functions $f \colon G \to V$, and

$$[\text{Ind}_K^G \sigma(g_1)f](g_2) = f(g_1^{-1}g_2), \qquad \text{for all } g_1, g_2 \in G \text{ and } f \in \text{Ind}_K^G V. \quad (1.32)$$

In the following, to simplify notation, we write $\lambda = \text{Ind}_K^G \sigma$.

Exercise 1.6.2 With the above notation, prove the following.

- λ is a representation. (In particular, check that $\lambda(g)f \in \text{Ind}_K^G V$ for all $g \in G$ and $f \in \text{Ind}_K^G V$).
- Suppose that V is equipped with a scalar product $\langle \cdot, \cdot \rangle_V$ and that σ is unitary. Define a scalar product in $\text{Ind}_K^G V$ by setting

$$\langle f_1, f_2 \rangle_{\text{Ind}_K^G V} = \frac{1}{|K|} \sum_{g \in G} \langle f_1(g), f_2(g) \rangle_V$$

for all $f_1, f_2 \in \text{Ind}_K^G V$. Show that λ is unitary.

Remark 1.6.3 The following yields an alternative approach to the definition of an induced representation. For $v \in V$ define $f_v \in V[G]$ by setting

$$f_v(g) = \begin{cases} \sigma(g^{-1})v & \text{if } g \in K \\ 0 & \text{otherwise.} \end{cases} \quad (1.33)$$

It is straightforward to check that $f_v \in \text{Ind}_K^G V$. Moreover, the set

$$\widetilde{V} = \{f_v : v \in V\} \subseteq \text{Ind}_K^G V$$

is a K-invariant subspace of $\text{Ind}_K^G V$ and $(\lambda|_K, \widetilde{V}) \sim (\sigma, V)$: indeed,

$$\lambda(k)f_v = f_{\sigma(k)v}$$

for all $k \in K$ and $v \in V$. Moreover, if $\mathscr{T} \subseteq G$ denotes a complete set of representatives for the left cosets of K in G, so that

$$G = \sqcup_{t \in \mathscr{T}} tK, \quad (1.34)$$

then we have

$$\text{Ind}_K^G V = \bigoplus_{t \in \mathscr{T}} \lambda(t)\widetilde{V}. \quad (1.35)$$

It follows immediately from (1.35) and the equality $|\mathscr{T}| = [G : K]$, the *index* of K in G, that

$$\dim \text{Ind}_K^G V = [G : K] \dim V. \quad (1.36)$$

Exercise 1.6.4 (Induction in stages)   Let $K \leq H \leq G$ be three groups and let (σ, V) be a K-representation. Then

$$\mathrm{Ind}_K^G \sigma \sim \mathrm{Ind}_H^G \mathrm{Ind}_K^H \sigma.$$

Hint: use the equivalence $(V[H])[G] \sim V[G \times H]$.

Example 1.6.5   Suppose that G acts transitively on a set X, let $x_0 \in X$, and set $K = \mathrm{Stab}_G(x_0)$. Recall (cf. Definition 1.2.50) that the permutation representation $(\lambda, L(X))$ is the G-representation defined by setting

$$[\lambda(g)f](x) = f(g^{-1}x)$$

for all $g \in G$, $x \in X$, and $f \in L(X)$. Then, denoting by $(\iota_K, \mathbb{C})$, the trivial representation of the group K, we have

$$\mathrm{Ind}_K^G \mathbb{C} = \{f \in L(G) : (\forall g \in G, k \in K) \, f(gk) = \iota_K(k)f(g) \equiv f(g)\} = L(G)^K.$$

So the map taking $f \in L(X)$ to $\widetilde{f} \in L(G)^K$ establishes an equivalence between $(\lambda, L(G/K))$ and $(\mathrm{Ind}_K^G \iota_K, \mathrm{Ind}_K^G \mathbb{C})$.

Example 1.6.6   Let $N \trianglelefteq G$ be a normal subgroup of G. Then, the corresponding homogeneous space $X = G/N$ has a natural structure of a group. Let $(\widetilde{\lambda_{G/N}}, L(G/N))$ be the G-representation defined by setting $\widetilde{\lambda_{G/N}}(g) = \lambda_{G/N}(gN)$ for all $g \in G$, where $\lambda_{G/N}$ denotes, as usual, the left regular representation of the group G/N. Then

$$(\lambda, L(X)) \equiv (\widetilde{\lambda_{G/N}}, L(G/N)).$$

Remark 1.6.7   More generally, if (σ, V) is a G/N-representation, its *inflation* $(\overline{\sigma}, V)$ is the G-representation defined by setting

$$\overline{\sigma}(g) = \sigma(gN) \tag{1.37}$$

for all $g \in G$.

Theorem 1.6.8 (Matrix coefficients)   *Let $K \leq G$ and let $\mathscr{T} \subseteq G$ as in* (1.34). *Given a K-representation (σ, V), take an orthonormal basis $\{e_1, e_2, \ldots, e_d\}$ of V. Then*

$$\{f_{t,j} := \lambda(t)f_{e_j} : t \in \mathscr{T}, j = 1, 2, \ldots, d\}$$

constitutes an orthonormal basis of $\mathrm{Ind}_K^G V$. Moreover, the corresponding matrix coefficients are given by

$$\langle \lambda(g)f_{t,j}, f_{s,i} \rangle_{\mathrm{Ind}_K^G V} = \begin{cases} \langle \sigma(s^{-1}gt)e_j, e_i \rangle_V & \text{if } s^{-1}gt \in K \\ 0 & \text{otherwise,} \end{cases}$$

for all $s, t \in \mathscr{T}$, $1 \leq i, j \leq d$, and $g \in G$.

Corollary 1.6.9 (Frobenius character formula) *With the notation of the above theorem, we have*

$$\chi^{\mathrm{Ind}_K^G \sigma}(g) = \sum_{\substack{t \in \mathscr{T}: \\ t^{-1} g t \in K}} \chi^\sigma(t^{-1} g t)$$

for all $g \in G$. □

Theorem 1.6.10 *Let (θ, W) be a G-representation and let (σ, V) be a K-representation, where $K \leq G$. Then*

$$W \otimes \mathrm{Ind}_K^G V \cong \mathrm{Ind}_K^G(\mathrm{Res}_K^G W \otimes V).$$

Proof Define $\phi \colon W \otimes \mathrm{Ind}_K^G V \to \mathrm{Ind}_K^G(\mathrm{Res}_K^G W \otimes V)$ by setting

$$\phi(w \otimes f)(g) = \theta(g^{-1}) w \otimes f(g)$$

for all $g \in G, w \in W$ and $f \in \mathrm{Ind}_K^G V$. We leave it as an exercise to check that ϕ is bijective, and furthermore that it is an intertwiner between $\theta \otimes \mathrm{Ind}_K^G \sigma$ and $\mathrm{Ind}_K^G(\mathrm{Res}_K^G \theta \otimes \sigma)$. □

From the above theorem, with $(\sigma, V) = (\iota_K, \mathbb{C})$ and setting $X = G/K$, we immediately deduce the following important relation between induction and restriction:

Corollary 1.6.11

$$W \otimes L(X) \cong \mathrm{Ind}_K^G \mathrm{Res}_K^G W.$$ □

1.6.2 Mackey theory

Theorem 1.6.12 (Frobenius reciprocity) *Let (θ, W) be a G-representation and let (σ, K) be a K-representation, where $K \leq G$. Then, as vector spaces,*

$$\mathrm{Hom}_G(W, \mathrm{Ind}_K^G V) \cong \mathrm{Hom}_K(\mathrm{Res}_K^G W, V).$$

Proof For $T \in \mathrm{Hom}_G(W, \mathrm{Ind}_K^G V)$, define $\widehat{T} \colon W \to V$ by setting $\widehat{T} w = [Tw](1_G)$ for all $w \in W$. We leave it as an exercise to check that $\widehat{T} \in \mathrm{Hom}_K(\mathrm{Res}_K^G W, V)$.

Vice versa, for $S \in \mathrm{Hom}_K(\mathrm{Res}_K^G W, V)$, we define $\check{S} \colon W \to V[G]$ by setting $[\check{S} w](g) = S(\theta(g^{-1} w))$ for all $g \in G$ and $w \in W$. Again, it is easy to check that $\check{S} \in \mathrm{Hom}_G(W, \mathrm{Ind}_K^G V)$. Moreover, $(\widehat{T})^{\check{}} = T$ and $(\check{S})^{\hat{}} = S$. These facts, and the obvious linearity of the maps $\hat{}$ and $\check{}$, end the proof. □

Remark 1.6.13 Let (ρ_1, V_1) and (ρ_2, V_2) be two G-representations and suppose that ρ_1 is irreducible. Then the multiplicity $m_{\rho_1}^{\rho_2}$ of ρ_1 in ρ_2 equals the dimension of $\mathrm{Hom}_G(V_1, V_2)$. See [19, Lemma 10.6.1.(i)].

From the remark and Frobenius reciprocity we immediately deduce:

Corollary 1.6.14   *Let* $(\theta, W) \in \widehat{G}$ *and let* $(\sigma, K) \in \widehat{K}$. *Then*

$$m_\theta^{\operatorname{Ind}_K^G \sigma} = m_\sigma^{\operatorname{Res}_K^G \theta}.$$

Let G be a group and let $H, K \leq G$ be two subgroups of G. Denote by $\mathscr{S}$ a complete set of representatives of the double cosets $H \backslash G / K$ so that $G = \sqcup_{s \in \mathscr{S}} H s K$. We suppose that $1_G \in \mathscr{S}$. For $s \in \mathscr{S}$ we set $G_s = H \cap sKs^{-1}$.

Exercise 1.6.15   Let $h_1, h_2 \in H$, $k_1, k_2 \in K$, and $s \in \mathscr{S}$. Show that $h_1 s k_1 = h_2 s k_2$ if and only if there exists $x \in G_s$ such that $h_2 = h_1 x$ and $k_2 = s^{-1} x^{-1} s k_1$. Deduce that $|HsK| = |H| \cdot |K| / |G_s|$.

Let (σ, V) be a K-representation and let (ν, U) be an H-representation. We define a G_s-representation (σ_s, V_s) by setting $V_s = V$ and $\sigma_s(x) = \sigma(s^{-1} xs)$ for all $x \in G_s$, and we set

$$\mathscr{S}_0 = \{ s \in \mathscr{S} : \operatorname{Hom}_{G_s}(\operatorname{Res}_{G_s}^H \nu, \sigma_s) \neq 0 \}.$$

We have the following fundamental results:

- **(Mackey's formula for invariants)**

$$\operatorname{Hom}_G(\operatorname{Ind}_H^G \nu, \operatorname{Ind}_K^G \sigma) \cong \bigoplus_{s \in \mathscr{S}_0} \operatorname{Hom}_{G_s}(\operatorname{Res}_{G_s}^H \nu, \sigma_s).$$

- **(Mackey's intertwining number theorem)**

$$\dim \operatorname{Hom}_G(\operatorname{Ind}_H^G \nu, \operatorname{Ind}_K^G \sigma) = \sum_{s \in \mathscr{S}_0} \dim \operatorname{Hom}_{G_s}(\operatorname{Res}_{G_s}^H \nu, \sigma_s).$$

- **(Mackey's irreducibility criterion)**

$$\operatorname{Ind}_K^G \sigma \text{ is irreducible} \Leftrightarrow \begin{cases} \sigma \text{ is irreducible and} \\ (\forall s \in \mathscr{S} \setminus \{1_G\}) \operatorname{Hom}_{G_s}(\operatorname{Res}_{G_s}^K \sigma, \sigma_s) = 0. \end{cases}$$

Remark 1.6.16   For a complete proof of Mackey's formula for invariants, see [19, Corollary 11.4.4]. When $H = K$ we shall revisit it in Section 1.6.5. Observe that it reduces to Frobenius reciprocity when $H = G$ and $\theta = \nu$. Moreover, both Mackey's intertwining number theorem and Mackey's irreducibility criterion are almost immediate consequences of Mackey's formula for invariants: we leave it to the reader to check the corresponding details.

Finally, we consider the counterpart of Corollary 1.6.11, namely the case when we restrict after inducing.

Theorem 1.6.17 (Mackey's lemma)

$$\operatorname{Res}_H^G \operatorname{Ind}_K^G \sigma \sim \bigoplus_{s \in \mathscr{S}} \operatorname{Ind}_{G_s}^H \sigma_s.$$

Proof See [19, Theorem 11.5.1]. $\qquad\qquad\qquad\qquad\qquad\qquad \square$

1.6.3 The little group method of Mackey and Wigner

In this section we present a method, due to Mackey and Eugene Paul Wigner, to obtain all irreducible representations of a group G admitting a normal abelian subgroup $A \trianglelefteq G$ and satisfying a suitable condition. We first observe that G acts on $\widehat{A}$ by conjugation: if $\chi \in \widehat{A}$ and $g \in G$ we define ${}^g\chi \in \widehat{A}$ by setting

$$ {}^g\chi(a) = \chi(g^{-1}ag) \tag{1.38}$$

for all $a \in A$ (we leave it to the reader to check that the map $(g, \chi) \mapsto {}^g\chi$ is an action).

Let $\chi \in \widehat{A}$. The *inertia group* of χ is the subgroup K_χ defined by $K_\chi = \operatorname{Stab}_G(\chi) = \{g \in G : {}^g\chi = \chi\}$. Note that since A is abelian we have $A \le K_\chi$. Moreover, an *extension* of χ to K_χ is a one-dimensional K_χ-representation $\widetilde{\chi}$ such that $\chi = \operatorname{Res}_A^{K_\chi} \widetilde{\chi}$.

We recall that given a $\psi \in \widehat{K_\chi/A}$, we denote by $\overline{\psi}$ its inflation (see (1.37)).

Theorem 1.6.18 *Suppose that $\chi \in \widehat{A}$ admits an extension $\widetilde{\chi}$ to K_χ. Then*

$$\operatorname{Ind}_A^{K_\chi} \chi = \bigoplus_{\psi \in \widehat{K_\chi/A}} d_\psi(\widetilde{\chi} \otimes \overline{\psi}). \tag{1.39}$$

Moreover, if every $\chi \in \widehat{A}$ admits an extension $\widetilde{\chi}$ to K_χ, then

$$\widehat{G} = \left\{ \operatorname{Ind}_{K_\chi}^G(\widetilde{\chi} \otimes \overline{\psi}) : \psi \in \widehat{K_\chi/A}, \chi \in X \right\}, \tag{1.40}$$

where X denotes a complete set of representatives of the orbits of G on $\widehat{A}$.

Proof

$$\operatorname{Ind}_A^{K_\chi} \chi = \operatorname{Ind}_A^{K_\chi}(\chi \otimes \iota_A)$$
$$= \operatorname{Ind}_A^{K_\chi}(\operatorname{Res}_A^{K_\chi} \widetilde{\chi} \otimes \iota_A)$$
$$= \widetilde{\chi} \otimes \operatorname{Ind}_A^{K_\chi} \iota_A = \widetilde{\chi} \otimes \overline{\lambda}$$

by Corollary 1.6.11, where $\overline{\lambda}$ denotes the inflation of the regular representation

$\lambda = \lambda_{K_\chi/A}$ of K_χ/A. By the Peter-Weyl Theorem (cf. Theorem 1.2.36), $\lambda = \bigoplus_{\psi \in \widehat{K_\chi/A}} d_\psi \psi$ so that

$$\overline{\lambda} = \bigoplus_{\psi \in \widehat{K_\chi/A}} d_\psi \overline{\psi},$$

from which (1.39) follows.

The proof of the other statement is more involved. Let $\mathscr{S}$ be a complete set of representatives for the double cosets $K_\chi \backslash G / K_\chi$ with $1_G \in \mathscr{S}$. Set $G_s = K_\chi \cap s K_\chi s^{-1}$ and $(\widetilde{\chi} \otimes \overline{\psi})_s(x) = (\widetilde{\chi} \otimes \overline{\psi})(s^{-1}xs)$ for all $s \in \mathscr{S}$ and $x \in G_s$. Since A is abelian, we have $(\widetilde{\chi} \otimes \overline{\psi})_s(a) = {}^s\chi(a)\psi(A)$ for all $a \in A$, so that

$$\mathrm{Res}_A^{G_s}(\widetilde{\chi} \otimes \overline{\psi})_s \sim d_\psi {}^s\chi$$

which in turn implies that, for $s \neq 1_G$ $\mathrm{Res}_{G_s}^{K_\chi}(\widetilde{\chi} \otimes \overline{\psi})$ and $(\widetilde{\chi} \otimes \overline{\psi})_s$ cannot have common irreducible subrepresentations (otherwise, restricting to A would give equivalent representations, violating the fact that ${}^s\chi \neq \chi$ if $s \neq 1_G$). From Mackey's irreducibility criterion, we deduce that $\mathrm{Ind}_{K_\chi}^G(\widetilde{\chi} \otimes \overline{\psi})$ is irreducible. Finally, from Mackey's lemma we deduce that

$$\mathrm{Res}_{K_\chi}^G \mathrm{Ind}_{K_\chi}^G(\widetilde{\chi} \otimes \overline{\psi}) \sim \bigoplus_{s \in \mathscr{S}} \mathrm{Ind}_{G_s}^{K_\chi}(\widetilde{\chi} \otimes \overline{\psi})_s.$$

We leave it as an exercise to deduce, from the above expression, that ψ is uniquely determined by $\mathrm{Ind}_{K_\chi}^G(\widetilde{\chi} \otimes \overline{\psi})$. $\qquad\square$

In the next theorem we apply the little group method in the case of a semi-direct product $G = A \rtimes H$, with A abelian. When both subgroups A and H are abelian, a simpler approach is presented in [21].

Theorem 1.6.19 *Let $G = A \rtimes H$ and suppose that A is abelian. For all $\chi \in \widehat{A}$, let $H_\chi = \mathrm{Stab}_H(\chi) = \{h \in H : {}^h\chi = \chi\}$. Then:*

(1) *the inertia group of χ is $K_\chi = A \rtimes H_\chi$;*
(2) *there exists an extension $\widetilde{\chi}$ of χ to $A \rtimes H_\chi$.*

Moreover,

$$\widehat{G} = \{\mathrm{Ind}_{A \rtimes H_\chi}^G(\widetilde{\chi} \otimes \overline{\psi}) : \chi \in X, \psi \in \widehat{H_\chi}\}, \tag{1.41}$$

where X denotes a complete set of representatives of the orbits of H on $\widehat{A}$.

Proof

(1) Given $a \in A$ and $h \in H$, we have ${}^{ah}\chi = \chi \Leftrightarrow {}^h\chi = \chi \Leftrightarrow h \in H_\chi$.
(2) Define $\widetilde{\chi} \colon A \rtimes H_\chi \to \mathbb{T}$ by setting $\widetilde{\chi}(ah) = \chi(a)$ for all $a \in A$ and $h \in H_\chi$.

We have

$$\widetilde{\chi}(a_1h_1 \cdot a_2h_2) = \widetilde{\chi}(a_1h_1a_2h_1^{-1} \cdot h_1h_2) = \chi(a_1h_1a_2h_1^{-1})$$
$$= \chi(a_1)\chi(h_1a_2h_1^{-1}) = \chi(a_1)\chi(a_2) = \widetilde{\chi}(a_1h_1)\widetilde{\chi}(a_2h_2),$$

showing that $\widetilde{\chi} \in \widehat{A \rtimes H_\chi}$. Finally, (1.41) follows immediately from (1.40). $\square$

1.6.4 Hecke algebras

This section is based on our recent work [20] (see also [15, Chapter 13] for the particular case when the K-representation is one-dimensional).

Let G be a finite group, let $K \leq G$ be a subgroup, and let (θ, V) be a K-representation. We set $\mathscr{H}(G, K, \theta)$ equal to

$$\{F: G \to \mathrm{End}(V) : F(k_1gk_2) = \theta(k_2^{-1})F(g)\theta(k_1^{-1}), \forall g \in G \text{ and } \forall k_1, k_2 \in K\}.$$

Given $F_1, F_2 \in \mathscr{H}(G, K, \theta)$ we define their *convolution product* $F_1 * F_2 : G \to \mathrm{End}(V)$ by setting

$$[F_1 * F_2](g) = \sum_{h \in G} F_1(h^{-1}g)F_2(h)$$

for all $g \in G$, and their scalar product as

$$\langle F_1, F_2 \rangle_{\widetilde{\mathscr{H}}(G, K, \theta)} = \sum_{g \in G} \langle F_1(g), F_2(g) \rangle_{\mathrm{End}(V)}.$$

Finally, for $F \in \widetilde{\mathscr{H}}(G, K, \theta)$ we define the *adjoint* $F^* : G \to \mathrm{End}(V)$ by setting

$$F^*(g) = [F(g^{-1})]^*$$

for all $g \in G$, where $[F(g^{-1})]^*$ is the adjoint of the operator $F(g^{-1}) \in \mathrm{End}(V)$.

Exercise 1.6.20 Let $F_1, F_2, F \in \widetilde{\mathscr{H}}(G, K, \theta)$. For $g \in G$, set

$$1_{\widetilde{\mathscr{H}}}(g) = \frac{1}{|K|}1_K(g)\theta(g^{-1}),$$

where $\mathbf{1}_K$ denotes the characteristic function of K. Show that

- $F_1 * F_2 \in \widetilde{\mathscr{H}}(G, K, \theta)$;
- $F^* \in \widetilde{\mathscr{H}}(G, K, \theta)$;
- $(F_1 * F_2)^* = F_2^* * F_1^*$;
- $1_{\widetilde{\mathscr{H}}} \in \widetilde{\mathscr{H}}(G, K, \theta)$, and $F * 1_{\widetilde{\mathscr{H}}} = 1_{\widetilde{\mathscr{H}}} * F = F$, for all $F \in \widetilde{\mathscr{H}}(G, K, \theta)$,

and deduce that $\widetilde{\mathscr{H}}$ is a unital $*$-algebra.

We refer to $\mathscr{H}(G,K,\theta)$ as the *Hecke algebra* associated with the triple (G,K,θ).

As in Section 1.6.2 (with $H = K$), we denote by $\mathscr{S} \subseteq G$ (with $1_G \in \mathscr{S}$) a complete set of representatives for the double K-cosets in G so that $G = \bigsqcup_{s \in \mathscr{S}} KsK$. For $s \in \mathscr{S}$ we set $K_s = K \cap sKs^{-1}$ and denote by (θ^s, V_s) the K_s-representation defined by setting $V_s = V$ and $\theta^s(x) = \theta(s^{-1}xs)$ for all $x \in K_s$. Finally, we set $\mathscr{S}_0 = \{s \in \mathscr{S} : \mathrm{Hom}_{K_s}(\mathrm{Res}_{K_s}^K \theta, \theta_s) \neq 0\}$.

Exercise 1.6.21 Choose $s \in \mathscr{S}$. For each $T \in \mathrm{Hom}_{K_s}(\mathrm{Res}_{K_s}^K \theta, \theta^s)$, define $\mathscr{L}_T : G \to \mathrm{End}(V)$ by setting

$$\mathscr{L}_T(g) = \begin{cases} \theta(k_2^{-1})T\theta(k_1^{-1}) & \text{if } g = k_1 s k_2 \text{ for some } k_1, k_2 \in K \\ 0 & \text{if } g \notin KsK. \end{cases}$$

Let $F \in \widetilde{\mathscr{H}}(G,K,\theta)$. Show that

(1) $\mathscr{L}_T$ is well defined and belongs to $\widetilde{\mathscr{H}}(G,K,\theta)$;
(2) $F(s) \in \mathrm{Hom}_{K_s}(\mathrm{Res}_{K_s}^K \theta, \theta^s)$ for all $s \in \mathscr{S}$;
(3) $F = \sum_{s \in \mathscr{S}_0} \mathscr{L}_{F(s)}$ and the nontrivial elements in this sum are linearly independent.

For $F \in \widetilde{\mathscr{H}}(G,K,\theta)$ we define $\xi(F) \colon \mathrm{Ind}_K^G V \to V[G]$ by setting

$$[\xi(F)f](g) = \sum_{h \in G} F(h^{-1}g)f(h)$$

for all $f \in \mathrm{Ind}_K^G V$ and $g \in G$. Also, for $T \in \mathrm{End}_G(\mathrm{Ind}_K^G V)$ we define $\Xi(T) \colon G \to \mathrm{End}(V)$ by setting

$$\Xi(T)(g)v = \frac{1}{|K|}[Tf_v](g),$$

for all $g \in G$ and $v \in V$, where f_v is as in (1.33).

Exercise 1.6.22 (1) Show that $\xi(F) \in \mathrm{End}_G(\mathrm{Ind}_K^G V)$ for all $F \in \widetilde{\mathscr{H}}(G,K,\theta)$.
(2) Show that $\xi(F_1 * F_2) = \xi(F_1)\xi(F_2)$ and $\xi(F^*) = \xi(F)^*$ for all $F_1, F_2, F \in \widetilde{\mathscr{H}}(G,K,\theta)$.
(3) Show that $\Xi(T) \in \widetilde{\mathscr{H}}(G,K,\theta)$ for all $T \in \mathrm{End}_G(\mathrm{Ind}_K^G V)$.
(4) Show that the normalized map $F \mapsto \frac{1}{\sqrt{|K|}}\xi(F)$ is an isometry.
(5) Show that

$$\frac{1}{|K|}\sum_{h \in G}[\lambda(h)f_{f(h)}] = f, \tag{1.42}$$

for all $f \in \mathrm{Ind}_K^G V$.

Theorem 1.6.23 *The map* $\xi\colon \widetilde{\mathscr{H}}(G,K,\theta)\to\mathrm{End}(\mathrm{Ind}_K^G V)$ *is a* $*$*-isomorphism of unital* $*$*-algebras with inverse the map* $\Xi\colon \mathrm{End}_G(\mathrm{Ind}_K^G V) \to \widetilde{\mathscr{H}}(G,K,\theta)$.

Proof Having established the results in Exercise 1.6.22, we only need to check that Ξ is a right-inverse of ξ.

Given $T \in \mathrm{End}_G(\mathrm{Ind}_K^G V)$, $f \in \mathrm{Ind}_K^G$, and $g \in G$, we have

$$
\begin{aligned}
[(\xi \circ \Xi(T)) f](g) &= \sum_{h \in G} [\Xi(T)(h^{-1}g)] f(h) \\
&= \tfrac{1}{|K|} \sum_{h \in G} \left[T f_{f(h)} \right](h^{-1}g) \\
&= \tfrac{1}{|K|} \sum_{h \in G} \left[\lambda(h) T f_{f(h)} \right](g) \\
&= \tfrac{1}{|K|} \sum_{h \in G} \left[T \lambda(h) f_{f(h)} \right](g) \\
&= \left[T \left(\tfrac{1}{|K|} \sum_{h \in G} \lambda(h) f_{f(h)} \right) \right](g) \\
&= [Tf](g)
\end{aligned}
$$

by (1.42). This shows that $\xi(\Xi(T)) = T$, as desired. $\qquad\square$

The Hecke algebra as a subalgebra of $L(G)$. Let (θ, V) be a K-representation as in the first part of this section, but we now assume that θ is *irreducible*. We fix $v \in V$ with $\|v\| = 1$ and we consider an orthonormal basis $\{v_1 = v, v_2, \ldots, v_{d_\theta}\}$ of V.

We define (everything depending on the choice of the fixed vector $v \in V$):

- $\psi \in L(G)$ by setting

$$
\psi(g) = \begin{cases} \dfrac{d_\theta}{|K|} \langle v, \theta(k) v \rangle_V & \text{if } g = k \in K \\ 0 & \text{otherwise;} \end{cases} \tag{1.43}
$$

- the convolution operator $P\colon L(G) \to L(G)$ by setting $Pf = T_\psi f = f * \psi$ for all $f \in L(G)$;
- the linear operator $T\colon \mathrm{Ind}_K^G V \to L(G)$ by setting

$$
[Tf](g) = \sqrt{d_\theta/|K|}\,\langle f(g), v \rangle_V
$$

for all $f \in \mathrm{Ind}_K^G V$ and $g \in G$, and denote its range by

$$
\mathscr{I}(G,K,\theta) = T\left(\mathrm{Ind}_K^G V\right) \subseteq L(G);
$$

- the map $S\colon \widetilde{\mathscr{H}}(G,K,\theta) \to L(G)$ by setting

$$
[SF](g) = d_\theta \langle F(g) v, v \rangle_V
$$

for all $F \in \widetilde{\mathscr{H}}(G,K,\theta)$ and $g \in G$;
- the subspace

$$
\mathscr{H}(G,K,\theta) = \{\psi * f * \psi : f \in L(G)\} \equiv \{f \in L(G) : f = \psi * f * \psi\} \leq L(G). \tag{1.44}
$$

Exercise 1.6.24 (1) Show that $T \in \mathrm{Hom}_G(\mathrm{Ind}_K^G V, L(G))$ and is an isometry.

(2) Deduce that $\mathscr{I}(G, K, \theta)$ is a λ_G-invariant subspace of $L(G)$, which is G-isomorphic to $\mathrm{Ind}_K^G V$.

(3) Show that $\psi * \psi = \psi$ and $\psi^* = \psi$ and deduce that P is the orthogonal projection of $L(G)$ onto $\mathscr{I}(G, K, \theta)$.

(4) Show that $S(F) \in \mathscr{H}(G, K, \theta)$ and $S(F_1 * F_2) = S(F_2)S(F_1)$ for all $F, F_1, F_2 \in \widetilde{\mathscr{H}}(G, K, \theta)$.

(5) Show that $\frac{1}{\sqrt{d_\theta}} S$ is an isometry.

(6) Show that every $f \in \mathscr{H}(G, K, \theta)$ is supported in $\bigsqcup_{s \in \mathscr{S}_0} KsK$.

Combining the results from the above exercise we establish the following:

Theorem 1.6.25 *$\mathscr{H}(G, K, \theta)$ is an involutive subalgebra of $L(G)$, and the map $S: \mathscr{H}(G, K, \theta) \to \mathscr{H}(G, K, \theta)$ is a $*$-anti-isomorphism of $*$-algebras.* $\qquad\square$

1.6.5 Multiplicity-free triples and their spherical functions

This section is also based on [20] and [15, Chapter 13].

Let G be a finite group, let $K \leq G$ be a subgroup, and let $(\theta, V) \in \widehat{K}$.

Definition 1.6.26 We say that (G, K, θ) is a *multiplicity-free triple* if the algebra $\mathscr{H}(G, K, \theta)$ (cf. (1.44) and Theorem 1.6.25) is commutative.

Theorem 1.6.27 *The following conditions are equivalent:*

(1) *(G, K, θ) is a multiplicity-free triple.*

(2) *$\mathrm{Ind}_K^G \theta$ decomposes without multiplicity.*

(3) *The algebra $\mathscr{H}(G, K, \theta)$ is commutative.*

(4) *The algebra $\mathrm{End}_G(\mathrm{Ind}_K^G(V))$ is commutative.*

(5) *The multiplicity of θ in $\mathrm{Res}_K^G \rho$ satisfies $m_\theta^{\mathrm{Res}_K^G \rho} \leq 1$, that is, we have*

$$\dim \mathrm{Hom}_K(V, \mathrm{Res}_K^G W) \leq 1$$

for every $(\rho, W) \in \widehat{G}$.

Proof The equivalences between (1), (2), and (3) follow from Theorem 1.6.23 and Theorem 1.6.25. The equivalence with (4) is obtained by arguing as in the proof of Theorem 1.2.60. The equivalence with the remaining condition follows from Frobenius reciprocity: we leave the details to the reader. $\qquad\square$

The following provides a simple condition guaranteeing multiplicity-freeness:

Proposition 1.6.28 *Suppose there exists an anti-automorphism τ of G such that $f(\tau(g)) = f(g)$ for all $f \in \mathcal{H}(G, K, \theta)$ and $g \in G$. Then (G, K, θ) is a multiplicity-free triple.*

Proof Let $f_1, f_2 \in \mathcal{H}(G, K, \theta)$ and $g \in G$. We have

$$
\begin{aligned}
[f_1 * f_2](g) &= \sum_{h \in G} f_1(gh) f_2(h^{-1}) \\
&= \sum_{h \in G} f_1(\tau(gh)) f_2(\tau(h^{-1})) \\
&= \sum_{h \in G} f_1(\tau(h)\tau(g)) f_2(\tau(h^{-1})) \\
&= \sum_{h \in G} f_2(\tau(h^{-1})) f_1(\tau(h)\tau(g)) \\
&= \sum_{t \in G} f_2(t^{-1}) f_1(t\tau(g)) \\
&= [f_2 * f_1](\tau(g)) = [f_2 * f_1](g).
\end{aligned}
$$

(setting $t = \tau(h)$ in the penultimate line). Thus $f_1 * f_2 = f_2 * f_1$, showing that $\mathcal{H}(G, K, \theta)$ is commutative. $\qquad\square$

Remark 1.6.29 When $\theta = \iota_K$, the trivial representation of the subgroup K, we recover the case of a Gelfand pair (see Theorem 1.2.60). Note that in this context the algebra $\mathcal{H}(G, K, \iota_K)$ coincides with the algebra $\widetilde{\mathcal{H}}(G, K, \iota_K)$ and the criterion in Proposition 1.6.28 reduces to the condition of a weakly symmetric Gelfand pair (see Exercise 1.2.55).

For the rest of this section we consider a multiplicity-free triple (G, K, θ). Generalizing the case of Gelfand pairs, we show that also in this setting it is possible to develop a complete theory of spherical functions.

Definition 1.6.30 A function $\phi \in \mathcal{H}(G, K, \theta)$ is *spherical* if

$$
\begin{cases}
\phi * f = \lambda_{\phi, f} \phi & \text{for all } f \in \mathcal{H}(G, K, \theta) \\
\phi(1_G) = 1.
\end{cases}
$$

The proof of the next results follow the same lines of the analogous results in the setting of Gelfand pairs (cf. Section 1.2.2) and we leave it to the reader (for more details, we refer to [20, Section 4.2]; see also [19, Section 13]).

Lemma 1.6.31 *A function $\phi \in L(G) \setminus \{0\}$ is spherical if and only if*

$$
\sum_{k \in K} \phi(gkh)\overline{\psi(k)} = \phi(g)\phi(h)
$$

for all $g, h \in G$, where $\psi \in L(G)$ is as in (1.43). $\qquad\square$

Theorem 1.6.32 *Let ϕ be a spherical function and define $\Phi \colon L(G) \to \mathbb{C}$ by setting $\Phi(f) = [f * \phi](1_G)$. Then Φ is a multiplicative functional on $\mathscr{H}(G, K, \theta)$. Conversely, every multiplicative functional on $\mathscr{H}(G, K, \theta)$ comes from a spherical function as above.* $\qquad\square$

We denote by $\mathscr{J} \subseteq \widehat{G}$ the set of all irreducible G-represetations that are contained in $\mathrm{Ind}_K^G \theta$. Note that $(\rho, W) \in \mathscr{J}$ if and only if $\dim \mathrm{Hom}_K(V, \mathrm{Res}_K^G W) = 1$ (cf. Theorem 1.6.27). We then have:

Corollary 1.6.33

$$|\{spherical\ functions\}| = |\mathscr{J}| = \dim \mathscr{H}(G, K, \theta).$$

Proposition 1.6.34 *Let ϕ, ϕ' be distinct spherical functions. Then*

- $\phi^* = \phi$;
- $\phi * \phi' = 0$;
- $\langle \lambda_G(g_1)\phi, \lambda_G(g_2)\phi' \rangle_{L(G)} = 0$ for all $g_1, g_2 \in G$. In particular $\phi \perp \phi'$. $\quad\square$

Theorem 1.6.35 *Let $U_\phi = \mathrm{span}\{\lambda_G(g)\phi : g \in G\}$. Then*

$$\mathscr{I}(G, K, \theta) = \bigoplus_{\phi \in \mathscr{J}} U_\phi. \qquad\square$$

We denote ϕ^σ the spherical function associated with $\sigma \in \mathscr{J}$.

Definition 1.6.36 The map $\mathscr{F} \colon \mathscr{H}(G, K, \theta) \to \mathbb{C}^{\mathscr{J}}$ defined by setting

$$[\mathscr{F}f](\sigma) = \langle f, \phi^\sigma \rangle_{L(G)} = [f * \phi^\sigma](1_G)$$

for all $f \in \mathscr{H}$ is the *spherical Fourier transform*.

Theorem 1.6.37 (Properties of the spherical Fourier transform) *The spherical Fourier transform is an algebra isomorphism between the commutative algebras $\mathscr{H}(G, K, \theta)$ and $\mathbb{C}^{\mathscr{J}}$. Moreover, for $f, f_1, f_2 \in \mathscr{H}(G, K, \theta)$ we have:*

- *(Convolution property)*

$$\mathscr{F}[f_1 * f_2] = \mathscr{F}(f_1) \cdot \mathscr{F}(f_2);$$

- *(Inversion formula)*

$$f = \frac{1}{|G|} \sum_{\sigma \in \mathscr{J}} d_\sigma [\mathscr{F}f](\sigma)\phi^\sigma;$$

- *(Parseval identity)*

$$\langle f_1, f_2 \rangle_{L(G)} = \frac{1}{|G|} \sum_{\sigma \in \mathscr{J}} d_\sigma [\mathscr{F}f_1](\sigma)\overline{[\mathscr{F}f_1](\sigma)}. \qquad\square$$

1.7 Representation theory of $\mathrm{GL}(2,\mathbb{F}_q)$

This final section is devoted to the study of the representation theory of the general linear group $\mathrm{GL}(2,\mathbb{F}_q)$ over a finite field with q elements. It is based on part IV of our monograph [19], which sheds light on the results and the calculations in the beautiful exposition of Piatetski–Shapiro [57] by framing them in a more comprehensive theory. We start with some elementary facts on finite fields and their characters, and determining, as intermediate steps, the irreducible representations of the affine group $\mathrm{Aff}(\mathbb{F}_q)$. We also describe the subgroup structure of $\mathrm{GL}(2,\mathbb{F}_q)$ by analyzing a few important subgroups, notably the *Borel subgroup B* and the *unipotent subgroup U*, as well as its *Bruhat decomposition*. We then determine all irreducible representations and their characters of $\mathrm{GL}(2,\mathbb{F}_q)$: these are of two types, parabolic (that can be obtained by inducing up characters of the Borel subgroup) and cuspidal (whose space of U-invariant vectors is trivial).

1.7.1 Finite fields and their characters

Let $\mathbb{F}$ be a finite field. We denote by $\mathbb{F}[x]$ the ring of all polynomials in the indeterminate x with coefficients in $\mathbb{F}$, and by $\partial p(x) := n$ the *degree* of a polynomial $p(x) = a_0 + a_1 x + \cdots + a_n x^n \in \mathbb{F}[x]$, $a_n \neq 0$.

Recall that the *characteristic* $\mathrm{char}(\mathbb{F})$ of $\mathbb{F}$, that is, the additive order of $1 \in \mathbb{F}$, is a prime number. Indeed, the map $\Phi \colon \mathbb{Z} \to \mathbb{F}$ defined by $\Phi(\pm n) = \pm(1 + 1 + \cdots + 1)$ for all $n \in \mathbb{N}$, is a ring homomorphism so that $\mathbb{Z}/\ker(\Phi)$ is isomorphic to $\Phi(\mathbb{Z}) \subset \mathbb{F}$. Now, $\Phi(\mathbb{Z})$, being a finite integral domain, it is itself a field (exercise). We deduce that $\ker(\Phi) = p\mathbb{Z}$ for a unique prime number p, and therefore $\mathrm{char}(\mathbb{F}) = p$.

An *extension* of $\mathbb{F}$ is a field $\mathbb{E}$ such that $\mathbb{F} \subset \mathbb{E}$. It then follows that $\mathbb{E}$ is a vector space over $\mathbb{F}$. We denote by $[\mathbb{E} : \mathbb{F}] = \dim_{\mathbb{F}} \mathbb{E}$ the *degree* of this extension.

Since $\mathbb{Z}/p\mathbb{Z} \cong \Phi(\mathbb{Z}) \subset \mathbb{F}$, we deduce that $|\mathbb{F}| = p^n$, where $n = [\mathbb{F} : \Phi(\mathbb{Z})]$.

Given an extension $\mathbb{F} \subset \mathbb{E}$, an element $\alpha \in \mathbb{E}$ is said to be *algebraic* over $\mathbb{F}$ if there exists a polynomial $p(x) \in \mathbb{F}[x]$ such that $p(\alpha) = 0$. If $\alpha \in \mathbb{E}$ is algebraic over $\mathbb{F}$, then the set $I_\alpha = \{p(x) \in \mathbb{F}[x] : p(\alpha) = 0\}$ is an ideal of $\mathbb{F}[x]$. Since $\mathbb{F}[x]$ is a *principal ideal domain*, there exists a *monic* polynomial $q(x) \in \mathbb{F}[x]$ such that $I_\alpha = q(x)\mathbb{F}[x]$. Such a polynomial $q(x)$, which is unique and irreducible (over $\mathbb{F}$), is called the *minimal polynomial* of α (over $\mathbb{F}$). Consider the ring homomorphism $\Phi \colon \mathbb{F}[x] \to \mathbb{E}$ defined by setting $\Phi(p(x)) = p(\alpha)$ for all $p(x) \in \mathbb{F}[x]$. Then $I_\alpha = \ker \Phi$ and

$$\mathbb{F}[x]/q(x)\mathbb{F}[x] \cong \Phi(\mathbb{F}[x]) = \mathbb{F}[\alpha] \leq \mathbb{E},$$

where $\mathbb{F}[\alpha]$, the subfield obtained by adjoining α to $\mathbb{F}$, satisfies $[\mathbb{F}[\alpha] : \mathbb{F}] = \partial q(x)$, the degree of the minimal polynomial of α.

Exercise 1.7.1 Let $\mathbb{F} \subset \mathbb{E}$ be an extension of fields. Show that if $[\mathbb{E} : \mathbb{F}] < \infty$, then every $\alpha \in \mathbb{E}$ is algebraic over $\mathbb{F}$.

Let $p(x) \in \mathbb{F}[x]$ of degree $\partial p(x) = n$. The smallest (i.e., of minimal degree) field extension $\mathbb{E}$ of $\mathbb{F}$ such that there exist $\alpha_1, \alpha_2, \ldots, \alpha_n \in \mathbb{E}$ and $c \in \mathbb{F}$ such that $p(x) = c(x - \alpha_1)(x - \alpha_2) \cdots (x - \alpha_n)$, is called a *splitting field* for $p(x)$ over $\mathbb{F}$.

Theorem 1.7.2 (Existence and uniqueness of finite fields)

(1) *The splitting field of any polynomial $p(x) \in \mathbb{F}[x]$ exists and is unique up to isomorphism.*

(2) *Suppose that $q = p^n$ for some integer $n \geq 1$. Then the splitting field of the polynomial $p(x) = x^q - x$ over $\mathbb{Z}/p\mathbb{Z}$ has exactly q elements, which consist of all the roots of $p(x)$.*

(3) *For every prime number p and integer $h \geq 1$ there exists a unique (up to isomorphism) finite field $\mathbb{F}_q$ of order $q = p^h$. It is isomorphic to*

$$\mathbb{F}_p[x]/\ell(x)\mathbb{F}_p[x],$$

where $\ell(x) = (x - \alpha)(x - \alpha^p)(x - \alpha^{p^2}) \cdots (x - \alpha^{p^{h-1}})$ and α is any generator of the cyclic group $\mathbb{F}_q^*$.

(4) *The (multiplicative) group $\mathbb{F}_q^*$ of invertible elements of $\mathbb{F}_q$ is cyclic (of order $q - 1$).*

Proof See, for instance [19, Theorem 1.1.21 and Theorem 6.3.3]. $\square$

The *Galois group* of an extension $\mathbb{F} \subset \mathbb{E}$ is the group

$$\mathrm{Gal}(\mathbb{E} : \mathbb{F}) = \{\xi \in \mathrm{Aut}(\mathbb{E}) : \xi(x) = x \text{ for all } x \in \mathbb{F}\}$$

of automorphisms of $\mathbb{E}$ fixing all elements of $\mathbb{F}$ pointwise.

Suppose that $\mathrm{char}(\mathbb{F}) = p$. Then the map $\sigma \colon \mathbb{F} \to \mathbb{F}$ defined by $\sigma(x) = x^p$ for all $x \in \mathbb{F}$ is an automorphism of $\mathbb{F}$, called the *Frobenius automorphism* of $\mathbb{F}$. Then if $|\mathbb{F}| = p^n$, the Galois group $\mathrm{Gal}(\mathbb{F} : \mathbb{F}_p)$ is cyclic of order n, indeed it is generated by the Frobenius automorphism, and equals $\mathrm{Aut}(\mathbb{F})$.

More generally, suppose $\mathbb{E} = \mathbb{F}_{q^h} = \mathbb{F}_{p^{nh}}$ and $\mathbb{F} = \mathbb{F}_q = \mathbb{F}_{p^n}$. Then $\mathrm{Gal}(\mathbb{E} : \mathbb{F})$ is cyclic of order h, indeed generated by $\overline{\sigma} = \sigma^n$ (thus $\overline{\sigma}(x) = x^{pn} = x^q$ for all $x \in \mathbb{E}$). The *trace* and the *norm* are the maps $\mathrm{Tr}_{\mathbb{E}/\mathbb{F}} \colon \mathbb{E} \to \mathbb{F}$ and $\mathrm{N}_{\mathbb{E}/\mathbb{F}} \colon \mathbb{E} \to \mathbb{F}$ given by

$$\mathrm{Tr}_{\mathbb{E}/\mathbb{F}}(\alpha) = \sum_{k=1}^{h} \overline{\sigma}^k(\alpha) \text{ and } \mathrm{N}_{\mathbb{E}/\mathbb{F}}(\alpha) = \prod_{k=1}^{h} \overline{\sigma}^k(\alpha) \qquad (1.45)$$

for all $\alpha \in \mathbb{E}$.

Exercise 1.7.3 Show that $\mathrm{Tr}_{\mathbb{E}/\mathbb{F}}(\alpha)$ (resp. $\mathrm{N}_{\mathbb{E}/\mathbb{F}}(\alpha)$) is indeed in $\mathbb{F}$ for every $\alpha \in \mathbb{E}$.

Theorem 1.7.4 (Hilbert Satz 90) (1) $\mathrm{Tr}_{\mathbb{E}/\mathbb{F}}$ *is a surjective $\mathbb{F}$-linear map from $\mathbb{E}$ onto $\mathbb{F}$ and*

$$\ker(\mathrm{Tr}_{\mathbb{E}/\mathbb{F}}) = \{\alpha - \overline{\sigma}(\alpha) : \alpha \in \mathbb{E}\}.$$

(2) $\mathrm{N}_{\mathbb{E}/\mathbb{F}}$ *yields (by restriction) a surjective homomorphism from the multiplicative group $\mathbb{E}^*$ of $\mathbb{E}$ into the multiplicative group $\mathbb{F}^*$ of $\mathbb{F}$ and*

$$\ker(\mathrm{N}_{\mathbb{E}/\mathbb{F}}) = \{\alpha\overline{\sigma}(\alpha)^{-1} : \alpha \in \mathbb{E}^*\}.$$

Quadratic extensions From now on we suppose that p is odd. An extension $\mathbb{F} \subset \mathbb{E}$ with $[\mathbb{E} : \mathbb{F}] = 2$ is called *quadratic*: it is a generalization of the familiar extension $\mathbb{R} \subset \mathbb{C}$ and the matrix representation $z = a + ib \leftrightarrow \begin{pmatrix} a & -b \\ b & a \end{pmatrix}$ for all $a, b \in \mathbb{R}$. Let $q = p^h$. Then $\mathrm{Gal}(\mathbb{F}_{q^2} : \mathbb{F}_q)$ is cyclic of order 2 and it is generated by σ, where $\overline{\sigma}(x) = x^q$ for all $x \in \mathbb{F}_{q^2}$. Moreover, there exists an irreducible monic polynomial of degree 2 over $\mathbb{F}_q$ (in fact, there are $(q^2 - q)/2$ such) $x^2 + ax + b$, say with roots α and β.

Exercise 1.7.5 With α and β as above, show that $\overline{\sigma}(\alpha) = \beta$ (and $\overline{\sigma}(\beta) = \alpha$).

Theorem 1.7.6 *Suppose that p is odd, and $q = p^h$. Let η be a generator of the cyclic group $\mathbb{F}_q^*$ and denote by $\pm i$ the square roots of η. Then $\pm i \notin \mathbb{F}_q$ and $\{1, i\}$ is a vector space basis for $\mathbb{F}_{q^2}$ over $\mathbb{F}_q$. Moreover, $\mathbb{F}_{q^2}$ is isomorphic (as an $\mathbb{F}_q$-algebra) to the algebra $\mathbf{M}_2(\mathbb{F}_q, \eta) \subseteq \mathbf{M}_2(\mathbb{F}_q)$ consisting of all matrices of the form*

$$\begin{pmatrix} \alpha & \eta\beta \\ \beta & \alpha \end{pmatrix}$$

with $\alpha, \beta \in \mathbb{F}_q$. The isomorphism is provided by the map $\mathbf{M}_2(\mathbb{F}_q, \eta) \to \mathbb{F}_{q^2}$ given by

$$\begin{pmatrix} \alpha & \eta\beta \\ \beta & \alpha \end{pmatrix} \mapsto \alpha + i\beta \qquad (1.46)$$

for all $\alpha, \beta \in \mathbb{F}_q$. Moreover $\overline{\sigma}(\alpha + i\beta) = \alpha - i\beta$ for all $\alpha, \beta \in \mathbb{F}_q$. $\square$

The *conjugate* of an element $\alpha \in \mathbb{F}_{q^2}$ is defined as $\overline{\alpha} = \overline{\sigma}(\alpha)$. Then

$$\mathrm{Tr}_{\mathbb{F}_{q^2}/\mathbb{F}_q}(\alpha) = \alpha + \overline{\alpha} \quad \text{and} \quad \mathrm{N}_{\mathbb{F}_{q^2}/\mathbb{F}_q}(\alpha) = \alpha\overline{\alpha}$$

for all $\alpha \in \mathbb{F}_{q^2}$. Moreover, $\alpha = \overline{\alpha}$ if and only if $\alpha \in \mathbb{F}_q$.

Characters of finite fields Let $\mathbb{F}_q$ be a finite field. An *additive character* of $\mathbb{F}_q$ is a character of the finite abelian group $(\mathbb{F}_q, +)$, that is, a map $\chi \colon \mathbb{F}_q \to \mathbb{T}$ (see Remark 1.2.29) such that $\chi(x+y) = \chi(x)\chi(y)$ for all $x, y \in \mathbb{F}_q$. The additive characters constitute a (multiplicative) abelian group, denoted by $\widehat{\mathbb{F}}_q$, called the *dual group* of $\mathbb{F}_q$.

We have the *orthogonality relations*:

$$\langle \chi, \xi \rangle_{L(\mathbb{F}_q)} \equiv \sum_{x \in \mathbb{F}_q} \chi(x)\overline{\xi(x)} = \begin{cases} q & \text{if } \chi = \xi \\ 0 & \text{otherwise,} \end{cases}$$

for all $\chi, \xi \in \widehat{\mathbb{F}}_q$.

The *principal* additive character of $\mathbb{F}_q$ is defined by setting, for all $x \in \mathbb{F}_q$,

$$\chi_{princ}(x) = \exp[2\pi i \mathrm{Tr}(x)/p], \tag{1.47}$$

where $\mathrm{Tr} = \mathrm{Tr}_{\mathbb{F}_q/\mathbb{F}_p}$ denotes the trace (cf. (1.45)) and we identify $\mathbb{F}_p$ with $\{0, 1, \ldots, p-1\}$ to compute the exponential. Since Tr is a surjective $\mathbb{F}_p$-linear map from $\mathbb{F}_q$ onto $\mathbb{F}_p$, the principal character χ_{princ} is indeed a nontrivial additive character.

Exercise 1.7.7 Let χ be a nontrivial additive character of $\mathbb{F}_q$. For each $y \in \mathbb{F}_q$ define $\chi_y \colon \mathbb{F}_q \to \mathbb{T}$ by setting

$$\chi_y(x) = \chi(xy)$$

for all $x \in \mathbb{F}_q$ (see Remark 1.2.29). Show that $\chi_y \in \widehat{\mathbb{F}}_q$, and that the map

$$\Psi \colon \quad \begin{array}{ccc} \mathbb{F}_q & \to & \widehat{\mathbb{F}}_q \\ y & \mapsto & \chi_y \end{array}$$

is a group isomorphism.

Exercise 1.7.8 Show that $\widehat{\mathbb{F}}_{q^2} = \{\chi_{s,t} : s, t \in \mathbb{F}_q\}$, where

$$\chi_{s,t}(x, y) = \chi_{princ}(sx + ty) \tag{1.48}$$

for all $s, t, x, y \in \mathbb{F}_q$.

A *multiplicative character* of $\mathbb{F}_q$ is a character of the finite cyclic group $(\mathbb{F}_q^*, \cdot)$, that is, a map

$$\psi \colon \mathbb{F}_q^* \to \mathbb{T}$$

such that $\psi(xy) = \psi(x)\psi(y)$ for all $x, y \in \mathbb{F}_q^*$. The set $\widehat{\mathbb{F}}_q^*$ of all multiplicative characters is a (multiplicative) cyclic group, called the *dual group* of $\mathbb{F}_q^*$.

We have the *orthogonality relations*:

$$\langle \psi, \phi \rangle = \sum_{x \in \mathbb{F}_q^*} \psi(x)\overline{\phi(x)} = \begin{cases} q-1 & \text{if } \psi = \phi \\ 0 & \text{otherwise.} \end{cases}$$

Let x be a generator of $\mathbb{F}_q^*$. The *principal multiplicative character* of $\mathbb{F}_q^*$ associated with x is the multiplicative character ψ_{princ} defined by setting

$$\psi_{princ}(x^k) = \exp\left(\frac{2\pi i k}{q-1}\right) \tag{1.49}$$

for all $k = 1, 2, \ldots, q-1$.

Exercise 1.7.9 Show that ψ_{princ} is a generator of $\widehat{\mathbb{F}_q^*}$.

Decomposable and indecomposable characters

Let v be a character of $\mathbb{F}_{q^2}^*$.

One says that v is *decomposable* if there exists a character ψ of $\mathbb{F}_q^*$ such that

$$v(\alpha) = \psi(\alpha\overline{\alpha}) \tag{1.50}$$

for all $\alpha \in \mathbb{F}_{q^2}^*$. If this is not the case, v is called *indecomposable*.

Moreover, the *conjugate* of v is the character $\overline{v}$ defined by $\overline{v}(\alpha) = v(\overline{\alpha})$ for all $\alpha \in \mathbb{F}_{q^2}^*$.

Exercise 1.7.10 A character $v \in \widehat{\mathbb{F}_{q^2}^*}$ is decomposable if and only if $v = \overline{v}$.

1.7.2 Representation theory of the affine group $\mathrm{Aff}(\mathbb{F}_q)$

Let p be a prime number and let $q = p^n$. The (*general*) *affine group* (*of degree one*) over $\mathbb{F}_q$ is the subgroup $\mathrm{Aff}(\mathbb{F}_q)$ of $\mathrm{GL}(2, \mathbb{F}_q)$ defined by

$$\mathrm{Aff}(\mathbb{F}_q) = \left\{ \begin{pmatrix} a & b \\ 0 & 1 \end{pmatrix} : a \in \mathbb{F}_q^*, b \in \mathbb{F}_q \right\}.$$

Exercise 1.7.11 Show that the action of $\mathrm{Aff}(\mathbb{F}_q)$ on the set $\left\{ \begin{pmatrix} x \\ 1 \end{pmatrix} : x \in \mathbb{F}_q \right\}$ by left multiplication is doubly transitve.

Consider the following abelian subgroups of $\mathrm{Aff}(\mathbb{F}_q)$:

$$A = \left\{ \begin{pmatrix} a & 0 \\ 0 & 1 \end{pmatrix} : a \in \mathbb{F}_q^* \right\} \cong \mathbb{F}_q^*$$

and

$$U = \left\{ \begin{pmatrix} 1 & b \\ 0 & 1 \end{pmatrix} : b \in \mathbb{F}_q \right\} \cong \mathbb{F}_q.$$

Exercise 1.7.12 (1) The inverse of $\begin{pmatrix} a & b \\ 0 & 1 \end{pmatrix} \in \mathrm{Aff}(\mathbb{F}_q)$ is

$$\begin{pmatrix} a & b \\ 0 & 1 \end{pmatrix}^{-1} = \begin{pmatrix} a^{-1} & -a^{-1}b \\ 0 & 1 \end{pmatrix};$$

(2) the subgroup U is normal and one has

$$\mathrm{Aff}(\mathbb{F}_q) \cong U \rtimes A \equiv \mathbb{F}_q \rtimes \mathbb{F}_q^*; \tag{1.51}$$

(3) the conjugacy classes of the group $\mathrm{Aff}(\mathbb{F}_q)$ are the following:

- $\mathscr{C}_0 = \left\{ \begin{pmatrix} 1 & 0 \\ 0 & 1 \end{pmatrix} \right\};$
- $\mathscr{C}_1 = \left\{ \begin{pmatrix} 1 & b \\ 0 & 1 \end{pmatrix} : b \in \mathbb{F}_q^* \right\};$
- $\mathscr{C}_a = \left\{ \begin{pmatrix} a & b \\ 0 & 1 \end{pmatrix} : b \in \mathbb{F}_q \right\}$, where $a \in \mathbb{F}_q^*, a \neq 1.$

Since $\mathrm{Aff}(\mathbb{F}_q)$ is a semidirect product with an abelian normal subgroup (cf. (1.51)), we can apply the little group method (Theorem 1.6.19) in order to get a complete list of all irreducible representations of $\mathrm{Aff}(\mathbb{F}_q)$.

Exercise 1.7.13 After identifying A with the multiplicative group $\mathbb{F}_q^*$ and U with the additive group $\mathbb{F}_q$, show that the conjugacy action (cf. (1.38)) of $A \equiv \mathbb{F}_q^*$ on $\widehat{U} \equiv \widehat{\mathbb{F}_q}$ is given by

$$^a\chi(b) = \chi(a^{-1}b) \tag{1.52}$$

for all $\chi \in \widehat{U}, b \in \mathbb{F}_q$, and $a \in \mathbb{F}_q^*$.

Exercise 1.7.14 Denote by $\chi_0 \equiv 1$ the trivial character of U.

(1) Show that the action of A on $\widehat{U}$ has exactly two orbits, namely $\{\chi_0\}$ and $\widehat{\mathbb{F}_q} \setminus \{\chi_0\}$.
(2) Show that the stabilizer of $\chi \in \widehat{U}$ is given by

$$\mathrm{Stab}_A(\chi) = \begin{cases} \{1_A\} & \text{if } \chi \neq \chi_0 \\ A & \text{if } \chi = \chi_0. \end{cases}$$

Theorem 1.7.15 *The group* $\mathrm{Aff}(\mathbb{F}_q)$ *has exactly* $q-1$ *one-dimensional representations, obtained by associating with each* $\psi \in \widehat{A}$ *the group homomorphism* $\Psi \colon \mathrm{Aff}(\mathbb{F}_q) \to \mathbb{T}$ *defined by*

$$\Psi\begin{pmatrix} a & b \\ 0 & 1 \end{pmatrix} = \psi(a) \tag{1.53}$$

for all $\begin{pmatrix} a & b \\ 0 & 1 \end{pmatrix} \in \mathrm{Aff}(\mathbb{F}_q)$, *and one* $(q-1)$-*dimensional irreducible represen-tation, given by*

$$\pi = \mathrm{Ind}_U^{\mathrm{Aff}(\mathbb{F}_q)} \chi, \tag{1.54}$$

where χ is any nontrivial character of U.

Proof By Exercise 1.7.14, the inertia group of the trivial character $\chi_0 \in \widehat{U}$ is $\mathrm{Aff}(\mathbb{F}_q)$. This provides the $q-1$ one-dimensional representations simply by taking any character $\psi \in \widehat{A}$. Moreover, the inertia group of any nontrivial character $\chi \in \widehat{U}$ is U since, by Exercise 1.7.14, $\mathrm{Stab}_A(\chi) = \{1_A\}$. We conclude by applying Theorem 1.6.19. $\qquad\qquad\qquad\qquad\qquad\qquad\qquad\qquad\square$

1.7.3 The general linear group $\mathrm{GL}(2, \mathbb{F}_q)$

Let $q = p^n$ with p an odd prime. We consider five important subgroups of $\mathrm{GL}(2, \mathbb{F}_q)$:

$$B = \left\{ \begin{pmatrix} \alpha & \beta \\ 0 & \delta \end{pmatrix} : \alpha, \delta \in \mathbb{F}_q^*, \beta \in \mathbb{F}_q \right\} \quad \text{(the *Borel* subgroup)}$$

$$D = \left\{ \begin{pmatrix} \alpha & 0 \\ 0 & \delta \end{pmatrix} : \alpha, \delta \in \mathbb{F}_q^* \right\} \quad \text{(the *diagonal subgroup*)}$$

$$U = \left\{ \begin{pmatrix} 1 & \beta \\ 0 & 1 \end{pmatrix} : \beta \in \mathbb{F}_q \right\} \quad \text{(the *unipotent subgroup*)}$$

$$Z = \left\{ \begin{pmatrix} \alpha & 0 \\ 0 & \alpha \end{pmatrix} : \alpha \in \mathbb{F}_q^* \right\} \quad \text{(the *center*)}$$

$$C = \left\{ \begin{pmatrix} \alpha & \eta\beta \\ \beta & \alpha \end{pmatrix} : \alpha, \beta \in \mathbb{F}_q, (\alpha, \beta) \neq (0,0) \right\} \quad \text{(the *Cartan subgroup*)},$$

where, as usual, $\mathbb{F}_q^*$ denotes the multiplicative subgroup of $\mathbb{F}_q$ consisting of all nonzero elements, and η is a generator of $\mathbb{F}_q^*$; cf. Theorem 1.7.6

We have the following:

- $B = U \rtimes D \cong \mathrm{Aff}(\mathbb{F}_q) \times Z$.
- $U = [B, B] = B'$ is abelian.
- Let $w = \begin{pmatrix} 0 & 1 \\ 1 & 0 \end{pmatrix} \in \mathrm{GL}(2, \mathbb{F}_q)$, then (*Bruhat decomposition*):

$$\mathrm{GL}(2, \mathbb{F}_q) = B \sqcup BwU = B \sqcup UwB = B \sqcup BwB.$$

- $\mathrm{Aff}(\mathbb{F}_q) = U \rtimes A$.

Exercise 1.7.16 Show that $|\mathrm{GL}(2, \mathbb{F}_q)| = (q^2 - 1)(q^2 - q) = q(q+1)(q-1)^2$.

Theorem 1.7.17   *The following table describes the conjugacy classes of the group* $\mathrm{GL}(2, \mathbb{F}_q)$.

TYPE	RE	NC	NE	NAME	C(RE)
(a)	$\begin{pmatrix} \lambda & 0 \\ 0 & \lambda \end{pmatrix}$, $\lambda \neq 0$	$q - 1$	1	central	$\mathrm{GL}(2, \mathbb{F}_q)$
(b$_1$)	$\begin{pmatrix} \lambda_1 & 0 \\ 0 & \lambda_2 \end{pmatrix}$, $\lambda_1 \neq \lambda_2$	$\dfrac{(q-1)(q-2)}{2}$	$q^2 + q$	hyperbolic	D
(b$_2$)	$\begin{pmatrix} \lambda & 1 \\ 0 & \lambda \end{pmatrix}$, $\lambda \neq 0$	$q - 1$	$q^2 - 1$	parabolic	ZU
(b$_3$)	$C \setminus Z$	$\dfrac{q(q-1)}{2}$	$q^2 - q$	elliptic	C

where

- *TYPE indicates* type *of the conjugacy class*
- *RE indicates* representative element*: for each (conjugacy) class we indicate a representative element;*
- *NC indicates* number of conjugacy classes*: this equals the number of representative elements;*
- *NE indicates the* number of elements *in each class;*
- *NAME indicates the* denomination *of this type of class;*
- *C(RE) indicates the* centralizer in $\mathrm{GL}(2, \mathbb{F}_q)$ *of the representative element.*

Proof   We leave it as an exercise. The main point is to observe that two matrices are conjugate if and only if they have the same *minimal* and *characteristic* polynomials (for *nonscalar* matrices, the characteristic polynomial suffices). □

The representation theory of the Borel subgroup B may be then easily deduced from Theorem 1.7.15 and the isomorphism

$$B = \mathrm{Aff}(\mathbb{F}_q) \times Z \cong \mathrm{Aff}(\mathbb{F}_q) \times \mathbb{F}_q^*,$$

which gives (see Theorem 1.2.46)

$$\widehat{B} = \widehat{\mathrm{Aff}(\mathbb{F}_q)} \boxtimes \widehat{Z} \cong \widehat{\mathrm{Aff}(\mathbb{F}_q)} \boxtimes \widehat{\mathbb{F}_q^*}.$$

Theorem 1.7.18   *The Borel subgroup B has:*

- $(q-1)^2$ one-dimensional representations, namely $\Psi_1 \boxtimes \Psi_2$, where Ψ_1 is a one-dimensional representation of $\mathrm{Aff}(\mathbb{F}_q)$ and $\Psi_2 \in \widehat{Z}$;
- $q-1$ irreducible $(q-1)$-dimensional representations, namely $\pi \boxtimes \Psi$, where π is the unique irreducible representation of $\mathrm{Aff}(\mathbb{F}_q)$ of dimension $q-1$ and $\Psi \in \widehat{Z}$.

Using the correspondence between characters of $\mathrm{Aff}(\mathbb{F}_q)$ and those of $\mathbb{F}_q^*$, given by (1.53), these representations are explicitly given by

$$(\Psi_1 \boxtimes \Psi_2) \begin{pmatrix} \alpha & \beta \\ 0 & \delta \end{pmatrix} = \psi_1(\alpha \delta^{-1})\psi_2(\delta) \quad \text{for all } \begin{pmatrix} \alpha & \beta \\ 0 & \delta \end{pmatrix} \in B,$$

with $\psi_1, \psi_2 \in \widehat{\mathbb{F}_q^*}$, and

$$(\pi \boxtimes \Psi) \begin{pmatrix} \alpha & \beta \\ 0 & \delta \end{pmatrix} = \pi \begin{pmatrix} \alpha \delta^{-1} & \beta \delta^{-1} \\ 0 & 1 \end{pmatrix} \psi(\delta) \quad \text{for all } \begin{pmatrix} \alpha & \beta \\ 0 & \delta \end{pmatrix} \in B,$$

with $\psi \in \widehat{\mathbb{F}_q^*}$.

Notation Rearranging the *parametrization* we set

$$\chi_{\psi_1, \psi_2} \begin{pmatrix} \alpha & \beta \\ 0 & \delta \end{pmatrix} = \psi_1(\alpha)\psi_2(\delta) \quad \text{and} \quad \chi_{\psi, \psi} = \psi(\det(b))$$

for all $\psi_1, \psi_2, \psi \in \widehat{\mathbb{F}_q^*}$ and $\begin{pmatrix} \alpha & \beta \\ 0 & \delta \end{pmatrix}, b \in B$. Also, we shall make no distinction between $\mathrm{Res}_D^B \chi_{\psi_1, \psi_2}$ and χ_{ψ_1, ψ_2}.

If $\chi \in \widehat{D}$, let $^w\chi$ be defined by $^w\chi(d) = \chi(wdw)$ for all $w \in D$. Then $^w\chi_{\psi_1, \psi_2} = \chi_{\psi_2, \psi_1}$.

1.7.4 Representations of $\mathrm{GL}(2, \mathbb{F}_q)$

In the first part of this section we determine the irreducible representations of $\mathrm{GL}(2, \mathbb{F}_q)$ that may be obtained by inducing up the characters of the Borel subgroup B. First we give a general principle.

Proposition 1.7.19 *Let G be a group and $N \trianglelefteq G$ a normal subgroup. Then the map $(\rho, U) \mapsto (\widetilde{\rho}, U)$ defined by*

$$\widetilde{\rho}(gN)u = \rho(g)u \tag{1.55}$$

for all $g \in G$ and $u \in U$, is a bijection between the set of all G-representations (ρ, U) such that $\mathrm{Res}_N^G \rho$ is trivial and the set of all G/N-representations.

Proof We leave it to the reader to check that $\widetilde{\rho}$ is well defined, and that the inverse map is given by the inflation. $\qquad\square$

Exercise 1.7.20 Let H be a finite group and denote by H' its derived subgroup. Deduce from the previous proposition that there exists a bijective correspondence between the set of all (irreducible) one-dimensional representations of H and the characters of H/H'.

Proposition 1.7.21 *Let (ρ, V) be an irreducible representation of a group G. Then, if $H \leq G$ and $V^{H'}$ denotes the subspace of all H'-invariant vectors in V (this is called the Jacquet module), we have*

$$V^{H'} \neq \{0\} \Leftrightarrow \text{ there exists } \chi \widehat{H} \text{ such that } d_\chi = 1 \text{ and } \rho \text{ is contained in } \text{Ind}_H^G \chi.$$

Proof The H-representation $(\text{Res}_H^G \rho, V^{H'})$ when restricted to H' is trivial. Therefore it yields a representation of the abelian group H/H' which is a direct sum of characters. By Exercise 1.7.20 it therefore corresponds to a direct sum of one-dimensional H-representations.

If $V^{H'} \neq 0$, then there exists a one-dimensional H-representation χ such that $(\chi, \mathbb{C}) \preceq (\text{Res}_H^G \rho, V^{H'}) \preceq (\text{Res}_H^G \rho, V)$. By Frobenius reciprocity we deduce that $\rho \preceq \text{Ind}_H^G \chi$. $\qquad\square$

Notation

- We set $G = \text{GL}(2, \mathbb{F}_q)$.
- If χ is a one-dimensional representation of B, we denote by $(\widehat{\chi}, V)$ the G-representation $(\text{Ind}_B^G \chi, \text{Ind}_B^G \mathbb{C})$ (note that $\dim V = q + 1$).
- Since $D = B/B'$, there exists a bijection between one-dimensional representations of B and characters of D: given $\chi \in \widehat{D}$, we denote by $^w\chi$ the one-dimensional representation of B corresponding to the character $\chi \in \widehat{D}$.

Proposition 1.7.22 *Let χ be a one-dimensional representation of B. Then*

$$(\text{Res}_B^G \widehat{\chi}, V^U) \sim (\chi \oplus{}^w\chi, \mathbb{C}^2).$$

Proof By our definitions, $V^U \leq \text{Ind}_B^G \mathbb{C}$ and, by the Bruhat decomposition, $f \in V^U$ only depends on $f(1_G)$ and $f(w)$. We then leave it to the reader to compute the corresponding matrix coefficients (for more details see [19, Proposition 14.5.5]) and complete the proof. $\qquad\square$

Notation For $\psi \in \widehat{\mathbb{F}_q^*}$, we define a one-dimensional G-representation by setting

$$\widehat{\chi}_\psi^0(g) = \psi(\det(g))$$

for all $g \in G$.

Theorem 1.7.23 (1) *Let $\psi_1, \psi_2, \xi_1, \xi_2 \in \widehat{\mathbb{F}_q^*}$. If $\psi_1 \neq \psi_2$, then $\widehat{\chi}_{\psi_1,\psi_2}$ is an irreducible G-representation of dimension $q + 1$. Moreover,*

$$\widehat{\chi}_{\psi_1,\psi_2} \sim \widehat{\chi}_{\xi_1,\xi_2} \Leftrightarrow \{\psi_1, \psi_2\} = \{\xi_1, \xi_2\}.$$

In particular,

$$\left\{ \widehat{\chi}_{\psi_1,\psi_2} : \psi_1 \neq \psi_2 \in \widehat{\mathbb{F}_q^*} \right\}$$

are $\frac{(q-1)(q-2)}{2}$ *pairwise nonequivalent irreducible representations of G.*

(2) *For each* $\psi \in \widehat{\mathbb{F}_q^*}$ *there exists an irreducible G-representation* $\widehat{\chi}_{\psi}^1$ *of dimension q such that*

$$\widehat{\chi}_{\psi,\psi} = \widehat{\chi}_{\psi}^0 \oplus \widehat{\chi}_{\psi}^1.$$

Moreover,

$$\left\{ \widehat{\chi}_{\psi}^1 : \psi \in \widehat{\mathbb{F}_q^*} \right\} \quad and \quad \left\{ \widehat{\chi}_{\psi}^0 : \psi \in \widehat{\mathbb{F}_q^*} \right\}$$

is a set of $(q-1)$ *pairwise nonequivalent q-dimensional G-representations, and the set of* all *one-dimensional G-representations, respectively.*

Proof By the Bruhat decomposition $G = B \sqcup BwB$ we have that $S = \{1_G, w\}$ is a complete set of representatives for the double coset $B\backslash G/B$. Moreover $G_w = B \cap wBw = D$ and Mackey's formula for invariants gives, for all one-dimensional representations χ, ξ of B:

$$\mathrm{Hom}_G(\widehat{\chi}, \widehat{\xi}) = \mathrm{Hom}_B(\chi, \xi) \oplus \mathrm{Hom}_D(\mathrm{Res}_D^B \chi, {}^w\xi) = \mathrm{Hom}_B(\chi, \xi) \oplus \mathrm{Hom}_D(\chi, {}^w\xi).$$

We deduce that

- if $\xi = \chi$ and $\chi \neq {}^w\chi$, then $\widehat{\chi}$ is irreducible;
- if $\chi \neq {}^w\chi$, $\xi \neq {}^w\xi$ and $\{\chi, {}^w\chi\} \neq \{\xi, {}^w\xi\}$, then $\widehat{\chi} \not\sim \widehat{\xi}$;
- if $\chi = {}^w\chi$, then $\dim\mathrm{Hom}_G(\widehat{\chi}, \widehat{\chi}) = 2$ so $\widehat{\chi} = \sigma_1 \oplus \sigma_2$, with $\sigma_1, \sigma_2 \in \widehat{G}$.

We observe that $\widehat{\chi}_{\psi}^0 \preceq \widehat{\chi}_{\psi,\psi}$: if $f(g) = \overline{\psi(\det g)}$, with $g \in G$, we have

$$f(gb) = \overline{\psi(\det gb)} = \overline{\psi(\det gb)} = \overline{\psi(\det g)\psi(\det b)} = \chi_{\psi,\psi}(b)f(g)$$

for all $b \in B$. As a consequence,

$$[\widehat{\chi}_{\psi,\psi}(g)f](g_0) = f(g^{-1}g_0) = \widehat{\chi}_{\psi}^0 f(g_0),$$

so there exists $\widehat{\chi}_{\psi}^1 \preceq \widehat{\chi}$ such that $\widehat{\chi} = \widehat{\chi}_{\psi}^0 \oplus \widehat{\chi}_{\psi}^1$. We leave it to the reader to check that if $\psi \neq \phi$, then $\widehat{\chi}_{\psi}^1 \not\sim \widehat{\chi}_{\phi}^1$. $\square$

Definition 1.7.24 A G-representation (ρ, V) is called *cuspidal* if the space $V^U = \{v \in V : \rho(u)v = v, \forall u \in U\}$ of all U-invariant vectors is trivial. We denote by $\mathrm{Cusp} = \mathrm{Cusp}(\mathrm{GL}(2, \mathbb{F}_q)) \subset \widehat{\mathrm{GL}(2, \mathbb{F}_q)}$ a complete set of pairwise nonequivalent *irreducible* cuspidal representations.

Theorem 1.7.25   *Let $\chi \in \widehat{U}$ be a nontrivial character. Then $\mathrm{Ind}_U^G \chi$ is multiplicity-free, does not depend on χ, and its decomposition is*

$$\mathrm{Ind}_U^G \chi = \left[\bigoplus_{\psi \in \widehat{\mathbb{F}_q^*}} \widehat{\chi}_\psi^1 \right] \oplus \left[\bigoplus_{\psi_1 \neq \psi_2 \in \widehat{\mathbb{F}_q^*}} \widehat{\chi}_{\psi_1, \psi_2} \right] \oplus \left[\bigoplus_{\rho \in \mathrm{Cusp}} \rho \right]$$

so that $\mathrm{Ind}_U^G \chi$ contains all the irreducible G-representations of dimension greater than one.

Proof   We have that $U \trianglelefteq B$ and $B = \sqcup_{d \in D} dU = \sqcup_{d \in D} U dU$ so that from the Bruhat decomposition we get

$$G = B \sqcup U w B = \left(\bigsqcup_{d \in D} U dU \right) \sqcup \left(\bigsqcup_{d \in D} U w dU \right).$$

We deduce the following facts.

- $\mathscr{S} = D \sqcup wD$ is a complete set of representatives for the double U-cosets in G.
- $dUd^{-1} \cap U = U$ and $wdUd^{-1}w \cap U = \{1_G\}$ for all $d \in D$.
- $\mathscr{S}_0 = Z \sqcup wD = \mathscr{S} \setminus (D \setminus Z)$.
- $f \in \mathscr{H}(G, K, \psi)$ vanishes on $\sqcup_{d \in D \setminus Z} dU$; equivalently f is supported in $\sqcup_{s \in Z \sqcup wD} U s U$.

Define $\tau \colon G \to G$ by setting

$$\tau \begin{pmatrix} \alpha & \beta \\ \gamma & \delta \end{pmatrix} = \begin{pmatrix} \delta & \beta \\ \gamma & \alpha \end{pmatrix}$$

for all $\begin{pmatrix} \alpha & \beta \\ \gamma & \delta \end{pmatrix} \in G$. It is immediate to check that τ is an involutive anti-automorphism of G. We claim that, if $f \in \mathscr{H}(G, U, \chi)$, then $f^\tau = f$, where $f^\tau(g) = f(\tau(g))$ for all $g \in G$.

Indeed, $\mathrm{supp}(f) \subseteq U(Z \sqcup wD)U$ and it obvious that $\tau|_U$ (resp. $\tau|_Z$) is the identity on U (resp. on Z). Since

$$\tau(wd) = \tau \left(\begin{pmatrix} 0 & 1 \\ 1 & 0 \end{pmatrix} \begin{pmatrix} \alpha & 0 \\ 0 & \beta \end{pmatrix} \right) = \tau \left(\begin{pmatrix} 0 & \beta \\ \alpha & 0 \end{pmatrix} \right) = \begin{pmatrix} 0 & \beta \\ \alpha & 0 \end{pmatrix} = wd$$

for all $d = \begin{pmatrix} \alpha & 0 \\ 0 & \beta \end{pmatrix} \in D$, the claim follows.

We deduce from Proposition 1.6.28, that the Hecke algebra $\mathscr{H}(G, U, \chi)$ is commutative and therefore (cf. Theorem 1.6.27) $\operatorname{Ind}_U^G \chi$ is multiplicity-free. By transitivity of induction we have

$$\operatorname{Ind}_U^G \chi = \operatorname{Ind}_{\operatorname{Aff}(\mathbb{F}_q)}^G \operatorname{Ind}_U^{\operatorname{Aff}(\mathbb{F}_q)} \chi = \operatorname{Ind}_{\operatorname{Aff}(\mathbb{F}_q)}^G \pi \qquad (1.56)$$

which implies that also $\operatorname{Ind}_{\operatorname{Aff}(\mathbb{F}_q)}^G \pi$ is multiplicity-free.

The explicit decomposition of $\operatorname{Ind}_U^G \chi$ follows from the following observations.

- The multiplicity of $\widehat{\chi}_\psi^1$ and $\widehat{\chi}_{\psi_1,\psi_2}$ in $\operatorname{Ind}_U^G \chi$ is one, for all $\psi_1, \psi_2, \psi \in \widehat{\mathbb{F}_q^*}$ (exercise; hint: use (1.56)).
- If ρ is a cuspidal representation, then $\operatorname{Res}_{\operatorname{Aff}(\mathbb{F}_q)}^G \rho$ cannot contain a one-dimensional representation of $\operatorname{Aff}(\mathbb{F}_q)$. Otherwise, the restriction to U (which equals the derived subgroup of $\operatorname{Aff}(\mathbb{F}_q)$) of a one-dimensional representation of $\operatorname{Aff}(\mathbb{F}_q)$ being trivial would provide nontrivial U-invariant vectors, contradicting ρ being cuspidal. It follows that $\operatorname{Res}_{\operatorname{Aff}(\mathbb{F}_q)}^G \rho = m\pi$ for some integer $m \geq 1$. But, by Frobenius reciprocity and (1.56),

$$1 \geq m_\rho^{\operatorname{Ind}_U^G \chi} = m \geq 1,$$

showing that the multiplicity of ρ in $\operatorname{Ind}_U^G \chi$ is exactly one. $\qquad\square$

References

[1] J. L. Alperin and R. B. Bell, *Groups and Representations*, Graduate Texts in Mathematics, **162**. Springer-Verlag, New York, 1995.

[2] R. A. Bailey, *Association Schemes: Designed Experiments, Algebra and Combinatorics*. Cambridge Studies in Advanced Mathematics **84**, Cambridge University Press, 2004.

[3] E. Bannai and T. Ito, *Algebraic Combinatorics*, Benjamin, Menlo Park, CA, 1984.

[4] E. Bannai, E. Bannai, T. Ito and R. Tanaka, *Algebraic Combinatorics*, De Gruyter Series in Discrete Mathematics and Applications volume 5, De Gruyter 2021.

[5] L. Bartholdi and R. I. Grigorchuk, On parabolic subgroups and Hecke algebras of some fractal groups. *Serdica Math. J.* **28** (2002), no. 1, 47–90.

[6] K. P. Bogart, An obvious proof of Burnside's lemma, *Amer. Math. Monthly* **98** (1991), no. 10, 927–928.

[7] A. Borodin and G. Olshanski, *Representations of the Infinite Symmetric Group*. Cambridge Studies in Advanced Mathematics, **160**. Cambridge University Press, Cambridge, 2017.

[8] A. E. Brouwer, A. M. Cohen and A. Neumaier, *Distance-Regular Graphs*. Ergebnisse der Mathematik und ihrer Grenzgebiete (3) [Results in Mathematics and Related Areas (3)], **18**. Springer-Verlag, Berlin, 1989.

[9] D. Bump and D. Ginzburg, Generalized Frobenius–Schur numbers, *J. Algebra* **278** (2004), no. 1, 294–313.

[10] W. Burnside, *Theory of Groups of Finite Order*, Cambridge University Press, 1897.

[11] P. J. Cameron, *Permutation Groups*. London Mathematical Society Student Texts, **45**. Cambridge University Press, Cambridge, 1999.

[12] P. J. Cameron and J. H. van Lint, *Designs, Graphs, Codes and their Links*. London Mathematical Society Student Texts, **22**. Cambridge University Press, Cambridge, 1991.

[13] T. Ceccherini-Silberstein, F. Scarabotti and F. Tolli, Trees, wreath products and finite Gelfand pairs, *Adv. Math.*, **206** (2006), no. 2, 503–537.

[14] T. Ceccherini-Silberstein, F. Scarabotti and F. Tolli, Finite Gelfand pairs and their applications to probability and statistics, *J. Math. Sci. (N.Y.)* **141** (2007), no. 2, 1182–1229.

[15] T. Ceccherini-Silberstein, F. Scarabotti and F. Tolli, *Harmonic Analysis on Finite Groups: Representation Theory, Gelfand Pairs and Markov Chains.* Cambridge Studies in Advanced Mathematics **108**, Cambridge University Press, Cambridge, 2008.

[16] T. Ceccherini-Silberstein, F. Scarabotti and F. Tolli, *Representation Theory of the Symmetric Groups: the Okounkov–Vershik Approach, Character Formulas, and Partition Algebras.* Cambridge Studies in Advanced Mathematics **121**, Cambridge University Press, Cambridge, 2010.

[17] T. Ceccherini-Silberstein, A. Machí, F. Scarabotti and F. Tolli, Induced representations and Mackey theory. Functional analysis. *J. Math. Sci. (N.Y.)* **156** (2009), no. 1, 11–28.

[18] T. Ceccherini-Silberstein, F. Scarabotti and F. Tolli, *Representation Theory and Harmonic Analysis of Wreath Products of Finite Groups*, London Mathematical Society Lecture Note Series **410**, Cambridge University Press, Cambridge, 2014.

[19] T. Ceccherini-Silberstein, F. Scarabotti and F. Tolli, *Discrete Harmonic Analysis: Representations, Number Theory, Expanders, and the Fourier Transform.* Cambridge Studies in Advanced Mathematics, **172**, Cambridge University Press, Cambridge, 2018.

[20] T. Ceccherini-Silberstein, F. Scarabotti and F. Tolli, *Gelfand Triples and their Hecke Algebras — harmonic analysis for multiplicity-free induced representations of finite groups. With a foreword by Eiichi Bannai*, Lecture Notes in Mathematics **2267**, Springer, Cham, 2020.

[21] T. Ceccherini-Silberstein, F. Scarabotti and F. Tolli, *Representation theory of finite group extensions. Clifford theory, Mackey obstruction, and the orbit method*, Springer Monographs in Mathematics, Springer, Cham, 2022.

[22] P. Corsini and V. Leoreanu, *Applications of Hyperstructure Theory*, Springer, 2003.

[23] D. D'Angeli and A. Donno, Self-similar groups and finite Gelfand pairs, *Algebra Discrete Math.* (2007), no. 2, 54–69.

[24] D. D'Angeli and A. Donno, A group of automorphisms of the rooted dyadic tree and associated Gelfand pairs, *Rend. Semin. Mat. Univ. Padova* **121** (2009), 73–92.

[25] Ph. Delsarte, *An algebraic approach to the association schemes of coding theory*, Philips Res. Rep. Suppl. No. 10 (1973).

[26] P. Diaconis, *Group Representations in Probability and Statistics.* IMS Hayward, CA, 1988.

[27] P. Diaconis and M. Shahshahani, Generating a random permutation with random transpositions, *Z. Wahrsch. Verw. Geb.*, **57** (1981), 159–179.

[28] P. Diaconis and M. Shahshahani, Time to reach stationarity in the Bernoulli–Laplace diffusion model, *SIAM J. Math. Anal.* **18** (1987), no. 1, 208–218.

[29] J. Dieudonné, *Treatise on Analysis* Vol. VI. Pure and Applied Mathematics, 10-VI. Academic Press, Inc. New York-London, 1978.

[30] Ch. F. Dunkl, The measure algebra of a locally compact hypergroup, *Trans. Amer. Math. Soc.* **179** (1973), 331–348.

[31] Ch. F. Dunkl, Structure hypergroups for measure algebras, *Pacific J. Math.* **47** (1973), 413–425.

[32] Ch. F. Dunkl, Spherical functions on compact groups and applications to special functions. Symposia Mathematica, Vol. XXII (Convegno sull'Analisi Armonica e Spazi di Funzioni su Gruppi Localmente Compatti, INDAM, Rome, 1976), pp. 145–161. Academic Press, London, 1977.

[33] Ch. F. Dunkl, Orthogonal functions on some permutation groups, *Proc. Symp. Pure Math.* **34**, Amer. Math. Soc., Providence, RI, (1979), 129–147.

[34] H. Dym and H. P. McKean, *Fourier Series and Integrals*, Probability and Mathematical Statistics, No. 14. Academic Press, New York-London, 1972.

[35] P. Ehrenfest and T. Ehrenfest, Über zwei bekannte Einwände gegen das Boltzmannsche H-Theorem. *Physikalische Zeitschrift* **8** (1907), 311–314.

[36] J. Faraut, Analyse harmonique sur les paires de Guelfand et les espaces hyperboliques, CIMPA lecture notes (1980).

[37] A. Figà-Talamanca, Note del Seminario di Analisi Armonica, A.A. 1990–91, Università di Roma "La Sapienza".

[38] A. Figà-Talamanca and C. Nebbia, *Harmonic Analysis and Representation Theory for Groups Acting on Homogeneous Trees*, London Mathematical Society Lecture Note Series **162**. Cambridge University Press, Cambridge, 1991.

[39] W. Fulton and J. Harris, *Representation Theory. A First Course*, Springer-Verlag, New York, 1991.

[40] I. M. Gelfand, Spherical functions in symmetric Riemann spaces, *Doklady Akad. Nauk SSSR (N.S.)* **70**, (1950); [Collected papers, Vol. II, Springer (1988) 31–35].

[41] C. D. Godsil, *Algebraic Combinatorics.* Chapman and Hall Mathematics Series. Chapman & Hall, New York, 1993.

[42] C. Godsil and G. Royle, *Algebraic Graph Theory.* Graduate Texts in Mathematics **207**, Springer-Verlag, New York, 2001.

[43] R. I. Grigorchuk, Just infinite branch groups, in *New Horizons in Pro-p Groups* (ed. Marcus de Sautoy, Dan Segal and Aner Shalev), 121–179, Progr. Math. **184**, Birkhäuser Boston, Boston, MA, 2000.

[44] S. Helgason, *Groups and Geometric Analysis: Integral Geometry, Invariant Differential Operators, and Spherical Functions.* Mathematical Surveys and Monographs, **83**. American Mathematical Society, Providence, RI, 2000.

[45] D. A. Holton and J. Sheehan, *The Petersen Graph.* Australian Mathematical Society Lecture Series **7**, Cambridge University Press, Cambridge, 1993.

[46] I. M. Isaacs, *Character Theory of Finite Groups*, Corrected reprint of the 1976 original [Academic Press, New York]. Dover Publications, Inc., New York, 1994.

[47] M. R. Jerrum, Computational Pólya theory, in *Surveys in Combinatorics, 1995* (P. Rowlinson, ed.), London Math. Soc. Lecture Notes **218**, Cambridge University Press, Cambridge, 1995, pp. 103–118.

[48] R. J. Jewett, Spaces with an abstract convolution of measures, *Advances in Math.* **18** (1975), no. 1, 1–101.

[49] J. H. van Lint and R.M. Wilson, *A Course in Combinatorics*. Second edition. Cambridge University Press, Cambridge, 2001.

[50] I. G. Macdonald, *Symmetric Functions and Hall Polynomials*. Second edition. With contributions by A. Zelevinsky. Oxford Mathematical Monographs. Oxford Science Publications. The Clarendon Press, Oxford University Press, New York, 1995.

[51] F. J. MacWilliams and N. J. A. Sloane, *The Theory of Error-Correcting Codes*, Vol. I and II. North-Holland Mathematical Library, Vol. 16. North-Holland Publishing Co., Amsterdam-New York-Oxford, 1977.

[52] M. A. Naimark and A. I. Stern, *Theory of Group Representations*, Springer-Verlag, New York, 1982.

[53] A. F. Nikiforov, S. K. Suslov and V. B. Uvarov, *Classical Orthogonal Polynomials of a Discrete Variable*. Springer Series in Computational Physics. Springer-Verlag, Berlin, 1991.

[54] P. M. Neumann, A lemma that is not Burnside's, *Math. Sci.* **4** (1979), no. 2, 133–141.

[55] A. Okounkov and A. M. Vershik, A new approach to representation theory of symmetric groups. *Selecta Math. (N.S.)* **2** (1996), no. 4, 581–605.

[56] F. Peter and H. Weyl, Die Vollständigkeit der primitiven Darstellungen einer geschlossenen kontinuierlichen Gruppe, *Math. Ann.*, **97** (1927), 737–755.

[57] I. I. Piatetski-Shapiro, *Complex Representations of* GL$(2,K)$ *for Finite Fields K*. Contemporary Mathematics, **16**. American Mathematical Society, Providence, R.I., 1983.

[58] M. Reed and B. Simon, *Methods of Modern Mathematical Physics. II. Fourier Analysis and Self-Adjointness*. Academic Press [Harcourt Brace Jovanovich, Publishers], New York-London, 1975.

[59] J. Saxl, On multiplicity-free permutation representations, in *Finite Geometries and Designs* (ed. P. J. Cameron, J. W. P. Hrschfeld and D. R. Hughes), pp. 337–353, London Math. Soc. Lecture Notes Series, **49**, Cambridge University Press, 1981.

[60] F. Scarabotti, Time to reach stationarity in the Bernoulli-Laplace diffusion model with many urns, *Adv. in Appl. Math.* **18** (1997), no. 3, 351–371.

[61] F. Scarabotti and F. Tolli, Harmonic analysis on a finite homogeneous space, *Proc. Lond. Math. Soc.*(3) **100** (2010), no. 2, 348–376.

[62] F. Scarabotti and F. Tolli, Fourier analysis of subgroup-conjugacy invariant functions on finite groups, *Monatsh. Math.* **170** (2013), 465–479.

[63] F. Scarabotti and F. Tolli, Hecke algebras and harmonic analysis on finite groups, *Rend. Mat. Appl.* (7) **33** (2013), no. 1-2, 27–51.

[64] F. Scarabotti and F. Tolli, Induced representations and harmonic analysis on finite groups, *Monatsh. Math.* **181** (2016), no. 4, 937–965.

[65] J. P. Serre, *Linear Representations of Finite Groups*, Graduate Texts in Mathematics **42**. Springer-Verlag, New York-Heidelberg, 1977.

[66] B. Simon, *Representations of Finite and Compact Groups*, American Math. Soc., 1996.

[67] D. Stanton, An introduction to group representations and orthogonal polynomials, in *Orthogonal Polynomials* (P. Nevai Ed.), 419–433, Kluwer Academic Dordrecht, 1990.

[68] S. Sternberg, *Group Theory and Physics*, Cambridge University Press, Cambridge, 1994.

[69] A. Terras, *Fourier Analysis on Finite Groups and Applications*. London Mathematical Society Student Texts **43**. Cambridge University Press, Cambridge, 1999.

[70] H. Wielandt, *Finite Permutation Groups*, Academic Press, New York-London, 1964.

[71] J. Wolf, *Harmonic Analysis on Commutative Spaces*. Mathematical Surveys and Monographs **142**. American Mathematical Society, Providence, RI, 2007.

[72] E. M. Wright, Burnside's lemma: a historical note, *J. Comb. Theory (B)*, **30** (1981), 89–90.

[73] P.-H. Zieschang, *Theory of Association Schemes*, Springer Monographs in Mathematics. Springer-Verlag, Berlin, 2005.

2

Quantum probability approach to spectral analysis of growing graphs

Nobuaki Obata

Abstract

These lecture notes provide quantum probabilistic concepts and methods for spectral analysis of graphs, in particular, for the study of asymptotic behavior of the spectral distributions of growing graphs. Quantum probability theory is an algebraic generalization of classical (Kolmogorovian) probability theory, where an element of a (not necessarily commutative) $*$-algebra is treated as a random variable. In this aspect the concepts and methods peculiar to quantum probability are applied to the spectral analysis of adjacency matrices of graphs. In particular, we focus on the method of quantum decomposition and the use of various concepts of independence. The former discloses the non-commutative nature of adjacency matrices and gives a systematic method of computing spectral distributions. The latter is related to various graph products and provides a unified aspect in obtaining the limit spectral distributions as corollaries of various central limit theorems.

Keywords: adjacency matrix, asymptotic combinatorics, central limit theorem, graph product, graph spectrum, growing graph, orthogonal polynomials, quantum decomposition, quantum probability, spectral distribution

Mathematics Subject Classification: Primary 05C50; Secondary 05A16, 05C30, 05C76, 33C45, 60F05, 81S25

2.1 Introduction

A graph is a common object which has been studied extensively in many branches of science. In recent years large graphs have attracted much attention from various points of view of analysis, combinatorics, probability as well as applications to complex and random networks [18, 23, 48]. Our work is inspired by the philosophy of *asymptotic combinatorics* proposed by Vershik [71, 72], where the main question is to explore the limit behavior of a combinatorial object when it grows. Accordingly, we will concentrate on a growing family of graphs rather than a single large graph, see Figure 2.1.

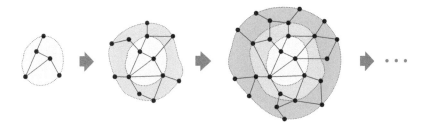

Figure 2.1 Illustration of growing graphs

In these lecture notes we employ *quantum probability theory* towards the spectral analysis of growing graphs. It is widely accepted that probability theory provides essential concepts and powerful tools for asymptotic problems in general. Quantum probability theory is an algebraic generalization of the classical (Kolmogorovian) probability theory, where an element of a (not necessarily commutative) ∗-algebra is treated as a random variable. Thus, the adjacency matrices of growing graphs are treated directly as random variables and their asymptotic properties are derived from their non-commutative nature that are not perceived by classical analysis.

Let us start with a simple example. Denote by C_n the cycle on n vertices and regard them as growing graphs as $n \to \infty$. The adjacency matrix of C_n is given by

$$
A_n = \begin{bmatrix}
0 & 1 & & & & & 1 \\
1 & 0 & 1 & & & & \\
& 1 & 0 & 1 & & & \\
& & \ddots & \ddots & \ddots & & \\
& & & & 1 & 0 & 1 \\
1 & & & & & 1 & 0
\end{bmatrix}.
$$

Since A_n is a real symmetric matrix, its eigenvalues are all real. The eigenvalues with multiplicities are described by a multi-set, called the *spectrum* of C_n. To avoid inessential argument we consider only C_{2n}. By elementary linear algebra we obtain

$$\text{Spec}(C_{2n}) = \left\{ -2(1), 2(1), 2\cos\frac{k\pi}{n}(2) \,; 1 \le k \le n-1 \right\}. \qquad (2.1)$$

We are interested in how the eigenvalues (with multiplicities) are distributed on the real line. An elementary approach of statistics is to visualize the distribution using a histogram, where the number of eigenvalues in each small interval is counted, see Figure 2.2.

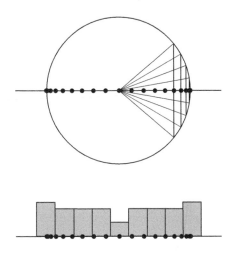

Figure 2.2　Spec(C_{2n}) and its distribution

When the cycle C_{2n} grows as $n \to \infty$, the total number of eigenvalues of C_{2n} as well as the number in each small interval diverges. Hence, to grasp the limit shape of the distribution we need to make the intervals finer as $n \to \infty$. For that purpose, it is essential to consider the *eigenvalue distribution*, namely, a probability distribution on $\mathbb{R}$ defined by

$$\mu_n = \frac{1}{2n}\left(\delta_{-2} + \delta_2 + 2\sum_{k=1}^{n-1} \delta_{2\cos\frac{k\pi}{n}} \right), \qquad (2.2)$$

where the multiplicities are taken into account. The limit of μ_n as $n \to \infty$ is obtained through integration. Take a continuous bounded function $f(x)$ on $\mathbb{R}$ and consider the integral:

$$\int_{-\infty}^{+\infty} f(x)\mu_n(dx) = \frac{1}{2n}\left\{ f(-2) + f(2) + 2\sum_{k=1}^{n-1} f\left(2\cos\frac{k\pi}{n}\right) \right\}.$$

Then, taking the definition of Riemann integral into account, we obtain

$$\lim_{n\to\infty} \int_{-\infty}^{+\infty} f(x)\mu_n(dx) = \lim_{n\to\infty} \frac{1}{n}\sum_{k=1}^{n-1} f\left(2\cos\frac{k\pi}{n}\right)$$
$$= \int_0^1 f(2\cos\pi t)dt = \int_{-2}^2 f(x)\frac{dx}{\pi\sqrt{4-x^2}}.$$

Thus, the limit of μ_n is given by the probability distribution with density function

$$\frac{1}{\pi\sqrt{4-x^2}}, \qquad -2 < x < 2, \tag{2.3}$$

which is known as the *arcsine law* (with mean 0 and variance 2), see Figure 2.3. We understand that the arcsine law presents a feature of the growing cycles C_{2n} as $n \to \infty$.

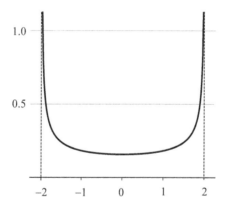

Figure 2.3 Arcsine law (with mean 0 and variance 2)

The above example shows our main question of asymptotic spectral analysis of growing graphs though the method there is entirely classical. The purpose of these lecture notes is to demonstrate the powerful methods of quantum probability theory, where the adjacency matrix itself is treated directly as a random variable. In fact, employing quantum probability we are able to obtain the eigenvalue distribution (2.2) without calculating the characteristic polynomial nor solving the eigenvalue problem associated to A_n. Moreover, we are able to get the limit distribution (2.3) without knowing Spec (C_{2n}) as mentioned explicitly in (2.1). We will come back to this example in Subsection 2.5.2.

These lecture notes are organized as follows. In Section 2.2 we review the basic concepts and notations in quantum probability theory, where we see how a classical random variable is replaced with an algebraic random variable. The concept of spectral distribution is introduced and the moment problem is mentioned.

In Section 2.3 we introduce an interacting Fock space (IFS) as a typical algebraic probability space. Combining the theory of orthogonal polynomials, we come to the concept of quantum decomposition of a real algebraic random variable. The *method of quantum decomposition* is a clue to disclose the non-commutative nature of a classical random variable and, in our context, is applied to the adjacency matrix of a graph.

In Section 2.4 we formulate the main question within the framework of quantum probability. We show an IFS structure emerges naturally in the stratification of a graph and illustrate how to compute the spectral distribution from the quantum decomposition of the adjacency matrix. Distance-regular graphs provide an important class of graphs which admit natural IFS structures. Our approach is efficient also to association schemes (Bose–Mesner algebras).

In Section 2.5 we study the asymptotic spectral distribution of growing regular graphs by means of quantum decomposition. It is noticeable that to obtain the limit distribution we do not need the spectral distribution of each graph of the growing graphs, but only need a simple set of geometric or combinatorial data describing how the graphs grow.

In Section 2.6 we focus on graph products. There is a remarkable relation between graph products and concepts of independence. In fact, quantum probability provides quite a few variants of "independence" thanks to non-commutativity and, as a result, the central limit theorem (CLT) yields different limit distributions. We see that repeated application of a certain graph product gives rise to an expression of the adjacency matrix as a sum of "independent" random variables. Then the limit distribution is obtained as a corollary of the corresponding CLT.

Although our study has been so far restricted to the adjacency matrices of graphs, our quantum probabilistic method is potentially applicable to other matrices such as Laplacian matrices, distance matrices, Q-matrices, and their modifications. However, our present method is not directly applicable to questions concerning complex spectrum. It is our hope to extend the quantum probability approach to spectral analysis of the unitary evolution matrices of quantum walks, adjacency matrices or Laplacian matrices of digraphs (directed graphs).

Acknowledgements: These lecture notes are based on my mini-course lecture in the International Conference and PhD–Masters Summer School on

Groups and Graphs, Designs and Dynamics (G2D2), held at China Three Gorges University in August 2019. I am thankful to the organizers and participants for creating the friendly and exciting atmosphere. I express my special appreciation to Professor Yaokun Wu who initiated our collaboration and provided warm encouragement. Also, I gratefully acknowledge the financial support from Three Gorges Mathematical Research Center.

2.2 Basic concepts of quantum probability

In the measure theoretical (Kolmogorovian) probability theory a random variable is a measurable function defined on a probability space $(\Omega, \mathscr{F}, P)$. Then the random variables are closed under addition, scalar multiplication and pointwise multiplication, namely, they form a commutative algebra. Quantum probability is an extension of the measure theoretical probability which allows a noncommutative algebra of "random variables." Thus, quantum probability is based on an algebraic probability space $(\mathscr{A}, \varphi)$ as is discussed in the following section.

These lecture notes being based on *quantum probability*, the measure theoretical probability theory is referred to as the *classical probability theory*. The quantum-classical correspondence will be discussed in Subsection 2.2.4.

2.2.1 Algebraic probability spaces

An algebra $\mathscr{A}$ over the complex number field $\mathbb{C}$ is called a *-algebra* if it is equipped with a *-operation $a \mapsto a^*$ such that

$$(a+b)^* = a^* + b^*, \quad (\lambda a)^* = \bar{\lambda} a^*, \quad (ab)^* = b^* a^*, \quad (a^*)^* = a,$$

where $a, b \in \mathscr{A}$ and $\lambda \in \mathbb{C}$. A *-algebra $\mathscr{A}$ is called *unital* if it has a multiplication unit $1_{\mathscr{A}}$. Identifying a complex number $\lambda \in \mathbb{C}$ with $\lambda 1_{\mathscr{A}}$ in $\mathscr{A}$, we write just λ for $\lambda 1_{\mathscr{A}}$. Accordingly, the multiplication unit is denoted by 1 simply. Obviously, $1^* = 1$.

Definition 2.2.1 (State)   Let $\mathscr{A}$ be a unital *-algebra. A map $\varphi : \mathscr{A} \to \mathbb{C}$ is called a *state* on $\mathscr{A}$ if it is

(1) linear, i.e., $\varphi(a+b) = \varphi(a) + \varphi(b)$ and $\varphi(\lambda a) = \lambda \varphi(a)$ for $a, b \in \mathscr{A}$ and $\lambda \in \mathbb{C}$;
(2) positive, i.e., $\varphi(a^*a) \geq 0$ for $a \in \mathscr{A}$;
(3) normalized, i.e., $\varphi(1) = 1$.

Definition 2.2.2 (Algebraic probability space)   A pair $(\mathscr{A}, \varphi)$, where $\mathscr{A}$ is a unital $*$-algebra and φ is a state on it, is called an *algebraic probability space*. Each element $a \in \mathscr{A}$ is called an *(algebraic) random variable*. It is called *real* if $a = a^*$.

Once we fix an algebraic probability space $(\mathscr{A}, \varphi)$, we simply call an element $a \in \mathscr{A}$ a *random variable* without specifying the state φ.

The statistical nature of an algebraic random variable a in $(\mathscr{A}, \varphi)$ is fully stored in its *mixed moments* defined by

$$\varphi(a^{\varepsilon_m} \cdots a^{\varepsilon_2} a^{\varepsilon_1}), \qquad \varepsilon_1, \varepsilon_2, \ldots, \varepsilon_m \in \{1, *\}.$$

Two algebraic random variables a in $(\mathscr{A}, \varphi)$ and b in $(\mathscr{B}, \psi)$ are called *stochastically equivalent* or *moment equivalent* if all of their mixed moments coincide:

$$\varphi(a^{\varepsilon_m} \cdots a^{\varepsilon_2} a^{\varepsilon_1}) = \psi(b^{\varepsilon_m} \cdots b^{\varepsilon_2} b^{\varepsilon_1}).$$

In that case we write

$$a \overset{\mathrm{m}}{=} b.$$

For a real random variable $a = a^* \in \mathscr{A}$ the mixed moments are reduced to the *moment sequence*:

$$\varphi(a^m), \qquad m = 1, 2, \ldots.$$

Then, two real random variables $a = a^*$ in $(\mathscr{A}, \varphi)$ and $b = b^*$ in $(\mathscr{B}, \psi)$ are moment equivalent $a \overset{\mathrm{m}}{=} b$ if and only if

$$\varphi(a^m) = \psi(b^m), \qquad m = 1, 2, \ldots.$$

Proposition 2.2.3   *Let $(\mathscr{A}, \varphi)$ be an algebraic probability space.*

(1) $\varphi(a^*) = \overline{\varphi(a)}$ *for any $a \in \mathscr{A}$.*
(2) *[Schwarz inequality]* $|\varphi(a^*b)|^2 \le \varphi(a^*a)\varphi(b^*b)$ *for any $a, b \in \mathscr{A}$.*

Proof   Exercise (consider $\varphi((a + \lambda 1)^*(a + \lambda 1)) \ge 0$).   □

Proposition 2.2.4   *Let $(\mathscr{A}, \varphi)$ be an algebraic probability space.*

(1) *If $\varphi(a^*a) = \varphi(aa^*) = 0$, then $a \overset{\mathrm{m}}{=} 0$.*
(2) *If $\varphi(a^*a) = \varphi(aa^*) = |\varphi(a)|^2$, then $a \overset{\mathrm{m}}{=} \varphi(a)1$.*

Proof   Exercise.   □

Remark 2.2.5   Note that $a \overset{\mathrm{m}}{=} b$ does not imply $a + c \overset{\mathrm{m}}{=} b + c$ in general. Of course, the implication is valid if $c = \lambda 1$ for $\lambda \in \mathbb{C}$.

The matrix algebra provides a typical example of an algebraic probability space. The $n \times n$ complex matrices

$$M(n, \mathbb{C}) = \{a = [a_{ij}] \, ; a_{ij} \in \mathbb{C}\}$$

becomes a unital $*$-algebra in an obvious manner. The most elementary states on $M(n, \mathbb{C})$ are the normalized trace and the vector states. The *normalized trace* is defined by

$$\varphi_{\mathrm{tr}}(a) = \frac{1}{n}\mathrm{Tr}\,(a) = \frac{1}{n}\sum_{i=1}^{n} a_{ii}, \qquad a = [a_{ij}] \in M(n, \mathbb{C}).$$

A *vector state* associated to a unit vector $\xi \in \mathbb{C}^n$ is defined by

$$\varphi(a) = \langle \xi, a\xi \rangle, \qquad a \in M(n, \mathbb{C}).$$

It is straightforward to verify that the normalized trace and the vector state are indeed states on $M(n, \mathbb{C})$.

In fact, there is a simple expression for a general state on $M(n, \mathbb{C})$. A matrix $\rho \in M(n, \mathbb{C})$ is called a *density matrix* if ρ is positive definite, i.e., $\langle \xi, \rho\xi \rangle \geq 0$ for all $\xi \in \mathbb{C}^n$, and $\mathrm{Tr}\,(\rho) = 1$.

Proposition 2.2.6   *For a state φ on $M(n, \mathbb{C})$ there exists a unique density matrix ρ such that*

$$\varphi(a) = \mathrm{Tr}\,(\rho a), \qquad a \in M(n, \mathbb{C}).$$

Moreover, the map $\varphi \mapsto \rho$ gives rise to a bijection from the set of states onto the set of density matrices.

Proof   Exercise.   □

2.2.2 Spectral distributions

Let μ be a probability distribution (or probability measure) on $\mathbb{R} = (-\infty, +\infty)$, where $\mathbb{R}$ is always equipped with the Borel σ-field. We say that a probability distribution μ on $\mathbb{R}$ has a finite moment of order $m = 1, 2, \ldots$ if

$$\int_{-\infty}^{+\infty} |x|^m \mu(dx) < \infty.$$

In that case the *mth moment* of μ is defined by

$$M_m(\mu) = \int_{-\infty}^{+\infty} x^m \mu(dx). \tag{2.4}$$

We set $M_0(\mu) = \mu(\mathbb{R}) = 1$. Recall that the mean value and variance of μ are given by

$$\text{mean}(\mu) = M_1(\mu), \qquad \text{var}(\mu) = M_2(\mu) - M_1(\mu)^2,$$

respectively. Let $\mathfrak{P}_{\text{fm}}(\mathbb{R})$ denote the set of probability distributions having finite moments of all orders.

The following result is fundamental.

Theorem 2.2.7 *Let $(\mathscr{A}, \varphi)$ be an algebraic probability space. For a real random variable $a = a^* \in \mathscr{A}$ there exists a probability distribution $\mu \in \mathfrak{P}_{\text{fm}}(\mathbb{R})$ such that*

$$\varphi(a^m) = M_m(\mu) = \int_{-\infty}^{+\infty} x^m \mu(dx), \qquad m = 1, 2, \dots.$$

Definition 2.2.8 (Spectral distribution) The probability distribution μ described in Theorem 2.2.7 is called the *spectral distribution* of a in the state φ. (As is remarked later, the spectral distribution is not necessarily uniquely determined.)

Outline of Proof We set $M_m = \varphi(a^m)$ and consider the so-called Hanckel determinants:

$$\Delta_m = \det \begin{bmatrix} M_0 & M_1 & \cdots & M_m \\ M_1 & M_2 & \cdots & M_{m+1} \\ \vdots & \vdots & & \vdots \\ M_m & M_{m+1} & \cdots & M_{2m} \end{bmatrix}, \qquad m = 0, 1, 2, \dots. \tag{2.5}$$

Then $\Delta_0 = M_0 = 1$ and one of the following two cases happens:

(1) [infinite type] $\Delta_m > 0$ for all $m = 0, 1, 2, \dots$;
(2) [finite type] there exists $d \geq 1$ such that $\Delta_0 > 0, \Delta_1 > 0, \dots, \Delta_{d-1} > 0$ and $\Delta_d = \Delta_{d+1} = \cdots = 0$.

Recall that the famous Hamburger theorem [16, 62, 64] says that the above condition (1) or (2) is necessary and sufficient for a real sequence $\{M_m\}$ to be a moment sequence of a probability distribution on $\mathbb{R}$. Our assertion is then a direct consequence of the Hamburger theorem. $\qquad \square$

In general, $\mu \in \mathfrak{P}_{\text{fm}}(\mathbb{R})$ is not uniquely determined by its moments. Namely, it may happen that two distinct probability distributions $\mu \neq \nu$ have a common moment sequence:

$$\int_{-\infty}^{+\infty} x^m \mu(dx) = \int_{-\infty}^{+\infty} x^m \nu(dx) = M_m, \qquad m = 0, 1, 2, \dots.$$

We say that μ *is the solution of a determinate moment problem* if μ is uniquely determined by its moment sequence. We only mention a simple and useful criterion for the uniqueness.

Proposition 2.2.9 (Carleman's moment test)   *Let $\{M_m\}$ be the moment sequence of a probability distribution in $\mathfrak{P}_{\mathrm{fm}}(\mathbb{R})$. If*

$$\sum_{m=1}^{\infty} M_{2m}^{-\frac{1}{2m}} = +\infty, \tag{2.6}$$

then there exists a unique probability distribution $\mu \in \mathfrak{P}_{\mathrm{fm}}(\mathbb{R})$ such that $M_m = M_m(\mu)$. (We understand that condition (2.6) is automatically satisfied when $M_{2m} = 0$ occurs for some m.)

Recall that the *support* of a probability distribution on $\mathbb{R}$ is a closed subset of $\mathbb{R}$ defined by

$$\mathrm{supp}\,\mu = \mathbb{R} \setminus \bigcup \{U \subset \mathbb{R}\,;\, \text{open set such that } \mu(U) = 0\}.$$

A *point mass* or δ-*measure* at $a \in \mathbb{R}$ is a probability distribution on $\mathbb{R}$ defined by

$$\delta_a(E) = \begin{cases} 1, & \text{if } a \in E, \\ 0, & \text{otherwise}, \end{cases}$$

where $E \subset \mathbb{R}$ is a Borel set. Obviously, $\mathrm{supp}\,\delta_a = \{a\}$. The integral with respect to the point mass δ_a gives rise to the evaluation at a as follows:

$$\int_{-\infty}^{+\infty} f(x)\delta_a(dx) = f(a).$$

Note that $\mathrm{supp}\,\mu$ consists of finitely many points if and only if μ is a finite sum of point masses.

Example 2.2.10   The following probability distributions are the solutions of determinate moment problems. Verification is by Carleman's moment test and is left to the readers.

(1) Any probability distribution having a finite support.
(2) Any probability distribution having a compact support.
(3) The normal law. We may use the moment sequence of the standard normal law:

$$M_{2m} = \frac{1}{\sqrt{2\pi}} \int_{-\infty}^{+\infty} x^{2m} e^{-x^2/2} dx = \frac{(2m)!}{2^m m!}, \qquad M_{2m+1} = 0,$$

of which derivation is by elementary calculus.

(4) The Poisson distribution with parameter $\lambda > 0$ defined by

$$e^{-\lambda} \sum_{n=0}^{\infty} \frac{\lambda^n}{n!} \delta_n.$$

Proposition 2.2.11 *Let $(\mathscr{A}, \varphi)$ be an algebraic probability space. Let $a = a^* \in \mathscr{A}$ be a real random variable and μ the spectral distribution. Let $\mathscr{B} \subset \mathscr{A}$ be the unital $*$-subalgebra generated by a, that is, the set of polynomials in a with complex coefficients. Then $\dim \mathscr{B} < \infty$ if and only if $|\operatorname{supp} \mu| < \infty$. In that case the spectral distribution is uniquely determined by a. Moreover, $\dim \mathscr{B} = |\operatorname{supp} \mu|$.*

Proof Exercise. □

2.2.3 Convergence of random variables

Let $(\mathscr{A}_1, \varphi_1), (\mathscr{A}_2, \varphi_2), \cdots$ and $(\mathscr{B}, \psi)$ be algebraic probability spaces. We say that a sequence of random variables $a_n \in \mathscr{A}_n$ converges to $b \in \mathscr{B}$ *in moments* if

$$\lim_{n \to \infty} \varphi_n(a_n^{\varepsilon_m} \cdots a_n^{\varepsilon_1}) = \psi(b^{\varepsilon_m} \cdots b^{\varepsilon_1})$$

for any choice of $m = 1, 2, \ldots$ and $\varepsilon_i \in \{1, *\}$. In that case we write

$$a_n \xrightarrow{\ \text{m}\ } b.$$

For real random variables $a_1 = a_1^*, a_2 = a_2^*, \cdots$ and $b = b^*$ we only need to consider the moment sequence. Hence $a_n \xrightarrow{\ \text{m}\ } b$ if and only if

$$\lim_{n \to \infty} \varphi_n(a_n^m) = \psi(b^m), \qquad m = 1, 2, \ldots. \tag{2.7}$$

Let $\mu_1, \mu_2, \ldots$ and ν be the spectral distributions of $a_1 = a_1^*, a_2 = a_2^*, \cdots$ and $b = b^*$, respectively. Then (2.7) is equivalent to

$$\lim_{n \to \infty} \int_{-\infty}^{+\infty} x^m \mu_n(dx) = \int_{-\infty}^{+\infty} x^m \nu(dx), \qquad m = 1, 2, \ldots. \tag{2.8}$$

In that case we say that μ_n *converges to ν in moments*.

In classical probability theory the concept of *weak convergence* (also called *vague convergence* in some literature) is common. Let $\mu_1, \mu_2, \ldots$ and ν be probability distributions on $\mathbb{R}$. We say that μ_n *converges weakly* to ν if

$$\lim_{n \to \infty} \int_{-\infty}^{+\infty} f(x) \mu_n(dx) = \int_{-\infty}^{+\infty} f(x) \nu(dx) \tag{2.9}$$

for any bounded continuous functions $f(x)$. Condition (2.8) being obtained simply by replacing $f(x)$ with x^m in (2.9), we should note that the monomial

x^m is a continuous function but not bounded. In this connection the following result is useful, for the proof see, for example, [17, Section 4.5].

Proposition 2.2.12   *Let $\mu_1, \mu_2, \ldots$ and ν be probability distributions in $\mathfrak{P}_{\mathrm{fm}}(\mathbb{R})$ and assume that μ_n converges to ν in moments. If ν is the solution to a determinate moment problem, then μ_n converges to ν weakly.*

The concept of convergence in moments is naturally extended to a sequence of random vectors. Let $(\mathscr{A}_1, \varphi_1), (\mathscr{A}_2, \varphi_2), \cdots$ and $(\mathscr{B}, \psi)$ be algebraic probability spaces. Consider k-dimensional vectors of random variables

$$(a_{n,1}, \ldots, a_{n,k}), \qquad a_{n,i} \in \mathscr{A}_n$$

and

$$(b_1, \ldots, b_k), \qquad b_i \in \mathscr{B}.$$

We say that $(a_{n,1}, \ldots, a_{n,k})$ *converges to* $(b_1, \ldots, b_k)$ *in moments* if every mixed moment formed by $a_{n,1}, \ldots, a_{n,k}$ converges to the mixed moment of the same type formed by $b_1, \ldots, b_k$. In that case we write

$$(a_{n,1}, \ldots, a_{n,k}) \xrightarrow{\mathrm{m}} (b_1, \ldots, b_k).$$

Note that there are many variants of mixed moments because the random variables under consideration do not necessarily commute. Moreover, we are immediately faced with a difficult problem of how to define the distribution of $(b_1, \ldots, b_k)$ even though all $b_1, \ldots, b_k$ are real.

2.2.4 Classical probability vs quantum probability

A classical (or Kolmogorovian) probability space is a triple $(\Omega, \mathscr{F}, P)$, where Ω is a non-empty set, $\mathscr{F}$ a σ-field of subsets of Ω, and P a probability measure on $\mathscr{F}$. A random variable X is a measurable function defined on Ω with values in $\mathbb{R}$ or in $\mathbb{C}$ (in this case X is called a complex random variable). The *mean value* or *expectation* of a complex random variable X is defined by

$$\mathbf{E}[X] = \int_\Omega X(\omega) P(d\omega), \tag{2.10}$$

whenever the integral converges absolutely.

Let $L^{\infty-}(\Omega, \mathscr{F}, P)$ denote the set of all complex random variables with finite moments of all orders, that is,

$$\mathbf{E}[|X|^m] = \int_\Omega |X(\omega)|^m P(d\omega) < \infty, \qquad m = 1, 2, \ldots.$$

It is verified that $L^{\infty-}(\Omega, \mathscr{F}, P)$ becomes a unital $*$-algebra over $\mathbb{C}$ equipped with pointwise operations, where the multiplication unit is just the random variable taking the constant value 1 and the $*$-operation is given by $X^*(\omega) = \overline{X(\omega)}$ (complex conjugate). Then (2.10) gives rise to a state on $L^{\infty-}(\Omega,\mathscr{F},P)$ and the pair $(L^{\infty-}(\Omega,\mathscr{F},P),\mathbf{E})$ becomes an algebraic probability space. The set of bounded complex random variables, denoted by $L^\infty(\Omega,\mathscr{F},P)$, becomes a unital $*$-subalgebra of $L^{\infty-}(\Omega,\mathscr{F},P)$ and hence the pair $(L^\infty(\Omega,\mathscr{F},P),\mathbf{E})$ is an algebraic probability space too. We call $(L^{\infty-}(\Omega,\mathscr{F},P),\mathbf{E})$ or $(L^\infty(\Omega,\mathscr{F},P),\mathbf{E})$ the algebraic probability space associated to the classical probability space $(\Omega,\mathscr{F},P)$.

In the algebraic probability space $(L^{\infty-}(\Omega,\mathscr{F},P),\mathbf{E})$, a real random variable is nothing but a real random variable in the classical sense. Following the classical theory, the *distribution* of X means a probability distribution μ_X on $\mathbb{R}$ uniquely specified by

$$P(X \le x) = \mu_X((-\infty,x]), \qquad x \in \mathbb{R}.$$

Then the mth moment of X becomes

$$\mathbf{E}[X^m] = \int_\Omega X(\omega)^m P(d\omega) = \int_{-\infty}^{+\infty} x^m \mu_X(dx), \qquad m = 1,2,\dots. \qquad (2.11)$$

Thus, the distribution μ_X of the classical random variable X coincides with the spectral distribution of X regarded as a real algebraic random variable. In this manner quantum probability theory generalizes classical probability theory, while keeping basic concepts consistent.

We note that all the statistical information of $(\Omega,\mathscr{F},P)$ is obtained from the associated algebraic probability space. In fact, the probability of each event $A \in \mathscr{F}$ is obtained from the indicator random variable defined by

$$1_A(\omega) = \begin{cases} 1, & \omega \in A, \\ 0, & \text{otherwise}, \end{cases}$$

in such a way that

$$P(A) = \int_A P(d\omega) = \int_\Omega 1_A(\omega) P(d\omega) = \mathbf{E}[1_A].$$

It is an essential nature of classical probability that the algebra generated by random variables becomes commutative. Moreover, under certain topological conditions one may construct a classical probability space from a commutative algebraic probability space $(\mathscr{A},\varphi)$ in such a way that every real random variable in $\mathscr{A}$ is realized as a classical random variable. For instance, if $\mathscr{A}$ is a commutative unital C^*-algebra, such a construction is possible due to the spectral theory of Gelfand, see, for example, Takesaki [68, Chapter I].

2.2.5 Notes

Quantum (also called non-commutative) probability theory traces back to von Neumann [75], where statistical questions of quantum mechanics were formulated in terms of selfadjoint operators in a Hilbert space and the method of computing probability of quantum mechanical events was established as an application of spectral analysis. Although the term "quantum probability" did not appear therein, von Nuemann's argument was actually based on the *-algebra of observables and states on it, which is nothing else but an algebraic probability space in our language.

The term "quantum probability" adopted in these lecture notes is attributed to Accardi [1], while it appeared occasionally in earlier literature with different usage, see, for example, Gudder [25, 26], Suppes [67]. Although the original motivation came from quantum physics, quantum probability theory has developed into a new mathematical paradigm for non-commutative analysis including our topics.

For more on basic concepts and results in quantum probability, see Hora and Obata [36] and Obata [59] on which these lecture notes are based, and see also Meyer [50] and Parthasarathy [60]. Free probability theory, regarded as a particular case of quantum probability, has developed mainly along with operator algebras and random matrices, see, for example, Hiai and Petz [33], Speicher [65], Voiculescu, Dykema and Nica [74]. It has an interesting connection with some combinatorial problems of partitions, see Nica and Speicher [56].

2.3 Quantum decomposition

2.3.1 Jacobi coefficients and interacting Fock spaces

Definition 2.3.1 (Jacobi coefficients)   A pair of sequences $(\{\omega_n\}, \{\alpha_n\})$ is called *Jacobi coefficients* if one of the following conditions is satisfied:

(1) both $\{\omega_n\} = \{\omega_1, \omega_2, \dots\}$ and $\{\alpha_n\} = \{\alpha_1, \alpha_2, \dots\}$ are infinite sequences such that $\omega_n > 0$ and $\alpha_n \in \mathbb{R}$ for all n;
(2) there exists $d \geq 1$ such that $\{\omega_n\} = \{\omega_1, \dots, \omega_{d-1}\}$ consists of $d-1$ positive numbers and $\{\alpha_n\} = \{\alpha_1, \dots, \alpha_d\}$ consists of d real numbers. (For $d = 1$ we tacitly understand that $\{\omega_n\}$ is an empty sequence.)

Jacobi coefficients satisfying condition (1) are called *of infinite type*, and those satisfying condition (2) are called *of finite type* with length d. For Jacobi coefficients of infinite type we set $d = \infty$.

Given Jacobi coefficients $(\{\omega_n\},\{\alpha_n\})$ of infinite type, we prepare a complex linear space Γ spanned by a linear basis $\{\Phi_0,\Phi_1,\Phi_2,\dots\}$ with an inner product uniquely determined by

$$\langle\Phi_m,\Phi_n\rangle = \delta_{mn} \quad \text{(Kronecker symbol).} \tag{2.12}$$

Then Γ becomes an infinite dimensional inner product space (pre-Hilbert space) with an orthonormal basis $\{\Phi_n\}$. A Hilbert space obtained by completing Γ is not used in these lecture notes though it becomes necessary when we go deep into functional analysis.

Suppose that the Jacobi coefficients $(\{\omega_n\},\{\alpha_n\})$ are of finite type with length d. Then let Γ be an inner product space spanned by a linear basis $\{\Phi_0,\Phi_1,\dots,\Phi_{d-1}\}$ with an inner product similar as in (2.12). In that case, we may take $\Gamma = \mathbb{C}^d$ with the canonical basis.

Having prepared an inner product space for Jacobi coefficients $(\{\omega_n\},\{\alpha_n\})$, we next define three linear operators A^+,A^-,A° by

$$A^+\Phi_n = \sqrt{\omega_{n+1}}\,\Phi_{n+1},$$
$$A^-\Phi_n = \sqrt{\omega_n}\,\Phi_{n-1},$$
$$A^\circ\Phi_n = \alpha_{n+1}\Phi_n,$$

where we tacitly understand that $A^-\Phi_0 = 0$ and that $A^+\Phi_{d-1} = 0$ if the Jacobi coefficients are of finite type of length d. These actions are illustrated in Figure 2.4.

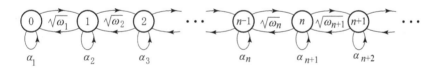

Figure 2.4 Actions of A^+,A^- and A°

Definition 2.3.2 (Interacting Fock space)   The quintuple $(\Gamma,\{\Phi_n\},A^+,A^-,A^\circ)$ described above is called an *interacting Fock space (IFS)* associated to Jacobi coefficients $(\{\omega_n\},\{\alpha_n\})$. We call A^+,A^- and A° the *creation, annihilation* and *conservation operators*, respectively.

Let $(\Gamma,\{\Phi_n\},A^+,A^-,A^\circ)$ be an IFS. Equipped with the $*$-operations uniquely determined by

$$(A^+)^* = A^-, \qquad (A^-)^* = A^+, \qquad (A^\circ)^* = A^\circ,$$

the algebra generated by A^+, A^-, A° becomes a $*$-algebra. The above $*$-operations are compatible to the inner product in the sense that

$$\langle \Phi_m, A^+ \Phi_n \rangle = \langle A^- \Phi_m, \Phi_n \rangle, \qquad \langle \Phi_m, A^\circ \Phi_n \rangle = \langle A^\circ \Phi_m, \Phi_n \rangle.$$

Unless otherwise stated, the $*$-algebra $\mathscr{A}$ is always regarded as an algebraic probability space equipped with the *vacuum state* defined by

$$\langle a \rangle = \langle \Phi_0, a \Phi_0 \rangle, \qquad a \in \mathscr{A}.$$

In this context the *canonical random variable*

$$A^+ + A^- + A^\circ$$

is of particular importance. The *vacuum spectral distribution* of $A^+ + A^- + A^\circ$ is a probability distribution $\mu \in \mathfrak{P}_{\mathrm{fm}}(\mathbb{R})$ determined by

$$\langle (A^+ + A^- + A^\circ)^m \rangle = \int_{-\infty}^{+\infty} x^m \mu(dx), \qquad m = 1, 2, \ldots.$$

2.3.2 Orthogonal polynomials

Let us start with a probability distribution $\mu \in \mathfrak{P}_{\mathrm{fm}}(\mathbb{R})$. Let $\mathbb{C}[x]$ denote the set of polynomials with complex coefficients and equip it with an inner product defined by

$$\langle f, g \rangle = \int_{-\infty}^{+\infty} \overline{f(x)} g(x) \mu(dx), \qquad f, g \in \mathbb{C}[x].$$

Although $\langle f, g \rangle$ is not necessarily an inner product because $\langle f, f \rangle = 0$ does not imply $f = 0$ in general, we use the term "inner product" for simplicity.

Applying the Gram–Schmidt orthogonalization to the sequence of monomials $1, x, x^2, \ldots, x^n, \ldots$, we obtain a sequence of polynomials inductively by

$$P_0(x) = 1, \tag{2.13}$$

$$P_n(x) = x^n - \sum_{k=0}^{n-1} \frac{\langle x^n, P_k \rangle}{\langle P_k, P_k \rangle} P_k(x), \qquad n = 1, 2, \ldots. \tag{2.14}$$

The above orthogonalization procedure stops at $n = d$ if $\langle P_d, P_d \rangle = 0$ happens for some $d \geq 1$. In that case we obtain just a finite sequence of polynomials $P_0(x), P_1(x), \ldots, P_{d-1}(x)$. At any rate the obtained sequence $\{P_n(x)\}$ is called the *orthogonal polynomials* associated to μ.

Proposition 2.3.3 *For a probability distribution $\mu \in \mathfrak{P}_{\mathrm{fm}}(\mathbb{R})$ the following conditions are equivalent:*

(1) *the Gram–Schmidt orthogonalization process stops at $n = d$;*

(2) $|\operatorname{supp}\mu| = d$.

In that case, letting $\operatorname{supp}\mu = \{\lambda_1,\dots,\lambda_d\}$, *we have*

$$P_d(x) = (x-\lambda_1)\cdots(x-\lambda_d).$$

Proof Exercise. □

Theorem 2.3.4 (Three-term recurrence relation) *Let* $\{P_n(x)\}$ *be the orthogonal polynomials associated to* $\mu \in \mathfrak{P}_{\mathrm{fm}}(\mathbb{R})$. *If* $|\operatorname{supp}\mu| = \infty$, *then there exist Jacobi coefficients* $(\{\omega_n\},\{\alpha_n\})$ *of infinite type such that*

$$P_0 = 1, \tag{2.15}$$

$$P_1 = x - \alpha_1, \tag{2.16}$$

$$xP_n = P_{n+1} + \alpha_{n+1}P_n + \omega_n P_{n-1}. \tag{2.17}$$

If $|\operatorname{supp}\mu| = d < \infty$, *then there exist Jacobi coefficients* $(\{\omega_n\},\{\alpha_n\})$ *of finite type of length d such that* (2.17) *holds up to* $n = d-1$.

Proof We give the proof for the case of $|\operatorname{supp}\mu| = \infty$, that is, $\{P_n(x)\}$ is an infinite sequence. The case of $|\operatorname{supp}\mu| = d < \infty$ requires simple additional treatment for $n = d-1$ and the details are left to the readers.

Obviously (2.15) holds by (2.13). Setting $n = 1$ in (2.14) we obtain

$$P_1(x) = x - \frac{\langle P_0, x\rangle}{\langle P_0, P_0\rangle}P_0(x) = x - \int_{\mathbb{R}}x\mu(dx),$$

and

$$\alpha_1 = \int_{\mathbb{R}}x\mu(dx) = \operatorname{mean}(\mu),$$

which shows (2.16). We will prove (2.17) for $n \geq 1$. Since $xP_n(x) = x^{n+1} + \cdots$ is a monic polynomial of degree $n+1$, we can write

$$xP_n(x) = P_{n+1}(x) + \sum_{k=0}^{n}c_{n,k}P_k(x). \tag{2.18}$$

Taking the inner product with $P_k(x)$, $0 \leq k \leq n-2$, we obtain

$$c_{n,k}\langle P_k, P_k\rangle = \langle P_k, xP_n\rangle = \langle xP_k, P_n\rangle = 0,$$

since $\deg xP_k(x) = k+1 \leq n-1$. Hence in the right-hand side of (2.18) the coefficients $c_{n,k} = 0$ except $k = n$ and $n-1$. Thus $xP_n(x)$ is written as in (2.17).

We will examine $\omega_n > 0$. Let $n \geq 1$. In view of (2.17) we have

$$\langle P_n, P_n\rangle = \langle P_n, xP_{n-1} - \alpha_n P_{n-1} - \omega_{n-1}P_{n-2}\rangle = \langle xP_n, P_{n-1}\rangle$$
$$= \langle P_{n+1} + \alpha_{n+1}P_n + \omega_n P_{n-1}, P_{n-1}\rangle = \omega_n\langle P_{n-1}, P_{n-1}\rangle.$$

Hence

$$\langle P_n, P_n \rangle = \omega_n \langle P_{n-1}, P_{n-1} \rangle = \cdots = \omega_n \omega_{n-1} \cdots \omega_1 \langle P_0, P_0 \rangle,$$

and applying $\langle P_0, P_0 \rangle = 1$, we obtain

$$\langle P_n, P_n \rangle = \omega_n \omega_{n-1} \cdots \omega_1 .$$

Then $\omega_n > 0$ follows from $\langle P_n, P_n \rangle > 0$.   □

The Jacobi coefficients described in Theorem 2.3.4 are called the Jacobi coefficients associated to $\mu \in \mathfrak{P}_{fm}(\mathbb{R})$. During the above proof we have established the following

Corollary 2.3.5   *Let $\mu \in \mathfrak{P}_{fm}(\mathbb{R})$. Let $\{P_n(x)\}$ and $(\{\omega_n\}, \{\alpha_n\})$ be the associated orthogonal polynomials and the Jacobi coefficients, respectively. Then we have*

$$\alpha_1 = \int_{-\infty}^{+\infty} x \mu(dx) = \mathrm{mean}(\mu),$$

$$\omega_1 = \int_{-\infty}^{+\infty} (x - \alpha_1)^2 \mu(dx) = \mathrm{var}(\mu),$$

$$\omega_n \omega_{n-1} \cdots \omega_1 = \int_{-\infty}^{+\infty} P_n(x)^2 \mu(dx).$$

An IFS structure emerges in the orthogonal polynomials. Define

$$\Phi_0(x) = P_0(x) = 1,$$

$$\Phi_n(x) = \frac{1}{\sqrt{\langle P_n, P_n \rangle}} P_n(x) = \frac{1}{\sqrt{\omega_n \omega_{n-1} \cdots \omega_1}} P_n(x), \qquad n \geq 1. \quad (2.19)$$

Then $\{\Phi_n(x)\}$ becomes an orthonormal set in $L^2(\mathbb{R}, \mu)$. Let Γ be the subspace of $L^2(\mathbb{R}, \mu)$ spanned by $\{\Phi_n(x)\}$. In accordance with the three-term recurrence relation (2.17) we define linear operators A^+, A°, A^- in Γ by

$$A^+ P_n = P_{n+1}, \qquad (2.20)$$

$$A^\circ P_n = \alpha_{n+1} P_n, \qquad (2.21)$$

$$A^- P_n = \omega_n P_{n-1}, \qquad (2.22)$$

where we understand $A^- P_0 = 0$ and $A^+ P_{d-1} = 0$ when the Jacobi coefficients are finite type of length d. We then see from (2.19) that

$$A^+ \Phi_n = \sqrt{\omega_{n+1}} \, \Phi_{n+1},$$

$$A^\circ \Phi_n = \alpha_{n+1} \Phi_n,$$

$$A^- \Phi_n = \sqrt{\omega_n} \, \Phi_{n-1}.$$

where $A^- \Phi_0 = 0$ and $A^+ \Phi_{d-1} = 0$ when the Jacobi coefficients are finite type of length d. We have thus obtained an IFS $(\Gamma, \{\Phi_n\}, A^+, A^\circ, A^-)$ associated to the Jacobi coefficients $(\{\omega_n\}, \{\alpha_n\})$.

We are interested in the vacuum spectral distribution of $A = A^+ + A^- + A^\circ$. It follows from (2.20)–(2.22) that

$$
\begin{aligned}
AP_n(x) &= A^+ P_n(x) + A^\circ P_n(x) + A^- P_n(x) \\
&= P_{n+1}(x) + \alpha_{n+1} P_n(x) + \omega_n P_{n-1}(x) \\
&= x P_n(x).
\end{aligned}
$$

Namely, A is the multiplication operator with the monomial x (or the coordinate function). In particular, for $\Phi_0(x) = P_0(x) = 1$ we have

$$
A^m \Phi_0(x) = x^m \Phi_0(x) = x^m
$$

and

$$
\langle \Phi_0, A^m \Phi_0 \rangle = \langle 1, x^m \rangle = \int_{-\infty}^{+\infty} x^m \mu(dx), \qquad m = 1, 2, \ldots.
$$

Therefore, μ coincides with the vacuum spectral distribution of $A = A^+ + A^\circ + A^-$.

Summing up,

Theorem 2.3.6 *Let $\mu \in \mathfrak{P}_{\mathrm{fm}}(\mathbb{R})$. Let $\{P_n(x)\}$ and $(\{\omega_n\}, \{\alpha_n\})$ be the associated orthogonal polynomials and the Jacobi coefficients, respectively. Let Γ be the linear space spanned by the orthonormal basis $\{\Phi_n\}$ as in (2.19), and A^+, A°, A^- the linear operators defined as in (2.20)–(2.22). Then the system $(\Gamma, \{\Phi_n\}, A^+, A^\circ, A^-)$ becomes an IFS associated to the Jacobi coefficients $(\{\omega_n\}, \{\alpha_n\})$ and the vacuum spectral distribution of $A^+ + A^\circ + A^-$ coincides with μ. Namely, we have*

$$
\langle \Phi_0, (A^+ + A^\circ + A^-)^m \Phi_0 \rangle = \int_{-\infty}^{+\infty} x^m \mu(dx), \qquad m = 1, 2, \ldots. \qquad (2.23)
$$

The above process is illustrated in the following diagram.

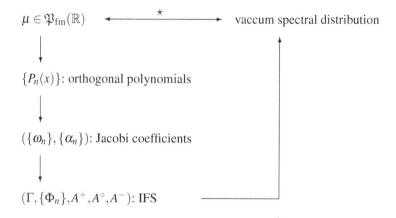

Since the vacuum spectral distribution is not uniquely determined in general, the relation $\overset{\star}{\longleftrightarrow}$ is understood in the sense of (2.23). When μ is the solution to a determinate moment problem, every arrow can be inverted.

Remark 2.3.7 Let $\mu \in \mathfrak{P}_{\mathrm{fm}}(\mathbb{R})$. Every polynomial being naturally regarded as a measurable function, we have a map $\iota : \mathbb{C}[x] \to L^2(\mathbb{R}, \mu)$. This map ι is injective if and only if $|\operatorname{supp}\mu| = \infty$. In that case the monomials $\{1, x, x^2, \dots\}$ are linearly independent in $L^2(\mathbb{R}, \mu)$. If $|\operatorname{supp}\mu| = d < \infty$, the monomials $\{1, x, x^2, \dots, x^{d-1}\}$ are linearly independent in $L^2(\mathbb{R}, \mu)$ and form a maximal linearly independent subset of $\{1, x, x^2, \dots\}$. In that case, setting $\operatorname{supp}\mu = \{\lambda_1, \dots, \lambda_d\}$, we see that $\operatorname{Ker}\iota$ is the ideal generated by $(x - \lambda_1)\cdots(x - \lambda_d)$.

Remark 2.3.8 It is noted that the polynomials are not necessarily dense in $L^2(\mathbb{R}, \mu)$ for an arbitrary $\mu \in \mathfrak{P}_{\mathrm{fm}}(\mathbb{R})$. Namely, the orthogonal polynomials $\{P_n(x)\}$ do not form an orthogonal *basis* of $L^2(\mathbb{R}, \mu)$ in general. It is known that they do so if μ is the solution of a determinate moment problem. But the converse is not true.

2.3.3 Quantum decomposition

We are now in a position to introduce one of the most fundamental concepts in quantum probability theory.

Let $(\mathscr{A}, \varphi)$ be an algebraic probability space and $a = a^* \in \mathscr{A}$ a real random variable. We take the spectral distribution of a, say $\mu \in \mathfrak{P}_{\mathrm{fm}}(\mathbb{R})$. Then we have

$$\varphi(a^m) = M_m(\mu) = \int_{-\infty}^{+\infty} x^m \mu(dx), \qquad m = 1, 2, \dots. \tag{2.24}$$

On the other hand, by Theorem 2.3.6 there exists an IFS $(\Gamma, \{\Phi_n\}, A^+, A^-, A^\circ)$ such that

$$\langle \Phi_0, (A^+ + A^\circ + A^-)^m \Phi_0 \rangle = \int_{-\infty}^{+\infty} x^m \mu(dx), \qquad m = 1, 2, \ldots. \qquad (2.25)$$

We then see from (2.24) and (2.25) that

$$\varphi(a^m) = \langle \Phi_0, (A^+ + A^\circ + A^-)^m \Phi_0 \rangle, \qquad m = 1, 2, \ldots,$$

which means that

$$a \stackrel{m}{=} A^+ + A^- + A^\circ. \qquad (2.26)$$

As a result, any real random variable in an algebraic probability space is decomposed into a sum of creation, annihilation and conservation operators in an IFS. The expression (2.26) is called a *quantum decomposition* of a, and A^+, A^- and A° are called its *quantum components*.

Recall that a classical random variable X having finite moments of all orders is regarded as an algebraic random variable. Then, as a particular case of (2.26) we obtain a quantum decomposition:

$$X \stackrel{m}{=} A^+ + A^- + A^\circ. \qquad (2.27)$$

It is noteworthy that (2.27) makes it possible to analyze a classical random variable by using the quantum components which are three mutually non-commutative operators A^+, A^- and A°. We will be convinced of its usefulness in some questions related to spectral analysis.

Example 2.3.9 (coin toss)   The outcome of tossing a fair coin is modelled by a classical random variable X such that

$$P(X = -1) = P(X = 1) = \frac{1}{2}.$$

The distribution of X is given by

$$\mu = \frac{1}{2} \delta_{-1} + \frac{1}{2} \delta_1, \qquad (2.28)$$

which is called the *Bernoulli distribution*. The moment sequence is given by

$$\mathbf{E}[X^m] = \begin{cases} 1, & \text{if } m \text{ is even}, \\ 0, & \text{if } m \text{ is odd}. \end{cases} \qquad (2.29)$$

As is easily verified, the orthogonal polynomials associated to μ in (2.28) are given by

$$P_0(x) = 1, \qquad P_1(x) = x.$$

In particular, $\alpha_1 = 0$. The rest of the Jacobi coefficients are obtained from the three-term recurrence relation

$$xP_1(x) = P_2(x) + \alpha_2 P_1(x) + \omega_1 P_0(x), \tag{2.30}$$

where $P_2(x) = (x+1)(x-1)$ since $\langle P_2, P_2 \rangle = 0$. Then (2.30) becomes

$$x^2 = (x+1)(x-1) + \alpha_2 x + \omega_1,$$

from which we obtain $\alpha_2 = 0$ and $\omega_1 = 1$. Consequently, the Jacobi coefficients of μ are given by

$$\omega_1 = 1, \qquad \alpha_1 = \alpha_2 = 0,$$

which are of finite type of length 2. Let $(\Gamma, \{\Phi_n\}, A^+, A^-)$ be the associated IFS, where we omitted A° for it is zero. (This IFS will be called the *fermion Fock space*, see Definition 2.3.19.) Since Γ is two-dimensional, it is convenient to take $\Gamma = \mathbb{C}^2$ with the canonical basis:

$$\Phi_0 = \begin{bmatrix} 0 \\ 1 \end{bmatrix}, \qquad \Phi_1 = \begin{bmatrix} 1 \\ 0 \end{bmatrix}.$$

Then the creation and annihilation operators are given in the matrix form:

$$A^+ = \begin{bmatrix} 0 & 1 \\ 0 & 0 \end{bmatrix}, \qquad A^- = \begin{bmatrix} 0 & 0 \\ 1 & 0 \end{bmatrix},$$

respectively. In fact,

$$A^+ \Phi_0 = \Phi_1, \quad A^+ \Phi_1 = 0; \qquad A^- \Phi_0 = 0, \quad A^- \Phi_1 = \Phi_0. \tag{2.31}$$

As is easily computed, we have

$$\langle \Phi_0, (A^+ + A^-)^m \Phi_0 \rangle = \left\langle \begin{bmatrix} 0 \\ 1 \end{bmatrix}, \begin{bmatrix} 0 & 1 \\ 1 & 0 \end{bmatrix}^m \begin{bmatrix} 0 \\ 1 \end{bmatrix} \right\rangle = \begin{cases} 1, & \text{if } m \text{ is even,} \\ 0, & \text{otherwise,} \end{cases} \tag{2.32}$$

which coincides with (2.29). We have thus proved that

$$X \overset{m}{=} A^+ + A^-. \tag{2.33}$$

This is the quantum decomposition of the fair coin toss X.

As a simple application of (2.33), we observe that the computation of moments $\mathbf{E}[X^m]$ is reduced to counting certain combinatorial numbers. In fact,

$$\mathbf{E}[X^m] = \langle \Phi_0, (A^+ + A^-)^m \Phi_0 \rangle = \sum \langle \Phi_0, A^{\varepsilon_m} \cdots A^{\varepsilon_1} \Phi_0 \rangle,$$

where the sum is taken over all possible $\varepsilon_1, \ldots, \varepsilon_m \in \{\pm\}$. It follows from (2.31) that $\langle \Phi_0, A^{\varepsilon_m} \cdots A^{\varepsilon_1} \Phi_0 \rangle$ is one for

$$\varepsilon_1 = +, \quad \varepsilon_2 = -, \quad \varepsilon_3 = +, \quad ,\ldots, \quad \varepsilon_{m-1} = +, \quad \varepsilon_m = -,$$

and zero otherwise, see Figure 2.5. Hence $\mathbf{E}[X^m] = 0$ for an odd m and $\mathbf{E}[X^m] = 1$ for an even m.

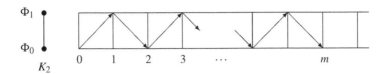

Figure 2.5 Calculating moments of coin toss

2.3.4 How to explicitly compute μ from $(\{\omega_n\}, \{\alpha_n\})$

For applications of quantum decomposition we need a method of computing a probability distribution from its Jacobi coefficients. We describe it as in the standard classical analysis.

For an arbitrary (not assuming finite moments) probability distribution μ on $\mathbb{R}$ the integral transform:

$$G_\mu(z) = \int_{-\infty}^{+\infty} \frac{\mu(dx)}{z - x}$$

is called the *Cauchy–Stieltjes transform* or simply the *Stieltjes transform*. The integral converges absolutely for any $z \in \mathbb{C} \backslash \mathrm{supp}\,\mu$.

Proposition 2.3.10 *Let $G(z) = G_\mu(z)$ be the Stieltjes transform of a probability distribution μ on $\mathbb{R}$.*

(1) *$G(z)$ is a holomorphic function in $\mathbb{C} \backslash \mathrm{supp}\,\mu$, in particular, in $\{\mathrm{Im}\,z > 0\}$.*
(2) *$\mathrm{Im}\,G(z) < 0$ for $\mathrm{Im}\,z > 0$ and $\mathrm{Im}\,G(z) > 0$ for $\mathrm{Im}\,z < 0$.*
(3) *$|G(z)| \leq |\mathrm{Im}\,z|^{-1}$ for $\mathrm{Im}\,z \neq 0$.*
(4) *$G(\bar{z}) = \overline{G(z)}$ for all $z \in \mathbb{C} \backslash \mathrm{supp}\,\mu$.*

The proofs are straightforward. Moreover, it is known that the Stieltjes transform determines a probability distribution uniquely. In fact, we have

Theorem 2.3.11 (Stieltjes inversion formula) *Let $G(z)$ be the Stieltjes transform of a probability distribution μ on $\mathbb{R}$. Then for any pair of real numbers $s < t$,*

$$-\frac{2}{\pi} \lim_{y \to +0} \int_s^t \mathrm{Im}\,G(x + iy)\,dx = \mu(\{s\}) + \mu(\{t\}) + 2\mu((s,t)).$$

Moreover,

$$\rho(x) = -\frac{1}{\pi} \lim_{y \to +0} \operatorname{Im} G(x + iy) \tag{2.34}$$

exists for a.e. $x \in \mathbb{R}$ and $\rho(x)dx$ is the absolutely continuous part of μ. (Here "a.e. x" means "for almost every x, that is, except on a null set.")

With Jacobi coefficients $(\{\omega_n\}, \{\alpha_n\})$ we associate a continued fraction of the form:

$$\cfrac{1}{z - \alpha_1 - \cfrac{\omega_1}{z - \alpha_2 - \cfrac{\omega_2}{z - \alpha_3 - \cfrac{\omega_3}{z - \alpha_4 - \cdots}}}}. \tag{2.35}$$

To save space, (2.35) is written as

$$\frac{1}{z - \alpha_1} - \frac{\omega_1}{z - \alpha_2} - \frac{\omega_2}{z - \alpha_3} - \frac{\omega_3}{z - \alpha_4} - \cdots \tag{2.36}$$

The continued fraction (2.36) converges by definition if the limit of the partial fractions exists:

$$\lim_{n \to \infty} \frac{1}{z - \alpha_1} - \frac{\omega_1}{z - \alpha_2} - \frac{\omega_2}{z - \alpha_3} - \cdots - \frac{\omega_{n-1}}{z - \alpha_n},$$

where the limit is taken only for $n \geq n_0$ if necessary.

Theorem 2.3.12 *Let $\mu \in \mathfrak{P}_{\mathrm{fm}}(\mathbb{R})$ and $(\{\omega_n\}, \{\alpha_n\})$ be its Jacobi coefficients. If μ is the unique solution to a determinate moment problem, the Stieltjes transform admits the continued fraction expansion:*

$$G_\mu(z) = \int_{-\infty}^{+\infty} \frac{\mu(dx)}{z - x} = \frac{1}{z - \alpha_1} - \frac{\omega_1}{z - \alpha_2} - \frac{\omega_2}{z - \alpha_3} - \frac{\omega_3}{z - \alpha_4} - \cdots,$$

where the right-hand side converges in $\{\operatorname{Im} z \neq 0\}$.

The detailed proofs of Theorems 2.3.11 and 2.3.12 can be found in [36, Chapter 1]. For uniqueness of μ we only mention the following criterion, for the proof see, for example, Shohat and Tamarkin [64], Wall [76, Chapter V]. In fact, Carleman's condition follows from his moment test but the converse implication is not true (so the former is strictly better than the latter).

Proposition 2.3.13 (Carleman's condition) *Let $\mu \in \mathfrak{P}_{\mathrm{fm}}(\mathbb{R})$ and $(\{\omega_n\}, \{\alpha_n\})$ its Jacobi coefficients. If*

$$\sum_{n=1}^{\infty} \frac{1}{\sqrt{\omega_n}} = +\infty, \tag{2.37}$$

then μ is a unique solution to the determinate moment problem. (We tacitly understand that (2.37) is satisfied for Jacobi coefficients of finite type. It is sometimes convenient to regard Jacobi coefficients of finite type as those of infinite type by concatenating zero sequence.)

Example 2.3.14 Consider Jacobi coefficients of finite type of length 4 given by

$$\omega_1 = 2, \quad \omega_2 = 1, \quad \omega_3 = 2, \quad \alpha_1 = \cdots = \alpha_4 = 0,$$

and let $(\Gamma, \{\Phi_n\}, A^+, A^-)$ be the associated IFS. Let us find the vacuum spectral distribution μ of $A^+ + A^-$. It follows from Theorem 2.3.12 that μ is uniquely determined by

$$G(z) = \int_{-\infty}^{+\infty} \frac{\mu(dx)}{z-x} = \frac{1}{z} - \frac{2}{z} - \frac{1}{z} - \frac{2}{z} - \frac{z^3 - 3z}{(z+1)(z-1)(z+2)(z-2)}.$$

Applying the partial fraction expansion, we obtain

$$G(z) = \int_{-\infty}^{+\infty} \frac{\mu(dx)}{z-x} = \frac{1/3}{z+1} + \frac{1/3}{z-1} + \frac{1/6}{z+2} + \frac{1/6}{z-2},$$

from which we have

$$\mu = \frac{1}{3}(\delta_{-1} + \delta_1) + \frac{1}{6}(\delta_{-2} + \delta_2).$$

Example 2.3.15 (Free Fock space) Consider the Jacobi coefficients of infinite type given by

$$\omega_1 = \omega_2 = \cdots = 1, \quad \alpha_1 = \alpha_2 = \cdots = 0.$$

Let $(\Gamma, \{\Phi_n\}, A^+, A^-)$ be the associated IFS (called the *free Fock space*) and μ the vacuum spectral distribution of $A^+ + A^-$. Since Carleman's condition (2.37) is satisfied obviously, the vacuum spectral distribution μ is uniquely determined and

$$G(z) = \int_{-\infty}^{+\infty} \frac{\mu(dx)}{z-x} = \frac{1}{z} - \frac{1}{z} - \frac{1}{z} - \frac{1}{z} - \cdots \tag{2.38}$$

holds, see Theorem 2.3.12. We need to calculate the continued fraction in (2.38). Writing

$$G(z) = \frac{1}{z - G(z)}$$

by periodicity, we obtain

$$\left(G(z) - \frac{z}{2}\right)^2 = \frac{1}{4}(z^2 - 4). \tag{2.39}$$

Here we need the square root of $z^2 - 4$. Let $\sqrt{z^2 - 4}$ be the holomorphic function in $\{\operatorname{Im} z > 0\}$ uniquely specified by

$$\sqrt{(iy)^2 - 4} = i\sqrt{y^2 + 4}, \qquad y > 0, \tag{2.40}$$

where $\sqrt{y^2 + 4}$ is the usual positive root of a positive number, for details see Remark 2.3.16 below. Then (2.39) is equivalent to

$$G(z) = \frac{z \pm \sqrt{z^2 - 4}}{2}. \tag{2.41}$$

The signature in front of $\sqrt{z^2 - 4}$ is chosen in such a way that

$$\operatorname{Im} G(iy) < 0, \qquad y > 0,$$

or

$$\lim_{y \to \infty} G(iy) = 0,$$

see Proposition 2.3.10. In view of (2.40) we should take

$$G(z) = \frac{z - \sqrt{z^2 - 4}}{2}. \tag{2.42}$$

Applying the Stieltjes inversion formula, we obtain

$$\rho(x) = -\frac{1}{\pi} \lim_{y \to +0} \operatorname{Im} G(x + iy) = \begin{cases} \dfrac{1}{2\pi}\sqrt{4 - x^2}, & \text{if } |x| \le 2, \\[2mm] 0, & \text{otherwise,} \end{cases} \tag{2.43}$$

for details see Remark 2.3.17 below. Finally, verifying that

$$\int_{-\infty}^{+\infty} \rho(x)\,dx = 1,$$

we conclude that the vacuum spectral distribution μ is explicitly given by

$$\mu(dx) = \frac{1}{2\pi}\sqrt{4 - x^2}\, 1_{[-2,2]}(x)\,dx,$$

which is known as the *Wigner semi-circle law*. By elementary calculus we obtain

$$\langle \Phi_0, (A^+ + A^-)^{2m}\Phi_0 \rangle = \frac{1}{2\pi}\int_{-2}^{2} x^{2m}\sqrt{4 - x^2}\,dx = \frac{1}{m+1}\binom{2m}{m}.$$

The last number is known as the *Catalan number*.

Remark 2.3.16 For the readers' convenience let us recall the definition of the square root $\sqrt{z^2 - 4}$. We start with $\sqrt{z}$. For $z \in \mathbb{C}\backslash[0, +\infty)$, writing

$$z = re^{i\theta}, \qquad r > 0, \quad 0 < \theta < 2\pi,$$

we define

$$g(z) = \sqrt{r}\, e^{i\theta/2},$$

where $\sqrt{r}$ is the usual positive root of $r > 0$. It is shown that $g(z)$ is a holomorphic function in $\mathbb{C}\backslash[0, +\infty)$ and satisfies $g(z)^2 = z$. Note that the holomorphic function $g(z)$ is uniquely determined by $g(-x) = i\sqrt{x}$ for $x > 0$. Upon the definition of homomorphic function $\sqrt{z}$, the choice of the domain is not unique. Our choice $\mathbb{C}\backslash[0, +\infty)$ is for later convenience.

Now let $z \in \{\mathrm{Im}\, z > 0\}$. Then $z^2 - 4 \in \mathbb{C}\backslash[-4, +\infty)$ and $f(z) = g(z^2 - 4)$ becomes a holomorphic function in $\{\mathrm{Im}\, z > 0\}$. Moreover, it is uniquely determined by

$$f(iy) = g(-y^2 - 4) = i\sqrt{y^2 + 4}, \qquad y > 0. \tag{2.44}$$

Since $f(z)$ satisfies $f(z)^2 = z^2 - 4$, we often write $f(z) = \sqrt{z^2 - 4}$. Strictly speaking, it does not make sense unless the domain and the branch are specified. However, in most cases, we may understand the meaning from the context.

Remark 2.3.17 Let $f(z) = \sqrt{z^2 - 4}$ be as in Remark 2.3.16, namely, with the domain $\{\mathrm{Im}\, z > 0\}$ and the branch as in (2.44). We show the calculation of

$$\lim_{y \to +0} f(x + iy), \qquad x \in \mathbb{R}.$$

For $x \in \mathbb{R}$ and $y > 0$ we define $r > 0$ and $0 < \theta < 2\pi$ uniquely by

$$re^{i\theta} = (x + iy)^2 - 4 = (x^2 - y^2 - 4) + 2xyi.$$

Then by definition

$$f(x + iy) = \sqrt{r}\, e^{i\theta/2}. \tag{2.45}$$

Note first that

$$\lim_{y \to +0} r = \lim_{y \to +0} \sqrt{(x^2 - y^2 - 4)^2 + (2xy)^2} = \sqrt{(x^2 - 4)^2} = |x^2 - 4|. \tag{2.46}$$

Suppose that $x > 2$. Then for sufficiently small $y > 0$ we have

$$\mathrm{Re}\,((x + iy)^2 - 4) = x^2 - y^2 - 4 > 0, \qquad \mathrm{Im}\,((x + iy)^2 - 4) = 2xy > 0.$$

Hence $0 < \theta \le \pi/2$ and

$$\lim_{y \to +0} \theta = 0. \tag{2.47}$$

It follows from (2.46) and (2.47) that

$$\lim_{y \to +0} f(x+iy) = \lim_{y \to +0} \sqrt{r}\,e^{i\theta/2} = \sqrt{x^2-4}. \tag{2.48}$$

Next suppose that $x < -2$. Then for sufficiently small $y > 0$ we have

$$\mathrm{Re}\left((x+iy)^2-4\right) = x^2 - y^2 - 4 > 0, \qquad \mathrm{Im}\left((x+iy)^2-4\right) = 2xy < 0.$$

Hence $3\pi/2 \le \theta < 2\pi$ and

$$\lim_{y \to +0} \theta = 2\pi. \tag{2.49}$$

It follows from (2.46) and (2.49) that

$$\lim_{y \to +0} f(x+iy) = \lim_{y \to +0} \sqrt{r}\,e^{i\theta/2} = \sqrt{x^2-4}\,e^{i\pi} = -\sqrt{x^2-4}. \tag{2.50}$$

Finally, consider the case of $|x| \le 2$. Since

$$\mathrm{Re}\left((x+iy)^2-4\right) = x^2 - y^2 - 4 < 0,$$

we have $\pi/2 < \theta < 3\pi/2$ and

$$\lim_{y \to +0} \theta = \pi. \tag{2.51}$$

It follows from (2.46) and (2.51) that

$$\lim_{y \to +0} f(x+iy) = \lim_{y \to +0} \sqrt{r}\,e^{i\theta/2} = \sqrt{4-x^2}\,e^{i\pi/2} = i\sqrt{4-x^2}. \tag{2.52}$$

Summing up, we have

$$\lim_{y \to +0} \sqrt{(x+iy)^2-4} = \begin{cases} \sqrt{x^2-4}, & \text{if } x > 2, \\ i\sqrt{4-x^2}, & \text{if } |x| \le 2, \\ -\sqrt{x^2-4}, & \text{if } x < -2. \end{cases}$$

2.3.5 Boson, fermion and free Fock spaces

Definition 2.3.18 (Boson Fock space)   An IFS associated to the Jacobi coefficients

$$\omega_n = n, \qquad \alpha_n = 0, \qquad n = 1, 2, \ldots,$$

is called the *boson Fock space*.

Definition 2.3.19 (Fermion Fock space)   An IFS associated to the Jacobi coefficients

$$\omega_1 = 1, \qquad \alpha_1 = \alpha_2 = 0,$$

which is of finite type of length 2, is called the *fermion Fock space*.

The boson and fermion Fock spaces are original concepts in quantum physics. The creation and annihilation operators of the boson Fock space fulfil the *canonical commutation relation (CCR)*:

$$A^- A^+ - A^+ A^- = I. \tag{2.53}$$

Similarly, the creation and annihilation operators of the fermion Fock space fulfil the *canonical anti-commutation relation (CAR)*:

$$A^- A^+ + A^+ A^- = I. \tag{2.54}$$

The vacuum spectral distribution of $A^+ + A^-$ in the boson Fock space is shown to be the standard normal distribution:

$$\mu(dx) = \frac{1}{\sqrt{2\pi}} e^{-x^2/2} dx.$$

In fact, by elementary calculus we easily obtain

$$\frac{1}{\sqrt{2\pi}} \int_{-\infty}^{+\infty} x^{2m-1} e^{-x^2/2} dx = 0, \qquad \frac{1}{\sqrt{2\pi}} \int_{-\infty}^{+\infty} x^{2m} e^{-x^2/2} dx = \frac{(2m)!}{2^m m!}. \tag{2.55}$$

On the other hand, verifying

$$\langle (A^+ + A^-)^{2m-1} \rangle = 0, \qquad \langle (A^+ + A^-)^{2m} \rangle = \frac{(2m)!}{2^m m!}, \tag{2.56}$$

we conclude that

$$\langle (A^+ + A^-)^m \rangle = \frac{1}{\sqrt{2\pi}} \int_{-\infty}^{+\infty} x^m e^{-x^2/2} dx, \qquad m = 1, 2, \ldots.$$

For the combinatorial proof of (2.56) by path counting, see [36, Section 1.7]. An alternative proof is by means of the Hermite polynomials, which are the orthogonal polynomials associated to the standard normal distribution.

The fermion Fock space is already discussed in Example 2.3.9. We know that the fair coin toss X admits the quantum decomposition

$$X \overset{\mathrm{m}}{=} A^+ + A^-.$$

Hence the vacuum spectral distribution of the fermion Fock space coincides with the *Bernoulli distribution*

$$\mu = \frac{1}{2} \delta_{-1} + \frac{1}{2} \delta_{+1}.$$

Definition 2.3.20 (Free Fock space)   An IFS associated to the Jacobi coefficients

$$\omega_n = 1, \qquad \alpha_n = 0, \qquad n = 1, 2, \ldots,$$

is called the *free Fock space*.

The free Fock space is already discussed in Example 2.3.15. The vacuum spectral distribution of $A^+ + A^-$ is given by the Wigner semi-circle law:

$$\mu(dx) = \frac{1}{2\pi}\sqrt{4-x^2}\,1_{[-2,2]}(x)dx.$$

We here mention the associated orthogonal polynomials.

The polynomials $\{U_n(x)\}$ defined by

$$U_n(\cos\theta) = \frac{\sin(n+1)\theta}{\sin\theta}, \qquad x=\cos\theta, \qquad n=0,1,2,\ldots,$$

are called the *Chebyshev polynomials of the second kind*. With the help of elementary knowledge of trigonometric functions we obtain

$$U_0(x)=1, \quad U_1(x)=2x, \quad U_{n+1}(x)-2xU_n(x)+U_{n-1}(x)=0.$$

We define

$$\tilde{U}_n(x) = U_n\left(\frac{x}{2}\right), \qquad n=0,1,2,\ldots.$$

Then $\tilde{U}_n(x)=x^n+\cdots$ becomes a monic polynomial. Moreover, the three-term recurrence relation:

$$\tilde{U}_0(x)=1, \quad \tilde{U}_1(x)=x, \quad x\tilde{U}_n(x)=\tilde{U}_{n+1}(x)+\tilde{U}_{n-1}(x), \tag{2.57}$$

and the orthogonal relation:

$$\int_{-2}^{2} \tilde{U}_m(x)\tilde{U}_n(x)\sqrt{4-x^2}\,dx = 0, \qquad m\neq n,$$

are verified easily. In other words, $\{\tilde{U}_n(x)\}$ are the orthogonal polynomials associated to the Wigner semi-circle law. We also obtain the Jacobi coefficients $(\{\omega_n\equiv 1\},\{\alpha_n\equiv 0\})$ from (2.57). In other words, the free Fock space structure emerges in the Chebyshev polynomials of the second kind.

As is easily verified, the creation and annihilation operators of the free Fock space fulfill

$$A^-A^+ = I, \tag{2.58}$$

which is often referred to as the *free commutation relation*. Thus, three commutation relations (2.53), (2.54) and (2.58) are unified in the *q-commutation relation*:

$$A^-A^+ - qA^+A^- = I. \tag{2.59}$$

In fact, for any $-1\leq q\leq 1$ the Jacobi coefficients are defined by

$$\omega_n = [n]_q = 1+q+q^2+\cdots+q^{n-1}, \qquad \alpha_n = 0, \qquad n=1,2,\ldots,$$

where $[n]_q$ is the so-called q-number of Gauss. The associated interacting Fock space is called the *q-Fock space*, where the q-commutation relation (2.59) holds. The vacuum spectral distribution of $A^+ + A^-$ is expressed in terms of the Jacobi theta function, see [12, 13, 69].

Remark 2.3.21 We are in a good position to mention the counterpart of the Chebyshev polynomials of the second kind. The polynomials $\{T_n(x)\}$ defined by

$$T_n(\cos\theta) = \cos n\theta, \qquad x = \cos\theta, \qquad n = 0, 1, 2, \dots,$$

are called the *Chebyshev polynomials of the first kind*. It follows from elementary knowledge of trigonometric function that

$$T_0(x) = 1, \quad T_1(x) = x, \quad T_{n+1}(x) - 2xT_n(x) + T_{n-1}(x) = 0.$$

We then define

$$\tilde{T}_0(x) = 1, \qquad \tilde{T}_n(x) = \left(\frac{1}{\sqrt{2}}\right)^{n-2} T_n\left(\frac{x}{\sqrt{2}}\right), \qquad n = 1, 2, \dots.$$

Note that $\tilde{T}_n(x) = x^n + \cdots$ becomes a monic polynomial of degree n. It is an easy task to derive the three-term recurrence:

$$\tilde{T}_0(x) = 1, \tag{2.60}$$
$$\tilde{T}_1(x) = x, \tag{2.61}$$
$$x\tilde{T}_1(x) = \tilde{T}_2(x) + \tilde{T}_0(x), \tag{2.62}$$
$$x\tilde{T}_n(x) = \tilde{T}_{n+1}(x) + \frac{1}{2}\tilde{T}_{n-1}(x), \qquad n = 2, 3, \dots, \tag{2.63}$$

and the orthogonal relation:

$$\int_{-\sqrt{2}}^{\sqrt{2}} \tilde{T}_m(x)\tilde{T}_n(x)\frac{dx}{\sqrt{2-x^2}} = 0, \qquad m \neq n.$$

Thus, $\{\tilde{T}_n(x)\}$ are the orthogonal polynomials associated to the arcsine law:

$$\mu(dx) = \frac{dx}{\pi\sqrt{2-x^2}} 1_{(-\sqrt{2},\sqrt{2})}(x)dx,$$

which is normalized to have mean zero and variance one. As is seen from (2.62) and (2.63), the Jacobi coefficients are given by

$$\omega_1 = 1, \quad \omega_2 = \omega_3 = \cdots = \frac{1}{2}, \qquad \alpha_1 = \alpha_2 = \cdots = 0.$$

2.3.6 Notes

An interacting Fock space (IFS) was originally introduced by Accardi for describing interacting quantum systems, see Accardi, Lu and Volovich [6]. In their context our IFS is a special case called *of one mode* and is almost useless in quantum physics, where infinite mode is essential. However, since the relation to orthogonal polynomials was found by Accardi and Bożejko [3], an IFS of one mode has become a central concept in quantum probability. In this line further study on the relationship between IFS's of *n* mode and orthogonal polynomials in *n* variables is highly desired.

The idea of quantum decomposition was proposed around 1998 as soon as the quantum probability approach to spectral analysis of graphs was launched by Hashimoto, Obata and Tabei [31]; see also Hashimoto [30] where the term "quantum decomposition" was first used explicitly. Since then the use of quantum decomposition has been spread over various topics such as quantum walks [44], random walks [40], resistor networks [39], Hecke algebras [29], and so forth; see also [4, 5, 63].

Among the enormous literature on orthogonal polynomials (in one variable), we refer to Chihara [16] for a concise introduction. During the long history of the moment problem, Akhiezer [7] and Shohat and Tamarkin [64] are renowned classics, see also Schmüdgen [62] and references cited therein. For analytic theory of continued fractions see Wall [76].

2.4 Spectral distributions of graphs

2.4.1 Adjacency matrix as a real random variable

Definition 2.4.1 (Graph) A *graph* is a pair $G = (V, E)$, where V is a nonempty set and E is a subset of the two-point set $\{\{x,y\}; x,y \in V,\ x \neq y\}$. An element of V is called a *vertex* and an element of E an *edge*. A graph $G = (V, E)$ is called *finite* if V is a finite set.

For two vertices $x, y \in V$ we write $x \sim y$ if $\{x,y\} \in E$, that is, if they are connected by an edge. The *degree* of $x \in V$ is defined by

$$\deg_G(x) = |\{y \in V; y \sim x\}|.$$

A graph is called *regular* if every vertex has a constant degree.

Definition 2.4.2 (Adjacency matrix) For a graph $G = (V, E)$ the *adjacency* *matrix $A = [a_{xy}]_{x,y \in V}$ is defined by*

$$a_{xy} = \begin{cases} 1, & x \sim y, \\ 0, & \text{otherwise.} \end{cases}$$

In these lecture notes we deal with both finite and infinite graphs. But an infinite graph is always assumed to be *locally finite*, namely,

$$\deg_G(x) < \infty \qquad \text{for all } x \in V.$$

This condition allows us to consider any powers of A and hence polynomials in A. The set of polynomials becomes a commutative unital $*$-algebra.

Definition 2.4.3 (Adjacency algebra) Let $G = (V, E)$ be a graph. The $*$-algebra of polynomials in A is called the *adjacency algebra* of G and is denoted by $\mathscr{A}(G)$.

Once $\mathscr{A}(G)$ is equipped with a state φ, the adjacency matrix A becomes a real random variable in the algebraic probability space $(\mathscr{A}(G), \varphi)$. We are mostly concerned with two states on the adjacency algebra $\mathscr{A}(G)$ as described below.

Let $C(V)$ be the space of $\mathbb{C}$-valued functions on V and $C_0(V)$ the subspace of those with finite supports. Of course, $C(V) = C_0(V)$ for a finite set V. With each $x \in V$ we associate $e_x \in C_0(V)$ defined by

$$e_x(y) = \begin{cases} 1, & \text{if } y = x, \\ 0, & \text{otherwise.} \end{cases}$$

The inner product on $C_0(V)$ is defined by

$$\langle e_x, e_y \rangle = \delta_{xy}, \qquad x, y \in V.$$

We call $\{e_x ; x \in V\}$ the canonical orthonormal basis of $C_0(V)$.

Let us start with a finite graph $G = (V, E)$. We identify $C(V)$ with $\mathbb{C}^{|V|}$ in a natural manner. The *tracial state* on $\mathscr{A}(G)$ is defined by

$$\varphi_{\mathrm{tr}}(a) = \langle a \rangle_{\mathrm{tr}} = \frac{1}{|V|} \operatorname{Tr}(a) = \frac{1}{|V|} \sum_{x \in V} \langle e_x, a e_x \rangle, \qquad a \in \mathscr{A}(G).$$

Since $\mathscr{A}(G)$ is of finite dimension, the spectral distribution of A in $\langle \cdot \rangle_{\mathrm{tr}}$ is uniquely determined by

$$\langle A^m \rangle_{\mathrm{tr}} = \int_{-\infty}^{+\infty} x^m \mu(dx), \qquad m = 1, 2, \ldots. \tag{2.64}$$

Since A is symmetric real matrix, its eigenvalues are all real. Let

$$\{\lambda_1(w_1), \lambda_2(w_2), \ldots, \lambda_s(w_s)\}$$

be the eigenvalues of A, where $\lambda_1, \lambda_2, \ldots, \lambda_s$ are mutually distinct and w_k stands for the multiplicities of λ_k. The *eigenvalue distribution* of G is defined by

$$\mu = \frac{1}{|V|} \sum_{k=1}^{s} w_k \delta_{\lambda_k} . \tag{2.65}$$

Lemma 2.4.4 *The spectral distribution of A in the tracial state $\langle \cdot \rangle_{\mathrm{tr}}$ coincides with the eigenvalue distribution of G.*

Proof Let μ be the eigenvalue distribution of G defined as in (2.65). Then we have

$$\langle A^m \rangle_{\mathrm{tr}} = \frac{1}{|V|} \mathrm{Tr}\,(A^m) = \frac{1}{|V|} \sum_{k=1}^{s} w_k \lambda_k^m = \int_{-\infty}^{+\infty} x^m \mu(dx),$$

which means that μ is the spectral distribution of A in $\langle \cdot \rangle_{\mathrm{tr}}$. □

Next we consider a finite or infinite graph $G = (V, E)$. For a fixed vertex $o \in V$ the *vacuum state at $o \in V$* is the vector state defined by

$$\langle a \rangle_o = \langle e_o, a e_o \rangle, \qquad a \in \mathscr{A}(G).$$

Lemma 2.4.5 *Let μ be the spectral distribution of A in the vacuum state at $o \in V$. Then we have*

$$\langle A^m \rangle_o = \langle e_o, A^m e_o \rangle = \int_{-\infty}^{+\infty} x^m \mu(dx) = |\{m\text{-step walks from } o \text{ to } o\}|.$$

Proof By definition we have

$$\langle e_o, A^m e_o \rangle = (A^m)_{oo} = \sum_{x_1, \ldots, x_{m-1} \in V} A_{ox_1} A_{x_1 x_2} \cdots A_{x_{m-1} o},$$

where

$$A_{ox_1} A_{x_1 x_2} \cdots A_{x_{m-1} o} = \begin{cases} 1, & \text{if } o \sim x_1 \sim x_2 \sim \cdots \sim x_{m-1} \sim o, \\ 0, & \text{otherwise.} \end{cases}$$

Therefore $\langle e_o, A^m e_o \rangle$ counts the number of m-step walks from o to itself. The rest of the assertion is obvious. □

2.4.2 IFS structure associated to graphs

Let $G = (V, E)$ be a graph with a distinguished vertex $o \in V$. Then we come to a natural stratification (distance partition):

$$V = \bigcup_{n=0}^{\infty} V_n, \qquad V_n = \{x \in V ; d(o, x) = n\}, \tag{2.66}$$

where $d(o,x)$ denotes the graph distance between two vertices o and x: see Figure 2.6. Note that $|V_n| < \infty$ by assumption of local finiteness.

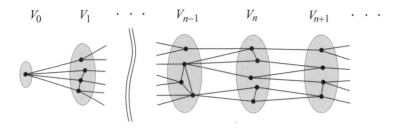

Figure 2.6 Stratification of a graph

As usual, let $\{e_x ; x \in V\}$ be the canonical basis of $C_0(V)$ and set

$$\Phi_n = |V_n|^{-1/2} \sum_{x \in V_n} e_x.$$

Let $\Gamma(G) \subset C_0(V)$ be the linear space spanned by $\{\Phi_n\}$, where we have

$$\langle \Phi_m, \Phi_n \rangle = \delta_{mn}.$$

For simplicity $\Gamma(G)$ is called the *Fock space associated to G* (and $o \in V$ to be precise). We next define A^+, A^- and A° respectively by

$$(A^\varepsilon)_{yx} = \begin{cases} 1, & x \in V_n, y \in V_{n+\varepsilon} \text{ and } x \sim y, \\ 0, & \text{otherwise,} \end{cases}$$

where $n + \varepsilon$ is a short-hand notation for $n+1$, $n-1$ and n according as $\varepsilon = +$, $\varepsilon = -$ and $\varepsilon = \circ$. See Figure 2.7.

Lemma 2.4.6 *It holds that*

$$A = A^+ + A^- + A^\circ, \tag{2.67}$$

and

$$(A^+)^* = A^-, \quad (A^\circ)^* = A^\circ.$$

Proof Exercise. Note that $(A^+)^* = A^-$ follows from $\langle e_x, A^+ e_y \rangle = \langle A^- e_x, e_y \rangle$. ∎

The expression in (2.67) is also called the *quantum decomposition* of A.

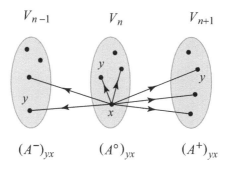

$$V_{n-1} \qquad V_n \qquad V_{n+1}$$

$$(A^-)_{yx} \qquad (A^\circ)_{yx} \qquad (A^+)_{yx}$$

Figure 2.7 Quantum decomposition $A = A^+ + A^- + A^\circ$

However, the quintuple $(\Gamma(G), \{\Phi_n\}, A^+, A^-, A^\circ)$ is not necessarily an IFS, because $\Gamma(G)$ is not necessarily invariant under the actions of A^+, A^- and A°. To describe condition for the invariance we use the convenient notation:

$$\omega_\varepsilon(x) = \{y \in V_{n+\varepsilon} ; y \sim x\}, \qquad x \in V_n, \quad \varepsilon \in \{+, -, \circ\}. \tag{2.68}$$

Proposition 2.4.7 *Let $G = (V, E)$ be a graph with a distinguished vertex $o \in V$ and $\Gamma(G)$ the associated Fock space. Then $\Gamma(G)$ is invariant under the actions of A^ε if and only if $\omega_\varepsilon(x)$ is constant on each V_n, where $\varepsilon \in \{+, -, \circ\}$. In that case, setting*

$$\omega_n = |\omega_-(y)|^2 \frac{|V_n|}{|V_{n-1}|}, \qquad y \in V_n, \tag{2.69}$$

$$\alpha_n = |\omega_\circ(y)|, \qquad y \in V_{n-1}, \tag{2.70}$$

we obtain

$$A^+\Phi_n = \sqrt{\omega_{n+1}}\,\Phi_{n+1}, \quad A^-\Phi_n = \sqrt{\omega_n}\,\Phi_{n-1}, \quad A^\circ\Phi_n = \alpha_{n+1}\Phi_n.$$

In other words, $(\Gamma(G), \{\Phi_n\}, A^+, A^-, A^\circ)$ is an IFS associated to the Jacobi coefficients $(\{\omega_n\}, \{\alpha_n\})$. Moreover,

$$\omega_1 = \deg(o), \qquad \alpha_1 = 0.$$

Proof First we consider the action of A^+. By definition we have

$$A^+ \Phi_n = \frac{1}{\sqrt{|V_n|}} \sum_{x \in V_n} A^+ e_x$$

$$= \frac{1}{\sqrt{|V_n|}} \sum_{x \in V_n} \sum_{y \in \omega_+(x)} e_y$$

$$= \frac{1}{\sqrt{|V_n|}} \sum_{y \in V_{n+1}} |\omega_-(y)| e_y.$$

The last vector is a constant multiple of Φ_{n+1} if and only if $|\omega_-(y)|$ is constant for $y \in V_{n+1}$. In that case we have

$$A^+ \Phi_n = \frac{|\omega_-(y)|}{\sqrt{|V_n|}} \sum_{y \in V_{n+1}} e_y = |\omega_-(y)| \frac{\sqrt{|V_{n+1}|}}{\sqrt{|V_n|}} \Phi_{n+1} = \sqrt{\omega_{n+1}} \Phi_{n+1},$$

where (2.69) is taken into account. Similarly we have

$$A^- \Phi_n = \frac{1}{\sqrt{|V_n|}} \sum_{y \in V_{n-1}} |\omega_+(y)| e_y.$$

Hence $A^- \Phi_n$ is a constant multiple of Φ_{n-1} if and only if $|\omega_+(y)|$ is constant for $y \in V_{n-1}$. In that case we have

$$A^- \Phi_n = |\omega_+(y)| \frac{\sqrt{|V_{n-1}|}}{\sqrt{|V_n|}} \Phi_{n-1}. \tag{2.71}$$

Counting the number of edges between V_{n-1} and V_n in two ways, we obtain

$$|\omega_+(y)||V_{n-1}| = |\omega_-(z)||V_n|, \qquad y \in V_{n-1}, \quad z \in V_n.$$

Then (2.71) becomes

$$A^- \Phi_n = |\omega_-(z)| \frac{\sqrt{|V_n|}}{\sqrt{|V_{n-1}|}} \Phi_{n-1} = \sqrt{\omega_n} \Phi_{n-1}.$$

Finally, for A° we have

$$A^\circ \Phi_n = \frac{1}{\sqrt{|V_n|}} \sum_{y \in V_n} |\omega_\circ(y)| e_y.$$

Hence $A^\circ \Phi_n$ is a constant multiple of Φ_n if and only if $|\omega_\circ(y)|$ is constant for $y \in V_n$. In that case, in view of (2.70) we have

$$A^\circ \Phi_n = |\omega_\circ(y)| \Phi_n = \alpha_{n+1} \Phi_n, \tag{2.72}$$

as desired. $\qquad\qquad\qquad\qquad\qquad\qquad\qquad\qquad\qquad\qquad\qquad\qquad\square$

Remark 2.4.8   A typical example satisfying the condition in Proposition 2.4.7 is a distance-regular graph, see Subsection 2.5.2 for detailed discussion; see also Definition 1.3.9 and the following discusson.

Example 2.4.9 (cycle C_6)   Consider a cycle on six vertices with a distinguished vertex o chosen arbitrarily. Let us calculate the spectral distribution μ of the adjacency matrix A in the vacuum state at o. Looking at Figure 2.8, we easily get

$$\omega_1 = 2, \qquad \omega_2 = 1, \qquad \omega_3 = 2, \qquad \alpha_1 = \cdots = \alpha_4 = 0.$$

The associated interacting Fock space is studied already in Example 2.3.14. The vacuum spectral distribution of $A = A^+ + A^-$ is given by

$$\mu = \frac{1}{3}(\delta_{+1} + \delta_{-1}) + \frac{1}{6}(\delta_{+2} + \delta_{-2}). \qquad (2.73)$$

As an application we get

$$\langle A^{2m+1} \rangle_o = \int_{-\infty}^{+\infty} x^{2m+1} \mu(dx) = 0,$$

$$\langle A^{2m} \rangle_o = \int_{-\infty}^{+\infty} x^{2m} \mu(dx)$$

$$= \frac{1}{3}((+1)^{2m} + (-1)^{2m}) + \frac{1}{6}((+2)^{2m} + (-2)^{2m}) = \frac{1}{3}(4^m + 2).$$

Recall that $\langle A^{2m} \rangle_o = \langle e_o, A^{2m} e_o \rangle$ coincides with the number of $2m$-step walks from the origin o to itself. It is noted that μ in (2.73) coincides with the eigenvalue distribution, as will be seen in Proposition 2.5.6.

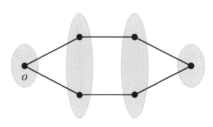

Figure 2.8 Stratification of C_6

Example 2.4.10 (Cube $K_2 \times K_2 \times K_2 = H(2,3)$)   Let us calculate the spectral distribution μ of the adjacency matrix A of the cube (Figure 2.9) in the vacuum state at an arbitrary fixed vertex o.

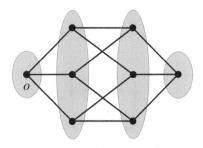

Figure 2.9 Stratification of the cube $K_2 \times K_2 \times K_2 = H(2,3)$

By easy observation we get

$$\omega_1 = 3, \qquad \omega_2 = 4, \qquad \omega_3 = 3, \qquad \alpha_1 = \cdots = \alpha_4 = 0.$$

Then the continued fraction becomes

$$G(z) = \cfrac{1}{z - \cfrac{3}{z - \cfrac{4}{z - \cfrac{3}{z}}}} = \frac{3/8}{z+1} + \frac{3/8}{z-1} + \frac{1/8}{z+3} + \frac{1/8}{z-3},$$

from which we obtain

$$\mu = \frac{3}{8}(\delta_{+1} + \delta_{-1}) + \frac{1}{8}(\delta_{+3} + \delta_{-3}).$$

The above μ coincides with the eigenvalue distribution, as will be seen in Proposition 2.5.6.

2.4.3 Homogeneous trees and Kesten distributions

Let T_κ denote the homogeneous tree of degree $\kappa \geq 2$ (see Figure 2.10), and fix a distinguished vertex arbitrarily. As usual we obtain the associated Fock space $\Gamma(T_\kappa)$ with the canonical basis $\{\Phi_n\}$. Let

$$A = A^+ + A^- + A^\circ$$

be the quantum decomposition of the adjacency matrix. By simple observation with Proposition 2.4.7 we obtain

$$A^+\Phi_0 = \sqrt{\kappa}\,\Phi_1, \quad A^+\Phi_n = \sqrt{\kappa-1}\,\Phi_{n+1}, \quad n \geq 1, \tag{2.74}$$

$$A^-\Phi_0 = 0, \quad A^-\Phi_1 = \sqrt{\kappa}\,\Phi_0, \quad A^-\Phi_n = \sqrt{\kappa-1}\,\Phi_{n-1}, \quad n \geq 2, \tag{2.75}$$

and

$$A^\circ = 0.$$

Then the Jacobi coefficients are given by

$$\omega_1 = \kappa, \quad \omega_2 = \omega_3 = \cdots = \kappa - 1, \qquad \alpha_1 = \alpha_2 = \cdots = 0. \qquad (2.76)$$

Figure 2.10 Homogeneous tree T_4

Theorem 2.4.11   *Let A be the adjacency matrix of a homogeneous tree T_κ, $\kappa \geq 2$. The vacuum spectral distribution of A at an arbitrary fixed vertex o is given by $\mu(dx) = \rho_\kappa(x)dx$ with*

$$\rho_\kappa(x) = \frac{\kappa}{2\pi} \frac{\sqrt{4(\kappa-1) - x^2}}{\kappa^2 - x^2}, \qquad |x| \leq 2\sqrt{\kappa - 1}. \qquad (2.77)$$

In other words, we have

$$\langle A^m \rangle_o = \langle e_o, A^m e_o \rangle = \int_{-2\sqrt{\kappa-1}}^{2\sqrt{\kappa-1}} x^m \rho_\kappa(x)dx, \qquad m = 1, 2, \ldots.$$

Proof   Let μ be the vacuum spectral distribution of the adjacency matrix A. Then μ is determined uniquely by Jacobi coefficients (2.76) and satisfies

$$G(z) = \int_{-\infty}^{+\infty} \frac{\mu(dx)}{z - x} = \frac{1}{z} - \frac{\kappa}{z} - \frac{\kappa-1}{z} - \frac{\kappa-1}{z} - \cdots.$$

The above continued fraction is computed in a similar manner as in Example 2.3.15 and we come to

$$G(z) = \frac{(\kappa-2)z - \kappa\sqrt{z^2 - 4(\kappa-1)}}{2(\kappa^2 - z^2)}, \qquad (2.78)$$

where $\sqrt{z^2 - 4(\kappa - 1)}$ is a holomorphic function on $\{\mathrm{Im}\, z > 0\}$ such that

$$\sqrt{(iy)^2 - 4(\kappa - 1)} = i\sqrt{y^2 + 4(\kappa - 1)}, \qquad y > 0.$$

Applying the Stieltjes inversion formula to (2.78) we obtain the density function as in (2.77). Verifying that

$$\int_{-2\sqrt{\kappa-1}}^{+2\sqrt{\kappa-1}} \rho_\kappa(x)\, dx = 1,$$

we conclude that the spectral distribution is $\mu(dx) = \rho_\kappa(x)dx$. Some examples are illustrated in Figure 2.11. $\square$

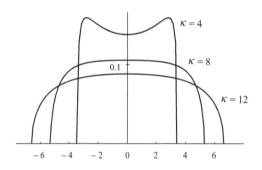

Figure 2.11 Spectral distribution of T_κ for $\kappa = 4, 8, 12$

Definition 2.4.12 (Kesten distribution) The *Kesten distribution* with parameters $p > 0$ and $q \geq 0$ is a probability distribution $\mu = \mu_{p,q}$ uniquely determined by

$$G(z) = \int_{-\infty}^{+\infty} \frac{\mu(dx)}{z - x} = \cfrac{1}{z -} \cfrac{p}{z -} \cfrac{q}{z -} \cfrac{q}{z -} \cdots. \qquad (2.79)$$

Accordingly, the vacuum spectral distribution of the adjacency matrix of a homogeneous tree T_κ of degree $\kappa \geq 2$ is the Kesten distribution with parameters $\kappa, \kappa - 1$. Note that the Kesten distribution with parameters $1, 1$ is nothing else but the Wigner semi-circle law.

An explicit expression for $\mu_{p,q}$ is obtained in a standard manner. Calculating the continued fraction in (2.79), we get

$$G(z) = -\frac{1}{2} \frac{(p - 2q)z + p\sqrt{z^2 - 4q}}{p^2 - (p - q)z^2}. \qquad (2.80)$$

Applying the Stieltjes inversion formula we obtain the absolutely continuous part:

$$\rho_{p,q}(x) = \begin{cases} \dfrac{p}{2\pi} \dfrac{\sqrt{4q - x^2}}{p^2 - (p-q)x^2}, & |x| \le 2\sqrt{q}, \\ 0, & |x| > 2\sqrt{q}. \end{cases} \tag{2.81}$$

On the other hand, by direct calculation we have

$$\frac{p}{2\pi} \int_{-2\sqrt{q}}^{2\sqrt{q}} \frac{\sqrt{4q - x^2}}{p^2 - (p-q)x^2} \, dx = \begin{cases} 1, & 0 < p \le 2q, \\ \dfrac{q}{p-q}, & 0 < 2q \le p. \end{cases}$$

When $0 < 2q \le p$, checking the poles and residues of $G(z)$ in (2.80) we find point masses. As a result, we have

$$\mu_{p,q}(dx) = \begin{cases} \rho_{p,q}(x)dx, & 0 < p \le 2q, \\ \rho_{p,q}(x)dx + \dfrac{p - 2q}{2(p-q)} \left(\delta_{-\frac{p}{\sqrt{p-q}}} + \delta_{\frac{p}{\sqrt{p-q}}} \right), & 0 < 2q \le p, \\ \dfrac{1}{2} \left(\delta_{-\sqrt{p}} + \delta_{\sqrt{p}} \right), & q = 0. \end{cases}$$

Example 2.4.13 (One-dimensional integer lattice $\mathbb{Z}$)   The homogeneous tree T_κ of degree $\kappa = 2$ coincides with the one-dimensional integer lattice $\mathbb{Z}$, in other words, the two-sided infinite path. By simple observation (see Figure 2.12) we obtain the Jacobi coefficients easily as

$$\omega_1 = 2, \qquad \omega_2 = \omega_3 = \cdots = 1, \qquad \alpha_1 = \alpha_2 = \cdots = 0,$$

and therefore the spectral distribution is the Kesten distribution $\mu_{1,2}$. Namely,

$$\mu_{1,2}(dx) = \frac{1}{\pi\sqrt{4 - x^2}} 1_{(-2,2)}(x)dx,$$

which coincides with the arcsine law with mean 0 and variance 2. By elementary calculus we obtain

$$\langle A^{2m} \rangle_o = \int_{-\infty}^{+\infty} x^{2m} \mu(dx) = \int_{-2}^{2} \frac{x^{2m}}{\pi\sqrt{4 - x^2}} \, dx = \binom{2m}{m},$$

which counts the number of $2m$-step walks from o to itself. Of course, the binomial coefficient is obtained from direct enumeration on $\mathbb{Z}$.

Example 2.4.14   Consider the "shoelace graph" as shown in Figure 2.13. Since

$$\omega_1 = 2, \qquad \omega_2 = \omega_3 = \cdots = 4, \qquad \alpha_1 = \alpha_2 = \cdots = 0,$$

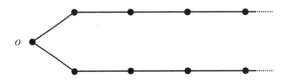

Figure 2.12 One-dimensional integer lattice $\mathbb{Z}$

which is obtained by easy observation, the vacuum spectral distribution at o is the Kesten distribution with parameters $2, 4$. It is given by $\mu_{2,4}(dx) = \rho(x)dx$ with the density function:

$$\rho(x) = \frac{1}{\pi} \frac{\sqrt{16 - x^2}}{2x^2 + 4}, \qquad |x| \le 4.$$

Figure 2.13 "Shoelace graph"

Example 2.4.15   We consider $\mathbb{Z}_+ = \{0, 1, 2, \dots\}$ as a (one-sided) infinite path. For $N \ge 1$ let $G^{(N)}$ be the graph obtained by joining N copies of $\mathbb{Z}_+$ at the end-vertex 0, and take the center as the distinguished vertex o, see Figure 2.14. This construction will be called the *star product*, see Subsection 2.6.2. By simple observation we obtain

$$\omega_1 = \deg(o) = N, \qquad \omega_2 = \omega_3 = \cdots = 1, \qquad \alpha_1 = \alpha_2 = \cdots = 0.$$

Therefore the vacuum spectral distribution at o is the Kesten distribution with parameters $N, 1$. For $N \ge 2$ the explicit form is given by

$$\mu_{N,1}(dx) = \rho_{N,1}(x)dx + \frac{N-2}{2(N-1)}\left(\delta_{-\frac{N}{\sqrt{N-1}}} + \delta_{\frac{N}{\sqrt{N-1}}}\right), \qquad (2.82)$$

where

$$\rho_{N,1}(x) = \begin{cases} \dfrac{N}{2\pi} \dfrac{\sqrt{4 - x^2}}{N^2 - (N-1)x^2}, & |x| \le 2, \\ 0, & |x| > 2. \end{cases}$$

It is noticeable that point masses appear in (2.82). In fact, there are two point masses with equal weights. Moreover, since

$$\lim_{N\to\infty} \int_{-\infty}^{+\infty} \rho_{N,1}(x)\,dx = \lim_{N\to\infty} \left(1 - \frac{N-2}{N-1}\right) = 0,$$

the point masses dominate the spectral distribution as $N \to \infty$. This phenomenon will be discussed in the more general context of the Boolean central limit theorem in Subsection 2.6.5.

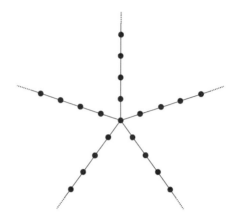

Figure 2.14 $G^{(5)}$: joining 5 copies of $\mathbb{Z}_+$ (star product)

2.5 Growing graphs

2.5.1 Formulation of question in general

By growing graphs we mean a set of graphs $G_\nu = (V_\nu, E_\nu)$, where ν runs over an infinite directed set and $|V_\nu| < |V_{\nu'}|$ for $\nu \prec \nu'$. We are interested in the spectral properties of G_ν as ν grows, say $\nu \to \infty$ for simplicity.

Given growing graphs $G_\nu = (V_\nu, E_\nu)$, we consider the adjacency matrix A_ν as a real random variable in an algebraic probability space $(\mathscr{A}(G_\nu), \langle\cdot\rangle_\nu)$ as usual. Let μ_ν be the spectral distribution of A_ν. Then we have

$$\langle A_\nu^m \rangle_\nu = \int_{-\infty}^{+\infty} x^m \mu_\nu(dx), \qquad m = 1, 2, \dots. \tag{2.83}$$

In what follows we occasionally omit the subscript ν to avoid messy symbols.

Recall that

$$\langle A_V \rangle = \mathrm{mean}(\mu_V), \qquad \Sigma^2(A_V) = \langle (A_V - \langle A_V \rangle)^2 \rangle = \mathrm{var}(\mu_V).$$

Definition 2.5.1 (Normalization) Let μ be a probability distribution on $\mathbb{R}$ with mean m and variance σ^2. The *normalization* of μ is a probability distribution $\tilde{\mu}$ defined by

$$\int_{-\infty}^{+\infty} f(x)\, \tilde{\mu}(dx) = \int_{-\infty}^{+\infty} f\!\left(\frac{x-m}{\sigma}\right) \mu(dx),$$

where $f(x)$ is a bounded continuous function on $\mathbb{R}$. Note that

$$\mathrm{mean}(\tilde{\mu}) = 0, \qquad \mathrm{var}(\tilde{\mu}) = 1.$$

Normalization is a standard concept in probability theory to study the asymptotic behaviors of distributions, for example, central limit theorem (CLT). The idea of normalization is also important to the study of asymptotic spectral distributions of growing graphs. In fact, for growing graphs G_V we study the limit of the normalized spectral distributions of A_V:

$$\mu = \lim_V \tilde{\mu}_V.$$

In other words, we are interested in a probability distribution μ characterized by

$$\lim_V \left\langle \left(\frac{A_V - \langle A_V \rangle}{\Sigma(A_V)} \right)^m \right\rangle = \int_{-\infty}^{+\infty} x^m \mu(dx), \qquad m = 1, 2, \dots.$$

A direct method would be to first find μ_V expicitly and then to calculate the normalized limit, but it is not employed here. We use the method of quantum decomposition which is significant in the sense that the limit distribution is obtained without knowing the explicit form of μ_V but from combinatorial data of how quantum components of A_V grow.

Example 2.5.2 (T_κ as $\kappa \to \infty$) Let us show the outline of the method of quantum decomposition to obtain the asymptotic spectral distribution of the homogeneous trees T_κ as $\kappa \to \infty$. Let A_κ be the adjacency matrix of T_κ and μ_κ the spectral distribution in the vacuum state at a fixed vertex of T_κ. As is suggested by the explicit expression (2.77), in the limit $\kappa \to \infty$ the support of μ_κ diverges and we cannot grasp anything meaningful. This is also seen from

$$\mathrm{var}\,\mu_\kappa = \Sigma^2(A_\kappa) = \langle A_\kappa^2 \rangle = \kappa \to \infty.$$

So we need normalization. Let $A_\kappa = A_\kappa^+ + A_\kappa^-$ be the quantum decomposition.

The normalization becomes

$$\frac{A_\kappa - \langle A_\kappa \rangle}{\Sigma(A_\kappa)} = \frac{A_\kappa}{\sqrt{\kappa}} = \frac{A_\kappa^+}{\sqrt{\kappa}} + \frac{A_\kappa^-}{\sqrt{\kappa}}.$$

We see from (2.74) and (2.75) that the actions of the normalized quantum components are given by

$$\frac{A_\kappa^+}{\sqrt{\kappa}}\Phi_0 = \Phi_1, \quad \frac{A_\kappa^+}{\sqrt{\kappa}}\Phi_n = \frac{\sqrt{\kappa-1}}{\sqrt{\kappa}}\Phi_{n+1}, \quad n=1,2,\ldots, \qquad (2.84)$$

$$\frac{A_\kappa^-}{\sqrt{\kappa}}\Phi_0 = 0, \quad \frac{A_\kappa^-}{\sqrt{\kappa}}\Phi_1 = \Phi_0, \quad \frac{A_\kappa^-}{\sqrt{\kappa}}\Phi_n = \frac{\sqrt{\kappa-1}}{\sqrt{\kappa}}\Phi_{n-1}, \quad n=2,3,\ldots.$$

$$(2.85)$$

It is emphasized here that the limit actions are immediate from (2.84) and (2.85). In fact, they are the same as the creation and annihilation operators in the free Fock space. Thus, letting $(\Gamma, \{\Psi_n\}, B^+, B^-)$ be the free Fock space, we claim that

$$\left(\frac{A_\kappa^+}{\sqrt{\kappa}}, \frac{A_\kappa^-}{\sqrt{\kappa}}\right) \xrightarrow{\text{m}} (B^+, B^-) \qquad \text{as } \kappa \to \infty. \qquad (2.86)$$

Strictly speaking, for (2.86) we need to show the convergence of mixed moments but the verification is easy. As a consequence of (2.86) we have

$$\lim_{\kappa \to \infty} \left\langle \left(\frac{A_\kappa}{\sqrt{\kappa}}\right)^m \right\rangle = \lim_{\kappa \to \infty} \left\langle \left(\frac{A_\kappa^+}{\sqrt{\kappa}} + \frac{A_\kappa^-}{\sqrt{\kappa}}\right)^m \right\rangle$$

$$= \langle \Psi_0, (B^+ + B^-)^m \Psi_0 \rangle$$

$$= \frac{1}{2\pi}\int_{-2}^{2} x^m \sqrt{4-x^2}\,dx, \qquad m=1,2,\ldots.$$

Consequently, the asymptotic spectral distribution of T_κ as $\kappa \to \infty$ is the Wigner semi-circle law.

Remark 2.5.3 In Theorem 2.4.11 we obtained the spectral distribution μ_κ of T_κ. It is the Kesten distribution given by $\mu_\kappa(dx) = \rho_\kappa(x)dx$, where the density function is known explicitly. Note that the normalization $\tilde{\mu}_\kappa$ is given by the normalized density function:

$$\tilde{\rho}_\kappa(x) = \sqrt{\kappa}\,\rho_\kappa(\sqrt{\kappa}x).$$

Then by elementary calculus one obtains the Wigner semi-circle law as the limit of $\tilde{\rho}_\kappa(x)$ as $\kappa \to \infty$. This is the conventional direct method of getting the asymptotic spectral distribution of T_κ. We are convinced by Example 2.5.2 that the quantum probability approach opens a new paradigm totally different from the conventional one.

2.5.2 Growing distance-regular graphs

Definition 2.5.4 (Distance-regular graph)   (See also Definition 1.3.9) A graph $G = (V,E)$ is called *distance-regular* if the intersection numbers:

$$p_{i,j}^k = |\{z \in V \,; d(x,z) = i, d(y,z) = j\}|,$$

are constant independently of the choice of a pair x, y with $d(x,y) = k$.

We start with a significant property of distance-regular graphs from the view point of quantum probability.

Proposition 2.5.5   *Let $G = (V,E)$ be a distance-regular graph with a distinguished vertex $o \in V$ chosen arbitrarily. Let $\Gamma(G)$ be the associated Fock space and $A = A^+ + A^- + A^\circ$ the quantum decomposition of the adjacency matrix. Then we have*

$$A^+\Phi_n = \sqrt{\omega_{n+1}}\,\Phi_{n+1}, \quad A^-\Phi_n = \sqrt{\omega_n}\,\Phi_{n-1}, \quad A^\circ\Phi_n = \alpha_{n+1}\Phi_n,$$

where

$$\omega_n = p_{1,n-1}^n p_{1,n}^{n-1}, \qquad \alpha_n = p_{1,n-1}^{n-1}.$$

In other words, $(\Gamma(G), \{\Phi_n\}, A^+, A^\circ, A^-)$ is an IFS associated to $(\{\omega_n\}, \{\alpha_n\})$. Moreover,

$$\mathrm{mean}(A) = \langle A \rangle_o = 0, \qquad \mathrm{var}(A) = \langle A^2 \rangle_o = \deg(o) = p_{11}^0$$

Proof   Exercise.   □

Proposition 2.5.6   *Let $G = (V,E)$ be a finite distance-regular graph and $\mathscr{A}(G)$ the adjacency algebra. Then the tracial state and the vacuum state at a vertex $o \in V$ coincide, that is,*

$$\langle a \rangle_{\mathrm{tr}} = \langle a \rangle_o = \langle e_o, ae_o \rangle, \qquad a \in \mathscr{A}(G).$$

Proof   From distance-regularity we see that

$$\langle e_x, A^m e_x \rangle = \langle e_o, A^m e_o \rangle, \qquad x \in V, \quad m = 1, 2, \dots.$$

Then for any $a \in \mathscr{A}(G)$ we have

$$\langle a \rangle_{\mathrm{tr}} = \frac{1}{|V|} \sum_{x \in V} \langle e_x, ae_x \rangle = \langle e_o, ae_o \rangle,$$

as desired.   □

We are interested in particular distance-regular graphs which form growing graphs, for example, Hamming graphs, Johnson graphs, odd graphs, homogeneous trees and so forth. Here we consider the class of Hamming graphs because it is interesting in itself and more importantly, it leads naturally to a general claim for growing distance-regular graphs.

For $d \geq 1$ and $N \geq 1$ we set

$$V = \{1,2,\ldots,N\}^d = \{x = (\xi_1,\ldots,\xi_d)\,;\, \xi_i \in \{1,2,\ldots,N\}\}.$$

An element $x \in V$ is often regarded as a word of length d with letters taken from $\{1,2,\ldots,N\}$. The Hamming distance on V is defined by

$$d(x,y) = |\{1 \leq i \leq d\,;\, \xi_i \neq \eta_i\}|, \qquad \begin{aligned} x &= (\xi_1,\ldots,\xi_d),\\ y &= (\eta_1,\ldots,\eta_d). \end{aligned}$$

Setting $E = \{\{x,y\}\,;\, x,y \in V, d(x,y) = 1\}$, we obtain a graph (V,E), which is called the *Hamming graph* and is denoted by $H(d,N)$. Note that the Hamming distance coincides with the graph distance.

In short, the Hamming graph $H(d,N)$ is the d-fold Cartesian power of the complete graph K_N, namely,

$$H(d,N) = K_N \times \cdots \times K_N \qquad (d \text{ factors}),$$

see Definition 2.6.7 below for a definition of product graph. Also, note that our Hamming graph $H(d,N)$ is exactly the graph $G_{d,N}^H$ introduced in Example 1.3.17.

By simple observation we see that the Hamming graph is distance-regular with intersection numbers:

$$p_{1,1}^0 = \deg(H(d,N)) = d(N-1),$$

$$p_{1,n-1}^n = n, \quad p_{1,n}^{n-1} = (d-n)(N-1), \quad p_{1,n-1}^{n-1} = (n-1)(N-2).$$

It then follows from Proposition 2.5.5 that $(\Gamma(H(d,N)), \{\Phi_n\}, A^+, A^-, A^\circ)$ becomes an interacting Fock space associated to Jacobi coefficients:

$$\omega_n = p_{1,n-1}^n p_{1,n}^{n-1} = n(d-n+1)(N-1), \qquad 1 \leq n \leq d,$$

$$\alpha_n = p_{1,n-1}^{n-1} = (n-1)(N-2), \qquad 1 \leq n \leq d+1.$$

Thus we have

$$A^+ \Phi_n = \sqrt{\omega_{n+1}}\, \Phi_{n+1} = \sqrt{(n+1)(d-n)(N-1)}\, \Phi_{n+1},$$

$$A^- \Phi_n = \sqrt{\omega_n}\, \Phi_{n-1} = \sqrt{n(d-n+1)(N-1)}\, \Phi_{n-1},$$

$$A^\circ \Phi_n = \alpha_{n+1} \Phi_n = n(N-2)\Phi_n.$$

Apparently, without normalization we cannot get reasonable actions in the limit as $N \to \infty$ and $d \to \infty$. In view of

$$\langle A \rangle = 0, \qquad \Sigma^2(A) = \langle A^2 \rangle = d(N-1),$$

the actions of normalized quantum components are given by

$$\frac{A^+}{\sqrt{d(N-1)}}\Phi_n = \sqrt{(n+1)\left(1-\frac{n}{d}\right)}\Phi_{n+1}, \qquad (2.87)$$

$$\frac{A^-}{\sqrt{d(N-1)}}\Phi_n = \sqrt{n\left(1-\frac{n-1}{d}\right)}\Phi_{n-1}, \qquad (2.88)$$

$$\frac{A^\circ}{\sqrt{d(N-1)}}\Phi_n = n\sqrt{\frac{N-2}{d}}\sqrt{\frac{N-2}{N-1}}\Phi_n. \qquad (2.89)$$

It is important to observe that, for reasonable limit actions, we need a proper scaling balance between $N \to \infty$ and $d \to \infty$. In fact, the limit actions are obtained only when

$$N \to \infty, \quad d \to \infty, \quad \frac{N}{d} \to \tau \geq 0. \qquad (2.90)$$

Let $(\Gamma, \{\Psi_n\}, B^+, B^-)$ be the boson Fock space. Since

$$B^+\Psi_n = \sqrt{n+1}\,\Psi_{n+1}, \quad B^-\Psi_n = \sqrt{n}\,\Psi_{n-1},$$

we have a useful formula:

$$B^+B^-\Psi_n = n\Psi_n.$$

Then the limit actions of (2.87)–(2.89) are given by B^+, B^- and $\sqrt{\tau}B^+B^-$, respectively. Verifying the convergence of mixed moments without difficulty, we come to the following assertion.

Theorem 2.5.7 (Quantum CLT for $H(d,N)$) *Let $A = A^+ + A^- + A^\circ$ be the quantum decomposition of the adjacency matrix of $H(d,N)$. Let its boson Fock space be $(\Gamma, \{\Psi_n\}, B^+, B^-)$. Then we have*

$$\left(\frac{A^+}{\sqrt{d(N-1)}}, \frac{A^-}{\sqrt{d(N-1)}}, \frac{A^\circ}{\sqrt{d(N-1)}}\right) \xrightarrow{m} (B^+, B^-, \sqrt{\tau}B^+B^-), \qquad (2.91)$$

where the limit is taken as in (2.90). Furthermore, for any $m = 1, 2, \ldots$ we have

$$\lim_{\substack{N\to\infty, d\to\infty \\ N/d\to\tau}} \left\langle\left(\frac{A}{\sqrt{d(N-1)}}\right)^m\right\rangle = \langle \Psi_0, (B^+ + B^- + \sqrt{\tau}B^+B^-)^m \Psi_0 \rangle. \qquad (2.92)$$

Let us examine the spectral distribution of $B^+ + B^- + \sqrt{\tau}B^+B^-$. We start with the case of $\tau = 0$. Since the vacuum spectral distribution of $B^+ + B^-$ is the standard normal distribution, we conclude that

$$\lim_{\substack{N\to\infty, d\to\infty \\ N/d\to 0}} \left\langle\left(\frac{A}{\sqrt{d(N-1)}}\right)^m\right\rangle = \frac{1}{\sqrt{2\pi}}\int_{-\infty}^{+\infty} x^m e^{-x^2/2}dx, \quad m = 1, 2, \ldots.$$

For $\tau > 0$ we need the following remarkable fact on Poisson distributions, see, for example, [36, Section 5.3], [59, Section 4.4].

Proposition 2.5.8 *Let $(\Gamma, \{\Psi_n\}, B^+, B^-)$ be the boson Fock space. For $\lambda > 0$ the vacuum spectral distribution of $(B^+ + \sqrt{\lambda})(B^- + \sqrt{\lambda})$ coincides with the Poisson distribution with parameter λ, namely,*

$$\langle \Psi_0, ((B^+ + \sqrt{\lambda})(B^- + \sqrt{\lambda}))^m \Psi_0 \rangle = \sum_{k=0}^{\infty} k^m \frac{\lambda^k}{k!} e^{-\lambda}, \qquad m = 1, 2, \ldots,$$

$$(2.93)$$

where the right-hand side is the mth moment of the Poisson distribution with parameter λ.

Writing the right-hand side of (2.92) as

$$\left\langle \Psi_0, \left(\sqrt{\tau} \left(B^+ + \frac{1}{\sqrt{\tau}} \right) \left(B^- + \frac{1}{\sqrt{\tau}} \right) - \frac{1}{\sqrt{\tau}} \right)^m \Psi_0 \right\rangle,$$

we see that the vacuum spectral distribution of $B^+ + B^- + \sqrt{\tau} B^+ B^-$ is an affine transformation of the Poisson distribution with parameter τ^{-1}. More precisely, it is a discrete distribution determined by

$$\mu_\tau = \sum_{k=0}^{\infty} p_k \delta_{\xi_k}, \qquad p_k = \frac{\tau^{-k}}{k!} e^{-1/\tau}, \qquad \xi_k = \sqrt{\tau} k - \frac{1}{\sqrt{\tau}}.$$

The above argument for the growing Hamming graphs is easily adapted to general growing distance-regular graphs. Let G_ν be growing distance-regular graphs with intersection numbers $p_{ij}^k(\nu)$ and let $A_\nu = A_\nu^+ + A_\nu^\circ + A_\nu^-$ be the quantum decomposition of the adjacency matrix. Using

$$\langle A_\nu \rangle = 0, \qquad \Sigma^2(A_\nu) = \langle A_\nu^2 \rangle = \deg(G_\nu) = p_{11}^0(\nu),$$

we obtain the normalization of A_ν as

$$\frac{A_\nu - \langle A_\nu \rangle}{\Sigma(A_\nu)} = \frac{A_\nu^+}{\sqrt{\deg(G_\nu)}} + \frac{A_\nu^\circ}{\sqrt{\deg(G_\nu)}} + \frac{A_\nu^-}{\sqrt{\deg(G_\nu)}}.$$

Then with the help of Proposition 2.5.5 we come to the following

Theorem 2.5.9 (Quantum CLT for growing distance-regular graphs) *Let G_ν be growing distance-regular graphs with intersection numbers $p_{i,j}^k(\nu)$ and set*

$$\omega_n(\nu) = p_{1,n-1}^n(\nu) p_{1,n}^{n-1}(\nu), \qquad \alpha_n(\nu) = p_{1,n-1}^{n-1}(\nu).$$

Assume that for all $n = 1, 2, \ldots$ the limits

$$\omega_n = \lim_\nu \frac{\omega_n(\nu)}{p_{1,1}^0(\nu)}, \qquad \alpha_n = \lim_\nu \frac{\alpha_n(\nu)}{\sqrt{p_{1,1}^0(\nu)}}, \qquad (2.94)$$

exist and $(\{\omega_n\}, \{\alpha_n\})$ *are Jacobi coefficients. Then, letting* $(\Gamma, \{\Phi_n\}, B^+, B^-,$ $B^\circ)$ *be the associated interacting Fock space, we have*

$$\left(\frac{A_v^+}{\sqrt{\deg(G_v)}}, \frac{A_v^-}{\sqrt{\deg(G_v)}}, \frac{A_v^\circ}{\sqrt{\deg(G_v)}}\right) \xrightarrow{\text{m}} (B^+, B^-, B^\circ).$$

Example 2.5.10 (Cycles C_k) The cycle C_k with $k \geq 3$ is distance-regular. We consider the growing cycles C_{2k} on $2k$ verices as $k \to \infty$. The argument is quite similar for cycles on odd numbers of vertices though small differences appear in the intersection numbers. By easy observation we obtain

$$p_{1,n-1}^n(2k) = \begin{cases} 1, & 1 \leq n \leq k-1, \\ 2, & n = k, \end{cases} \qquad p_{1,n}^{n-1}(2k) = \begin{cases} 2, & n = 1, \\ 1, & 2 \leq n \leq k, \end{cases}$$

$$p_{1,n-1}^{n-1}(2k) = 0, \qquad n = 1, 2, \ldots, k+1.$$

Note that $p_{1,1}^0(2k) = \deg(C_{2k}) = 2$. Then (2.94) becomes

$$\omega_n = \lim_{k \to \infty} \frac{p_{1,n-1}^n(2k) p_{1,n}^{n-1}(2k)}{p_{1,1}^0(2k)} = \begin{cases} 1, & n = 1, \\ 1/2, & n = 2, 3, \ldots, \end{cases}$$

and

$$\alpha_n = \lim_{k \to \infty} \frac{p_{1,n-1}^{n-1}(2k)}{\sqrt{p_{1,1}^0(2k)}} = 0, \qquad n = 1, 2, \ldots.$$

It is known (see Remark 2.3.21) that the above Jacobi coefficients correspond to the normalized arcsine law. Thus we conclude that the asymptotic spectral distribution of C_{2k} as $k \to \infty$ is the normalized arcsine law. For C_{2k+1} as $k \to \infty$ the final result is the same.

Remark 2.5.11 In the Introduction we demonstrated the conventional method of calculating the limit of eigenvalue distributions of C_{2k}. Since the eigenvalue distribution and vacuum spectral distribution coincide for C_{2k}, the argument in Example 2.5.10 gives a quantum probability approach to the same problem. It is instructive to turn our attention to a difference between two approaches. The conventional approach in the Introduction does not require normalization but the argument in Example 2.5.10 does. Nevertheless, we obtained the same limit distributions (with different variances). This is because $\deg(C_{2k}) = 2$ stays constant as $k \to \infty$.

Example 2.5.12 (Complete graphs) The complete graph K_n on n vertices is distance-regular. By simple observation we obtain

$$\omega_1(n) = n - 1, \quad \alpha_1(n) = 0, \quad \alpha_2(n) = n - 2. \tag{2.95}$$

Note that

$$\lim_{n\to\infty} \frac{\omega_1(n)}{p_{1,1}^0(n)} = \lim_{n\to\infty} \frac{n-1}{n-1} = 1$$

exists, however,

$$\lim_{n\to\infty} \frac{\alpha_2(n)}{\sqrt{p_{1,1}^0(n)}} = \lim_{n\to\infty} \frac{n-2}{\sqrt{n-1}} = \infty.$$

Hence our general result in Theorem 2.5.9 is not applicable to the growing complete graphs K_n as $n \to \infty$.

Remark 2.5.13   From (2.95) we obtain the vacuum spectral distribution of K_n easily. In fact, we have

$$\begin{aligned} G_n(z) &= \frac{1}{z} \frac{n-1}{z-(n-2)} \\ &= \frac{z-(n-2)}{(z+1)(z-(n-1))} = \frac{(n-1)/n}{z+1} + \frac{1/n}{z-(n-1)}, \end{aligned}$$

from which the spectral distribution is obtained as

$$\mu_n = \frac{n-1}{n}\delta_{-1} + \frac{1}{n}\delta_{n-1}.$$

Of course, the above μ_n coincides with the eigenvalue distribution of K_n. Since mean $\mu_n = 0$ and var $\mu_n = n-1$, the normalization is given by

$$\tilde{\mu}_n = \frac{n-1}{n}\delta_{-\frac{1}{\sqrt{n-1}}} + \frac{1}{n}\delta_{\frac{n-1}{\sqrt{n-1}}}.$$

As is suggested by Example 2.5.12, the moments of $\tilde{\mu}_n$ do not converge. In fact, for the third moment we obtain

$$M_3(\tilde{\mu}_n) = \frac{n-1}{n}\left(-\frac{1}{\sqrt{n-1}}\right)^3 + \frac{1}{n}\left(\frac{n-1}{\sqrt{n-1}}\right)^3 \to \infty, \qquad \text{as } n \to \infty.$$

On the other hand, for any bounded continuous function $f(x)$ we have

$$\begin{aligned} \int_{-\infty}^{+\infty} f(x)\tilde{\mu}_n(dx) &= \frac{n-1}{n} f\left(-\frac{1}{\sqrt{n-1}}\right) + \frac{1}{n} f\left(\frac{n-1}{\sqrt{n-1}}\right) \\ &\to f(0) = \int_{-\infty}^{+\infty} f(x)\delta_0(dx) \end{aligned}$$

which means that $\tilde{\mu}_n$ converges weakly to δ_0. Thus, a difference between convergence in moments and weak convergence is crucial in this example.

2.5.3 Growing regular graphs

The method of quantum decomposition for growing graphs G_v is still useful when $\Gamma(G_v)$ is not necessarily invariant under the quantum components of the adjacency matrix but an interacting Fock space emerges in the limit. We begin with an example.

Consider the integer lattice $\mathbb{Z}^N$ as $N \to \infty$. Note that $\mathbb{Z}^N$ is not distance-regular except for $N = 1$ but it is a regular graph with $\deg(\mathbb{Z}^N) = 2N$. Taking a distinguished vertex o, say the origin of the lattice, we obtain the associated Fock space $\Gamma(\mathbb{Z}^N)$. Let

$$A_N = A_N^+ + A_N^-$$

be the quantum decomposition of the adjacency matrix of $\mathbb{Z}^N$, where $A_N^\circ = 0$ is clear. Now note that $\Gamma(\mathbb{Z}^N)$ is not invariant under A^+ and A^-. This is seen, for example, by counting the number of outgoing edges; this number is not constant on each stratum: see Figure 2.15.

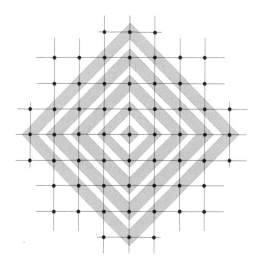

Figure 2.15 Stratification of $\mathbb{Z}^N$ with $N = 2$

A slightly more careful observation leads

$$A^+ \Phi_n = \sqrt{2N} \sqrt{n+1}\, \Phi_{n+1} + O(1), \tag{2.96}$$

$$A^- \Phi_n = \sqrt{2N} \sqrt{n}\, \Phi_{n-1} + O(N^{-1/2}). \tag{2.97}$$

Using

$$\langle A_N \rangle = 0, \qquad \Sigma^2(A_V) = \langle A_N^2 \rangle = \deg(\mathbb{Z}^N) = 2N,$$

the actions of normalized quantum components become

$$\frac{A_N^+}{\sqrt{2N}} \Phi_n = \sqrt{n+1}\, \Phi_{n+1} + O(N^{-1/2}), \qquad (2.98)$$

$$\frac{A_N^-}{\sqrt{2N}} \Phi_n = \sqrt{n}\, \Phi_{n-1} + O(N^{-1}), \qquad (2.99)$$

from which the limit actions are easily obtained. Let $(\Gamma, \{\Psi_n\}, B^+, B^-)$ be the boson Fock space. Then we have

$$\left(\frac{A_N^+}{\sqrt{2N}}, \frac{A_N^-}{\sqrt{2N}} \right) \xrightarrow{\mathrm{m}} (B^+, B^-).$$

In particular,

$$\lim_{N \to \infty} \left\langle \left(\frac{A_N}{\sqrt{2N}} \right)^m \right\rangle = \langle \Psi_0, (B^+ + B^-)^m \Psi_0 \rangle$$

$$= \frac{1}{\sqrt{2\pi}} \int_{-\infty}^{+\infty} x^m e^{-x^2/2} dx, \qquad m = 1, 2, \dots.$$

Thus the asymptotic spectral distribution of the integer lattice $\mathbb{Z}^N$ as $N \to \infty$ is the standard normal distribution.

The key of the above argument is that the actions of the normalized quantum components do not define an interacting Fock space as is seen in (2.98) and (2.99), but in the limit they do. Formalizing this situation, we may obtain a general statement.

We consider growing regular graphs $G_\nu = (V^{(\nu)}, E^{(\nu)})$, where

(1) $\lim_\nu \deg(G_\nu) = \infty$.

We need some statistical quantities obtained from the stratification of G_ν. For $\varepsilon \in \{+, -, \circ\}$ and $n = 0, 1, 2, \dots$ we set

$$M(\omega_\varepsilon | V_n) = \frac{1}{|V_n|} \sum_{x \in V_n} |\omega_\varepsilon(x)|,$$

$$\Sigma^2(\omega_\varepsilon | V_n) = \frac{1}{|V_n|} \sum_{x \in V_n} \{|\omega_\varepsilon(x)| - M(\omega_\varepsilon | V_n)\}^2,$$

$$L(\omega_\varepsilon | V_n) = \max\{|\omega_\varepsilon(x)| ; x \in V_n\},$$

where $\omega_\varepsilon(x)$ is defined in (2.68). We prepare conditions to control these statistics in the limit as $\nu \to \infty$.

(2) For each $n = 1, 2, \ldots,$

$$\omega_n = \lim_v M(\omega_- | V_n^{(v)}) < \infty, \qquad (2.100)$$

$$\lim_v \Sigma^2(\omega_- | V_n^{(v)}) = 0, \qquad \sup_v L(\omega_- | V_n^{(v)}) < \infty.$$

(3) For each $n = 0, 1, 2, \ldots,$

$$\alpha_n = \lim_v \frac{M(\omega_\circ | V_n^{(v)})}{\sqrt{\deg(G_v)}} < \infty, \qquad (2.101)$$

$$\lim_v \frac{\Sigma^2(\omega_\circ | V_n^{(v)})}{\deg(G_v)} = 0, \qquad \sup_v \frac{L(\omega_\circ | V_n^{(v)})}{\sqrt{\deg(G_v)}} < \infty.$$

Under the above conditions it is proved that $\omega_n > 0$ for all n even if G_v are finite graphs. In other words, $(\{\omega_n\}, \{\alpha_n\})$ become Jacobi coefficients of infinite type. Note that $\omega_1 = 1$ and $\alpha_1 = 0$ by definition.

Theorem 2.5.14 (Quantum CLT for growing regular graphs) *Let G_v be growing regular graphs satisfying conditions (1)–(3). Let $(\Gamma, \{\Psi_n\}, B^+, B^-, B^\circ)$ be the interacting Fock space associated to the Jacobi coefficients $(\{\omega_n\}, \{\alpha_n\})$ defined in (2.100) and (2.101). Then we have*

$$\left(\frac{A_v^+}{\sqrt{\deg(G_v)}}, \frac{A_v^-}{\sqrt{\deg(G_v)}}, \frac{A_v^\circ}{\sqrt{\deg(G_v)}} \right) \xrightarrow{\mathrm{m}} (B^+, B^-, B^\circ). \qquad (2.102)$$

In particular, the asymptotic spectral distribution of the normalized A_v in the vacuum state is the probability distribution associated to $(\{\omega_n\}, \{\alpha_n\})$.

Outline of Proof The complete proof is found in [36, Chapter 7] and [37]. We set $\kappa(v) = \deg(G_v)$ and often omit indicating the growing parameter v. The actions of the normalized quantum components of the adjacency matrix $A = A_v$ are written in the form:

$$\frac{A^\varepsilon}{\sqrt{\kappa}} \Phi_n = \gamma_{n+\varepsilon}^\varepsilon \Phi_{n+\varepsilon} + S_{n+\varepsilon}^\varepsilon, \qquad \varepsilon \in \{+, -, \circ\}, \quad n = 0, 1, 2, \ldots,$$

where

$$\gamma_n^+ = M(\omega_- | V_n) \left(\frac{|V_n|}{\kappa |V_{n-1}|} \right)^{1/2},$$

$$\gamma_n^- = M(\omega_+ | V_n) \left(\frac{|V_n|}{\kappa |V_{n+1}|} \right)^{1/2},$$

$$\gamma_n^\circ = \frac{M(\omega_\circ | V_n)}{\sqrt{\kappa}}.$$

Moreover, one may check that

$$|V_n| = \left\{ \prod_{k=1}^{n} M(\omega_- |V_k) \right\}^{-1} \kappa^n + O(\kappa^{n-1}).$$

It then follows from conditions (1)–(3) that

$$\lim_{\nu} \gamma_n^+ = \sqrt{\omega_n}, \qquad \lim_{\nu} \gamma_n^- = \sqrt{\omega_{n+1}}, \qquad \lim_{\nu} \gamma_n^\circ = \alpha_{n+1}.$$

Repeated application of the normalized quantum components gives rise to

$$\frac{A^{\varepsilon_m}}{\sqrt{\kappa}} \cdots \frac{A^{\varepsilon_1}}{\sqrt{\kappa}} \Phi_n = \gamma_{n+\varepsilon_1}^{\varepsilon_1} \gamma_{n+\varepsilon_1+\varepsilon_2}^{\varepsilon_2} \cdots \gamma_{n+\varepsilon_1+\cdots+\varepsilon_m}^{\varepsilon_m} \Phi_{n+\varepsilon_1+\cdots+\varepsilon_m}$$

$$+ \sum_{k=1}^{m} \underbrace{\gamma_{n+\varepsilon_1}^{\varepsilon_1} \cdots \gamma_{n+\varepsilon_1+\cdots+\varepsilon_{k-1}}^{\varepsilon_{k-1}}}_{(k-1) \text{ factors}} \underbrace{\frac{A^{\varepsilon_m}}{\sqrt{\kappa}} \cdots \frac{A^{\varepsilon_{k+1}}}{\sqrt{\kappa}}}_{(m-k) \text{ factors}} S_{n+\varepsilon_1+\cdots+\varepsilon_k}^{\varepsilon_k}.$$

The "error terms" are estimated by conditions (1)–(3) and we may show that

$$\lim_{\nu} \left\langle \Phi_j, \frac{A^{\varepsilon_m}}{\sqrt{\kappa(\nu)}} \cdots \frac{A^{\varepsilon_{k+1}}}{\sqrt{\kappa(\nu)}} S_{n+\varepsilon_1+\cdots+\varepsilon_k}^{\varepsilon_k} \right\rangle = 0$$

for any $j = 0, 1, 2, \ldots$. Finally, it is shown that the actions of normalized quantum components of the adjacency matrix coincide with the ones of the interacting Fock space associated to $(\{\omega_n\}, \{\alpha_n\})$. Thus (2.102) follows. $\square$

2.5.4 Notes

The study of eigenvalues of the adjacency matrix of a graph is one of the main subjects of *spectral graph theory*. Among the large literature, we refer to Biggs [10], Brouwer and Haemers [15], Cvetković, Rowlinson and Simić [19, 20]. Our main purpose is to investigate "universal shapes" of the eigenvalue distributions or, more generally, spectral distributions of large graphs. For that purpose we consider not a single large graph but a family of growing graphs. In this aspect our study contains a trend different from the traditional spectral graph theory.

The method of quantum decomposition for spectral analysis of growing graphs was first introduced by Hashimoto, Obata and Tabei [31], where the limit distributions of growing Hamming graphs are obtained. A similar idea was applied to other growing graphs, for example, certain Cayley graphs by Hashimoto [30], Johnson graphs by Hora [35] and Hashimoto, Hora and Obata [32], odd graphs and spidernets by Igarashi and Obata [38]. And the systematic approach to growing distance-regular graphs was established by Hora and

Obata [36] thanks to fruitful communications with experts in algebraic combinatorics. We here mention only Bannai and Ito [9] and Brouwer, Cohen and
Neumaier [14] for references. The quantum CLT for growing regular graphs
was proved by Hora and Obata [37]. For more recent topics see Koohestani,
Obata and Tanaka [45] and Leyva [47].

A multi-variate generalization of the method of quantum decomposition is
one of the current topics where an interesting relation to orthogonal polynomials in several variables is expected. Motivated by the limit argument of growing
Hamming graphs due to Hora [34], Morales, Obata and Tanaka [51] study an
example of bivariate extension arising from growing pairs of strongly regular
graphs though the quantum decomposition is not yet discussed. For a relevant
result see also Dhahri, Obata and Yoo [21].

The spectral analysis of homogeneous trees traces back to the famous paper
by Kesten [42], after whom the spectral distribution is named. It is noticeable that the derivation of the Kesten distribution is highly simplified by using
the method of quantum decomposition. There is a larger family of probability
distributions including the Kesten distributions, called the *free Meixner distributions*. These appear in the spectral analysis of spidernets obtained by adding
large cycles to homogeneous trees, see Hora and Obata [36] for details.

2.6 Concepts of independence and graph products

2.6.1 From classical to commutative independence

The classical central limit theorem (CLT) describes the asymptotic behavior of
the sum of independent, identically distributed random variables:

$$S_N = X_1 + X_2 + \cdots + X_N.$$

The most traditional case is the coin toss, where X_n stands for the result of the
nth coin toss which obeys

$$P(X_n = 1) = P(X_n = 0) = \frac{1}{2}.$$

Then S_N counts the number of heads during the N trials and we know that S_N
obeys the binomial distribution $B(N, 1/2)$. Moreover, the classical de Moivre–
Laplace theorem says that $B(N, 1/2)$ approaches the normal distribution with
mean $N/2$ and variance $N/4$. The classical central limit theorem claims that
a similar statement remains valid whatever the distribution of X_n is. Namely,
the distribution of S_N approaches the normal distribution with mean Nm and

variance $N\sigma^2$, where $m = \mathbf{E}[X_n]$ and $\sigma^2 = \mathbf{V}[X_n]$. A precise statement is given as follows.

Theorem 2.6.1 (Classical CLT) *Let $X_1, X_2, \ldots$ be a sequence of independent, identically distributed random variables. Assume that they are normalized in such a way that $\mathbf{E}[X_n] = 0$ and $\mathbf{V}[X_n] = 1$. Then*

$$\lim_{N \to \infty} P\left(\frac{1}{\sqrt{N}} \sum_{n=1}^{N} X_n \le a \right) = \frac{1}{\sqrt{2\pi}} \int_{-\infty}^{a} e^{-x^2/2} dx, \qquad a \in \mathbb{R},$$

or equivalently, for any bounded continuous function $f(x)$,

$$\lim_{N \to \infty} \mathbf{E}\left[f\left(\frac{1}{\sqrt{N}} \sum_{n=1}^{N} X_n \right) \right] = \frac{1}{\sqrt{2\pi}} \int_{-\infty}^{+\infty} f(x) e^{-x^2/2} dx. \qquad (2.103)$$

The standard proof is based on the theory of characteristic functions (Fourier transforms) and is quite analytic. But when X_n admits finite moments of all orders, a combinatorial proof is possible.

A Combinatorial Proof of Classical CLT   We start with the expansion:

$$\mathbf{E}\left[\left(\frac{1}{\sqrt{N}} \sum_{n=1}^{N} X_n \right)^m \right] = \frac{1}{N^{m/2}} \sum_{n_1, \ldots, n_m = 1}^{N} \mathbf{E}[X_{n_1} X_{n_2} \cdots X_{n_m}]. \qquad (2.104)$$

The key step is to pick up the essential mixed moments

$$\mathbf{E}[X_{n_1} X_{n_2} \cdots X_{n_m}] \qquad (2.105)$$

that contributes to the limit as $N \to \infty$. We note that the mixed moment (2.105) is determined by the "pattern" of the arrangement of $X_1, X_2, \ldots, X_N$ because of their being independent and identically distributed.

Suppose first that there exists some X_i which appears just once in the product $X_{n_1} X_{n_2} \cdots X_{n_m}$. Such a factor is called a *singleton*. Then, by independence we have

$$\mathbf{E}[X_{n_1} X_{n_2} \cdots X_{n_m}] = \mathbf{E}[X_i] \mathbf{E}[\cdots \text{rest} \cdots] = 0. \qquad (2.106)$$

Consequently, the mixed moments containing a singleton do not contribute to the sum in the right-hand side of (2.104) and are ignored. Next we consider mixed moments (2.105) without singletons. Let s be the number of distinct factors appearing therein. Of course, $1 \le s \le [m/2]$ and (2.104) becomes

$$\mathbf{E}\left[\left(\frac{1}{\sqrt{N}} \sum_{n=1}^{N} X_n \right)^m \right] = \frac{1}{N^{m/2}} \sum_{s=1}^{[m/2]} \sum \mathbf{E}[X_{n_1} X_{n_2} \cdots X_{n_m}], \qquad (2.107)$$

where the inner sum is taken over $(n_1, \ldots, n_m)$ such that s distinct numbers appear and each of them appears more than once. The number of such arrangements $(n_1, \ldots, n_m)$ is given by

$$\binom{N}{s} \times W(s,m),$$

where $W(s,m)$ is the number of "words" of length m with s distinct "letters" each of which appears more than once. By assumption of being identically distributed, the mixed moments of $X_1, X_2, \ldots$ of order m are bounded by some constant $C_m > 0$. Thus the inner sum in (2.107) is bounded as

$$\sum |\mathbf{E}[X_{n_1} X_{n_2} \cdots X_{n_m}]| \le \binom{N}{s} \times W(s,m) C_m \sim c(s,m) N^s, \qquad (2.108)$$

where $c(s,m)$ is a constant independent of N. If m is odd, then $s \le [m/2] < m/2$ and, comparing with the denominator $N^{m/2}$ in (2.107), we obtain

$$\lim_{N \to \infty} \mathbf{E}\left[\left(\frac{1}{\sqrt{N}} \sum_{n=1}^{N} X_n \right)^m \right] = 0. \qquad (2.109)$$

If m is even, only the terms corresponding to $s = [m/2] = m/2$ contribute to the limit of (2.107). Hence, setting $m = 2s$ we have

$$\lim_{N \to \infty} \mathbf{E}\left[\left(\frac{1}{\sqrt{N}} \sum_{n=1}^{N} X_n \right)^{2s} \right] = \lim_{N \to \infty} \frac{1}{N^s} \sum \mathbf{E}[X_{n_1} X_{n_2} \cdots X_{n_{2s}}], \qquad (2.110)$$

where the sum is taken over $(n_1, \ldots, n_{2s})$ such that s distinct numbers appear and each of them appears exactly twice. The number of such $(n_1, \ldots, n_{2s})$ is given by

$$\binom{N}{s} \frac{(2s)!}{2^s} \qquad (2.111)$$

and for such a $(n_1, \ldots, n_{2s})$ we have

$$\mathbf{E}[X_{n_1} X_{n_2} \cdots X_{n_{2s}}] = \mathbf{E}[X_{i_1}^2 X_{i_2}^2 \cdots X_{i_s}^2]$$
$$= \mathbf{E}[X_{i_1}^2] \mathbf{E}[X_{i_2}^2] \cdots \mathbf{E}[X_{i_s}^2] = 1. \qquad (2.112)$$

Consequently, (2.110) becomes

$$\lim_{N \to \infty} \mathbf{E}\left[\left(\frac{1}{\sqrt{N}} \sum_{n=1}^{N} X_n \right)^{2s} \right] = \lim_{N \to \infty} \frac{1}{N^s} \binom{N}{s} \frac{(2s)!}{2^s} = \frac{(2s)!}{2^s s!}. \qquad (2.113)$$

Since the results in (2.109) and (2.113) coincide with the moments of the standard normal distribution (see, for example, (2.55)), we conclude that

$$\lim_{N \to \infty} \mathbf{E}\left[\left(\frac{1}{\sqrt{N}} \sum_{n=1}^{N} X_n \right)^m \right] = \frac{1}{\sqrt{2\pi}} \int_{-\infty}^{+\infty} x^m e^{-x^2/2} dx, \qquad m = 1, 2, \ldots.$$

In other words, the distribution of

$$\frac{1}{\sqrt{N}} \sum_{n=1}^{N} X_n$$

converges to the standard normal distribution in moments. Since the standard normal distribution is the solution to a determinate moment problem, the convergence is also valid in the weak sense, see Proposition 2.2.12. This completes the proof of (2.103).   □

Remark 2.6.2   A collection ϑ of non-empty subsets of a finite set S is called a *partition* if $S = \bigcup_{v \in \vartheta} v$ and $v \cap v' = \emptyset$ for any $v, v' \in S$ with $v \neq v'$. A partition ϑ is called a *pair partition* if $|v| = 2$ for any $v \in \vartheta$. For $S = \{1, 2, \ldots, 2s\}$ the set of pair partitions is denoted by PP(s). We then see easily that

$$|\mathrm{PP}(s)| = \frac{(2s)!}{2^s s!},$$

which appeared in (2.113). In fact, the number in (2.111) is understood as

$$\binom{N}{s} \frac{(2s)!}{2^s} = \binom{N}{s} |\mathrm{PP}(s)| \times s!$$

and (2.113) as

$$\lim_{N \to \infty} \mathbf{E}\left[\left(\frac{1}{\sqrt{N}} \sum_{n=1}^{N} X_n \right)^{2s} \right] = \lim_{N \to \infty} \frac{1}{N^s} \binom{N}{s} |\mathrm{PP}(s)| \times s! = |\mathrm{PP}(s)|.$$

Thus, the $2s$-th moment converges to the number of pair partitions. This combinatorial viewpoint is essential in the study of other concepts of independence, in particular free independence, see, for example, [56, 65].

Motivated by the classical CLT, we are interested in the asymptotics of

$$\frac{1}{\sqrt{N}} \sum_{n=1}^{N} a_n,$$

where $a_n = a_n^*$ are real random variables in an algebraic probability space $(\mathscr{A}, \varphi)$ that are "independent" and identically distributed. In the combinatorial proof of CLT, upon calculating or estimating the mixed moments we use factorization as in (2.106) and (2.112) to reduce the degree of the mixed moments. This aspect is essential to introduce various concepts of independence. In fact, due to non-commutativity many variants of factorization rules are introduced.

We start with a direct generalization of classical independence. A subset $\mathscr{B}$ of a unital $*$-algebra $\mathscr{A}$ is called a $*$-*subalgebra* if it is closed under the algebraic operations and the involution, and is called a *unital ∗-subalgebra* if it is a ∗-subalgebra containing the multiplication unit of $\mathscr{A}$ (not of $\mathscr{B}$).

Definition 2.6.3 Let $(\mathscr{A}, \varphi)$ be an algebraic probability space. A family $\{\mathscr{A}_\lambda\}$ of unital ∗-subalgebras of $\mathscr{A}$ is called *commutative independent* or *tensor independent* if

$$\varphi(a_1 \cdots a_m), \qquad a_i \in \mathscr{A}_{\lambda_i}, \qquad \lambda_1 \neq \lambda_2 \neq \cdots \neq \lambda_m,$$

admits the following reduction process:

$$\varphi(a_1 \cdots a_m) = \begin{cases} \varphi(a_1)\varphi(a_2 \cdots a_m), & \lambda_1 \notin \{\lambda_2, \ldots, \lambda_m\}, \\ \varphi(a_2 \cdots a_{r-1}(a_1 a_r)a_{r+1} \cdots a_m), & \text{otherwise,} \end{cases}$$

where $r \geq 3$ is the smallest number such that $\lambda_1 = \lambda_r$. Note that neither $\mathscr{A}_\lambda$ nor $\mathscr{A}$ is assumed to be commutative.

Definition 2.6.4 (Commutative independence) A sequence of random variables $a_1, a_2, \ldots$ in an algebraic probability space $(\mathscr{A}, \varphi)$ is called *commutative independent* if the unital ∗-subalgebras $\mathscr{A}_n$ generated by a_n are commutative independent in the sense of Definition 2.6.3.

Theorem 2.6.5 (Commutative CLT) *Let $a_n = a_n^*$ be a sequence of real random variables in an algebraic probability space $(\mathscr{A}, \varphi)$, normalized as $\varphi(a_n) = 0$ and $\varphi(a_n^2) = 1$. Assume that $\{a_n\}$ has uniformly bounded mixed moments, i.e.,*

$$C_m = \sup\{|\varphi(a_{n_1} \cdots a_{n_m})|; n_1, \ldots, n_m \geq 1\} < \infty, \qquad m = 1, 2, \ldots. \quad (2.114)$$

If $\{a_n\}$ are commutative independent, we have

$$\lim_{N \to \infty} \varphi\left(\left(\frac{1}{\sqrt{N}} \sum_{n=1}^{N} a_n\right)^m\right) = \frac{1}{\sqrt{2\pi}} \int_{\mathbb{R}} x^m e^{-x^2/2} dx, \qquad m = 1, 2, \ldots.$$

Proof We need only to repeat the combinatorial proof of the classical CLT, for the complete argument see [36, Chapter 8]. □

Remark 2.6.6 In the classical case, the uniform boundedness of mixed moments (2.114) follows from the condition of being independent and identically distributed. In Theorem 2.6.5, $a_1, a_2, \ldots$ being not necessarily commutative, we need (2.114) instead of being identically distributed. On the other hand, we only assume that $a_1, a_2, \ldots$ have common mean $\varphi(a_n) = 0$ and common variance $\varphi(a_n^2) = 1$, and do not require coincidence of higher order moments.

2.6.2 Graph products

A binary operation of graphs $G_1 \# G_2$ is reasonably called a *graph product* if it satisfies associativity: $(G_1 \# G_2) \# G_3 = G_1 \# (G_2 \# G_3)$; see Figure 2.16. In these lecture notes we will restrict ourselves to the case where the vertex set of the product graph $G_1 \# G_2$ is given by the direct product of the vertex sets of G_1 and G_2.

Figure 2.16 Illustration of graph product

To be precise, let $G_1 = (V_1, E_1)$ and $G_2 = (V_2, E_2)$ be two graphs. Their adjacency matrices A_1 and A_2 act on the vector spaces $C_0(V_1)$ and $C_0(V_2)$, respectively. A graph product $G_1 \# G_2$ will be defined as a graph on the direct product set $V = V_1 \times V_2$ with certain adjacency relation on it, and hence the adjacency matrix $A = A[G_1 \# G_2]$ acts on $C_0(V)$. On the other hand, the canonical isomorphism between $C_0(V) = C_0(V_1 \times V_2)$ and $C_0(V_1) \otimes C_0(V_2)$ is given by the correspondence between the orthonormal basis

$$e_{(x,y)} \leftrightarrow e_x \otimes e_y, \qquad x \in V_1, \quad y \in V_2.$$

Then A will be expressed by A_1 and A_2.

Choose vertices $o_1 \in V_1$ and $o_2 \in V_2$ arbitrarily. Let φ be the vacuum state at $(o_1, o_2) \in V$, that is, a state on $\mathscr{A}(G_1 \# G_2)$ defined by

$$\varphi(a) = \langle e_{(o_1, o_2)}, a e_{(o_1, o_2)} \rangle, \qquad a \in \mathscr{A}(G_1 \# G_2).$$

If a admits a product form $a = b_1 \otimes b_2$ according to the canonical isomorphism $C_0(V) \cong C_0(V_1) \otimes C_0(V_2)$, we have

$$\begin{aligned}
\varphi(a) &= \langle e_{(o_1, o_2)}, a e_{(o_1, o_2)} \rangle \\
&= \langle e_{o_1} \otimes e_{o_2}, (b_1 \otimes b_2)(e_{o_1} \otimes e_{o_2}) \rangle \\
&= \langle e_{o_1} \otimes e_{o_2}, (b_1 e_{o_1}) \otimes (b_2 e_{o_2}) \rangle \\
&= \langle e_{o_1}, b_1 e_{o_1} \rangle \langle e_{o_2}, b_2 e_{o_2} \rangle.
\end{aligned}$$

In this sense, φ is called a *product state*. We regard $A = A[G_1 \# G_2]$ as a random variable in $(\mathscr{A}(G_1 \# G_2), \varphi)$.

Definition 2.6.7 (Cartesian product graph)   The *Cartesian product* or *direct product* of G_1 and G_2, denoted by $G_1 \times G_2$, is a graph on $V = V_1 \times V_2$ with adjacency relation:

$$(x_1, y_1) \sim (x_2, y_2) \quad \Longleftrightarrow \quad \begin{cases} x_1 = x_2 \\ y_1 \sim y_2 \end{cases} \text{ or } \begin{cases} x_1 \sim x_2 \\ y_1 = y_2. \end{cases}$$

See Figure 2.17.

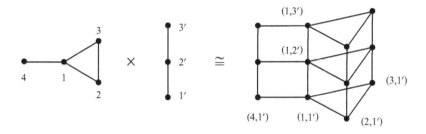

Figure 2.17 Cartesian product graph $G_1 \times G_2$

Lemma 2.6.8   *Let $G_1 = (V_1, E_1)$ and $G_2 = (V_2, E_2)$ be two graphs with adjacency matrices A_1 and A_2, respectively. Then*

$$A[G_1 \times G_2] = A_1 \otimes I_2 + I_1 \otimes A_2, \tag{2.115}$$

where I_1 and I_2 are the identity matrices on $C_0(V_1)$ and $C_0(V_2)$, respectively.

Proof   Exercise.   □

Definition 2.6.9 (Hierarchical product graph)   Let $G_1 = (V_1, E_1)$ and $G_2 = (V_2, E_2)$ be two graphs. We fix $o_2 \in V_2$ as a distinguished vertex. The *comb product* or the *hierarchical product* of G_1 and G_2, denoted by $G_1 \rhd_{o_2} G_2$, is a graph on $V = V_1 \times V_2$ with adjacency relation:

$$(x_1, y_1) \sim (x_2, y_2) \quad \Longleftrightarrow \quad \begin{cases} x_1 = x_2 \\ y_1 \sim y_2 \end{cases} \text{ or } \begin{cases} x_1 \sim x_2 \\ y_1 = y_2 = o_2. \end{cases} \tag{2.116}$$

See Figure 2.18.

We see from definition (or Lemma 2.6.10 below) that the comb product is associative but not commutative.

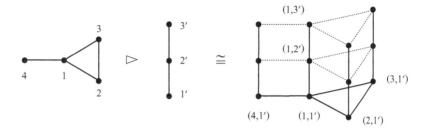

Figure 2.18 Comb product graph $G_1 \rhd_{o_2} G_2$ with $o_2 = 1'$

Lemma 2.6.10   *Let $G_1 = (V_1, E_1)$ and $G_2 = (V_2, E_2)$ be two graphs with adjacency matrices A_1 and A_2, respectively. Let $o_2 \in V_2$ be a distinguished vertex chosen arbitrarily. Then*

$$A[G_1 \rhd_{o_2} G_2] = A_1 \otimes P_2 + I_1 \otimes A_2, \qquad (2.117)$$

where I_1 is the identity matrix on $C_0(V_1)$ and P_2 the rank one projection from $C_0(V_2)$ onto the space spanned by $e_{o_2} \in C_0(V_2)$.

Proof   Let $x_1, x_2 \in V_1$ and $y_1, y_2 \in V_2$. Note first that

$$
\begin{aligned}
&(A_1 \otimes P_2 + I_1 \otimes A_2)_{(x_1, y_1), (x_2, y_2)} \\
&= (A_1 \otimes P_2)_{(x_1, y_1), (x_2, y_2)} + (I_1 \otimes A_2)_{(x_1, y_1), (x_2, y_2)} \\
&= (A_1)_{x_1, x_2} (P_2)_{(y_1, y_2)} + (I_1)_{x_1, x_2} (A_2)_{(y_1, y_2)}.
\end{aligned}
\qquad (2.118)
$$

Since $(P_2)_{(y_1, y_2)} = \langle e_{y_1}, P_2 e_{y_2} \rangle = \delta_{y_1, o_2} \delta_{y_2, o_2}$ by definition, the expression in (2.118) is equal to one if and only if the condition in (2.116) is satisfied. Therefore $A_1 \otimes P_2 + I_1 \otimes A_2$ is the adjacency matrix of $G_1 \rhd_{o_2} G_2$ as desired.   □

Definition 2.6.11 (Star product graph)   Let $G_1 = (V_1, E_1)$ and $G_2 = (V_2, E_2)$ be two graphs with distinguished vertices $o_1 \in V_1$ and $o_2 \in V_2$, respectively. The *star product* of G_1 and G_2, denoted by $G_1 \,_{o_1}\!\star_{o_2} G_2$, is a graph on $V = V_1 \times V_2$ with adjacency relation:

$$
(x_1, y_1) \sim (x_2, y_2) \quad \Longleftrightarrow \quad
\begin{cases} x_1 = x_2 = o_1 \\ y_1 \sim y_2 \end{cases}
\text{ or }
\begin{cases} x_1 \sim x_2 \\ y_1 = y_2 = o_2. \end{cases}
\qquad (2.119)
$$

See Figure 2.19.

Lemma 2.6.12   *Let $G_1 = (V_1, E_1)$ and $G_2 = (V_2, E_2)$ be two graphs with ad-*

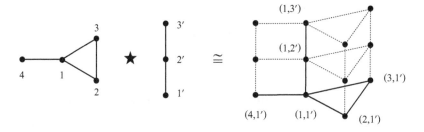

Figure 2.19 Star product graph $G_1{}_{o_1}\star_{o_2} G_2$ with $o_1 = 1, o_2 = 1'$

jacency matrices A_1 and A_2, respectively. Let $o_1 \in V_1$ and $o_2 \in V_2$ be distinguished vertices chosen arbitrarily. Then,

$$A[G_1{}_{o_1}\star_{o_2} G_2] = A_1 \otimes P_2 + P_1 \otimes A_2, \qquad (2.120)$$

where P_1 and P_2 are rank one projections from $C_0(V_1)$ onto the space spanned by $e_{o_1} \in C_0(V_1)$ and from $C_0(V_2)$ onto the space spanned by $e_{o_2} \in C_0(V_2)$, respectively.

Proof Exercise. □

We often write $G_1 \star G_2$ for $G_1{}_{o_1}\star_{o_2} G_2$ whenever there is no confusion. Define a subset of $V_1 \times V_2$ by

$$V_1 \star V_2 = \{(x, o_2) ; x \in V_1\} \cup \{(o_1, y) ; y \in V_2\}.$$

Note that every vertex (x, y) of the star product $G_1 \star G_2$ which does not belong to $V_1 \star V_2$ is isolated. In fact, we are mostly concerned with the induced subgraph spanned by $V_1 \star V_2$, which is also denoted by $G_1 \star G_2$ for simplicity. Obviously, the latter $G_1 \star G_2$ is a graph obtained simply by joining two vertices o_1 and o_2.

We mention three more graph products in terms of adjacency matrices. The *lexicographic product* $G_1 \rhd_L G_2$ is defined by

$$A[G_1 \rhd_L G_2] = A_1 \otimes J_2 + I_1 \otimes A_2, \qquad (2.121)$$

where J_2 denotes the so-called all-one matrix, that is the entries are all one. The *Kronecker product* $G_1 \times_K G_2$ is defined by

$$A[G_1 \times_K G_2] = A_1 \otimes A_2. \qquad (2.122)$$

The *strong product graph* $G_1 \times_S G_2$ is defined by

$$A[G_1 \times_S G_2] = A_1 \otimes I_2 + I_1 \otimes A_2 + A_1 \otimes A_2. \qquad (2.123)$$

Note that the names of graph products are not unified in the literature.

2.6.3 Central Limit Theorem for Cartesian powers

Theorem 2.6.13   *Let $G_1 = (V_1, E_1)$ and $G_2 = (V_2, E_2)$ be two graphs with adjacency matrices A_1 and A_2, respectively. Then the adjacency matrix of the Cartesian product $G_1 \times G_2$ is decomposed into a sum of commutative independent random variables:*

$$A[G_1 \times G_2] = A_1 \otimes I_2 + I_1 \otimes A_2 \qquad (2.124)$$

with respect to the vacuum state at (o_1, o_2), where $o_1 \in V_1$ and $o_2 \in V_2$ are distinguished vertices chosen arbitrarily.

Proof   Relation (2.124) is already shown in Lemma 2.6.8. Let φ be the vacuum state at (o_1, o_2). We prove that $A_1 \otimes I_2$ and $I_1 \otimes A_2$ are commutative independent random variables in $(\mathscr{A}, \varphi)$. Let $\mathscr{A}_1$ and $\mathscr{A}_2$ be the unital $*$-subalgebras generated by $A_1 \otimes I_2$ and $I_1 \otimes A_2$, respectively. We need to verify the condition mentioned in Definition 2.6.3. Upon checking the factorization of

$$\varphi(a_1 \cdots a_m), \quad a_i \in \mathscr{A}_{\lambda_i}, \quad \lambda_i \in \{1, 2\}, \quad \lambda_1 \neq \lambda_2 \neq \cdots \neq \lambda_m,$$

by linearity it is sufficient to take a_i to be a power of $A_1 \otimes I_2$ or $I_1 \otimes A_2$. Suppose that $\lambda_1 \notin \{\lambda_2, \ldots, \lambda_m\}$. There are a few cases and here we only consider the case of $m = 2$, $\lambda_1 = 1$ and $\lambda_2 = 2$. Using $a_1 a_2 = (A_1 \otimes I_2)^p (I_1 \otimes A_2)^q = A_1^p \otimes A_2^q$, we obtain

$$\varphi(a_1 a_2) = \langle e_{o_1} \otimes e_{o_2}, (A_1^p \otimes A_2^q)(e_{o_1} \otimes e_{o_2}) \rangle = \langle e_{o_1}, A_1^p e_{o_1} \rangle \langle e_{o_2}, A_2^q e_{o_2} \rangle.$$

On the other hand, we have

$$\varphi(a_1) = \langle e_{o_1} \otimes e_{o_2}, (A_1 \otimes I_2)^p (e_{o_1} \otimes e_{o_2}) \rangle = \langle e_{o_1}, A_1^p e_{o_1} \rangle,$$
$$\varphi(a_2) = \langle e_{o_1} \otimes e_{o_2}, (I_1 \otimes A_2)^q (e_{o_1} \otimes e_{o_2}) \rangle = \langle e_{o_2}, A_2^q e_{o_2} \rangle.$$

Consequently,

$$\varphi(a_1 a_2) = \varphi(a_1) \varphi(a_2),$$

as desired. We next suppose that $\lambda_1 \in \{\lambda_2, \ldots, \lambda_m\}$. Among a few similar cases, we consider here the case of $\lambda_1 = 1, \lambda_2 = 2$ and $\lambda_3 = 1$. Since $A_1 \otimes I_2$ and $I_1 \otimes A_2$ are commutative, we obtain

$$\varphi(a_1 \cdots a_m) = \varphi(a_2 (a_1 a_3) a_4 \cdots a_m).$$

Thus, the condition mentioned in Definition 2.6.3 is verified. $\qquad\square$

We consider the Cartesian powers. Let $G = (V, E)$ be a graph with adjacency matrix A. For $N \geq 1$ let

$$G^{(N)} = G \times \cdots \times G \qquad (N \text{ times})$$

denote the N-fold Cartesian power of G and $A^{(N)}$ the adjacency matrix of $G^{(N)}$. By repeated application of Lemma 2.6.8 we have

$$A^{(N)} = \sum_{n=1}^{N} \overbrace{I \otimes \cdots \otimes I}^{n-1} \otimes A \otimes \overbrace{I \otimes \cdots \otimes I}^{N-n}. \qquad (2.125)$$

Fix a distinguished vertex $o \in V$ and consider the vacuum state at $(o, \dots, o) \in V^N$, denoted by $\langle \cdot \rangle$ for simplicity. Since $\langle \cdot \rangle$ is a product state, we have

$$\langle I^{\otimes(n-1)} \otimes A \otimes I^{\otimes(N-n)} \rangle = \langle e_o, A e_o \rangle = 0, \qquad (2.126)$$

$$\langle (I^{\otimes(n-1)} \otimes A \otimes I^{\otimes(N-n)})^2 \rangle = \langle e_o, A^2 e_o \rangle = \deg(o). \qquad (2.127)$$

We set

$$B_n = \frac{1}{\sqrt{\deg(o)}} \overbrace{I \otimes \cdots \otimes I}^{n-1} \otimes A \otimes \overbrace{I \otimes \cdots \otimes I}^{N-n}.$$

Then B_n is normalized in such a way that $\langle B_n \rangle = 0$ and $\langle B_n^2 \rangle = 1$. As is easily seen, the mixed moments of $B_1, B_2, \dots$ satisfy the uniform boundedness condition (2.114). Moreover, in a similar manner as in the proof of Theorem 2.6.13, we see that

$$\frac{A_N}{\sqrt{\deg(o)}} = \sum_{n=1}^{N} B_n$$

is a sum of commutative independent random variables with respect to the vacuum state $\langle \cdot \rangle$. Finally, applying the commutative CLT (Theorem 2.6.5) we come to the following result.

Theorem 2.6.14 (CLT for Cartesian powers) *Let $G = (V, E)$ be a graph with adjacency matrix A. For $N \geq 1$ let $G^{(N)}$ be the N-fold Cartesian power of G and $A^{(N)}$ the adjacency matrix. Then as a random variable with respect to the vacuum state at $(o, \dots, o)$, where $o \in V$ is a distinguished vertex chosen arbitrarily, the normalized adjacency matrix*

$$\frac{A^{(N)}}{\sqrt{N}\sqrt{\deg(o)}}$$

converges to the standard normal distribution in moments. Namely,

$$\lim_{N\to\infty}\left\langle\left(\frac{A^{(N)}}{\sqrt{N}\sqrt{\deg(o)}}\right)^m\right\rangle = \frac{1}{\sqrt{2\pi}}\int_{-\infty}^{+\infty} x^m e^{-x^2/2}dx, \quad m=1,2,\dots.$$

Remark 2.6.15 In Theorem 2.6.14 we take $\langle\cdot\rangle$ to be the vacuum state at $(o,\dots,o)$. A similar argument is applicable whenever $\langle\cdot\rangle$ is a product state of the form: $\varphi\otimes\cdots\otimes\varphi$ where φ is an arbitrary state on $\mathscr{A}(G)$. In that case, the normalized adjacency matrix of $G^{(N)}$ is given by

$$\frac{A^{(N)} - N\varphi(A)}{\sqrt{N}\sqrt{\varphi(A^2) - \varphi(A)^2}}$$

and converges to the standard normal distribution in moments.

2.6.4 Monotone independence and comb product

Let $(\Lambda, <)$ be a totally ordered set. In a finite sequence

$$\lambda_1 \neq \lambda_2 \neq \dots \neq \lambda_s \neq \dots \neq \lambda_m, \quad \lambda_i \in \Lambda, \quad m \geq 2, \tag{2.128}$$

λ_s is called a *peak* if (i) $1 < s < m$, $\lambda_{s-1} < \lambda_s$ and $\lambda_s > \lambda_{s+1}$; or (ii) $s=1$ and $\lambda_1 > \lambda_2$; or (iii) $s=m$ and $\lambda_{m-1} < \lambda_m$.

Definition 2.6.16 (Monotone independence) Let $(\mathscr{A}, \varphi)$ be an algebraic probability space and let $\{\mathscr{A}_\lambda\}$ be a set of $*$-subalgebras of $\mathscr{A}$ indexed by a totally ordered set Λ. We say that $\{\mathscr{A}_\lambda\}$ is *monotone independent* if

$$\varphi(a_1\cdots a_s\cdots a_m) = \varphi(a_s)\varphi(a_1\cdots\check{a}_s\cdots a_m), \tag{2.129}$$

for any $a_i \in \mathscr{A}_{\lambda_i}$, whenever $\lambda_1 \neq \lambda_2 \neq \dots \neq \lambda_m$ and λ_s is a peak. (Here $\check{a}_s$ means that the term a_s is omitted from the sequence.)

Definition 2.6.17 (Monotone independent sequence) A sequence of random variables $a_1, a_2,\dots$ in an algebraic probability space $(\mathscr{A}, \varphi)$ is called *monotone independent* if the $*$-subalgebras $\mathscr{A}_n$ generated by a_n are monotone independent in the sense of Definition 2.6.16. (Unless otherwise stated, the natural numbers appearing as an index set are equipped with the usual order.)

Remark 2.6.18 By definition a $*$-subalgebra of $\mathscr{A}$ does not necessarily contain the multiplication unit $1_\mathscr{A}$. If it contains $1_\mathscr{A}$, then it is called a *unital* $*$-subalgebra. Note that monotone independence for unital $*$-subalgebras falls into a trivial situation. For example, suppose that $\mathscr{A}_1$ and $\mathscr{A}_2$ are monotone independent $*$-subalgebras and $1_\mathscr{A} \in \mathscr{A}_2$. Then for any $a \in \mathscr{A}_1$ we have $\varphi(a^*a) = \varphi(a^*1_\mathscr{A}a) = \varphi(a^*)\varphi(a)$ and $\varphi(aa^*) = \varphi(a1_\mathscr{A}a^*) = \varphi(a^*)\varphi(a)$, which imply that $a \overset{m}{=} \varphi(a)1_{\mathscr{A}}$. This is why the monotone independence is defined in terms of *-subalgebras.

Example 2.6.19 We illustrate how the factorization rule (2.129) works for computing monotone independent random variables. For monotone independent random variables $a_1, a_2, \ldots$ let us consider the mixed moment:

$$\varphi(a_2 a_1 a_4 a_3 a_4 a_3 a_6 a_6 a_4 a_3 a_3 a_5) = \varphi(214343664335),$$

where the right-hand side is just for simplicity. Looking at the peaks, we have

$$= \varphi(2)\varphi(4)\varphi(4)\varphi(66)\varphi(5)\varphi(133433),$$

and continue similar factorization to get

$$= \varphi(2)\varphi(4)\varphi(4)\varphi(66)\varphi(5)\varphi(4)\varphi(13333)$$
$$= \varphi(2)\varphi(4)\varphi(4)\varphi(66)\varphi(5)\varphi(4)\varphi(3333)\varphi(1),$$

see Figure 2.20. On the other hand, if $a_1, a_2, \ldots$ are commutative independent, we have

$$\varphi(214343664335) = \varphi(1)\varphi(2)\varphi(3333)\varphi(444)\varphi(5)\varphi(66).$$

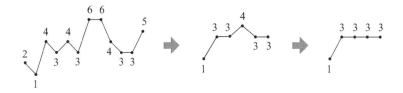

Figure 2.20 Factorization procedure for monotone independence

Theorem 2.6.20 *Let $G_1 = (V_1, E_1)$ and $G_2 = (V_2, E_2)$ be two graphs with adjacency matrices A_1 and A_2, respectively. Let $o_2 \in V_2$ be a distinguished vertex. Then the adjacency matrix of the comb product $G_1 \triangleright_{o_2} G_2$ is decomposed into a sum of monotone independent random variables:*

$$A[G_1 \triangleright_{o_2} G_2] = A_1 \otimes P_2 + I_1 \otimes A_2 \qquad (2.130)$$

with respect to the vacuum state at (o_1, o_2) with arbitrarily chosen $o_1 \in V_1$, where P_2 is the rank one projection from $C_0(V_2)$ onto the space spanned by $e_{o_2} \in C_0(V_2)$.

More generally,

Theorem 2.6.21 *For $n = 1, 2, \ldots, N$ let $G_n = (V_n, E_n)$ be a graph with adjacency matrix A_n. Let $o_n \in V_n$ be a distinguished vertex. Then the adjacency matrix of the comb product*

$$G = G_1 \rhd_{o_2} G_2 \rhd_{o_3} \cdots \rhd_{o_N} G_N$$

is decomposed into a sum of monotone independent random variables:

$$A[G] = \sum_{n=1}^{N} I_1 \otimes I_2 \otimes \cdots \otimes I_{n-1} \otimes A_n \otimes P_{n+1} \otimes \cdots \otimes P_N \tag{2.131}$$

with respect to the vacuum state at $(o_1, \ldots, o_N)$, where P_n is the rank one projection from $C_0(V_n)$ onto the space spanned by $e_{o_n} \in C_0(V_n)$.

Relation (2.130) is already shown in Lemma 2.6.10 and its repeated application gives rise to (2.131) easily. Note also that $G_1 \rhd_{o_2} G_2 \rhd_{o_3} G_3$ is a short-hand notation for

$$(G_1 \rhd_{o_2} G_2) \rhd_{o_3} G_3 \cong G_1 \rhd_{(o_2, o_3)} (G_2 \rhd_{o_3} G_3).$$

For the monotone independence we give a general result.

Let $\mathscr{A}_n$ be the matrix algebra $M(d_n, \mathbb{C})$ and choose a unit vector $\xi_n \in \mathbb{C}^{d_n}$, where $d_n \geq 2$ but will be not used explicitly. Let φ_n be the vector state on $\mathscr{A}_n$ associated to ξ_n, that is,

$$\varphi_n(b) = \langle \xi_n, b\xi_n \rangle, \qquad b \in \mathscr{A}_n.$$

Let p_n be the projection defined by

$$p_n \eta = \langle \xi_n, \eta \rangle \xi_n \qquad \eta \in \mathbb{C}^{d_n}.$$

Then for $b, b' \in \mathscr{A}_n$ we have

$$\varphi_n(bp_nb') = \langle \xi_n, bp_nb'\xi_n \rangle = \langle \xi_n, b\langle \xi_n, b'\xi_n \rangle \xi_n \rangle = \langle \xi_n, b\xi_n \rangle \langle \xi_n, b'\xi_n \rangle$$

and hence

$$\varphi_n(bp_nb') = \varphi_n(b)\varphi_n(b'), \qquad b, b' \in \mathscr{A}_n. \tag{2.132}$$

In particular,

$$\varphi_n(p_n) = \langle \xi_n, p_n\xi_n \rangle = \langle \xi_n, \xi_n \rangle = 1. \tag{2.133}$$

Now fixing $N \geq 1$ we consider the matrix algebra $\mathscr{A} = \mathscr{A}_1 \otimes \cdots \otimes \mathscr{A}_N$ equipped with the product state $\varphi = \varphi_1 \otimes \cdots \otimes \varphi_N$. Given $b_n \in \mathscr{A}_n$ we define a sequence of random variables in $(\mathscr{A}, \varphi)$ by

$$a_1 = b_1 \otimes p_2 \otimes p_3 \otimes \cdots \otimes p_{N-1} \otimes p_N,$$
$$a_2 = 1 \otimes b_2 \otimes p_3 \otimes \cdots \otimes p_{N-1} \otimes p_N,$$
$$\cdots$$
$$a_n = \overbrace{1 \otimes 1 \otimes \cdots \otimes 1}^{n-1} \otimes b_n \otimes \overbrace{p_{n+1} \otimes \cdots \otimes p_N}^{N-n},$$
$$\cdots$$
$$a_N = 1 \otimes 1 \otimes 1 \otimes \cdots \otimes 1 \otimes b_N,$$

In view of (2.133) we have

$$\varphi(a_n) = \varphi_1(1)\varphi_2(1) \cdots \varphi_n(b_n)\varphi_{n+1}(p_{n+1}) \cdots \varphi_N(p_N) = \varphi_n(b_n). \quad (2.134)$$

Theorem 2.6.22 *Notation and assumptions being as above, the random variables $a_1, a_2, \ldots, a_N$ in $(\mathscr{A}, \varphi)$ are monotone independent.*

Proof The formal proof is routine. Here we only illustrate the factorization rule by an example $\varphi(a_1 a_2 a_1)$. Note first that

$$a_1 a_2 a_1 = (b_1 \cdot 1 \cdot b_1) \otimes (p_2 \cdot b_2 \cdot p_2) \otimes (p_3 \cdot p_3 \cdot p_3) \otimes \cdots \otimes (p_N \cdot p_N \cdot p_N).$$

Then by (2.132) and (2.133) we have

$$\varphi(a_1 a_2 a_1) = \varphi_1(b_1^2)\varphi_2(p_2 b_2 p_2)\varphi_3(p_3) \cdots \varphi_N(p_N) = \varphi_1(b_1^2)\varphi_2(b_2).$$

Finally, by (2.134) we have

$$\varphi(a_1 a_2 a_1) = \varphi(a_1^2)\varphi(a_2),$$

as desired. The essense of the above argument is that in the product $a_1 a_2 a_1$ the peak is attained by a_2 and it corresponds to $p_2 b_2 p_2$. $\qquad \square$

Theorem 2.6.23 (Monotone CLT) *Let $a_n = a_n^*$ be a sequence of real random variables in an algebraic probability space $(\mathscr{A}, \varphi)$, normalized as $\varphi(a_n) = 0$ and $\varphi(a_n^2) = 1$. Assume that $\{a_n\}$ has uniformly bounded mixed moments, see (2.114). If $\{a_n\}$ are monotone independent, we have*

$$\lim_{N \to \infty} \varphi\left(\left(\frac{1}{\sqrt{N}}\sum_{n=1}^{N} a_n\right)^m\right) = \frac{1}{\pi}\int_{-\sqrt{2}}^{\sqrt{2}} \frac{x^m}{\sqrt{2-x^2}}\,dx, \quad m = 1, 2, \ldots, \quad (2.135)$$

where the probability distribution in the right-hand side is the normalized arcsine law.

Outline of proof   Just like the combinatorial proof of the classical CLT, we start with the expansion:

$$\varphi\left(\left(\frac{1}{\sqrt{N}}\sum_{n=1}^{N}a_n\right)^m\right)=\frac{1}{N^{m/2}}\sum_{n_1,\ldots,n_m=1}^{N}\varphi(a_{n_1}a_{n_2}\cdots a_{n_m}).$$

It is easy to see that $\varphi(a_{n_1}a_{n_2}\cdots a_{n_m})=0$ if the mixed moment contains a singleton. We next observe that $\varphi(a_{n_1}a_{n_2}\cdots a_{n_m})$ contributes to the limit only if the number s of distinct factors appearing therein is $s=[m/2]$. In particular,

$$\lim_{N\to\infty}\varphi\left(\left(\frac{1}{\sqrt{N}}\sum_{n=1}^{N}a_n\right)^{2m-1}\right)=0.$$

For $m=2s$ we evaluate $\varphi(a_{n_1}a_{n_2}\cdots a_{n_{2s}})$ with distinct factors appearing exactly twice and we come to

$$\lim_{N\to\infty}\varphi\left(\left(\frac{1}{\sqrt{N}}\sum_{n=1}^{N}a_n\right)^{2s}\right)=\frac{(2s)!}{2^s s!s!}=\frac{1}{\pi}\int_{-\sqrt{2}}^{\sqrt{2}}\frac{x^{2s}}{\sqrt{2-x^2}}\,dx,$$

for $s=1,2,\ldots$. Thus (2.135) is proved. For the complete proof see, for example, Hora and Obata [36, Section 8.2], Saigo [61].   □

Theorem 2.6.24 (CLT for comb powers)   *Let $G=(V,E)$ be a graph with adjacency matrix A. Let $o\in V$ be a distinguished vertex chosen arbitrarily. For $N\geq 1$ let $G^{(N)}=G\triangleright_o G\triangleright_o\cdots\triangleright_o G$ be the N-fold comb power of G and $A^{(N)}$ the adjacency matrix. Then as a random variable with respect to the vacuum state at $(o,\ldots,o)$, the normalized adjacency matrix*

$$\frac{A^{(N)}}{\sqrt{N}\sqrt{\deg(o)}}$$

converges to the normalized arcsine law in moments. Namely,

$$\lim_{N\to\infty}\left\langle\left(\frac{A^{(N)}}{\sqrt{N}\sqrt{\deg(o)}}\right)^m\right\rangle=\int_{-\sqrt{2}}^{+\sqrt{2}}x^m\frac{dx}{\pi\sqrt{2-x^2}},\quad m=1,2,\ldots.$$

Remark 2.6.25   Monotone independence appears also in the lexicographic product (2.121). In fact, $A[G_1\triangleright_L G_2]=A_1\otimes J_2+I_1\otimes A_2$ is a sum of monotone independent random variables with respect to the product state $\varphi_1\otimes\varphi_2$, where φ_1 is an arbitrary state on $\mathscr{A}(G_1)$ and φ_2 is the state on $\mathscr{A}(G_2)$ defined by the density matrix $J_2/|V_2|$. For the details see [58, 59].

2.6.5 Boolean independence and star product

Definition 2.6.26 (Boolean independence)   Let $(\mathscr{A}, \varphi)$ be an algebraic probability space. A family $\{\mathscr{A}_\lambda\}$ of $*$-subalgebras of $\mathscr{A}$ is called *Boolean independent* if

$$\varphi(a_1 \cdots a_m) = \varphi(a_1)\varphi(a_2 \cdots a_m)$$

for any $a_i \in \mathscr{A}_{\lambda_i}$, $i = 1, \ldots, m$, whenever $\lambda_1 \neq \lambda_2 \neq \ldots \neq \lambda_m$.

Definition 2.6.27 (Boolean independent sequence)   A sequence of random variables $a_1, a_2, \ldots$ in an algebraic probability space $(\mathscr{A}, \varphi)$ is called *Boolean independent* if the $*$-subalgebras $\mathscr{A}_n$ generated by a_n are Boolean independent in the sense of Definition 2.6.26.

Remark 2.6.28   As in the case of monotone independence, Boolean independence for unital $*$-subalgebras falls into a trivial situation. For example, suppose that $\mathscr{A}_1$ and $\mathscr{A}_2$ are Boolean independent $*$-subalgebras and $1_{\mathscr{A}} \in \mathscr{A}_2$. Then for any $a \in \mathscr{A}_1$ we have $\varphi(a^*a) = \varphi(a^* 1_{\mathscr{A}} a) = \varphi(a^*)\varphi(a)$ and $\varphi(aa^*) = \varphi(a 1_{\mathscr{A}} a^*) = \varphi(a)\varphi(a^*)$, which imply that $a \overset{m}{=} \varphi(a)1_{\mathscr{A}}$.

Example 2.6.29   Suppose $a_1, a_2, \ldots$ are Boolean independent. Then we have

$$\varphi(a_2 a_1 a_4 a_3 a_4 a_3 a_6 a_6 a_4 a_3 a_3 a_5)$$
$$= \varphi(214343664335)$$
$$= \varphi(2)\varphi(1)\varphi(4)\varphi(3)\varphi(4)\varphi(3)\varphi(66)\varphi(4)\varphi(33)\varphi(5),$$

which shows a significant difference from monotone or commutative independence, see Example 2.6.19.

Theorem 2.6.30   *Let $G_1 = (V_1, E_1)$ and $G_2 = (V_2, E_2)$ be two graphs with adjacency matrices A_1 and A_2, respectively. Let $o_1 \in V_1$ and $o_2 \in V_2$ be distinguished vertices. Then the adjacency matrix of the star product $G_{1\,o_1}\star_{o_2} G_2$ is decomposed into a sum of Boolean independent random variables:*

$$A[G_{1\,o_1}\star_{o_2} G_2] = A_1 \otimes P_2 + P_1 \otimes A_2 \qquad (2.136)$$

with respect to the vacuum state at (o_1, o_2), where P_1 and P_2 are the rank one projections from $C_0(V_1)$ onto the space spanned by $e_{o_1} \in C_0(V_1)$ and from $C_0(V_2)$ onto the space spanned by $e_{o_2} \in C_0(V_2)$, respectively.

More generally,

Theorem 2.6.31   *For $n = 1, 2, \ldots, N$ let $G_n = (V_n, E_n)$ be a graph with adjacency matrix A_n. Let $o_n \in V_n$ be a distinguished vertex. Then the adjacency* *matrix of the star product*

$$G = G_{1\ o_1 \star o_2} G_{2\ o_2 \star o_3} \cdots {}_{o_{N-1} \star o_N} G_N$$

is decomposed into a sum of Boolean independent random variables:

$$A[G] = \sum_{n=1}^{N} P_1 \otimes \cdots \otimes P_{n-1} \otimes A_n \otimes P_{n+1} \otimes \cdots \otimes P_N \qquad (2.137)$$

with respect to the vacuum state at $(o_1, \ldots, o_N)$, where P_n is the rank one projection from $C_0(V_n)$ onto the space spanned by $e_{o_n} \in C_0(V_n)$.

Relation (2.136) is already shown in Lemma 2.6.10 and its repeated application gives rise to (2.137) immediately. Note also that $G_{1\ o_1 \star o_2} G_{2\ o_2 \star o_3} G_3$ is a short-hand notation for

$$\left(G_{1\ o_1 \star o_2} G_2\right)_{(o_1, o_2) \star o_3} G_3 \cong G_{1\ o_1 \star (o_2, o_3)} \left(G_{2\ o_2 \star o_3} G_3\right).$$

For the Boolean independence we state a slightly more general result. Let $(\mathscr{A}_n, \varphi_n)$, p_n and $(\mathscr{A}, \varphi)$ be the same as in the paragraph just before Theorem 2.6.22. Given $b_n \in \mathscr{A}_n$ we define a sequence of random variables in $(\mathscr{A}, \varphi)$ by

$$a_1 = b_1 \otimes p_2 \otimes p_3 \otimes \cdots \otimes p_{N-1} \otimes p_N,$$
$$a_2 = p_1 \otimes b_2 \otimes p_3 \otimes \cdots \otimes p_{N-1} \otimes p_N,$$
$$\cdots$$
$$a_n = p_1 \otimes p_2 \otimes \cdots \otimes p_{n-1} \otimes b_n \otimes p_{n+1} \otimes \cdots \otimes p_N,$$
$$\cdots$$
$$a_N = p_1 \otimes p_2 \otimes p_3 \otimes \cdots \otimes p_{N-1} \otimes b_N.$$

Then by similar consideration as in the case of monotone independence, we have the following result.

Theorem 2.6.32 *Notation and assumptions being as above, the random variables $a_1, a_2, \ldots, a_N$ in $(\mathscr{A}, \varphi)$ are Boolean independent.*

Theorem 2.6.33 (Boolean CLT) *Let $a_n = a_n^*$ be a sequence of real random variables in an algebraic probability space $(\mathscr{A}, \varphi)$, normalized as $\varphi(a_n) = 0$ and $\varphi(a_n^2) = 1$. Assume that $\{a_n\}$ has uniformly bounded mixed moments, see (2.114). If $\{a_n\}$ are Boolean independent, we have*

$$\lim_{N \to \infty} \varphi\left(\left(\frac{1}{\sqrt{N}} \sum_{n=1}^{N} a_n\right)^m\right) = \frac{1}{2} \int_{-\infty}^{\infty} x^m (\delta_{-1} + \delta_{+1})(dx), \quad m = 1, 2, \ldots,$$

where the probability distribution in the right-hand side is the normalized Bernoulli distribution.

Outline of Proof By similar argument as in the proofs of the commutative and monotone CLTs, we can obtain

$$\lim_{N \to \infty} \varphi\left(\left(\frac{1}{\sqrt{N}} \sum_{n=1}^{N} a_n\right)^{2m-1}\right) = 0,$$

$$\lim_{N \to \infty} \varphi\left(\left(\frac{1}{\sqrt{N}} \sum_{n=1}^{N} a_n\right)^{2m}\right) = 1,$$

which coincides with the moments of the Bernoulli distribution. $\square$

Theorem 2.6.34 (CLT for star powers) *Let* $G = (V, E)$ *be a graph with adjacency matrix A. Let $o \in V$ be a distinguished vertex chosen arbitrarily. For $N \geq 1$ let $G^{(N)} = G \, {}_o{\star}_o \, G \, {}_o{\star}_o \cdots {}_o{\star}_o \, G$ be the N-fold star power of G and $A^{(N)}$ the adjacency matrix. Then as a random variable with respect to the vacuum state at $(o, \ldots, o)$, the normalized adjacency matrix*

$$\frac{A^{(N)}}{\sqrt{N}\sqrt{\deg(o)}}$$

converges to the normalized Bernoulli distribution in moments. Namely,

$$\lim_{N \to \infty} \left\langle \left(\frac{A^{(N)}}{\sqrt{N}\sqrt{\deg(o)}}\right)^m \right\rangle = \frac{1}{2} \int_{-\infty}^{\infty} x^m (\delta_{-1} + \delta_{+1})(dx), \quad m = 1, 2, \ldots.$$

Example 2.6.35 In Example 2.4.15 we obtained explicitly the vacuum spectral distribution μ_N of the N-fold star power of $\mathbb{Z}_+$ and observed that μ_N has two point masses with the same weight $(N-2)/2(N-1)$ which tends to $1/2$ as $N \to \infty$. After normalization we see that $\tilde{\mu}_N$ converges to the Bernoulli distribution. This phenomenon is now understood as a special case of the Boolean CLT.

2.6.6 Convolutions of spectral distributions

For two probability distributions μ_1 and μ_2 the *(classical) convolution*, denoted by $\mu_1 * \mu_2$, is defined to be a unique probability distribution μ characterized by

$$\int_{-\infty}^{+\infty} f(z)\mu(dz) = \int_{-\infty}^{+\infty} \int_{-\infty}^{+\infty} f(x+y)\mu_1(dx)\mu_2(dy),$$

where $f(x)$ is a bounded continuous function. If μ_1 and μ_2 admit density functions, so does $\mu = \mu_1 * \mu_2$. In fact, for $\mu_1(dx) = \rho_1(x)dx$ and $\mu_2(dx) = \rho_2(x)dx$ we have $\mu(dx) = \rho(x)dx$ with

$$\rho(x) = \int_{-\infty}^{+\infty} \rho_1(x-y)\rho_2(y)dy = \int_{-\infty}^{+\infty} \rho_1(y)\rho_2(x-y)dy.$$

If μ_1 and μ_2 admit finite moments of all orders, so does $\mu = \mu_1 * \mu_2$. In fact, we have

$$M_m(\mu_1 * \mu_2) = \sum_{k=0}^{m} \binom{m}{k} M_k(\mu_1) M_{m-k}(\mu_2). \qquad (2.138)$$

Recall that the classical convolution appears as the distribution of a sum of independent random variables. Let X and Y be classical random variables with distributions μ_1 and μ_2, and assume that X and Y are independent. Let μ be the distribution of $X + Y$. Then by definition of characteristic function we have

$$\mathbf{E}[e^{it(X+Y)}] = \int_{-\infty}^{+\infty} e^{itz} \mu(dz), \qquad t \in \mathbb{R}. \qquad (2.139)$$

On the other hand, by independence the left-hand side becomes

$$\begin{aligned} \mathbf{E}[e^{it(X+Y)}] &= \mathbf{E}[e^{itX}]\mathbf{E}[e^{itY}] \\ &= \int_{-\infty}^{+\infty} e^{itx} \mu_1(dx) \int_{-\infty}^{+\infty} e^{ity} \mu_2(dy) \\ &= \int_{-\infty}^{+\infty} \int_{-\infty}^{+\infty} e^{it(x+y)} \mu_1(dx) \mu_2(dy). \end{aligned} \qquad (2.140)$$

From (2.139) and (2.140) we see that $\mu = \mu_1 * \mu_2$. The last analytic argument is not directly applicable to algebraic random variables, but the formula (2.138) suggests some possibility.

Now let $(\mathscr{A}, \varphi)$ be an algebraic probability space. Let $a = a^*$ and $b = b^*$ be two real random variables with spectral distributions μ_1 and μ_2, respectively. Consider the sum $a + b$ and its spectral distribution μ. We expect that the spectral distribution μ is obtained from μ_1 and μ_2 when a and b satisfy some independence condition. If so, this procedure gives rise to a binary operation $\mu_1 \# \mu_2$ for two probability distributions μ_1 and μ_2.

The classical convolution naturally appears in connection with commutative independence.

Theorem 2.6.36 *Let $a = a^*$ and $b = b^*$ be commutative independent random variables of an algebraic probability space $(\mathscr{A}, \varphi)$. Let μ_1 and μ_2 be the spectral distributions of a and b, respectively. Then the spectral distribution of $a + b$ is given by the classical convolution $\mu_1 * \mu_2$.*

Proof By expansion we have

$$\varphi((a+b)^m) = \sum_{x_1,\ldots,x_m \in \{a,b\}} \varphi(x_1 \cdots x_m).$$

By commutative independence we see that

$$\varphi(x_1 \cdots x_m) = \varphi(a^k)\varphi(b^{m-k}) = M_k(\mu_1) M_{m-k}(\mu_2),$$

if a appears k times (hence b appears $m-k$ times) in the product $x_1 \cdots x_m$. Then we see from (2.138) that

$$\varphi((a+b)^m) = \sum_{k=0}^m \binom{m}{k} M_k(\mu_1) M_{m-k}(\mu_2) = M_m(\mu_1 * \mu_2),$$

which means that the spectral distribution of $a+b$ is given by $\mu_1 * \mu_2$. $\qquad\square$

Theorem 2.6.37 *Let $G_1 = (V_1, E_1)$ and $G_2 = (V_2, E_2)$ be two graphs, and μ_1 and μ_2 the spectral distributions of the adjacency matrices A_1 and A_2 in the vacuum states at $o_1 \in V_1$, $o_2 \in V_2$, respectively. Let $G = G_1 \times G_2$ be the Cartesian product and A the adjacency matrix. Then the spectral distributions of A in the vacuum states at (o_1, o_2) is given by the classical convolution $\mu_1 * \mu_2$.*

Proof Direct consequence of Theorems 2.6.13 and 2.6.36. $\qquad\square$

Example 2.6.38 Let $d \geq 1$ and $N \geq 1$. The Hamming graph $H(d,N)$ is the d-fold Cartesian power of the complete graph K_N, that is,

$$H(d,N) = K_N \times \cdots \times K_N \qquad \text{(d factors)}.$$

Let A and B be the adjacency matrices of $H(d,N)$ and K_N, respectively. The spectral distribution of B in the vacuum state at an arbitrarily chosen vertex is given by

$$\nu = \frac{1}{N} \delta_{N-1} + \frac{N-1}{N} \delta_{-1},$$

which coincides with the eigenvalue distribution of K_N because it is distance-regular. Then it follows from Theorem 2.6.37 that the spectral distribution of A in the vacuum state at an arbitrarily chosen vertex is given by the d-fold convolution power of ν, that is,

$$\mu = \nu^{*d} = \nu * \cdots * \nu \qquad \text{(d times)}.$$

Using the simple formula

$$\delta_a * \delta_b = \delta_{a+b}, \qquad a, b \in \mathbb{R},$$

and linearity we obtain

$$\mu = \left(\frac{1}{N} \delta_{N-1} + \frac{N-1}{N} \delta_{-1} \right)^{*d}$$
$$= \sum_{k=0}^d \binom{d}{k} \left(\frac{1}{N} \right)^k \left(\frac{N-1}{N} \right)^{d-k} \delta_{kN-d}. \qquad (2.141)$$

In fact, (2.141) coincides with the eigenvalue distribution of $H(d,N)$ because it is distance-regular. Needless to say, the eigenvalues of the adjacency matrix A of $H(d,N)$ can be obtained by an elementary combinatorial argument, see Bannai and Ito [9], Brouwer, Cohen and Neumaier [14, Section 9.2].

Example 2.6.39 Let $G_1 = (V_1, E_1)$ and $G_2 = (V_2, E_2)$ be two graphs, and μ_1 and μ_2 the spectral distributions of the adjacency matrices A_1 and A_2 in the vacuum states at $o_1 \in V_1$, $o_2 \in V_2$, respectively. Let $G = G_1 \times G_2$ be the Cartesian product and A the adjacency matrix. Let μ be the spectral distributions of A in the vacuum states at (o_1, o_2). Recall that the spectral distribution of the adjacency matrix in the vacuum state at a vertex is related to the number of walks from the vertex to itself. That is,

$$M_m(\mu) = \langle e_{(o_1,o_2)}, A^m e_{(o_1,o_2)} \rangle = |\{m\text{-step walks from } (o_1, o_2) \text{ to itself}\}|.$$

Then Theorem 2.6.37 gives rise to a formula of counting walks. Namely, we have

$$\langle e_{(o_1,o_2)}, A^m e_{(o_1,o_2)} \rangle = \sum_{k=0}^{m} \binom{m}{k} \langle e_{o_1}, A_1^m e_{o_1} \rangle \langle e_{o_2}, A_2^m e_{o_2} \rangle. \tag{2.142}$$

For monotone independence the following result is fundamental. For the proof we refer to Muraki [52, 53] and Hasebe [28].

Lemma 2.6.40 Let $a_1 = a_1^*$ and $a_2 = a_2^*$ be real random variables in an algebraic probability space $(\mathcal{A}, \varphi)$ with spectral distributions μ_1 and μ_2, respectively. If a_1 and a_2 are monotone independent with respect to the usual order $1 < 2$, the spectral distribution of $a_1 + a_2$ is a probability distribution μ specified by

$$H_\mu(z) = H_{\mu_1}(H_{\mu_2}(z)), \qquad \operatorname{Im} z > 0, \tag{2.143}$$

where $H_\mu(z)$ is the reciprocal Stieltjes transform defined by $H_\mu(z) = 1/G_\mu(z)$.

Given two probability distributions μ_1 and μ_2, the probability distribution μ determined by (2.143) is called the *monotone convolution* of μ_1 and μ_2 and denoted by $\mu_1 \triangleright \mu_2$. Note that the monotone convolution is not commutative.

Theorem 2.6.41 Let $G_1 = (V_1, E_1)$ and $G_2 = (V_2, E_2)$ be two graphs with adjacency matrices A_1 and A_2, respectively. Let μ_1 and μ_2 be the spectral distributions of A_1 and A_2 in the vacuum state at $o_1 \in V_1$ and $o_2 \in V_2$, respectively. Then the spectral distribution of the adjacency matrix of the comb product $G_1 \triangleright_{o_2} G_2$ in the vacuum state at (o_1, o_2) is given by the monotone convolution $\mu_1 \triangleright \mu_2$.

Proof Direct consequence from Theorem 2.6.20 and the definition of monotone convolution. $\square$

Example 2.6.42 Let $G_1 = K_3$ and $G_2 = K_2$. Choose arbitrary vertices o_1 of G_1 and o_2 of G_2, and consider the comb product $G_1 \triangleright G_2 = G_1 \triangleright_{o_2} G_2$, see Figure 2.21 where (o_1, o_2) is indicated by double circle. The spectral distribution of K_3 in the vacuum state at a vertex coincides with the eigenvalue distribution and is given by

$$\mu_1 = \frac{2}{3}\delta_{-1} + \frac{1}{3}\delta_2 .$$

Similarly, the spectral distribution of K_2 in the vacuum state at a vertex coincides with the eigenvalue distribution and is given by

$$\mu_2 = \frac{1}{2}\delta_{-1} + \frac{1}{2}\delta_1 .$$

The Stieltjes transforms of μ_1 and μ_2 are given respectively by

$$G_1(z) = \frac{2/3}{z+1} + \frac{1/3}{z-2} = \frac{z-1}{z^2-z-2},$$

$$G_2(z) = \frac{1/2}{z+1} + \frac{1/2}{z-1} = \frac{z}{z^2-1},$$

and hence

$$H_1(z) = \frac{1}{G_1(z)} = \frac{z^2-z-2}{z-1},$$

$$H_2(z) = \frac{1}{G_2(z)} = \frac{z^2-1}{z}.$$

For the computation of monotone convolution we need to compute $H_1(H_2(z))$ and its reciprocal:

$$G(z) = \frac{1}{H_1(H_2(z))} = \frac{z^3 - z^2 - z}{(z^2+z-1)(z^2-2z-1)}$$

$$= \frac{2z/3}{z^2+z-1} + \frac{z/2}{z^2-2z-1}$$

$$= \frac{p_1}{z+(1+\sqrt{5})/2} + \frac{p_2}{z+(1-\sqrt{5})/2} + \frac{p_3}{z-(1+\sqrt{2})} + \frac{p_4}{z-(1-\sqrt{2})},$$

where

$$p_1 = \frac{5+\sqrt{5}}{15}, \quad p_2 = \frac{5-\sqrt{5}}{15}, \quad p_3 = \frac{2+\sqrt{2}}{12}, \quad p_4 = \frac{2-\sqrt{2}}{12}.$$

Thus, the spectral distribution of $K_3 \triangleright K_2$ in the vacuum state at (o_1, o_2) is given by

$$\mu = \mu_1 \triangleright \mu_2 = p_1\delta_{-(1+\sqrt{5})/2} + p_2\delta_{-(1-\sqrt{5})/2} + p_3\delta_{1+\sqrt{2}} + p_4\delta_{1-\sqrt{2}}.$$

As an application, the mth moment:

$$M_m(\mu) = p_1\left(-\frac{1+\sqrt{5}}{2}\right)^m + p_2\left(\frac{1-\sqrt{5}}{2}\right)^m + p_3(1+\sqrt{2})^m + p_4(1-\sqrt{2})^m$$

coincides with the number of m-step walks from (o_1, o_2) to itself.

Figure 2.21   The comb product $K_3 \triangleright K_2$

We next discuss the Boolean independence. In general, for $\mu \in \mathfrak{P}_{\mathrm{fm}}(\mathbb{R})$ the *moment generating function* is defined by

$$M(z; \mu) = \sum_{m=0}^{\infty} M_m(\mu)z^m = 1 + \sum_{m=1}^{\infty} M_m(\mu)z^m,$$

where $M_m(\mu)$ is the mth moment of μ. We first mention the fundamental result without proof.

Lemma 2.6.43   *For $\mu_1, \mu_2 \in \mathfrak{P}_{\mathrm{fm}}(\mathbb{R})$ there exists $\mu \in \mathfrak{P}_{\mathrm{fm}}(\mathbb{R})$ such that*

$$M(z; \mu) = \frac{M(z; \mu_1)M(z; \mu_2)}{M(z; \mu_1) + M(z; \mu_2) - M(z; \mu_1)M(z; \mu_2)}. \qquad (2.144)$$

If μ in Lemma 2.6.43 is the solution to a determinate moment problem, it is called the *Boolean convolution* of μ_1 and μ_2, and is denoted by $\mu_1 \uplus \mu_2$.

For a real random variable $a = a^*$ in an algebraic probability space $(\mathscr{A}, \varphi)$ the moment generating function is defined by

$$M(z; a) = \sum_{m=0}^{\infty} \varphi(a^m)z^m.$$

Lemma 2.6.44   *Let $a = a^*$ and $b = b^*$ be two random variables in an algebraic probability space $(\mathscr{A}, \varphi)$ and, $M(z; a)$ and $M(z; b)$ the moment generating functions of a and b, respectively. If a and b are Boolean independent, then* *we have*

$$M(z;a+b) = \frac{M(z;a)M(z;b)}{M(z;a)+M(z;b)-M(z;a)M(z;b)}.$$

In other words, the spectral distribution of $a+b$ is given by the Boolean convolution of the spectral distributions of a and b.

Theorem 2.6.45 *Let $G_1 = (V_1,E_1)$ and $G_2 = (V_2,E_2)$ be two graphs with adjacency matrices A_1 and A_2, respectively. Let μ_1 and μ_2 be the spectral distributions of A_1 and A_2 in the vacuum state at $o_1 \in V_1$ and $o_2 \in V_2$, respectively. Then the spectral distribution of the adjacency matrix of the star product $G_1\,{}_{o_1}\!\star_{o_2} G_2$ in the vacuum state at (o_1,o_2) is given by the Boolean convolution $\mu_1 \uplus \mu_2$.*

Proof Immediate consequence from Theorem 2.6.30 and the definition of Boolean convolution. □

It is interesting and instructive to apply the Boolean convolution to counting walks. Let $G_1 = (V_1,E_1)$ and $G_2 = (V_2,E_2)$ be two graphs with distinguished vertices $o_1 \in V_1$ and $o_2 \in V_2$, respectively. Let μ_1 be the spectral distribution of the adjacency matrix A_1 of G_1 in the vacuum state at o_1. Let μ_2 be the similar one for G_2. Let $G = G_1\,{}_{o_1}\!\star_{o_2} G_2$ be the star product and A the adjacency matrix. Let μ be the spectral distribution of A in the vacuum state at (o_1,o_2).

Recall that the moment $M_m(\mu)$ coincides with the number of m-step walks in G from (o_1,o_2) to itself. Along with the idea of the first return of a Markov chain, we will derive an important functional relation for the moment generating function:

$$M(z;\mu) = \sum_{m=0}^{\infty} M_m(\mu)z^m. \tag{2.145}$$

For $m \geq 1$ let $F_m(G,(o_1,o_2))$ denote the number of m-step walks from (o_1,o_2) to itself which do not come back to (o_1,o_2) before the mth step. We define

$$F(z;G,(o_1,o_2)) = \sum_{m=1}^{\infty} F_m(G,(o_1,o_2))z^m. \tag{2.146}$$

For now, we write $F_m = F_m(G,(o_1,o_2))$ and $F(z) = F(z;G,(o_1,o_2))$ for brevity. We start with

$$M_m(\mu) = \sum_{k=1}^{m} F_k M_{m-k}(\mu), \qquad m = 1,2,\dots, \tag{2.147}$$

which is easily verified by definition. Then multiplying both sides of (2.147) by z^m and summing up, we obtain

$$
\begin{aligned}
M(z;\mu) - 1 &= \sum_{m=1}^{\infty} M_m(\mu) z^m = \sum_{m=1}^{\infty} \sum_{k=1}^{m} F_k M_{m-k}(\mu) z^m \\
&= \sum_{k=1}^{\infty} \sum_{m=k}^{\infty} F_k M_{m-k}(\mu) z^m = \sum_{k=1}^{\infty} \sum_{m=0}^{\infty} F_k z^k M_m(\mu) z^m \\
&= F(z) M(z;\mu).
\end{aligned}
$$

We thus come to important functional relations:

$$
F(z;G,(o_1,o_2)) = 1 - \frac{1}{M(z;\mu)}, \qquad M(z;\mu) = \frac{1}{1 - F(z;G,(o_1,o_2))}. \tag{2.148}
$$

We now take into account that $G = G_1 {}_{o_1}\!\star_{o_2} G_2$ is the star product. Let $F_m(G_1,o_1)$ be the number of m-step walks from o_1 to itself in G_1 which do not come back to o_1 before the mth step. Let $F_m(G_2,o_2)$ be the similar number for G_2. An m-step walk from (o_1,o_2) to itself in G which does not come back to (o_1,o_2) before the mth step must stay in either V_1 or V_2. Therefore, we have the additive formula:

$$
F_m(G,(o_1,o_2)) = F_m(G_1,o_1) + F_m(G_2,o_2), \qquad m = 1,2,\ldots,
$$

and hence

$$
F(z;G,(o_1,o_2)) = F(z;G_1,o_1) + F(z;G_2,o_2), \tag{2.149}
$$

where

$$
F(z;G_1,o_1) = \sum_{m=1}^{\infty} F_m(G_1,o_1) z^m, \qquad F(z;G_2,o_2) = \sum_{m=1}^{\infty} F_m(G_2,o_2) z^m.
$$

Applying (2.148) to (2.149), we obtain

$$
1 - \frac{1}{M(z;\mu)} = \left\{ 1 - \frac{1}{M(z;\mu_1)} \right\} + \left\{ 1 - \frac{1}{M(z;\mu_2)} \right\},
$$

that is,

$$
M(z;\mu) = \frac{M(z;\mu_1)M(z;\mu_2)}{M(z;\mu_1) + M(z;\mu_2) - M(z;\mu_1)M(z;\mu_2)}.
$$

Thus (2.144) is derived.

Remark 2.6.46 For a general $\mu \in \mathfrak{P}_{\mathrm{fm}}(\mathbb{R})$ a sequence $\{F_k\}$ is defined by (2.147). We call $F_k = F_k(\mu)$ the kth *Boolean cumulant* of μ. It is readily shown that

$$
F_k(\mu_1 \uplus \mu_2) = F_k(\mu_1) + F_k(\mu_2), \qquad \mu_1, \mu_2 \in \mathfrak{P}_{\mathrm{fm}}(\mathbb{R}).
$$

Finally, we briefly mention the Kronecker product. As is mentioned in (2.122) the Kronecker product of two graphs $G_1 = (V_1, E_1)$ and $G_2 = (V_2, E_2)$, denoted by $G_1 \times_K G_2$, is defined through their adjacency matrices:

$$A = A[G_1 \times_K G_2] = A_1 \otimes A_2.$$

If $G_1 = K_1$, that is $|V_1| = 1$, then the Kronecker product $K_1 \times_K G_2$ becomes a graph on V_2 with no edges (the empty graph on V_2). If G_1 and G_2 are connected with $|V_1| \geq 2$ and $|V_2| \geq 2$, then $G_1 \times_K G_2$ has at most two connected components.

Given distinguished vertices $o_1 \in V_1$ and $o_2 \in V_2$, let μ_1 and μ_2 be the vacuum spectral distributions of A_1 at o_1 and of A_2 at o_2, respectively. Moreover, let μ be the vacuum spectral distribution of A at (o_1, o_2). Then we have

$$
\begin{aligned}
M_m(\mu) &= \langle e_{(o_1, o_2)}, A^m e_{(o_1, o_2)} \rangle \\
&= \langle e_{o_1} \otimes e_{o_2}, (A_1 \otimes A_2)^m e_{o_1} \otimes e_{o_2} \rangle \\
&= \langle e_{o_1}, A_1^m e_{o_1} \rangle \langle e_{o_2}, A_2^m e_{o_2} \rangle \\
&= M_m(\mu_1) M_m(\mu_2).
\end{aligned}
$$

Hence, in integral forms we have

$$
\begin{aligned}
\int_{-\infty}^{+\infty} x^m \mu(dx) &= \int_{-\infty}^{+\infty} x_1^m \mu_1(dx_1) \int_{-\infty}^{+\infty} x_2^m \mu_2(dx_2) \\
&= \int_{-\infty}^{+\infty} \int_{-\infty}^{+\infty} (x_1 x_2)^m \mu_1(dx_1) \mu_2(dx_2).
\end{aligned}
$$

In general, a probability distribution μ determined by the above relation is called the *Mellin convolution* and is denoted by $\mu = \mu_1 *_M \mu_2$. Thus, we come to the following result.

Theorem 2.6.47 *Let $G_1 = (V_1, E_1)$ and $G_2 = (V_2, E_2)$ be two graphs with adjacency matrices A_1 and A_2, respectively. Let μ_1 and μ_2 be the spectral distributions of A_1 and A_2 in the vacuum state at $o_1 \in V_1$ and $o_2 \in V_2$, respectively. Then the spectral distribution of the adjacency matrix of the Kronecker product $G_1 \times_K G_2$ in the vacuum state at (o_1, o_2) is given by the Mellin convolution $\mu_1 *_M \mu_2$.*

Example 2.6.48 As is verified from Figure 2.22, we see that

$$K_4 \times_K K_2 \cong K_2 \times K_2 \times K_2 = H(3, 2).$$

It follows from Theorem 2.6.47 that the vacuum spectral distribution of the graph $K_4 \times_K K_2$ at an arbitrarily fixed (o_1, o_2) is given by

$$\left(\frac{3}{4} \delta_{-1} + \frac{1}{4} \delta_3 \right) *_M \left(\frac{1}{2} \delta_{-1} + \frac{1}{2} \delta_1 \right).$$

By using the formula:

$$\delta_a *_M \delta_b = \delta_{ab}, \qquad a,b \in \mathbb{R},$$

which is verified easily by definition, and the linearity we obtain

$$\left(\frac{3}{4}\delta_{-1} + \frac{1}{4}\delta_3\right) *_M \left(\frac{1}{2}\delta_{-1} + \frac{1}{2}\delta_1\right) = \frac{1}{8}\delta_{-3} + \frac{3}{8}\delta_{-1} + \frac{3}{8}\delta_1 + \frac{1}{8}\delta_3.$$

Of course, this coincides with the spectral distribution of the Hamming graph $H(3,2)$, see Example 2.6.38. On the other hand, from $K_2 \times K_2 \times K_2 = H(3,2)$ we see that

$$\left(\frac{1}{2}\delta_{-1} + \frac{1}{2}\delta_1\right)^{*3} = \frac{1}{8}\delta_{-3} + \frac{3}{8}\delta_{-1} + \frac{3}{8}\delta_1 + \frac{1}{8}\delta_3.$$

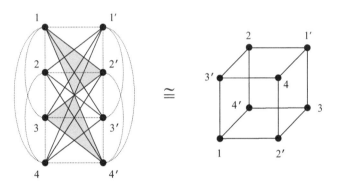

Figure 2.22 The Kronecker product $K_4 \times_K K_2$

Remark 2.6.49 Let $f_1(x)$ and $f_2(x)$ be density functions and consider $\mu_1(dx) = f_1(x)dx$ and $\mu_2(dx) = f_2(x)dx$. If $f_1(-x) = f_1(x)$ and $f_2(-x) = f_2(x)$, then $\mu_1 *_M \mu_2$ admits a symmetric density function $2f_1 \star f_2(x)$ defined by

$$f_1 \star f_2(x) = \int_0^\infty f_1(y)f_2\left(\frac{x}{y}\right)\frac{dy}{y} = \int_0^\infty f_1\left(\frac{x}{y}\right)f_2(y)\frac{dy}{y}, \quad x > 0. \quad (2.150)$$

In fact, (2.150) is the standard convolution of the multiplicative group $\mathbb{R}_{>0}$.

For further applications of the Kronecker product, in particular, to counting walks in restricted lattices, see [59].

2.6.7 Notes

Quite a few concepts of independence have been introduced in quantum probability: free independence by Voiculescu [73], monotone independence by Lu [49] and Muraki [52], and Boolean independence by Speicher and Woroudi [66] also by Bożejko [11] implicitly, and others. For attempts to unify and axiomatize concepts of independence, see Lenczewski [46] and Muraki [54, 55], among others.

In connection with spectral analysis of graphs, Accardi, Ben Ghorbal and Obata [2] first observed that the adjacency matrix of a comb product graph is decomposed into a sum of monotone independent random variables. Similarly, it was found by Obata [57] that Boolean independence appears in a star product graph.

Although very interesting, the free product graph is out of our scope because the vertex set is not the direct product but the so-called free product. For example, the homogeneous tree T_κ is the free product of κ copies of K_2 and hence the adjacency matrix is decomposed into a sum of free independent random variables. For recent topics in this line see, for example, Vargas and Kulkarni [70] and references cited therein.

References

[1] L. Accardi, *Topics in quantum probability. New stochastic methods in physics*, Phys. Rep. **77** (1981), 169–192.

[2] L. Accardi, A. Ben Ghorbal and N. Obata, Monotone independence, comb graphs and Bose-Einstein condensation, *Infin. Dimens. Anal. Quantum Probab. Relat. Top.* **7** (2004), 419–435.

[3] L. Accardi and M. Bożejko, Interacting Fock spaces and Gaussianization of probability measures, *Infin. Dimens. Anal. Quantum Probab. Relat. Top.* **1** (1998), 663–670.

[4] L. Accardi, H. Rebei and A. Riahi, The quantum decomposition of random variables without moments, *Infin. Dimens. Anal. Quantum Probab. Relat. Top.* **16** (2013), 1350012.

[5] L. Accardi, A. Riahi and H. Rebei, The quantum decomposition associated with the Lévy white noise processes without moments, *Probab. Math. Stat.* **34** (2014), 337–362.

[6] L. Accardi, Y. G. Lu and I. Volovich, *Quantum Theory and its Stochastic Limit*, Springer-Verlag, Berlin, 2002.

[7] N. I. Akhiezer, *The Classical Moment Problem and Some Related Questions in Analysis*, Hafner Publishing Co., New York, 1965.

[8] O. Arizmendi and T. Gaxiola, On the spectral distribution of distance-k graph of free product graphs, *Infin. Dimens. Anal. Quantum Probab. Relat. Top.* **19** (2016), 1650017.

[9] E. Bannai and T. Ito: *Algebraic Combinatorics. I. Association Schemes*, The Benjamin Cummings Publ. Co., Menlo Park, CA, 1984.

[10] N. Biggs, *Algebraic Graph Theory* (2nd Ed.), Cambridge University Press, Cambridge, 1993.

[11] M. Bożejko, Positive definite functions on the free group and the noncommutative Riesz product, *Boll. Un. Mat. Ital.* A (6) **5** (1986), 13–21.

[12] M. Bożejko, B. Kümmerer and R. Speicher, q-Gaussian processes: Noncommutative and classical aspects, *Commun. Math. Phys.* **185** (1997), 129–154.

[13] M. Bożejko and R. Speicher, An example of a generalized Brownian motion, *Commun. Math. Phys.* **137** (1991), 519–531.

[14] A. E. Brouwer, A. M. Cohen and A. Neumaier, *Distance-Regular Graphs*, Springer-Verlag, Berlin, 1989.

[15] A. E. Brouwer and W. H. Haemers, *Spectra of Graphs*, Springer, Berlin, 2012.

[16] T. S. Chihara, *An Introduction to Orthogonal Polynomials*, Gordon and Breach Science Publishers, New York, 1978.

[17] K. L. Chung, *A Course in Probability Theory* (3rd Ed.), Academic Press, San Diego, 2001.

[18] F. Chung and L. Lu, *Complex Graphs and Networks*, Amer. Math. Soc., Providence, 2006.

[19] D. Cvetković, P. Rowlinson and S. Simić, *Eigenspaces of Graphs*, Cambridge University Press, Cambridge, 1997.

[20] D. Cvetković, P. Rowlinson and S. Simić, *An Introduction to the Theory of Graph Spectra*, Cambridge University Press, Cambridge, 2010.

[21] A. Dhahri, N. Obata and H. J. Yoo, Multivariate orthogonal polynomials: quantum decomposition, deficiency rank and support of measure, *J. Math. Anal. Appl.* **485** (2020), 123775.

[22] R. Durrett, *Probability; Theory and Examples* (4th Ed.), Cambridge University Press, Cambridge, 2010.

[23] R. Durrett, *Random Graph Dynamics*, Cambridge University Press, Cambridge, 2010.

[24] C. D. Godsil and B. D. McKay, A new graph product and its spectrum, *Bull. Austral. Math. Soc.* **18** (1978), 21–28.

[25] S. P. Gudder, Quantum probability spaces, *Proc. Amer. Math. Soc.* **21** (1969), 296–302.

[26] S. P. Gudder, *Quantum Probability*, Academic Press, Boston, 1988.

[27] R. Hammack, W. Imrich and S. Klavžar, *Handbook of Product Graphs* (2nd Ed.), CRC Press, Boca Raton, 2011.

[28] T. Hasebe, Monotone convolution and monotone infinite divisibility from complex analytic viewpoint, *Infin. Dimens. Anal. Quantum Probab. Relat. Top.* **13** (2010), 111–131.

[29] T. Hasegawa, H. Saigo, S. Saito and S. Sugiyama, A quantum probabilistic approach to Hecke algebras for $\mathfrak{p}$-adic PGL_2, *Infin. Dimens. Anal. Quantum Probab. Relat. Top.* **21** (2018), 1850015.

[30] Y. Hashimoto, Quantum decomposition in discrete groups and interacting Fock spaces, *Infin. Dimens. Anal. Quantum Probab. Relat. Top.* **4** (2001), 277–287.

[31] Y. Hashimoto, N. Obata and N. Tabei, A quantum aspect of asymptotic spectral analysis of large Hamming graphs, in *Quantum Information III* (ed. T. Hida and K. Saitô), pp. 45–57, World Scientific Publ., River Edge, 2001.

[32] Y. Hashimoto, A. Hora and N. Obata, Central limit theorems for large graphs: method of quantum decomposition, *J. Math. Phys.* **44** (2003), 71–88.

[33] F. Hiai and D. Petz, *The Semicircle Law, Free Random Variables and Entropy*, Amer. Math. Soc., Providence, 2000.

[34] A. Hora, Central limit theorems and asymptotic spectral analysis on large graphs, *Infin. Dimens. Anal. Quantum Probab. Relat. Top.* **1** (1998), 221–246.

[35] A. Hora, Scaling limit for Gibbs states for Johnson graphs and resulting Meixner classes, *Infin. Dimens. Anal. Quantum Probab. Relat. Top.* **6** (2003), 139–143.

[36] A. Hora and N. Obata, *Quantum Probability and Spectral Analysis of Graphs*, Springer, 2007.

[37] A. Hora and N. Obata, Asymptotic spectral analysis of growing regular graphs, *Trans. Amer. Math. Soc.* **360** (2008), 899–923.

[38] D. Igarashi and N. Obata, Asymptotic spectral analysis of growing graphs: Odd graphs and spidernets, Banach Center Publications **73** (2006), 245–265.

[39] S. Jafarizadeh, R. Sufiani and M. A. Jafarizadeh, Evaluation of effective resistances in pseudo-distance-regular resistor networks, *J. Stat. Phys.* **139** (2010), 177–199.

[40] Y. Kang, Quantum decomposition of random walk on Cayley graph of finite group, *Physica A* **458** (2016), 146–156.

[41] S. Karlin and J. McGregor, Random walks, *Illinois J. Math.* **3** (1959), 66–81.

[42] H. Kesten, Symmetric random walks on groups, *Trans. Amer. Math. Soc.* **92** (1959), 336–354.

[43] A. N. Kolmogorov, *Grundbegriffe der Wahrscheinlichkeitsrechnung*, Julius Springer, Berlin, 1933.

[44] N. Konno, N. Obata and E. Segawa, Localization of the Grover walks on spidernets and free Meixner laws, *Commun. Math. Phys.* **322** (2013), 667–695.

[45] M. Koohestani, N. Obata and H. Tanaka, Scaling limits for the Gibbs states on distance-regular graphs with classical parameters, *SIGMA* **17** (2021), 104, 22 pages.

[46] R. Lenczewski, Unification of independence in quantum probability, *Infin. Dimens. Anal. Quantum Probab. Relat. Top.* **1** (1998), 383–405.

[47] M. T. G. Leyva, Analysis of growing graphs and quantum probability, Thesis, Centro de Investigación en Matemáticas, Mexico, 2017.

[48] L. Lovász, *Large Networks and Graph Limits*, Amer. Math. Soc., Providence, RI, 2012.

[49] Y. G. Lu, An interacting free Fock space and the arcsine law, *Probab. Math. Stat.* **17** (1997), 149–166.

[50] P.-A. Meyer, *Quantum Probability for Probabilists*, Lect. Notes in Math. Vol. 1538, Springer, 1993.

[51] J. V. S. Morales, N. Obata and H. Tanaka, Asymptotic joint spectra of Cartesian powers of strongly regular graphs and bivariate Charlier-Hermite polynomials, *Colloq. Math.* **162** (2020), 1–22.

[52] N. Muraki, Noncommutative Brownian motion in monotone Fock space, *Commun. Math. Phys.* **183** (1997), 557–570.

[53] N. Muraki, Monotonic independence, monotonic central limit theorem and monotonic law of small numbers, *Infin. Dimens. Anal. Quantum Probab. Relat. Top.* **4** (2001), 39–58.

[54] N. Muraki, The five independences as natural products, *Infin. Dimens. Anal. Quantum Probab. Relat. Top.* **6** (2003), 337–371.

[55] N. Muraki, A simple proof of the classification theorem for positive natural products, *Probab. Math. Statist.* **33** (2013), 315–326.

[56] A. Nica and R. Speicher, *Lectures on the Combinatorics of Free Probability*, Cambridge University Press, Cambridge, 2006.

[57] N. Obata, Quantum probabilistic approach to spectral analysis of star graphs, *Interdiscip. Inform. Sci.* **10** (2004), 41–52.

[58] N. Obata, Quantum probability aspects to lexicographic and strong products of graphs, *Interdiscip. Inform. Sci.* **22** (2016), 143–146.

[59] N. Obata, *Spectral Analysis of Growing Graphs. A Quantum Probability Point of View*, Springer, 2017.

[60] K. R. Parthasarathy, *An Introduction to Quantum Stochastic Calculus*, Birkhäuser Verlag, Basel, 1992.

[61] H. Saigo, A simple proof for monotone CLT, *Infin. Dimens. Anal. Quantum Probab. Relat. Top.* **13** (2010), 339–343.

[62] K. Schmüdgen, *The Moment Problem*, Springer, 2017.

[63] R. Schott and G. S. Staples, Connected components and evolution of random graphs: an algebraic approach, *J. Alg. Comb.* **35** (2012), 141–156.

[64] J. A. Shohat and J. D. Tamarkin, *The Problem of Moments*, Amer. Math. Soc., New York, 1943.

[65] R. Speicher, *Free Probability Theory*, Oxford University Press, Oxford, 2011.

[66] R. Speicher and R. Woroudi, Boolean convolution, in *Free Probability Theory* (ed. D. Voiculescu), pp. 267–279, Amer. Math. Soc., Providence, 1997.

[67] P. Suppes, The probabilistic argument for a non-classical logic of quantum mechanics, *Philosophy of Science* **33** (1966), 14–21.

[68] M. Takesaki, *Theory of Operator Algebra I*, Springer, Berlin, 2002.

[69] H. van Leeuwen and H. Maassen, A q deformation of the Gauss distribution, *J. Math. Phys.* **36** (1995), 4743–4756.

[70] J. G. Vargas and A. Kulkarni, Spectra of infinite graphs via freeness with amalgamation, *Canad. J. Math.* **75** (2023), 1633–1684.

[71] A. M. Vershik, Asymptotic combinatorics and algebraic analysis, Proc. ICM, Zürich (ed. S. D. Chatterji), Vol. 2, pp. 1384–1394, Birkhäuser, Basel, 1995.

[72] A. M. Vershik, Between "very large" and "infinite": The asymptotic representation theory, *Probab. Math. Stat.* **33** (2013), 467–476.

[73] D. Voiculescu, Symmetries of some reduced free product C^*-algebras, in *Operator Algebras and Their Connections with Topology and Ergodic Theory* (ed. H. Araki, C. C. Moore, Ş. Stratila, D. Voiculescu and Gr. Arsene), pp. 556–588, Lect. Notes in Math., Vol. 1132, Springer, Berlin, 1985.

[74] D. V. Voiculescu, K. J. Dykema and A. Nica, *Free Random Variables*, Amer. Math. Soc., Providence, 1992.

[75] J. von Neumann, *Mathematische Grundlagen der Quantenmechanik*, Springer, Berlin, 1932.

[76] H. S. Wall, *Analytic Theory of Continued Fractions*, D. Van Nostrand Company, New York, 1948.

3

Laplacian eigenvalues and optimality

R. A. Bailey and Peter J. Cameron

Abstract

Eigenvalues of the Laplacian matrix of a graph have been widely used in studying connectivity and expansion properties of networks, and also in analysing random walks on a graph.

Independently, statisticians introduced various optimality criteria in experimental design, the goal being to obtain more accurate estimates of quantities of interest in an experiment. It turns out that the most popular of these optimality criteria for block designs are determined by the Laplacian eigenvalues of the concurrence graph, or of the Levi graph, of the design.

The most important optimality criteria, called A (average), D (determinant) and E (extreme), are related to the conductance of the graph as an electrical network, the number of spanning trees, and the isoperimetric properties of the graphs, respectively.

The number of spanning trees is also an evaluation of the Tutte polynomial of the graph, and is the subject of the Merino–Welsh conjecture relating it to acyclic and totally cyclic orientations, of interest in their own right.

This chapter ties these ideas together, building on the work in [4, 5].

Keywords: block design, optimality, resistance distance, spanning trees

Mathematics Subject Classification: 05B05, 05C25, 05C50, 20B25, 20B30, 62K05, 62K10

3.1 Block designs in experiments

Pure mathematicians and statisticians often use very different language to describe block designs. A mathematician typically begins with a set of *points*, and each *block* is regarded as a subset of this. On the other hand, a statistician usually begins with a set of *treatments* which are going to be compared in an experiment. Each treatment is allocated to several *experimental units*, which might be plots in an agricultural field, or sessions in a laboratory, or people-month combinations. If the experimental units are not all alike, it is a good idea to group them into homogenous *blocks*: then the design question becomes "How should we allocate the treatments to blocks?"

R. A. Fisher and Frank Yates did pioneering work on this at Rothamsted Experimental Station in the UK in the 1920s, 1930s and 1940s. Many of their methods used finite abelian groups, even if they did not say so explicitly. In the 1930s, R. C. Bose, who was working at the Indian Statistical Institute in Calcutta, introduced some further methods based on finite fields. (This means "fields" in the algebraic sense, not the agricultural one.)

The following quotation is taken from [10], in a special issue of the journal *Sankhyā* devoted to Bose's memory.

There is a very famous joke about Bose's work in Giridh. Professor Mahalanobis wanted Bose to visit the paddy fields and advise him on sampling problems for the estimation of yield of paddy. Bose did not very much like the idea, and he used to spend most of the time at home working on combinatorial problems using Galois fields. The workers of the ISI used to make a joke about this. Whenever Professor Mahalanobis asked about Bose, his secretary would say that Bose is working in fields, which kept the Professor happy.

This section gives a brief introduction to block designs as used in experiments. It includes the notation and statistical models usually assumed by statisticians, and summarizes these in matrix form. In the rest of this chapter, we will use the terms "point" and "treatment" interchangeably.

3.1.1 Experiments in blocks

Here are some examples of experiments in practice. They are all taken from [2].

Example 3.1.1 We have six varieties of cabbage to compare in the field with 24 plots shown in Figure 3.1. There is a patch of stony ground, which may affect the growth of the cabbages. There are also some trees at one end of the field: in some experiments here in previous years the crows nesting in the trees damaged the crops closest to them.

How do we avoid bias in this experiment?

The solution is to partition the experimental units into homogeneous blocks. One block consists of the six plots on stony ground. A second block can then

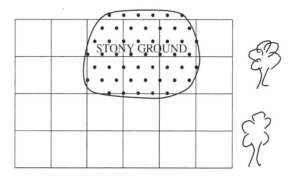

Figure 3.1 Experimental field in Example 3.1.1

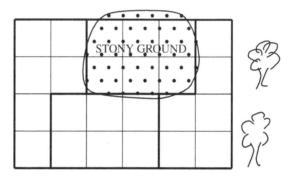

Figure 3.2 Blocking in the experimental field in Example 3.1.1

consist of the remaining six plots closest to the trees. The remaining plots can be divided into two blocks of six nearby plots in an obvious way. The result is shown in Figure 3.2. Then each variety can be planted on one plot in each block.

Example 3.1.2 This experiment on people was reported in the newspaper *The Weekend Australian* on 28 August 2004.

Several studies have suggested that drinking red wine gives some protection against heart disease, but it is not known whether the effect is caused by the alcohol or by some other ingredient of red wine. To investigate this, medical scientists enrolled 40 volunteers into a trial lasting 28 days.

For the first 14 days, half the volunteers drank two glasses of red wine per day, while the other half had two standard drinks of gin. For the remaining 14 days the drinks were reversed: those who had been drinking red wine changed to gin, while those who had been drinking gin changed to red wine.

On days 14 and 28, the scientists took a blood sample from each volunteer and measured the amount of inflammatory substance in the blood.

In this experiment, each experimental unit consists of one volunteer for 14 days. So there are 80 experimental units. Each volunteer forms a block of size two. The treatments are the two types of drink.

Example 3.1.3 This experiment on diffusion of proteins was conducted by a post-doctoral researcher in Biology.

She added from zero to four extra green fluorescent proteins to cells of *Escherichia coli*, adding zero to each of ten cells, adding one to each of ten further cells, and so on. Then she measured the rate of diffusion of proteins in each of the 50 cells.

She conducted the whole experiment in a single week, by dealing with ten cells each day. Figure 3.3 shows what she did.

Day	Monday	Tuesday	Wednesday	Thursday	Friday
Added cells	0000000000	1111111111	2222222222	3333333333	4444444444

Figure 3.3 An experiment on diffusion of proteins

After the experiment, the post-doc took her data to a statistician, who had to point out (gently) that the experiment had been badly designed. Any perceived differences in the rate of diffusion might indeed have been caused by different numbers of added cells. On the other hand, the post-doc might simply have got better at preparing the samples as the week wore on. There is no way of distinguishing these two causes. Likewise, there might have been environmental changes in the lab that could have contributed to the differences.

The statistician said that it would have been better to regard each day as a block, and use each added number twice in each day, as shown in Figure 3.4.

Day	Monday	Tuesday	Wednesday	Thursday	Friday
Added cells	0011223344	0011223344	0011223344	0011223344	0011223344

Figure 3.4 Suggested blocking in the experiment on diffusion of proteins

There may still be systematic differences within each day, so the statistician suggested a further improvement: randomize the order within each day. One possible outcome is shown in Figure 3.5.

Day	Monday	Tuesday	Wednesday	Thursday	Friday
Added cells	1040223134	2230110443	1421324030	4420013312	3204320411

Figure 3.5 Randomized version of the block design in Figure 3.4

Example 3.1.4   In a consumer experiment, twelve housewives volunteer to test new detergents in their own washing machines. (This was 40 years ago, when most homemakers in the UK were female.) There are 16 new detergents to compare, but it is not realistic to ask any one volunteer to compare this many detergents.

Each housewife tests one detergent per washload for each of four washloads, and assesses the cleanliness of each washload.

In this experiment, the experimental units are the washloads. The house-wives form 12 blocks of size four. The treatments are the 16 new detergents.

Notation 3.1.5   In general, we shall use the following notation. There are v treatments that we want to compare. There are b blocks, with k plots in each block. Here we write "plot" as short for "experimental unit", even if the experiment does not take place in a field.

The preceding examples are summarized in Table 3.1.

Example	Blocks	b	k	Treatments	v
3.1.1	contiguous plots	4	6	cabbage varieties	6
3.1.2	volunteers	40	2	drinks	2
3.1.3	days	5	10	numbers of cells	5
3.1.4	housewives	12	4	detergents	16

Table 3.1 *Summary of preceding examples*

Given the numbers v, b and k, the statistician has to ask herself the following questions.

- How should I choose a block design?
- How should I randomize it?
- How should I analyse the data after the experiment?

These questions are all interlinked, and they suggest the following important question:

- What makes a block design good?

3.1.2 Complete-block designs

We begin with so-called *complete-block designs*, as a gentle introduction to the statistical ideas involved.

Definition 3.1.6 (Complete-block design)   In a complete-block design, there are v treatments, and b blocks of size v (so that $k = v$).

Construction and randomization

The construction is very simple: each treatment occurs on one plot per block. The designs in Examples 3.1.1 and 3.1.2 are complete-block designs.

   The purpose of blocking is to remove foreseen differences that are not caused by the treatments. To allow for other, unforeseen, differences, each design is randomized before it is used. For a complete-block design, this is the appropriate method of randomization.

• Within each block independently, randomize the order of the treatments.

Statistical Model

Notation 3.1.7   We use the following notation. Individual plots are denoted by lower-case Greek letters. Let

$$f(\omega) = \text{treatment on plot } \omega;$$
$$g(\omega) = \text{block containing plot } \omega;$$
$$Y_\omega = \text{response on plot } \omega.$$

We assume that the response Y_ω is a random variable satisfying

$$Y_\omega = \tau_{f(\omega)} + \beta_{g(\omega)} + \varepsilon_\omega, \tag{3.1}$$

where τ_i is a constant depending on treatment i, β_j is a constant depending on block j, and the ε_ω are independent (normal) random variables with zero mean and (unknown) variance σ^2.

   For any constant c, we can replace τ_i and β_j by $\tau_i + c$ and $\beta_j - c$ without changing the model. So we cannot estimate $\tau_1, \ldots, \tau_v$. But we can estimate treatment differences $\tau_i - \tau_l$, and we can estimate sums $\tau_i + \beta_j$.

Estimating treatment differences

Definition 3.1.8 (Properties of estimators)   An estimator for $\tau_1 - \tau_2$ is called

• *best* if it has minimum variance subject to any other specified conditions;
• *linear* if it is a linear combination of $Y_1, Y_2, \ldots, Y_{bk}$;
• *unbiased* if its expectation is equal to the (unknown) true value $\tau_1 - \tau_2$.

Why do these properties matter? It is because the estimator for $\tau_1 - \tau_2$ is a random variable. On any one occasion it may not give the correct value, but if it is unbiased then it will give the correct value on average, if many similar experiments are performed. The spread of the possible values for the estimator depends on its variance: the smaller the variance, the smaller the spread, so the more likely it is that the observed value from one experiment is close to the average value. Thus if the estimator is unbiased and has low variance then the value obtained from this one experiment is likely to be close to the true value.

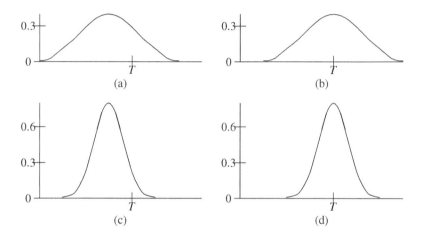

Figure 3.6 Probability density functions for four possible estimators for the true value T

Figure 3.6 compares the probability distibutions for four hypothetical estimators for the true value T. The two distributions in Figures 3.6(a) and (c) tend to underestimate T, but those in Figures 3.6(b) and (d) are unbiased. The variance of those in Figures 3.6(c) and (d) is one quarter of the variance of those in Figures 3.6(a) and (b), so the values in each of the former are closer together. Given a choice amongst these four methods of estimation, the one in Figure 3.6(d) would be preferred.

For a complete-block design, standard theory of linear models (for example, see [17, 18, 29]) shows that the best linear unbiased estimator (which is often abbreviated to BLUE) of $\tau_1 - \tau_2$ is

(average response on treatment 1) − (average response on treatment 2).

Moreover, the variance of this estimator is $2\sigma^2/b$.

Residuals

For a complete-block design, the best linear unbiased estimator of $\tau_i + \beta_j$ is

(average response on treatment i) + (average response on block j)

$-$(average response overall).

Notation 3.1.9 Statisticians write this estimated value as $\hat{\tau}_i + \hat{\beta}_j$.

Definition 3.1.10 (Residual and residual sum of squares) The *residual* on experimental unit ω is defined to be

$$Y_\omega - \hat{\tau}_{f(\omega)} - \hat{\beta}_{g(\omega)}.$$

Then the *residual sum of squares* RSS is defined to be $\sum_\omega (Y_\omega - \hat{\tau}_{f(\omega)} - \hat{\beta}_{g(\omega)})^2$.

It can be shown that RSS is equal to

$$\sum_\omega Y_\omega^2 - \sum_{i=1}^v \frac{(\text{total on treatment } i)^2}{b} - \sum_{j=1}^b \frac{(\text{total on block } j)^2}{k} + \frac{\left(\sum_\omega Y_\omega\right)^2}{bk}.$$

Notation 3.1.11 If Z is any random variable, its expectation is denoted by $\mathbb{E}(Z)$.

Theorem 3.1.12 *In a complete-block design,* $\mathbb{E}(\text{RSS}) = (b-1)(v-1)\sigma^2$.

Hence

$$\frac{\text{RSS}}{(b-1)(v-1)}$$

is an unbiased estimator of σ^2.

Remark 3.1.13 We are not usually interested in the block parameters β_j.

Remark 3.1.14 If $k = vs$ and each treatment occurs s times in each block, then estimation is similar. (Example 3.1.3 is like this, with $s = 2$.) Then the variance of the best linear unbiased estimator of $\tau_i - \tau_j$ is $2\sigma^2/bs$.

Remark 3.1.15 In particular, if there is a single block and each treatment occurs r times then the variance of the best linear unbiased estimator of $\tau_i - \tau_j$ is $2\sigma^2/r$.

3.1.3 Incomplete-block designs

For an incomplete-block design, there are v treatments, and b blocks of size k, where $2 \le k < v$.

Construction and randomization

There are many methods of construction, but how do we choose a suitable design? In order to do that, we have to know what makes an incomplete-block design good. Both of these are discussed in more detail later in this section and in Section 3.3.

The appropriate randomization is slightly more complicated than that for a complete-block design. There are two steps.

- Randomize the order of the blocks, because they do not all have the same treatments.
- Within each block independently, randomize the order of the treatments.

Examples and terminology

Here we show a few examples, in order to introduce some definitions. Figures 3.7–3.9 show incomplete-block designs with $b = 7$ and $k = 3$. We use the conventions that columns are blocks; the order of treatments within each block is irrelevant, and the order of blocks is irrelevant, because both of these may be changed by randomization.

Example 3.1.16   Figure 3.7 shows two incomplete-block designs for 15 treatments in seven blocks of size three.

(a)   (b)

Figure 3.7 Two block designs with $v = 15$, $b = 7$ and $k = 3$

Definition 3.1.17 (Replication; equireplicate; queen-bee)   The *replication* of a treatment is its number of occurrences. A design is *equireplicate* if all treatments have the same replication. A design is a *queen-bee* design if there is a treatment that occurs in every block.

Remark 3.1.18   It is always true that the average replication $\bar{r}$ is given by $\bar{r} = bk/v$. For the two designs in Figure 3.7, $\bar{r} = 1.4$.

Remark 3.1.19   Neither of the designs in Figure 3.7 is equireplicate.

Remark 3.1.20   The design in Figure 3.7(b) is a queen-bee design. On the other hand, the replications of any two treatments in the design in Figure 3.7(a) differ by no more than one.

Theorem 3.1.21 *If every treatment is replicated r times then vr = bk.*

Proof Count the number of experimental units in two different ways. □

Statisticians tend to prefer equireplicate designs, for reasons which will be discussed in Section 3.3. On the other hand, biologists tend to prefer queen-bee designs, because the treatment which occurs in every block can be some sort of standard treatment, which may be referred to as the "control".

Example 3.1.22 Figure 3.8 shows two incomplete-block designs for five treatments in seven blocks of size three. In both designs, the average replication $\bar{r}$ is equal to 4.2.

1	1	1	1	2	2	2
2	3	3	4	3	3	4
3	4	5	5	4	5	5

(a)

1	1	1	1	2	2	2
1	3	3	4	3	3	4
2	4	5	5	4	5	5

(b)

Figure 3.8 Two block designs with $v = 5$, $b = 7$ and $k = 3$

Definition 3.1.23 (Binary)   A block design is *binary* if no treatment occurs more than once in any block.

The block design in Figure 3.8(a) is binary; the block design in Figure 3.8(b) is non-binary.

We shall not consider any design in which there is any block having the same treatment on every experimental unit. The responses on the plots in any such block give no information about the differences between treatments.

Example 3.1.24 Figure 3.9 shows two incomplete-block designs for seven treatments in seven blocks of size three. Both designs are equireplicate, with replication $r = 3$.

1	2	3	4	5	6	7
2	3	4	5	6	7	1
4	5	6	7	1	2	3

(a)

1	2	3	4	5	6	7
2	3	4	5	6	7	1
3	4	5	6	7	1	2

(b)

Figure 3.9 Two block designs with $v = 7$, $b = 7$ and $k = 3$

Definition 3.1.25 (Balance)   A binary incomplete-block design is *balanced* if every pair of distinct treatments occurs together in the same number of blocks. (These are also called 2-*designs*.)

The design in Figure 3.9(a) is balanced, but the design in Figure 3.9(b) is not. The first of the two is the celebrated *Fano plane*, shown in geometric form in Figure 3.10.

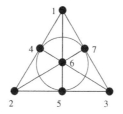

Figure 3.10 The Fano plane

Theorem 3.1.26   *If a binary design is balanced, with every pair of distinct treatments occuring together in λ blocks, then the design is equireplicate and $r(k-1) = \lambda(v-1)$.*

Proof   Suppose that treatment i has replication r_i, for $i = 1, \ldots, v$. The design is binary, so treatment i occurs in r_i blocks. Each of these blocks has $k-1$ other experimental units, each with a treatment other than i. Each other treatment must occur on λ of these experimental units. There are $v-1$ other treatments, and so

$$r_i(k-1) = \lambda(v-1).$$

Hence $r_i = r = \lambda(v-1)/(k-1)$ for $i = 1, \ldots, v$.   □

3.1.4 Matrix formulae

The statistical model assumed is the one given in Equation (3.1), using the notation in that section.

When the data are collected, they are usually written in a column vector of length bk:

$$Y = \begin{bmatrix} Y_1 \\ Y_2 \\ \vdots \\ Y_{bk} \end{bmatrix}.$$

(Statisticians typically use column vectors rather than row vectors because data are stored in a column in a spreadsheet or a csv file.) Similarly, define column vectors

$$\tau = \begin{bmatrix} \tau_1 \\ \tau_2 \\ \vdots \\ \tau_v \end{bmatrix}, \quad \beta = \begin{bmatrix} \beta_1 \\ \beta_2 \\ \vdots \\ \beta_b \end{bmatrix} \quad \text{and} \quad \varepsilon = \begin{bmatrix} \varepsilon_1 \\ \varepsilon_2 \\ \vdots \\ \varepsilon_{bk} \end{bmatrix}.$$

The model in Equation (3.1) can now be rewritten in vector form as follows:

$$Y = X\tau + Z\beta + \varepsilon, \tag{3.2}$$

where

$$X_{\omega i} = \begin{cases} 1 & \text{if } f(\omega) = i \\ 0 & \text{otherwise,} \end{cases} \quad \text{and} \quad Z_{\omega j} = \begin{cases} 1 & \text{if } g(\omega) = j \\ 0 & \text{otherwise.} \end{cases}$$

Here the matrix X has bk rows (labelled by the experimental units) and v columns (labelled by the treatments); while the matrix Z has bk rows (labelled by the experimental units) and b columns (labelled by the blocks).

Further matrices are defined in terms of X and Z.

Notation 3.1.27 The "same block" indicator matrix B is defined by $B = ZZ^{\top}$.

Thus the rows and columns of B are labelled by the experimental units, and its entries are given by

$$B_{\alpha\omega} = \begin{cases} 1 & \text{if } \alpha \text{ and } \omega \text{ are in the same block} \\ 0 & \text{otherwise.} \end{cases}$$

On the other hand, it is clear that $Z^{\top}Z = kI_b$, where I_b is the $b \times b$ identity matrix.

Notation 3.1.28 Similarly, $X^{\top}X$ is the $v \times v$ diagonal matrix R of treatment replications r_1 to r_v.

Definition 3.1.29 (Incidence matrix) The $v \times b$ *incidence matrix* N is defined by $N = X^{\top}Z$. Thus the entry N_{ij} is equal to the number of times that treatment i occurs in block j.

Definition 3.1.30 (Concurrence; concurrence matrix) The $v \times v$ *concurrence matrix* Λ is defined by $\Lambda = NN^{\top}$, so that the entry λ_{ij} is equal to the number of occurrences of treatments i and j in the same block, which is called the *concurrence* of treatments i and j.

Definition 3.1.31 (Laplacian matrix; information matrix) Two further relevant $v \times v$ matrices are the *Laplacian matrix* L and the *information matrix* C. They are defined by $L = kR - \Lambda$ and $C = k^{-1}L$.

Example 3.1.32 Here is a small example to demonstrate the notation. Figure 3.11 shows an incomplete-block design with $v = 8$, $b = 4$ and $k = 3$. The blocks are labelled B1 to B4, and the treatments are labelled 1 to 8.

B1	B2	B3	B4
1	2	3	4
2	3	4	1
5	6	7	8

Figure 3.11   Block design in Example 3.1.32

Label the experimental units 1 to 12, starting with the top experimental unit in block B1, working down to the bottom experimental unit in the block, then repeating the process in blocks B2 to B4. Then the matrices X and Z are as follows:

$$
X =
\begin{array}{c@{\quad}}
\begin{array}{cccccccc}
1 & 2 & 3 & 4 & 5 & 6 & 7 & 8
\end{array} \\
\left[
\begin{array}{cccccccc}
1 & 0 & 0 & 0 & 0 & 0 & 0 & 0 \\
0 & 1 & 0 & 0 & 0 & 0 & 0 & 0 \\
0 & 0 & 0 & 0 & 1 & 0 & 0 & 0 \\
0 & 1 & 0 & 0 & 0 & 0 & 0 & 0 \\
0 & 0 & 1 & 0 & 0 & 0 & 0 & 0 \\
0 & 0 & 0 & 0 & 0 & 1 & 0 & 0 \\
0 & 0 & 1 & 0 & 0 & 0 & 0 & 0 \\
0 & 0 & 0 & 1 & 0 & 0 & 0 & 0 \\
0 & 0 & 0 & 0 & 0 & 0 & 1 & 0 \\
0 & 0 & 0 & 1 & 0 & 0 & 0 & 0 \\
1 & 0 & 0 & 0 & 0 & 0 & 0 & 0 \\
0 & 0 & 0 & 0 & 0 & 0 & 0 & 1
\end{array}
\right]
\end{array}
\quad \text{and} \quad
Z =
\begin{array}{c@{\quad}}
\begin{array}{cccc}
\text{B1} & \text{B2} & \text{B3} & \text{B4}
\end{array} \\
\left[
\begin{array}{cccc}
1 & 0 & 0 & 0 \\
1 & 0 & 0 & 0 \\
1 & 0 & 0 & 0 \\
0 & 1 & 0 & 0 \\
0 & 1 & 0 & 0 \\
0 & 1 & 0 & 0 \\
0 & 0 & 1 & 0 \\
0 & 0 & 1 & 0 \\
0 & 0 & 1 & 0 \\
0 & 0 & 0 & 1 \\
0 & 0 & 0 & 1 \\
0 & 0 & 0 & 1
\end{array}
\right]
\end{array}
.
$$

It follows that

$$B = ZZ^\top = \begin{bmatrix} 1 & 1 & 1 & 0 & 0 & 0 & 0 & 0 & 0 & 0 & 0 & 0 \\ 1 & 1 & 1 & 0 & 0 & 0 & 0 & 0 & 0 & 0 & 0 & 0 \\ 1 & 1 & 1 & 0 & 0 & 0 & 0 & 0 & 0 & 0 & 0 & 0 \\ 0 & 0 & 0 & 1 & 1 & 1 & 0 & 0 & 0 & 0 & 0 & 0 \\ 0 & 0 & 0 & 1 & 1 & 1 & 0 & 0 & 0 & 0 & 0 & 0 \\ 0 & 0 & 0 & 1 & 1 & 1 & 0 & 0 & 0 & 0 & 0 & 0 \\ 0 & 0 & 0 & 0 & 0 & 0 & 1 & 1 & 1 & 0 & 0 & 0 \\ 0 & 0 & 0 & 0 & 0 & 0 & 1 & 1 & 1 & 0 & 0 & 0 \\ 0 & 0 & 0 & 0 & 0 & 0 & 1 & 1 & 1 & 0 & 0 & 0 \\ 0 & 0 & 0 & 0 & 0 & 0 & 0 & 0 & 0 & 1 & 1 & 1 \\ 0 & 0 & 0 & 0 & 0 & 0 & 0 & 0 & 0 & 1 & 1 & 1 \\ 0 & 0 & 0 & 0 & 0 & 0 & 0 & 0 & 0 & 1 & 1 & 1 \end{bmatrix}.$$

Furthermore, $Z^\top Z = 3I_4$, and

$$X^\top X = R = \begin{bmatrix} 2 & 0 & 0 & 0 & 0 & 0 & 0 & 0 \\ 0 & 2 & 0 & 0 & 0 & 0 & 0 & 0 \\ 0 & 0 & 2 & 0 & 0 & 0 & 0 & 0 \\ 0 & 0 & 0 & 2 & 0 & 0 & 0 & 0 \\ 0 & 0 & 0 & 0 & 1 & 0 & 0 & 0 \\ 0 & 0 & 0 & 0 & 0 & 1 & 0 & 0 \\ 0 & 0 & 0 & 0 & 0 & 0 & 1 & 0 \\ 0 & 0 & 0 & 0 & 0 & 0 & 0 & 1 \end{bmatrix}.$$

The incidence matrix N is given by

$$N = X^\top Z = \begin{bmatrix} 1 & 0 & 0 & 1 \\ 1 & 1 & 0 & 0 \\ 0 & 1 & 1 & 0 \\ 0 & 0 & 1 & 1 \\ 1 & 0 & 0 & 0 \\ 0 & 1 & 0 & 0 \\ 0 & 0 & 1 & 0 \\ 0 & 0 & 0 & 1 \end{bmatrix},$$

and the concurrence matrix Λ is given by

$$\Lambda = NN^\top = \begin{bmatrix} 2 & 1 & 0 & 1 & 1 & 0 & 0 & 1 \\ 1 & 2 & 1 & 0 & 1 & 1 & 0 & 0 \\ 0 & 1 & 2 & 1 & 0 & 1 & 1 & 0 \\ 1 & 0 & 1 & 2 & 0 & 0 & 1 & 1 \\ 1 & 1 & 0 & 0 & 1 & 0 & 0 & 0 \\ 0 & 1 & 1 & 0 & 0 & 1 & 0 & 0 \\ 0 & 0 & 1 & 1 & 0 & 0 & 1 & 0 \\ 1 & 0 & 0 & 1 & 0 & 0 & 0 & 1 \end{bmatrix}.$$

The Laplacian matrix L is given by

$$L = kR - \Lambda = \begin{bmatrix} 4 & -1 & 0 & -1 & -1 & 0 & 0 & -1 \\ -1 & 4 & -1 & 0 & -1 & -1 & 0 & 0 \\ 0 & -1 & 4 & -1 & 0 & -1 & -1 & 0 \\ -1 & 0 & -1 & 4 & 0 & 0 & -1 & -1 \\ -1 & -1 & 0 & 0 & 2 & 0 & 0 & 0 \\ 0 & -1 & -1 & 0 & 0 & 2 & 0 & 0 \\ 0 & 0 & -1 & -1 & 0 & 0 & 2 & 0 \\ -1 & 0 & 0 & -1 & 0 & 0 & 0 & 2 \end{bmatrix}.$$

In general, the concurrence λ_{ij} of treatments i and j is given by $\lambda_{ij} = \sum_{m=1}^{b} N_{im} N_{jm}$, which is the number of ordered pairs of experimental units (α, ω) with $g(\alpha) = g(\omega)$ (they are in the same block) and $f(\alpha) = i$ and $f(\omega) = j$ (they have treatments i and j).

If the design is binary, then λ_{ij} is simply equal to the number of blocks in which treatments i and j both occur: in particular, $\lambda_{ii} = r_i$ for $i = 1, \ldots, v$.

Whether or not the design is binary, counting pairs (α, ω) with $g(\alpha) = g(\omega)$ and $f(\alpha) = i$ shows that

$$r_i k = \sum_{j=1}^{v} \lambda_{ij} = \lambda_{ii} + \sum_{j \neq i} \lambda_{ij}.$$

Since $L = kR - \Lambda$, this shows that

$$L_{ii} = r_i k - \lambda_{ii} = \sum_{j \neq i} \lambda_{ij}.$$

On the other hand, if $j \neq i$ then $L_{ij} = -\lambda_{ij}$. This proves the following theorem.

Theorem 3.1.33   *The entries in each row of the Laplacian matrix sum to zero.*

This gives a simple way to calculate the Laplacian by hand. Simply write in the off-diagonal entries as the negatives of the appropriate concurrences (of course, $\lambda_{ij} = \lambda_{ji}$), and then put in the diagonal entries in such a way as to make the entries in each row sum to zero.

The Laplacian of the binary block design in Figure 3.8(a) is

$$\begin{bmatrix} 8 & -1 & -3 & -2 & -2 \\ -1 & 8 & -3 & -2 & -2 \\ -3 & -3 & 10 & -2 & -2 \\ -2 & -2 & -2 & 8 & -2 \\ -2 & -2 & -2 & -2 & 8 \end{bmatrix}, \tag{3.3}$$

while that of the non-binary block design in Figure 3.8(b) is

$$\begin{bmatrix} 8 & -2 & -2 & -2 & -2 \\ -2 & 8 & -2 & -2 & -2 \\ -2 & -2 & 8 & -2 & -2 \\ -2 & -2 & -2 & 8 & -2 \\ -2 & -2 & -2 & -2 & 8 \end{bmatrix}. \tag{3.4}$$

The second of these Laplacians is *completely symmetric* in the sense that all diagonal entries are the same and all off-diagonal entries are the same. Some people find it surprising that a non-binary block design can have a "nicer" Laplacian matrix than a binary block design with the same parameters.

3.1.5 Eigenspaces of real symmetric matrices

Before finishing our introduction to the use of matrices in the investigation of incomplete-block designs, we recall some useful facts about the eigenspaces of real symmetric matrices.

Notation 3.1.34 Suppose that M is an $n \times n$ real symmetric matrix. Denote by Θ its set of distinct eigenvalues. These are all real. For θ in Θ, denote by W_θ the set consisting of the zero vector and all eigenvectors of M with eigenvalue θ. Then W_θ is a subspace of $\mathbb{R}^n$.

We use the usual inner product on $\mathbb{R}^n$, in which vectors x and z are orthogonal to each other if $x^\top z = 0$. Two subspaces are orthogonal to each other if every vector in one is orthogonal to every vector in the other.

When M is symmetric, the eigenspaces W_θ and W_ϕ are orthogonal to each other when $\theta \neq \phi$. Furthermore, M is *diagonalizable* in the sense that the whole space $\mathbb{R}^n$ has a basis of eigenvectors, so it is the direct sum of the eigenspaces:

$$\mathbb{R}^n = \bigoplus_{\theta \in \Theta} W_\theta. \tag{3.5}$$

Definition 3.1.35 (Matrix of orthogonal projection)   The matrix P_θ of *orthogonal projection* onto W_θ is the $n \times n$ real matrix for which $P_\theta x = x$ if $x \in W_\theta$ and $P_\theta z = 0$ if z is orthogonal to W_θ.

It follows from Equation (3.5) that

$$I_n = \sum_{\theta \in \Theta} P_\theta. \tag{3.6}$$

Moreover, $P_\theta P_\phi = 0$ if $\theta \neq \phi$, while $P_\theta^2 = P_\theta$, so that P_θ is *idempotent*. Furthermore,

$$M = \sum_{\theta \in \Theta} \theta P_\theta. \tag{3.7}$$

Definition 3.1.36 (Spectral decomposition)   The sum on the right-hand side of Equation (3.7) is called the *spectral decomposition* of M.

The minimal polynomial of M has no repeated factors. Hence

$$P_\theta = \prod_{\phi \in \Theta \setminus \{\theta\}} \frac{1}{\theta - \phi} (M - \phi I).$$

This can be checked by verifying that $P_\theta x = x$ if $x \in W_\theta$ and $P_\theta x$ is the zero vector if $x \in W_\phi$ with $\phi \neq \theta$.

Definition 3.1.37 (Non-negative definite)   The matrix M is called *nonnegative definite* if $\theta \geq 0$ for all θ in Θ.

(Such matrices are also called *positive semidefinite* in the literature.)

We can write any vector x in $\mathbb{R}^n$ as $\sum_\theta x_\theta$, where $x_\theta = P_\theta x \in W_\theta$. Then $x^\top M x = \sum_\theta x_\theta^\top M x_\theta = \sum_\theta \theta x_\theta^\top x_\theta$. It follows that if M is nonnegative definite then $x^\top M x \geq 0$ for all vectors x in $\mathbb{R}^n$.

3.1.6 Fisher's Inequality

We can use the ideas from Section 3.1.5 to prove a famous inequality which holds for all balanced incomplete-block designs. This was first noticed by R. A. Fisher.

Theorem 3.1.38 (Fisher's Inequality)   *If an incomplete-block design with v treatments and b blocks is balanced, then $b \geq v$.*

Proof   By definition, the design is binary, so

$$\Lambda = rI_v + \lambda(J_v - I_v) = (r - \lambda)\left(I_v - v^{-1}J_v\right) + [\lambda(v-1) + r]v^{-1}J_v, \tag{3.8}$$

where I_v is the $v \times v$ identity matrix and J_v is the $v \times v$ all-1 matrix. The matrices

$I_v - v^{-1}J_v$ and $v^{-1}J_v$ are mutually orthogonal idempotents of ranks $v - 1$ and 1 respectively. Hence, comparing Equation (3.8) with Equation (3.7) shows that the eigenvalues of Λ are $r - \lambda$ and $\lambda(v - 1) + r$. Theorem 3.1.26 shows that $r(k - 1) = \lambda(v - 1)$; moreover, $k < v$, and so it follows that $\lambda < r$, and thus that $r - \lambda > 0$. It also follows from Theorem 3.1.26 that $\lambda(v - 1) + r = rk > 0$. Hence both of these eigenvalues are nonzero. Therefore

$$v = \mathrm{rank}(\Lambda) = \mathrm{rank}(NN^\top) = \mathrm{rank}(N^\top N) \le b. \qquad \square$$

3.1.7 Constructions

Here are three very standard constructions of equireplicate binary incomplete-block designs.

Cyclic designs

This construction works if $b = v$. First, label the treatments by the integers modulo v. Choose a so-called *initial* block with treatment-set $\{i_1, i_2, \ldots, i_k\}$. The next block is $\{i_1 + 1, i_2 + 1, \ldots, i_k + 1\}$, and so on, with all arithmetic done modulo v. The two block designs with $b = v = 7$ and $k = 3$ in Figure 3.9 are both cyclic, with initial blocks $\{1, 2, 4\}$ and $\{1, 2, 3\}$ respectively.

The initial block defines a $k \times k$ *table of differences*. The rows and columns are labelled by the elements of the initial block, in the same order. The entry in row ℓ and column m is the difference $i_\ell - i_m$ modulo v. In particular, all of the diagonal entries are zero. It is straightforward to show that, for treatments i and j, the concurrence λ_{ij} is equal to the number of occurrences of $i - j$ in the table of differences. Hence the design is balanced if and only if every nonzero integer modulo v occurs equally often in the table of differences.

Table 3.2 shows the tables of differences for the two cyclic designs in Figure 3.9. These confirm that the design in Figure 3.9(a) is balanced while that in Figure 3.9(b) is not.

$-$	1	2	4
1	0	6	4
2	1	0	5
4	3	2	0

(a)

$-$	1	2	3
1	0	6	5
2	1	0	6
3	2	1	0

(b)

Table 3.2 *Tables of differences for the cyclic block designs in Figure 3.9*

Square lattice designs

This construction works if $v = k^2$ and there are $r - 2$ mutually orthogonal Latin squares of order v.

Write out the treatments in a $k \times k$ square array. In the first replicate, the rows are blocks. In the second replicate, the columns are blocks. If a third replicate is needed, write out a $k \times k$ Latin square, superimpose it on the square array, and use its letters as blocks. For a fourth replicate, use a Latin square orthogonal to the first one, and so on.

All concurrences in the resulting design are in $\{0, 1\}$. When $r = k + 1$, the design is balanced. In this case, the design is also called an *affine plane* of order k.

Example 3.1.39   When $v = 9$ and $k = 3$, the ingredients for the construction are shown in Figure 3.12.

1	2	3
4	5	6
7	8	9

A	B	C
B	C	A
C	A	B

A	B	C
C	A	B
B	C	A

Figure 3.12 Ingredients for the construction of a square lattice design for nine treatments: from the left, the square array of treatments, the first Latin square, the second Latin square

For $r \in \{2, 3, 4\}$, these ingredients give the block design shown in the first r replicates of Figure 3.13.

1	4	7
2	5	8
3	6	9

1	2	3
4	5	6
7	8	9

1	2	3
6	4	5
8	9	7

1	2	3
5	6	4
9	7	8

Figure 3.13 Square lattice designs for nine treatments in two, three or four replicates: for each r, use the first r replicates from the left

A square lattice design would be suitable for the experiment in Example 3.1.4.

Projective planes

This construction works if $v = b = (k - 1)^2 + k$ and there are $k - 2$ mutually orthogonal Latin squares of order $k - 1$.

Start with a square lattice design for $(k - 1)^2$ treatments in $k(k - 1)$ blocks of size $k - 1$, in other words, an affine plane. Add a new treatment to every block in the first replicate. Add a second new treatment to every block in the second replicate. Continue like this for every replicate. Then add an extra block containing all the new treatments.

The resulting design is balanced, with all concurrences equal to 1.

Example 3.1.40 For example, when $v = b = 13$ and $k = 4$ we may apply this method to the block design in Figure 3.13 to obtain the projective plane in Figure 3.14.

1	4	7		1	2	3		1	2	3		1	2	3		10
2	5	8		4	5	6		6	4	5		5	6	4		11
3	6	9		7	8	9		8	9	7		9	7	8		12
10	10	10		11	11	11		12	12	12		13	13	13		13

Figure 3.14 A projective plane for 13 treatments

When $v = b = 7$ and $k = 3$ this method gives the Fano plane in Figure 3.10 as the smallest projective plane.

3.1.8 Partially balanced designs

An important subclass of equireplicate incomplete-block designs consists of the so-called *partially balanced* designs. Their definition depends on the concept of an association scheme. We define this briefly here. For more details, see [1]. The definition given in Section 1.4 is rather more general.

Definition 3.1.41 (Association scheme; partially balanced block design) An *association scheme* on the treatments is a partition of the set of v^2 ordered pairs of treatments into $s + 1$ associate classes, labelled $0, 1, \ldots, s$, subject to the four conditions listed below. For the m-th associate class, define the $v \times v$ matrix A_m to have (i, j)-entry equal to

$$\begin{cases} 1 & \text{if } i \text{ and } j \text{ are } m\text{-th associates} \\ 0 & \text{otherwise.} \end{cases}$$

The four conditions are defined in terms of these matrices, as follows.

(1) $A_0 = I_v$;
(2) $A_0, A_1, \ldots, A_s$ are all symmetric;
(3) $A_0 + A_1 + \cdots + A_s = J_v$;
(4) $A_l A_m$ is a linear combination of $A_0, \ldots, A_s$, for $0 \leq l \leq s$ and $0 \leq m \leq s$.

A block design is *partially balanced* (with respect to this association scheme) if Λ is a linear combination of $A_0, \ldots, A_s$.

Because it is symmetric, with zero diagonal, each matrix $A_1, \ldots, A_s$ can be regarded as the adjacency matrix of a graph whose vertices are labelled by the treatments. When $m = 2$, the foregoing conditions are equivalent to saying that the graphs defined by A_1 and A_2 are *strongly regular*. More generally, if a connected graph has maximum distance s between any pair of vertices, then we can partition ordered pairs of vertices into distance classes according to the distance between them. If these classes form an association scheme, then the graph is called *distance-regular*. See Section 3.2.2 for the standard definition of these graphs. Section 3.4.1 gives an example of a distance-regular graph. Strongly regular graphs are covered in more detail in Section 1.3.2, but note that the definition there excludes disconnected graphs, which we do not. Distance-regular graphs are covered in more detail in Sections 1.3.1 and 2.5.2.

We have already seen some examples of partially balanced incomplete-block designs. Cyclic designs are partially balanced with respect to the *cyclic* association scheme, which has $s = \lfloor v/2 \rfloor$. Treatments i and j are m-th associates if $i - j = \pm m$ modulo v.

Square lattice designs are partially balanced with respect to the *Latin-square-type* association scheme, which has $s = 2$. Treatments i and j are first associates if $\lambda_{ij} = 1$; they are second associates otherwise.

Here is another type of partially balanced incomplete-block design which has $s = 2$. Suppose that $v = gh$ and that the treatments are partitioned into g groups of size h. (Here the word "group" simply means a subset, with no algebraic connotations.) In the *group-divisible* association scheme, distinct treatments in the same group are first associates; treatments in different groups are second associates.

Example 3.1.42 For example, let $v = 6$, $g = 3$ and $h = 2$, with groups $\{1,4\}$, $\{2,5\}$ and $\{3,6\}$. Figure 3.15 shows a group-divisible design with $b = 4$ and $k = 3$. If treatments i and j are first associates then $\lambda_{ij} = 0$. If treatments i and j are second associates then $\lambda_{ij} = 1$.

1	1	2	3
2	5	4	4
3	6	6	5

Figure 3.15 A group-divisible block design for six treatments

Remark 3.1.43 Here are some warnings about terminology. "Partially balanced" does not mean "not balanced". Balanced incomplete-block designs are in fact the special case of partially balanced incomplete-block designs with

$s = 1$. Moreover, if an incomplete-block design is not balanced then this does not imply that it is partially balanced.

3.1.9 Laplacian matrix and information matrix

In Section 3.1.4 we defined the $bk \times bk$ matrix B by putting $B = ZZ^\top$. It follows that

$$B^2 = ZZ^\top ZZ^\top = Z(Z^\top Z)Z^\top = Z(kI_b)Z^\top = kB.$$

Therefore the matrix $k^{-1}B$ is idempotent. It is also symmetric, and has rank b.

Put $Q = I_{bk} - k^{-1}B$. Then Q is also idempotent and symmetric. It has rank $b(k-1)$. Therefore

$$X^\top QX = X^\top Q^2 X = X^\top Q^\top QX = (QX)^\top (QX),$$

which is nonnegative definite.

It turns out that we have seen this matrix before, because

$$X^\top QX = X^\top \left(I_{bk} - \frac{1}{k}B \right) X = X^\top X - \frac{1}{k}X^\top ZZ^\top X = R - \frac{1}{k}\Lambda = \frac{1}{k}L = C,$$

where L is the Laplacian matrix and C is the information matrix. It follows that L and C are both nonnegative definite, so their nonzero eigenvalues are all positive.

By Theorem 3.1.33, all row-sums of L are zero, so L has 0 as an eigenvalue on the all-1 vector.

Definition 3.1.44 (Connected block design)   A block design is defined to be *connected* if 0 is a simple eigenvalue of L.

From now on, we always assume connectivity. Call the remaining eigenvalues of L *nontrivial*. They are all nonnegative.

Under the assumption of connectivity, the null space of L is spanned by the all-1 vector. The matrix $v^{-1}J_v$ is the orthogonal projector onto this null space. Hence we can define the *Moore–Penrose generalized inverse* L^- of L by

$$L^- = \left(L + \frac{1}{v}J_v \right)^{-1} - \frac{1}{v}J_v.$$

Roughly speaking, L^- inverts L where it can, in the sense that $LL^-L = L$ and $L^-LL^- = L^-$. Moreover, $C = k^{-1}L$ and so $C^- = kL^-$. There are more details in Section 3.2.6.

3.1.10 Estimation and variance

Covariance matrices

Definition 3.1.45 (Variance–covariance matrix)   In general, if

$$
U = \begin{bmatrix} U_1 \\ U_2 \\ \vdots \\ U_n \end{bmatrix}
$$

is a random vector of length n, then its *variance–covariance matrix* $\mathrm{Cov}(U)$ is the $n \times n$ real symmetric matrix whose diagonal entries are the variances $\mathrm{Var}(U_1), \ldots, \mathrm{Var}(U_n)$ and whose (i, j)-off-diagonal entry is the covariance $\mathrm{Cov}(U_i, U_j)$.

Variance–covariance matrices are always nonnegative definite.

The following theorem gives a standard result from elementary probability theory.

Theorem 3.1.46   *If M is an $m \times n$ real matrix then MU is a random vector of length m and $\mathrm{Cov}(MU) = M \mathrm{Cov}(U) M^\top$.*

Equation (3.2) still gives our assumption for the vector Y of responses in the experiment as $Y = X\tau + Z\beta + \varepsilon$. All the entries in $X\tau$ and $Z\beta$ are constants, and so

$$
\mathrm{Cov}(Y) = \mathrm{Cov}(\varepsilon) = I_{bk}\sigma^2,
$$

as we assumed in Section 3.1.2.

Estimation

Now we generalize the methods and results of Section 3.1.2 to general block designs. Since $Q = I_{bk} - k^{-1}B$, we have

$$
QZ = Z - k^{-1}(ZZ^\top)Z = Z - k^{-1}Z(kI_b) = 0.
$$

Hence

$$
QY = Q(X\tau + Z\beta + \varepsilon) = QX\tau + QZ\beta + Q\varepsilon = QX\tau + Q\varepsilon. \qquad (3.9)
$$

By Theorem 3.1.46, $\mathrm{Cov}(Q\varepsilon) = Q\mathrm{Cov}(\varepsilon)Q^\top = Q\sigma^2$, which is essentially scalar for vectors in the column-space of Q. Therefore standard least squares theory shows that the best linear unbiased estimators are obtained by ignoring the random part of Equation (3.9), pre-multiplying everything by the transpose of the matrix in front of τ, and replacing the unknown τ with its estimate $\hat{\tau}$. This gives $(QX)^\top QY = (QX)^\top QX\hat{\tau}$, so that

$$X^\top QY = X^\top QX\hat{\tau} = C\hat{\tau}. \tag{3.10}$$

We want to estimate *contrasts* in the treatment parameters, which are linear combinations of the form $\sum_i x_i \tau_i$ with $\sum_i x_i = 0$. In particular, we want to estimate all the simple differences $\tau_i - \tau_j$. If x is a contrast and the design is connected then x is in the column space of C. Hence there is another contrast u such that $Cu = x$. Then $\sum_i x_i \tau_i = x^\top \tau = u^\top C\tau$. Pre-multiplying Equation (3.10) by $u^\top$ gives

$$u^\top X^\top QY = u^\top C\hat{\tau} = x^\top \hat{\tau} = \sum_i x_i \hat{\tau}_i.$$

Variance

Theorem 3.1.46 shows that the variance of this estimator is

$$u^\top X^\top Q(I_{bk}\sigma^2)QXu = u^\top X^\top QXu\sigma^2 = u^\top Cu\sigma^2$$
$$= u^\top x\sigma^2 = x^\top C^- x\sigma^2 = (x^\top L^- x)k\sigma^2.$$

In particular,

$$\mathrm{Var}(\hat{\tau}_i - \hat{\tau}_j) = (L_{ii}^- + L_{jj}^- - 2L_{ij}^-)k\sigma^2. \tag{3.11}$$

It is interesting to see that the entries in the Moore–Penrose inverse of the Laplacian matrix give such a simple formula for this variance. We should like all such variances to be as small as possible.

If the block design is balanced then $r(k-1) = \lambda(v-1)$ and

$$\begin{aligned}
L = krI_v - \Lambda &= krI_v - (rI_v + \lambda(J_v - I_v)) \\
&= r(k-1)I_v - \lambda(J_v - I_v) \\
&= \lambda(v-1)I_v - \lambda(J_v - I_v) \\
&= v\lambda \left(I_v - \frac{1}{v}J_v\right).
\end{aligned}$$

Hence

$$L^- = \frac{1}{v\lambda}\left(I_v - \frac{1}{v}J_v\right)$$

and all variances of estimators of pairwise differences are the same, namely

$$\frac{2k}{v\lambda}\sigma^2 = \frac{2k(v-1)}{vr(k-1)}\sigma^2.$$

If there were no need for blocking, then Section 3.1.2 shows that this variance would be $2\sigma^2/r$, so we see that the variance in a balanced incomplete-block design is

$$\frac{k(v-1)}{(k-1)v} \times \text{value in unblocked case.}$$

If the block design is partially balanced then L is a linear combination of A_0, ..., A_s. Then the conditions for an association scheme show that L^- is also a linear combination of A_0, ..., A_s, so there is a single pairwise variance for all pairs in the same associate class.

In particular, if $s = 2$ then there are precisely two concurrences and two pairwise variances, and all pairs with the same concurrence have the same pairwise variance. It can be shown that the smaller concurrence corresponds to the larger variance: see [3].

Remark 3.1.47 This simple pattern does not hold for arbitrary block designs. In general, pairs with the same concurrence may have different pairwise variances. There are some designs where some pairs with low concurrence have smaller pairwise variance than some pairs with high concurrence.

Remark 3.1.48 Matrix inversion was not easy in the pre-computer age. One reason for the introduction of balanced incomplete-block designs and partially balanced incomplete-block designs was that it was relatively easy to calculate L^- and hence to calculate the pairwise variances.

Residuals
Residuals and the residual sums of squares are defined exactly as they were in Section 3.1.2.

Theorem 3.1.49 *If the block design is connected then the expected value of the residual sume of squares is given by* $\mathbb{E}(\text{RSS}) = (bk - b - v + 1)\sigma^2$.

Hence

$$\frac{\text{RSS}}{bk - b - v + 1}$$

is an unbiased estimator of σ^2.

3.1.11 Reparametrization
Equation (3.1) is the standard way to write the model assumed for the expectation of the responses in a block design. Here we introduce a non-standard reparametrization of blocks by putting $\gamma_j = -\beta_j$ for $j = 1, \ldots, b$. Then

$$Y_\omega = \tau_{f(\omega)} - \gamma_{g(\omega)} + \varepsilon_\omega.$$

Now we can add the same constant to every τ_i and every γ_j without changing the model. So we cannot estimate $\tau_1, \ldots, \tau_v$, or $\gamma_1, \ldots, \gamma_b$. But we can aspire to estimate differences such as $\tau_i - \tau_l$, $\gamma_j - \gamma_m$ and $\tau_i - \gamma_j$.

In matrix form,

$$Y = X\tau - Z\gamma + \varepsilon, \tag{3.12}$$

where $\gamma = -\beta$. We can put the matrices X and $-Z$ into a single $bk \times (v+b)$ matrix $[X \mid -Z]$, and combine the column vectors τ and γ into a single column vector

$$\begin{bmatrix} \tau \\ \gamma \end{bmatrix}.$$

This enables us to rewrite Equation (3.12) as

$$Y = [X \mid -Z] \begin{bmatrix} \tau \\ \gamma \end{bmatrix} + \varepsilon.$$

Now the same theory as before shows that the best linear unbiased estimates of contrasts in $(\tau_1, \ldots, \tau_v, \gamma_1, \ldots, \gamma_b)$ satisfy

$$[X \mid -Z]^\top Y = [X \mid -Z]^\top [X \mid -Z] \begin{bmatrix} \hat{\tau} \\ \hat{\gamma} \end{bmatrix} = \tilde{L} \begin{bmatrix} \hat{\tau} \\ \hat{\gamma} \end{bmatrix},$$

where

$$\tilde{L} = \begin{bmatrix} X^\top \\ -Z^\top \end{bmatrix} [X \mid -Z] = \begin{bmatrix} X^\top X & -X^\top Z \\ -Z^\top X & Z^\top Z \end{bmatrix} = \begin{bmatrix} R & -N \\ -N^\top & kI_b \end{bmatrix}.$$

(Recall that R is the diagonal matrix of treatment replications and that N is the incidence matrix.)

Now let x be a contrast vector in $\mathbb{R}^{v+b}$. If $\tilde{L}u = x$ then

$$x^\top \begin{bmatrix} \hat{\tau} \\ \hat{\gamma} \end{bmatrix} = u^\top \tilde{L} \begin{bmatrix} \hat{\tau} \\ \hat{\gamma} \end{bmatrix} = u^\top \begin{bmatrix} X^\top \\ -Z^\top \end{bmatrix} Y.$$

Moreover, the variance of this estimator is $(x^\top \tilde{L}^- x)\sigma^2$. In particular,

$$\mathrm{Var}(\hat{\tau}_i - \hat{\tau}_j) = (\tilde{L}^-_{ii} + \tilde{L}^-_{jj} - 2\tilde{L}^-_{ij})\sigma^2. \tag{3.13}$$

It is interesting to compare this with with Equation (3.11).

3.1.12 Exercises

Exercise 3.1.1 Construct a square lattice design for 16 treatments in 12 blocks of size 4.

Exercise 3.1.2 For the design in Exercise 3.1.1, put $A = \Lambda - R$.

(i)   Express AJ_{16} as a linear combination of A, I_{16} and J_{16}.
      (This is possible because the design is equireplicate.)

(ii) Express A^2 as a linear combination of A, I_{16} and J_{16}.
      (This is possible because the design is partially balanced.)

Exercise 3.1.3   For the matrix A in Exercise 3.1.2, find its eigenvalues and their multiplicities. Hence find the eigenvalues, with their multiplicities, for the Laplacian matrix L for the design in Exercise 3.1.1.

Exercise 3.1.4   Let M be a real symmetric matrix. Give proofs of the following statements made in Section 3.1.5.

(i)   All eigenvalues of M are real.

(ii) M is diagonalizable.

(iii) If W_θ and W_ϕ are the eigenspaces of M corresponding to distinct eigenvalues θ and ϕ then W_θ is orthogonal to W_ϕ.

Exercise 3.1.5   By expressing the matrix L in Exercise 3.1.3 either as a linear combination of the symmetric idempotent matrices projecting onto its eigenspaces or as a linear combination of the matrices A, I_{16} and J_{16} in Exercise 3.1.2 (or otherwise, if you know of another method), find L^- as a linear combination of A, I_{16} and J_{16}. Hence find the variance V_{ij} of the best linear unbiased estimator of the treatment difference $\tau_i - \tau_j$. (These variances are not all the same.)

Exercise 3.1.6   An agricultural scientist is planning an experiment to compare 6 varieties of cabbage. He has enough cabbages to plant 10 plots of each variety. He has a choice of two fields. One field has 10 blocks of 6 plots each; the second field has 15 blocks of 4 plots each.

(i)   Construct a suitable design for each field.

(ii) The variance in the first field is σ_1^2, and the variance in the second field is σ_2^2. For each field, calculate the variance of the estimator of the difference between two varieties, using the design which you have constructed.

(iii) Which field would you recommend using if $\sigma_1^2 = 3\sigma_2^2/2$?

3.2 Laplacian matrices and their eigenvalues

This section will be about the Laplacian matrix of a graph and its eigenvalues, and their relation to some graph parameters.

This is not a complete account of the theory, but concentrates mainly on the things that are most relevant for experimental design.

For further reading, we recommend the surveys by Bollobás [9, Chapters II and IX] and Mohar [27].

Note that, as our matrices are usually symmetric, it doesn't matter whether we use row or column vectors; but, as we have seen, statisticians use column vectors for data, so we will follow this convention.

3.2.1 Which graph is best?

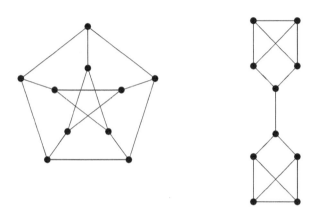

Figure 3.16 Which is the better network?

Example 3.2.1 Which of the two graphs in Figure 3.16 is better to use as a network? Of course, the question is not well defined. But which would you choose for a network, if you were concerned about connectivity, reliability, etc.?

We will use these graphs frequently as examples throughout this section. The first is the celebrated *Petersen graph*, probably the most famous finite graph, which has an entire book devoted to it [21]. The second has the same number of vertices and the same valency, but you would probably not choose it. We will investigate reasons for this.

So what makes a good network?
We might want the following conditions to be satisfied.

- No two vertices should be too far apart.
- There should be several alternative routes between two vertices. (But should these routes be disjoint?)
- There should be no "bottlenecks".

- Loss of a small part of the network should not result in disconnection.

Of course, we are resource-limited, otherwise we would just put multiple edges between any two vertices. So we might want to do as well as possible on these criteria if the number of edges allowed is bounded, say.

We will return to this after introducing some graph-theoretic terminology. This is mostly standard and can be found in any graph theory textbook.

3.2.2 Graph terminology

Definition 3.2.2 (Graph)   A *graph* has a set of *vertices*, some pairs of which are *adjacent*, or joined by *edges*. In this chapter, an edge will never connect a vertex to itself (that is, *loops* are forbidden), but we will allow multiple edges. A graph is *simple* if there are no multiple edges and no loops. We denote by $V(G)$ and $E(G)$ the vertex set and edge set of the graph G.

Definition 3.2.3 (Weighted graph)   Sometimes we consider a more general concept, where edges have nonnegative real *weights*; we speak of a *weighted graph*. If the weights are positive integers, we can think of an edge of weight k as being the same as k simple edges joining the same two vertices. In some contexts (but not here), vertices may also be weighted, and the graphs we consider are described as *edge-weighted graphs*.

Definition 3.2.4 (Connected graph)   A graph is *connected* if there is a path between any two of its vertices. In general, being joined by a path is an equivalence relation whose equivalence classes are the *connected components* of the graph. Moreover, in a connected graph, there is a metric on the set of vertices, called the *path metric*, where the distance between two vertices is the number of edges in a shortest path joining them. We denote the distance from v to w by $d(v,w)$.

Definition 3.2.5 (Regular graph)   A graph is *regular* with *valency d* if every vertex lies on d edges. The graphs in Figure 3.16 are each regular with valency 3.

Definition 3.2.6 (Strongly regular graph)   A simple graph is *strongly regular* with *parameters* (n,k,λ,μ) if the following hold:

- there are n vertices;
- the graph is regular with valency k;
- the number of common neighbours of two vertices i and j is λ if i and j are joined, or μ if they are not joined.

For example, the Petersen graph (Example 3.2.1) is strongly regular, with parameters $(10,3,0,1)$. We met the notion of *association scheme* in Definition 3.1.41. Note that a strongly regular graph defines an association scheme with two associate classes (edges and non-edges).

Distance-regular graphs are a generalization of strongly regular graphs, and occur in several places in this volume (for example, Sections 1.3.1 and 2.5.2).

Definition 3.2.7 (Distance-regular graph)   A connected graph G which has diameter d is *distance-regular* if there are numbers c_i, a_i, b_i for $0 \leq i \leq d$ such that, for any pair (x,y) of vertices at distance i, the number of vertices adjacent to y and at distance j from x is

$$\begin{cases} c_i, & \text{if } j = i-1, \\ a_i, & \text{if } j = i, \\ b_i, & \text{if } j = i+1. \end{cases}$$

The number is clearly zero if $j \notin \{i-1, i, i+1\}$. It is also zero if $i = 0$ and $j \neq 1$, or if $i = d$ and $j = d+1$. The mnemonic for the parameters is $c =$ "closer", $a =$ "across", $b =$ "beyond".

We can represent the parameters of a distance-regular graph by a diagram consisting of $d+1$ circles arranged in a line, with edges joining adjacent circles and loops at each circle after the first. For $0 \leq i \leq d$, the ith circle contains the number k_i of vertices at distance i from a fixed vertex x (this is independent of the choice of x), and the edge leaving the ith circle for the jth is labelled with c_i, a_i or b_i according as $j = i-1$, i or $i+1$. (So an edge can get different labels at each end.) The sum of the labels on edges leaving any circle is the valency of the graph. See Figure 3.17.

Figure 3.17 Parameters of a distance-regular graph

For example, Figure 3.18 shows the diagram for the Petersen graph (Example 3.2.1). In this case, the loop on the vertex corresponding to distance 1 is omitted, since $a_1 = 0$.

We will see a beautiful distance-regular graph, the *Sylvester graph*, in Section 3.4.1.

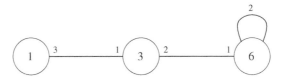

Figure 3.18 The diagram for the Petersen graph

For more information on distance-regular graphs, including the algebraic theory and a survey of some families of examples, we refer to Sections 1.3.1 and 2.5.2.

Definition 3.2.8 (Tree; forest) A *forest* is a graph without cycles. A *tree* is a connected forest.

A tree on n vertices has $n-1$ edges. More generally, a forest on n vertices which has k connected components has $n-k$ edges, since each connected component has one fewer edge than vertices.

Definition 3.2.9 (Spanning subgraph; spanning tree) Given a graph G, a *spanning subgraph* of G is a graph consisting of all the vertices and some of the edges of G. If it is a tree (or forest), we call it a *spanning tree* (or *spanning forest*) of G.

Every connected graph has a spanning tree. *Cayley's Theorem* says that the complete graph K_n has n^{n-2} spanning trees.

Connectivity

We return to our comparison of the two graphs in Figure 3.16. Here are some ways we might measure the "connectivity" of a graph. One of these is the isoperimetric number, which we first define.

Definition 3.2.10 (Isoperimetric number) The *isoperimetrc number* of a graph G is defined to be

$$\iota(G) = \min\left\{\frac{|\partial S|}{|S|} : S \subseteq V(G), 0 < |S| \leq n/2\right\},$$

where $n = |V(G)|$ and, for a set S of vertices, ∂S is the set of edges from S to its complement. The isoperimetric number is nonzero if and only if the graph is connected. Large isoperimetric number means that there are many edges out of any set of vertices.

- *Spanning trees.* How many spanning trees does it have? The more spanning trees, the better connected. (In particular, the number of spanning trees is nonzero if and only if the graph is connected.) The first graph has 2000 spanning trees, the second has 576.
- *Electrical resistance.* Imagine that the graph is an electrical network with each edge being a 1-ohm resistor. Now calculate the effective resistance R between each pair of terminals, by connecting a battery of voltage V to these vertices and measuring the current I that flows. By Ohm's law (see Section 3.2.7), $V = IR$. Now sum the effective resistances over all pairs of vertices. The lower the total, the better connected the graph. In particular, the average resistance is finite if and only if the graph is connected. In the first graph, the sum is 33; in the second, it is $206/3$, more than twice as large. (We will see later how to compute this number for a graph.)
- *Isoperimetric number.* The isoperimetric number for the first graph is 1 (there are just five edges between the inner and outer pentagons), that of the second graph is $1/5$ (there is just one edge between the top and bottom pieces).

So the Petersen graph wins on all three comparisons.

From now on, all our graphs will be connected unless we say otherwise.

3.2.3 The Laplacian of a graph

Let G be a graph on n vertices. (Multiple edges are allowed but loops are not.)

Definition 3.2.11 (Laplacian matrix of a graph)   The *Laplacian matrix* of G is the $n \times n$ matrix $L = L(G)$ whose (i, i) entry is the number of edges containing vertex i, while for $i \neq j$ the (i, j) entry is the negative of the number of edges joining vertices i and j.

This is a real symmetric matrix; its eigenvalues are the *Laplacian eigenvalues* of G. Note that its row sums are zero.

The definition can be extended to weighted graphs. Suppose that we have positive weights $w(e)$ on the edges of G. Then the *weighted Laplacian* has the (i, i) entry the sum of weights of edges containing i, and its (i, j) entry for $i \neq j$ is minus the sum of the weights of edges joining i to j.

If the weights are rational, then we may multiply them by the least common multiple of the denominators to make them integers. Then replace an edge of weight w by w edges, to obtain a multigraph with the same Laplacian.

For general real weights, we can replace them first by rational approximations.

We will not consider weighted Laplacians.

Relation to classical Laplacian

The classical Laplacian is a second-order differential operator defined on functions on a manifold, closely related to potential theory, the wave equation, etc.

A manifold can be *triangulated*, that is, approximated by a graph drawn in it. If we take the weight of an edge to be inversely proportional the square of its length, then the weighted Laplacian of the graph is an approximation to the Laplacian of the manifold.

In the other direction, given a graph, we can build a manifold reflecting its structure. Given a d-valent vertex, take a sphere with d holes; then glue spheres corresponding to the vertices of an edge together along the corresponding holes.

We won't pursue this any further.

Adjacency matrix and Laplacian

Definition 3.2.12 (Adjacency matrix) The usual *adjacency matrix* $A(G)$ of a graph G on n vertices has rows and columns indexed by vertices; the (i, j) entry is the number of edges connecting i to j.

Note that we can allow loops in this definition (though it is not clear whether a loop should contribute 1 or 2 to the corresponding diagonal entry). However, as said earlier, our graphs will not have loops.

If G is a regular graph with valency d, then $L(G) = dI - A(G)$; so the Laplacian eigenvalues are obtained by subtracting the adjacency matrix eigenvalues from d.

If G is not regular, there is no such simple relationship between the eigenvalues of the two matrices.

Non-negative definiteness

We give a more general form of Definition 3.1.37.

Definition 3.2.13 (Non-negative definite) A square matrix A is *nonnegative definite* (also called *positive semidefinite*) if $v^\top Av \geq 0$ for all column vectors v. If A is symmetric, this is equivalent to saying that all its eigenvalues are nonnegative.

Theorem 3.2.14 *The Laplacian of a graph is nonnegative definite.*

Proof L is the sum of submatrices $\begin{bmatrix} +1 & -1 \\ -1 & +1 \end{bmatrix}$, one for each edge (this 2×2 matrix in the positions indexed by the two vertices of the edge, with zeros elsewhere). This matrix is nonnegative definite (its eigenvalues are 2 and 0).

Now a sum of nonnegative definite matrices is nonnegative definite; for if $A = \sum A_e$ and $v^\top A_e v \ge 0$ for all e, then $v^\top A v \ge 0$. $\qquad\square$

We'll see another proof of this theorem after Theorem 3.2.22.
It follows that the eigenvalues of L are all nonnegative.

The multiplicity of zero

Theorem 3.2.15 *The multiplicity of 0 as an eigenvalue of L is equal to the number of connected components of the graph.*

Proof An eigenvector with zero eigenvalue is a function on the vertices whose value at i is the weighted average of its values on the neighbours of i, each neighbour weighted by the number of edges joining it to i. (If you know about harmonic functions, you will recognise this!) Let i be a vertex at which the maximum value of such a function f is achieved. Then $f(i) = f(j)$ for all vertices j adjacent to i. Hence by an easy induction, f is constant on the connected component containing i. So the eigenvector is constant on connected components. Conversely, any vector which is constant on the connected components is a Laplacian eigenvector with eigenvalue 0. $\qquad\square$

In particular, if the graph is connected (as we usually assume), the zero eigenvalue (called "trivial") has multiplicity 1; the other eigenvalues are *nontrivial*. The eigenvectors for the trivial eigenvalue are the constant vectors.

Eigenvalues

Since the Laplacian is nonnegative definite, its eigenvalues are all nonnegative.

The sum of the eigenvalues is the trace of L, which is the sum of the vertex valencies, or twice the number of edges. So the average of the nontrivial eigenvalues is $2|E(G)|/(|V(G)| - 1)$; it depends just on the numbers of vertices and edges, and the detailed structure of the graph has no effect.

We'll see that other means, in particular the geometric and harmonic means, of the nontrivial eigenvalues, give us important information!

Examples

In Example 3.2.1, The Petersen graph is strongly regular; its adjacency matrix A satisfies $A^2 + A - 2I = J$, where J is the all-1 matrix; its eigenvalues are 3, 1 and -2, and so the Laplacian eigenvalues are 0, 2 and 5, with multiplicities 1, 5 and 4 respectively.

For the other graph in Figure 3.16, the Laplacian eigenvalues are 0, 2, 3 (multiplicity 2), 4 (multiplicity 2), 5, and the zeros of $x^3 - 9x^2 + 20x - 4$ (which are approximately 0.2215, 3.2892, and 5.4893).

3.2.4 Isoperimetric number

Recall from Section 3.1.5 that eigenvectors corresponding to distinct eigenvalues of a symmetric matrix are orthogonal. The Rayleigh Principle characterizes eigenvectors by a minimization procedure.

Theorem 3.2.16 (Rayleigh Principle) *Let θ_1, θ_2 be the smallest and second-smallest eigenvalues of the real symmetric matrix A, and suppose that θ_1 is a simple eigenvalue with eigenvector u. Let v be any nonzero vector orthogonal to u. Then*

$$\frac{v^\top A v}{v^\top v} \geq \theta_2,$$

with equality if and only if v is an eigenvector with eigenvalue θ_2.

This is obvious when v is expressed as a linear combination of eigenvectors of A.

There is an extension to any eigenvalue.

We use this to establish a relation between cutsets and Laplacian eigenvalues.

Theorem 3.2.17 (Cutset Lemma) *Let G be a connected graph on n vertices, and E a set of m edges whose removal disconnects G into vertex sets of sizes n_1 and n_2, with $n_1 + n_2 = n$. Let μ be the smallest nontrivial eigenvalue of L. Then $m \geq \mu n_1 n_2/(n_1 + n_2)$.*

Proof For let V_1 and V_2 be the vertex sets in the theorem, and let v be the vector with value n_2 on vertices in V_1, and $-n_1$ on vertices in V_2. These values are chosen so that v is orthogonal to the all-1 vector (the trivial eigenvector). Clearly, $v^\top v = n_1 n_2^2 + n_2 n_1^2 = (n_1 + n_2)n_1 n_2$.

We claim that $v^\top L v = m(n_1 + n_2)^2$. Recall that L is the sum of submatrices corresponding to edges; we have to add the contributions of these. An edge within one of the parts contributes 0; an edge between the parts contributes $(n_1 + n_2)^2$. The claim follows.

The theorem now follows from the Rayleigh Principle. □

Theorem 3.2.18 *Let G be a connected graph whose smallest nontrivial Laplacian eigenvalue is μ. Then the isoperimetric number $\iota(G)$ is at least $\mu/2$.*

Proof Let S be a set of at most half the vertices; let $|S| = n_1$, $|V(G) \setminus S| = n_2$, and $|\partial(S)| = m$. Then by the Cutset Lemma,

$$\frac{|\partial(S)|}{|S|} = \frac{m}{n_1} \geq \frac{\mu n_2}{n_1 + n_2} \geq \frac{\mu}{2}.$$

□

So, on one of our criteria, a good network is one whose smallest nontrivial Laplacian eigenvalue is as large as possible.

Examples

In Example 3.2.1, the smallest nontrivial Laplacian eigenvalues are 2 (for the Petersen graph) and 0.2215 (for the other graph).

The Petersen graph has isoperimetric number 1, meeting the bound of half the least nontrivial eigenvalue. So the vector which is $+1$ on the outer pentagon and -1 on the inner pentagon is a Laplacian eigenvector corresponding to the smallest nontrivial eigenvalue.

In the other graph, the isoperimetric number is a little more than half the smallest nontrivial eigenvalue.

Expanders

Loosely, an expander is a regular connected graph whose smallest nontrivial Laplacian eigenvalue is large. The above result shows that expanders have large isoperimetric numbers. More precisely:

Definition 3.2.19 (Expanders)   A sequence $(G_n : n \in \mathbb{N})$ of graphs is said to be a sequence of *expanders* if there is a constant $c > 0$ such that the smallest nontrivial Laplacian eigenvalue of every graph G_n is at least c.

It is known that a random regular graph is an expander with high probability; but explicit constructions are more difficult. The first constructions were given by Margulis and by Lubotzky, Phillips and Sarnak, and depend on substantial number-theoretic and group-theoretic background.

Cheeger's Inequality

Cheeger's Inequality is a result about Laplacians of manifolds; it has a discrete analogue. It gives a bound in the other direction between the isoperimetric number and the smallest nontrivial Laplacian eigenvalue.

Theorem 3.2.20 (Discrete version of Cheeger's Inequality)   *Let G be a connected graph; let Δ be the maximum valency of G, and μ the smallest nontrivial Laplacian eigenvalue. Then*

$$\iota(G) \leq \sqrt{2\Delta\mu}.$$

Mohar improved the upper bound to $\sqrt{(2\Delta - \mu)\mu}$ if the graph is connected but not complete. We refer to Mohar's article [27] for this.

3.2.5 Signed incidence matrix

Definition 3.2.21 (Signed incidence matrix)   Choose a fixed but arbitrary orientation of the edges of the graph G, so that each edge has a *head* and a *tail*, and is directed from the tail to the head. We define the edge-vertex *signed incidence matrix* Q to have rows indexed by edges, columns by vertices, and (e, v) entry $+1$ if v is the head of the edge e, -1 if v is the tail of e, and 0 otherwise. Thus Q depends on the chosen orientation. See Figure 3.19.

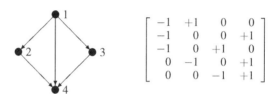

Figure 3.19 The signed incidence matrix

Theorem 3.2.22   *Let G have signed incidence matrix Q and Laplacian L. Then $Q^\top Q = L$.*

Proof   The (i, i) entry of $Q^\top Q$ is the number of edges with either head or tail at i; and the (i, j) entry is the sum of -1 for all edges with head at i and tail at j or *vice versa*.   □

This gives another proof of Theorem 3.2.14 that L is nonnegative definite, since

$$v^\top (Q^\top Q)v = (Qv)^\top Qv \geq 0.$$

And note that the chosen orientation doesn't affect the formula for L.

3.2.6 Generalized inverse; Moore–Penrose inverse

Definition 3.2.23 (Generalized inverse)   The notion of generalized inverse makes sense in any system with an associative multiplication (that is, any *semigroup*); for us, that means matrices. A *generalized inverse* of an element A is an element B such that $ABA = A$. (Note that, if there is an identity element I and B is an inverse of A, satisfying $AB = I$, then B is a generalized inverse of A.) A generalized inverse is sometimes referred to as a *von Neumann inverse*. In semigroup theory, an element with a generalized inverse is called *regular*, and a semigroup is *regular* if all its elements are regular.

In general, we lose no generality by assuming that also $BAB = B$, that is, A is also a generalized inverse of B. For suppose that $ABA = A$, and let $C = BAB$. Then

$$ACA = ABABA = ABA = A,$$
$$CAC = BABABAB = BABAB = BAB = C,$$

so we can use C instead of B.

Now the main facts about generalized inverses of matrices are given in the next two theorems.

Theorem 3.2.24 *Every matrix over a field has a generalized inverse.*

Proof Let A be a matrix. Choose vectors $v_1, \ldots, v_r$ spanning the image of A, and let $w_1, \ldots, w_r$ be pre-images of $v_1, \ldots, v_r$. Choose B mapping v_i to w_i for $i = 1, \ldots, r$. Then $ABA = A$. $\qquad\square$

Here we are mostly interested in square matrices. The foregoing result shows that the semigroup of all $n \times n$ matrices over a given field is regular.

Theorem 3.2.25 *For a square matrix A over a field, the following three conditions are equivalent:*

(1) *A has a generalized inverse that commutes with A;*
(2) *A has a generalized inverse which is a polynomial in A;*
(3) *0 is not a multiple zero of the minimal polynomial of A.*

Proof (2) implies (1): clear.

(3) implies (2): Suppose that 0 is not a multiple zero of the minimal polynomial of A. Then there is a polynomial f with zero constant term and nonzero coefficient of x which is satisfied by A. (If 0 is a zero of the minimal polynomial, use this polynomial; otherwise use the minimal polynomial multiplied by x.) After multiplying by a nonzero scalar, we can write $f(x) = x - x^2 h(x)$. Then $h(A)$ is a generalized inverse of A.

(1) implies (3): Suppose that $ABA = A$ and $AB = BA$. Then $BA^2 = A$. But, if 0 is a repeated zero of the minimum polynomial of A, there is a vector v mapped to 0 by A^2 but not by A, and applying the above equation to v gives a contradiction. $\qquad\square$

Corollary 3.2.26 *If the matrix A is diagonalizable, then it has a generalized inverse which is a polynomial in A. In particular, this holds for real symmetric matrices.*

Now we reconsider a specific generalized inverse of a real symmetric matrix, the *Moore–Penrose inverse*, defined below.

Definition 3.2.27 (Spectral decomposition)   Let A be a real symmetric matrix. Then we have a *spectral decomposition* of A:

$$A = \sum_{\theta \in \Theta} \theta P_\theta,$$

where Θ is the set of eigenvalues of A, and P_θ is the orthogonal projector onto the space of eigenvectors with eigenvalue θ.

Definition 3.2.28 (Moore–Penrose inverse)   We define the *Moore–Penrose inverse* of A by

$$A^- = \sum_{\theta \in \Theta \setminus \{0\}} \theta^{-1} P_\theta.$$

In other words, we invert A where we can.

The Moore–Penrose inverse is a generalized inverse of A: that is,

$$A^- A A^- = A^-, \qquad A A^- A = A.$$

We will see a lot of the matrix L^-, where L is the Laplacian of a graph G on n vertices.

If G is connected, then the projector onto the trivial eigenspace of L is $n^{-1}J_n$, where J_n is the all-1 matrix of order n. So adding this to L changes the trivial eigenvalue to 1, and subtracting it takes the 1 off again.

In other words,

$$L^- = (L + n^{-1}J_n)^{-1} - n^{-1}J_n.$$

Finally, we remark that the Moore–Penrose inverse of A does commute with A, so it is a polynomial in A. This is because both AA^- and A^-A have eigenvalues 0 on the null space of A and 1 on the image of A.

3.2.7 Electrical networks

As mentioned on page 207, we regard the graph G as an electrical network, where we regard each edge as a one-ohm resistor. Given any two vertices i and j, the effective resistance $R(i, j)$ between i and j is the voltage of a battery which, when connected to the two vertices, causes a current of 1 ampere to flow.

In order to calculate $R(i, j)$, we need to calculate the currents and potentials in the network. This requires some basic results from 19th century physics: Ohm's and Kirchhoff's laws.

In the more general context of weighted graphs, we interpret m edges between i and j as a single edge with resistance $1/m$ (by the parallel law for resistance). Thus it is better in this case to think of the weights as the *conductances*, where conductance is the reciprocal of resistance.

Ohm's law and Kirchhoff's laws

- *Ohm's law*: the potential drop in each edge is the product of the current and the resistance (and so is equal to the current since we have set all edge resistances to 1).
- *Kirchhoff's voltage law*: the sum of the potential drops on any path between vertices i and j is independent of the choice of path.
- *Kirchhoff's current law*: if vertex i is not connected to the battery, then the sum of the currents flowing into i is equal to the sum of the currents flowing out.

Resistance distance

We begin by showing that effective resistance defines a metric on the vertex set of a connected graph.

Theorem 3.2.29 *R is a metric on the vertex set of the graph.*

The proof depends on a lemma:

Lemma 3.2.30 *If we connect the terminals of a battery to two vertices i and j, then the potential of any other vertex lies between the potentials of i and j.*

Proof of the lemma First observe that, if h is any vertex other than a terminal, then the net current into h is zero; by Ohm's law, this implies that the potential on h is the average of the potentials of its neighbours.

Let h be a vertex with smallest potential, and suppose that it is a non-terminal vertex. Then all neighbours of h must have the same potential as h.

By induction and connectedness, all vertices have the same potential, a contradiction.

So the smallest potential is realized at a terminal, as required. The same argument applies also to the largest potential. □

Proof of the theorem Clearly $R(i, j)$ is nonnegative, and zero only if $i = j$; also it is symmetric in i and j. We have to prove the triangle inequality: $R(i, j) + R(j, k) \geq R(i, k)$.

Let $r_1 = R(i, j)$, and $r_2 = R(j, k)$. We define two current flows in the graph as follows. For the first one, we connect a battery with voltage r_1 to i and j (with i at the higher potential). This causes unit current to flow out of i and into j. If P_1 denotes the corresponding potential, then $P_1(i) - P_1(j) = r_1$, and $P_1(j) - P_1(k) \leq 0$, by the lemma. So $P_1(i) - P_1(k) \leq r_1$.

Similarly, a battery of voltage r_2 connected to j and k causes a unit current to flow out of j and into k; the potential satisfies $P_2(j) - P_2(k) = r_2$, and $P_2(i) - P_2(j) \leq 0$. So $P_2(i) - P_2(k) \leq r_2$.

Since Ohm's and Kirchhoff's laws are linear in the potentials and currents, we can add these two solutions and obtain another solution. The result has a unit current flowing out of i and into k. With $P = P_1 + P_2$, we thus have $P(i) - P(k) = R(i,k)$, and so

$$R(i,k) = (P_1(i) - P_1(k)) + (P_2(i) - P_2(k)) \leq r_1 + r_2. \qquad \square$$

Definition 3.2.31 (Resistance distance)   In view of this theorem, we refer to the metric $R(i,j)$ as the *resistance distance*.

This is an important metric, with many applications. Michael Kagan's conference talk at G2D2 gave much further information from the points of view of both mathematics and physics: see [22]. His talk also included other methods of computing the metric, and applications to a process called *resistance distance transform* of a graph, related to Weisfeiler–Lehman stabilization and the Graph Isomorphism Problem.

Resistance distance and the Laplacian

We now show how the resistance distance can be calculated from the Laplacian matrix of the graph.

Theorem 3.2.32   *Let G be a connected graph with Laplacian L. Then the effective resistance between i and j is*

$$L_{ii}^- + L_{jj}^- - L_{ij}^- - L_{ji}^-,$$

where L^- is the Moore–Penrose inverse of L.

Proof   Kirchhoff's voltage law and Ohm's law are taken care of if we take a vector p of potentials with components indexed by vertices, and require that the current on the edge e is equal to the potential difference between its ends. (As before, we take a fixed orientation of each edge, and take the current to be negative if it flows from head to tail of the edge.) Note that p is defined up to adding a constant vector.

This is expressed by the requirement that Qp is the vector of currents in the edges, where Q is the signed incidence matrix with respect to this orientation.

Then $Q^\top Qp = Lp$ is a vector whose ith entry is the sum of the signed currents into the vertex i. So Kirchhoff's current law says that $Q^\top Qp$ has all entries zero except at the two vertices connected to the battery. If the total current flowing from the battery is 1 amp, the entries in $Lp = Q^\top Qp$ are $+1$ and -1 on these two vertices. Let us write $Lp = f_i - f_j$, where f_i is the unit basis vector corresponding to vertex i.

Now $f_i - f_j$ is orthogonal to the all-1 vector, so $L^-(f_i - f_j) = p$. This gives the vector of potentials. The potential difference between i and j is the dot product of this vector with $f_i - f_j$, that is, $x^\top L^- x$, where $x = f_i - f_j$. This is the potential difference required to make a current of 1 amp flow; hence it is the effective resistance between i and j. This can be written

$$R(i,j) = L_{ii}^- + L_{jj}^- - L_{ij}^- - L_{ji}^-,$$

as required. □

The average pairwise resistance

One of our criteria for a good network is that the average pairwise resistance between two vertices should be small. The next theorem, the resistance theorem, shows that this is equivalent to maximizing the *harmonic mean* of the nontrivial Laplacian eigenvalues.

Theorem 3.2.33 *The average pairwise resistance is equal to 2 divided by the harmonic mean of the nontrivial Laplacian eigenvalues.*

Proof The sum of the resistances between all ordered pairs of vertices is

$$\sum_{i \neq j} R(i,j) = 2(n-1)\operatorname{Trace}(L^-) - 2\sum_{i \neq j} L_{ij}^- = 2n\operatorname{Trace}(L^-),$$

since the sum of all elements of L^- is zero (as the all-1 vector is an eigenvector with eigenvalue 0).

So the average pairwise resistance is $2\operatorname{Trace}(L^-)/(n-1)$.

Now the trace of L^- is the sum of the reciprocals of the nonzero eigenvalues of L, and so we are done. □

Examples

We return to Example 3.2.1. For the Petersen graph, the harmonic mean of the nontrivial eigenvalues is

$$(((5 \cdot 1/2) + (4 \cdot 1/5))/9)^{-1} = 30/11,$$

so the average resistance is $11/15$.

For the other graph, a similar calculation gives $135/103$, so the average resistance is $206/135$.

As we will see, it is often the case that the resistances can be calculated without computing the Moore–Penrose inverse of the Laplacian.

For the Petersen graph, we can exploit symmetry to calculate the resistance between two terminals. In Figure 3.20, we use the two lowest vertices as terminals. There is a symmetry group of order 4 fixing these vertices; its equivalence classes or orbits on the remaining vertices are labelled a, b, c in the left-hand side of the figure. Two vertices equivalent under a symmetry fixing the terminals must be at the same potential, and so they can be identified with each other, and edges between them can be neglected. This gives the graph shown on the right, with two or four parallel edges in place of the edges labelled 2 and 4. Then the path round the top of the pentagon has resistance $1/2 + 1/4 + 1/4 + 1/2 = 3/2$. This is in parallel with the edge at the bottom, so the overall resistance is $1/(1 + 2/3) = 3/5$.

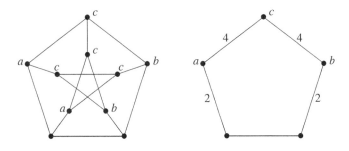

Figure 3.20 Using symmetry in the Petersen graph

Similar but slightly more complicated arguments give the resistance between non-adjacent terminals as $4/5$. (Exercise: prove this using symmetry.)

So the total is $15 \cdot 3/5 + 30 \cdot 4/5 = 33$, and the average is $33/45 = 11/15$, agreeing with the eigenvalue calculation.

Another technique for sparse graphs is given in the next section.

3.2.8 The Matrix-Tree Theorem

Definition 3.2.34 (Minor; cofactor)   Let A be an $n \times n$ matrix. The (i, j) *minor* is the determinant of the matrix obtained by deleting the ith row and jth column of A; the (i, j) *cofactor* is obtained by multiplying this minor by $(-1)^{i+j}$.

Theorem 3.2.35 (Matrix-Tree Theorem)   *Let G be a connected graph on n vertices. Then the following three quantities are equal:*

(1) *the number of spanning trees of G;*

(2) $(\theta_2 \cdots \theta_n)/n$, where $\theta_2, \ldots, \theta_n$ are the nontrivial Laplacian eigenvalues of G;

(3) any cofactor of $L(G)$.

Since one of our criteria for a good network is a large number of spanning trees, this is equivalent to maximizing the geometric mean of the nontrivial Laplacian eigenvalues.

The proof depends on the *Cauchy–Binet formula* , which says the following:

Theorem 3.2.36 (Cauchy–Binet formula)   *Let A be a $p \times q$ matrix, and B a $q \times p$ matrix, where $p < q$. Then*

$$\det(AB) = \sum_X \det(A(p,X)) \det(B(X,p)),$$

where X ranges over all p-element subsets of $\{1,\ldots,q\}$. Here $A(p,X)$ is the $p \times p$ matrix whose columns are the columns of A with index in X, and $B(X,p)$ is the $p \times p$ matrix whose rows are the rows of B with index in X.

The proof is an exercise.

Proof of the Matrix-Tree Theorem   Let Q be the signed incidence matrix of G, so that $Q^\top Q = L$. Let i be any vertex of G, and let $N = Q_i$ be the matrix obtained by deleting the column of Q indexed by i.

If X is a set of $n-1$ edges, then $\det(N(X, n-1)) = \det(N^\top(n-1, X))$ is ± 1 if X is the edge set of a spanning tree, and is 0 otherwise. (The proof of this is also an exercise.)

By the Cauchy–Binet formula, $\det(N^\top N)$ is equal to the number of spanning trees. But $N^\top N$ is the principal cofactor of $L(G)$ obtained by deleting the row and column indexed by i.

To finish the proof, let A be any square matrix with row and column sums zero, and let $B = A + J$, where J is the all-1 matrix of the same size as A. We evaluate $\det(B)$.

- Replace the first row by the sum of all the rows; this makes the entries in the first row n and doesn't change the other entries; the determinant is unchanged.
- Replace the first column by the sum of all the columns. This makes the first entry n^2, and the other entries in this column n, and doesn't change the other entries of the matrix; the determinant is unchanged.
- Subtract $1/n$ times the first row from each other row. The elements of the first column, other than the first, become 0; we subtract 1 from all elements

not in the first row or column of B, leaving the entries of A; and the determinant is unchanged.

We conclude that $\det(B)$ is n^2 times the $(1,1)$ cofactor of A.

It is easily checked that the argument works for any cofactor of A. So all cofactors of A are equal.

Finally, the all-1 vector is an eigenvector of B with eigenvalue n, while its other eigenvalues are the same as those of A. Thus $\det(B)$ is n times the product of the nontrivial eigenvalues of A.   □

We give now a simpler proof that all the cofactors of the Laplacian matrix are equal. The basic idea in this proof is due to Paul Glendinning, but we have made a simplification. First we remind the reader of the adjoint of a square matrix and its basic property.

Definition 3.2.37 (Adjoint)   The *adjoint* of A, $\mathrm{Adj}(A)$, is the matrix whose (i,j) entry is the (j,i) cofactor of A.

Lemma 3.2.38   *For any square matrix A,*

$$A \cdot \mathrm{Adj}(A) = \det(A)I.$$

Proof   The (i,i) entry of $A \cdot \mathrm{Adj}(A)$ is equal to

$$\sum_{j=1}^{n} a_{ij}\,\mathrm{Adj}(A)_{ji} = \sum_{j=1}^{n} a_{ij}C_{ij} = \det(A),$$

where C_{ij} is the (i,j) cofactor, since this is the expansion of the determinant using the ith row.

For $k \neq i$, the (k,i) entry is

$$\sum_{j=1}^{n} a_{kj}C_{ij}.$$

This is the expansion using the ith row of the determinant of the matrix B obtained by replacing the ith row of A by the kth. (This doesn't alter cofactors of elements in the ith row.) But B has two equal rows, so $\det(B) = 0$.   □

Proposition 3.2.39   *Let A be a real $n \times n$ matrix with all column sums 0. Then all columns of $\mathrm{Adj}(A)$ are equal.*

Proof   If the rank of A is less than $n - 1$, then all cofactors are zero, and there is nothing to prove. So assume that the rank is $n - 1$. Let j denote the all-1 row vector. By hypothesis, $jA = 0$; so the row null space of A is spanned by j. But $\mathrm{Adj}(A)A = 0$; so the rows of $\mathrm{Adj}(A)$ are multiples of j, which implies that all columns are equal.   □

Now let A be a square matrix whose row sums and column sums are all 0. Then all columns of $\mathrm{Adj}(A)$ are equal, and dually all rows of $\mathrm{Adj}(A)$ are equal; so all entries of $\mathrm{Adj}(A)$ are equal, as required.

Cayley's formula

The Matrix-Tree Theorem gives us a simple proof of the famous formula of Cayley:

Theorem 3.2.40 (Cayley's formula)   *The number of spanning trees in the complete graph on n vertices is equal to n^{n-2}.*

For the Laplacian of the complete graph is $nI - J$, where J is the all-1 matrix; its nontrivial Laplacian eigenvalues are all equal to n, and so the number of spanning trees is $n^{n-1}/n = n^{n-2}$.

In our two examples, the number of spanning trees are 2000 and 576 respectively. (Exercise: prove this using our earlier calculations.)

The Jacobian group

Definition 3.2.41 (Jacobian group)   The *Jacobian group* $\mathrm{Jac}(G)$ of a graph G on n vertices (also known as the *Picard group*, *critical group*, or *sandpile group*) is defined to be the quotient of $\mathbb{Z}^n$ by the subgroup spanned by the rows of $L(G)$ over $\mathbb{Z}$.

$\mathrm{Jac}(G)$ is a finitely generated abelian group. If G is connected, then $L(G)$ has nullity 1, and so $\mathrm{Jac}(G)$ has one factor which is the infinite cyclic group $\mathbb{Z}$. So it has the form $\mathbb{Z} \oplus A$, where A is the torsion subgroup of $\mathrm{Jac}(G)$, and is a finite abelian group whose order is $T(G)$, the number of spanning trees of G. (The torsion subgroup is called the *reduced* Jacobian group.)

For example, $\mathrm{Jac}(K_n)$ is the direct sum of $\mathbb{Z}$ and $n-2$ copies of the cyclic group $\mathbb{Z}/(n)$ of order n; the order of the torsion subgroup is thus n^{n-2}, in agreement with Cauchy's formula.

We see that $\mathrm{Jac}(G)$ carries more information than $T(G)$, since different graphs may have the same number of spanning trees but different structures of the group $\mathrm{Jac}(G)$ (as direct sums of cyclic groups).

In his invited lecture at the G2D2 conference, Alexander Mednykh explained this material, and gave details of the calculation of the structure of $\mathrm{Jac}(G)$ for a number of graphs G, including circulants and cyclic covers with finite valency. See also [26].

However, we do not know an application of the Jacobian group in the theory of optimal design. Perhaps one of our readers can find one ...

3.2.9 Markov chains

See also Section 1.5.1.

Definition 3.2.42 (Markov chain; transition matrix)   A *Markov chain* on a finite state space S is a sequence of random variables with values in S which has no memory: the state at time $n + 1$ depends only on the state at time n.

A Markov chain is defined by a *transition matrix* P, with rows and columns indexed by S, where p_{ij} is the probability of moving from state i to state j in one time step.

As usual, the entries of P are nonnegative and the row sums are equal to 1.

Random walks

An important example of a Markov chain is the *random walk* on a graph G. The state space is the vertex set $V(G)$. At time n, if the process is at vertex i, it chooses at random (with equal probabilities) an edge containing i, and at stage $n + 1$ moves to the other end of this edge.

If the graph has no loops, then the probability of moving from i to j is $-L_{ij}/L_{ii}$, where L is the Laplacian. In particular, if the graph is regular with degree d, then $P = I - L/d$.

More generally, $P = I - D^{-1}L$, where D is the diagonal matrix whose (i,i) entry is the number of edges incident with i.

Theory of Markov chains

In this section, we suspend temporarily our usual convention that vectors are column vectors.

If a Markov chain has transition matrix P, then the (i, j) entry of P^m is the probability of moving from i to j in m steps.

The Markov chain is *connected* if, for any i and j, there exists m such that $(P^m)_{ij} \neq 0$; it is *aperiodic* if the the greatest common divisors of the values of m for which $(P^m)_{ii} \neq 0$ for some i is 1.

A random walk on a graph G is connected if and only if G is connected, and is aperiodic if and only if G is not bipartite.

Theorem 3.2.43   *A connected aperiodic Markov chain has a unique limiting distribution, to which it converges from any starting distribution.*

Since the row sums of P are all 1, we see that $Pp = p$, where p is the all-1 column vector; our assumptions imply that the multiplicity of 1 as eigenvalue is 1. Now left and right eigenvalues are equal, so there is a row vector $q \neq 0$ such that $qP = q$. It can be shown that the entries of q are nonnegative; we can normalize it so that their sum is 1. Then q is a probability distribution which is fixed by P, so it is the unique stationary distribution.

Suppose that P is symmetric. Then we can write $P = \sum \theta P_\theta$ where θ runs over the eigenvalues, and P_θ is the projection onto the θ eigenspace. Then $P^m = \sum \theta^m P_\theta$.

It is also true, by the *Perron–Frobenius theorem*, that every eigenvalue θ satisfies $|\theta| \leq 1$. If the Markov chain is irreducible and aperiodic, then 1 is a simple eigenvalue, and all other eigenvalues have modulus strictly less than 1.

Now let x be any nonnegative row vector whose coordinates sum to 1. (We suspend our usual convention here.) We can regard x as the initial probability distribution. Then we have

$$xP^m = \sum_\theta \theta^m x P_\theta \to x P_1$$

as $m \to \infty$. So $xP_1 = q$ is the limiting distribution, and the rate of convergence to q is like μ^m where μ is the second-largest modulus of an eigenvalue. So the convergence is exponential if μ is not close to 1.

Random walks revisited

For a random walk, we have $P = I - D^{-1}L$. Then

$$D^{1/2}PD^{-1/2} = I - D^{-1/2}LD^{-1/2}.$$

This matrix is symmetric, and the displayed equation shows that it is similar to P; so P is indeed diagonalizable. However, the analysis is a bit more complicated, and not given here.

Its eigenvalues are $1 - \theta$ for each eigenvalue θ of the nonnegative definite matrix $D^{-1/2}LD^{-1/2}$, so for rapid convergence we require that the smallest positive eigenvalue of this matrix should be as large as possible.

Thus the problem is a twisted version of the usual problem about the smallest nontrivial Laplacian eigenvalue. If the graph is regular, so that $D = dI$, then it reduces exactly to the former problem.

Other results

The smallest nontrivial Laplacian eigenvalue μ of a graph G is an important parameter which occurs in many other situations.

For example, a result of Krivelevich and Sudakov [24] asserts that, in a regular graph of valency d on n vertices, if μ is sufficiently large in terms of n and d, then the graph is Hamiltonian.

Summing up

We saw that three important parameters of a connected graph, which are determined by its Laplacian spectrum, are:

- the harmonic mean of the nontrivial Laplacian eigenvalues, which tells us about the average resistance between pairs of vertices;
- the geometric mean of the nontrivial Laplacian eigenvalues, which tells us about the number of spanning trees;
- the smallest nontrivial Laplacian eigenvalue, which is related to the isoperimetric number and the rate of convergence of the random walk on the graph.

In the next section, we will see that these are also important parameters in experimental design!

3.2.10 Exercises

Exercise 3.2.1(i)   Prove the Cauchy–Binet formula (page 219).

(ii)   Prove the assertion in the proof of the Matrix-Tree Theorem (page 219), that if X is a set of $n-1$ edges of a graph with n vertices and signed incidence matrix N then $\det(N(X))$ is ± 1 if X is the set of edges of a spanning tree, and is zero otherwise.

Exercise 3.2.2   Show that the adjacency matrix A of the Petersen graph satisfies

$$A^2 + A - 2I = J,$$

where I and J are the identity and all-1 matrices of order 10. Hence find the eigenvalues of A, and the Laplacian eigenvalues of the Petersen graph.

Exercise 3.2.3   Show that a distance-regular graph of diameter 2 is the same thing as a connected strongly regular graph, with $k = b_0$, $\lambda = a_1$, $\mu = c_2$.

Exercise 3.2.4   Let G be a connected graph of diameter d on the vertex set V. Show that the partition of V^2 into $d+1$ classes $C_0,\ldots,C_d$, where C_i is the set of pairs of vertices at distance i, is an association scheme if and only if G is distance-regular.

Exercise 3.2.5   A connected graph is said to be *distance-transitive* if, given vertices x,y,u,v with $d(x,y) = d(u,v)$, there is a graph automorphism carrying the pair (x,y) to the pair (u,v). Prove that a distance-transitive graph is distance-regular.

Exercise 3.2.6(i)   Prove that the numbers of spanning trees of the two graphs in Figure 3.16 are 2000 and 576 respectively.

(ii)   Show that the isoperimetric number of the Petersen graph is equal to 1.

Exercise 3.2.7   Use symmetry to show that the resistance between any two non-adjacent vertices of the Petersen graph is equal to $4/5$.

Exercise 3.2.8   Let A be the adjacency matrix of a graph G. Show that the following conditions are equivalent:

(i)   there is a polynomial F such that $F(A) = J$;
(ii) G is regular and connected.

Exercise 3.2.9   Why is it true that, for any real matrix A, the matrix $A^\top A$ is nonnegative definite?

3.3 Designs, graphs and optimality

This section returns to the topic of incomplete-block designs introduced in Section 3.1. We introduce two graphs associated with a block design, and then show how the ideas and results of Section 3.2 can be used to find block designs that are good for experiments.

Recall that a block design Γ consists of

- a set of bk experimental units (also called plots), partitioned into b blocks of size k;
- a set of v treatments;
- a function f from the experimental units onto the set of treatments, so that $f(\omega)$ denotes the treatment applied to experimental unit ω.

The block containing experimental unit ω is denoted $g(\omega)$, and N_{ij} denotes the number of occurrences of treatment i in block j.

3.3.1 Two graphs associated with a block design

Here we introduce two different graphs, $\tilde{G}$ and G, associated with a block design Γ.

Levi graph

Definition 3.3.1 (Levi graph)   The *Levi graph* $\tilde{G}$ of a block design Γ has

- one vertex for each treatment,
- one vertex for each block,
- one edge for each experimental unit, with edge ω joining vertex $f(\omega)$ to vertex $g(\omega)$.

It is a bipartite graph, with N_{ij} edges between treatment-vertex i and block-vertex j.

Historical note Friedrich W. Levi was a German Jewish mathematician who had to leave Nazi Germany in the 1930s. He moved to India, where he worked in the Mathematics Department at the University of Calcutta, and interacted with R. C. Bose. He invented this graph to describe a block design. Some later authors named it after him. Some other authors call it the *incidence graph*. In 1952, he returned to Germany, working at the Frei Universität Berlin.

Example 3.3.2 Figure 3.21(a) shows a block design with $v = 4$ and $b = k = 3$. Its Levi graph is in Figure 3.21(b). Here the vertices shown as black filled circles represent the treatments, and they are labelled accordingly. The vertices shown as white squares represent the blocks. Although they are not labelled, there is enough information in the Levi graph for us to reconstruct the design. The block labelled by the left-most white square contains treatments 1, 3 and 4, and so it must be the first block in Figure 3.21(a). The block labelled by the right-most white square contains treatments 2, 3 and 4, and so it must be the middle block in Figure 3.21(a). The remaining block is represented by the middle white square. Treatment 2 occurs twice in this block, and so there are two edges between this white square and the vertex for treatment 2. The remaining edge at this white square joins it to the vertex for treatment 1, which occurs once in this block.

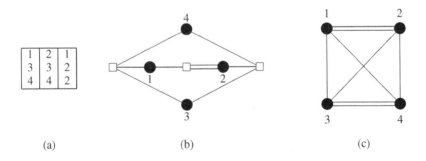

1	2	1
3	3	2
4	4	2

(a)   (b)   (c)

Figure 3.21 Block design, Levi graph and concurrence graph in Example 3.3.2

Example 3.3.3 Figure 3.22(a) shows a block design with $v = 8$, $b = 4$ and

$k = 3$. Because this design is binary, its Levi graph, shown in Figure 3.22(b), has no repeated edges.

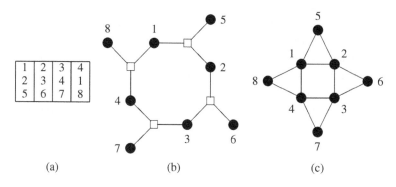

(a) (b) (c)

Figure 3.22 Block design, Levi graph and concurrence graph in Example 3.3.3

Up to naming of the blocks, a block design can always be recovered from its Levi graph.

Concurrence graph

Definition 3.3.4 (Concurrence graph) The *concurrence graph G* of a block design Γ has

- one vertex for each treatment,
- one edge for each unordered pair α, ω, with $\alpha \neq \omega$, $g(\alpha) = g(\omega)$ and $f(\alpha) \neq f(\omega)$: this edge joins vertices $f(\alpha)$ and $f(\omega)$.

There are no loops.

If $i \neq j$ then the number of edges between vertices i and j is

$$\lambda_{ij} = \sum_{s=1}^{b} N_{is}N_{js}.$$

As we saw in Section 3.1.4, this is called the concurrence of i and j, and is the (i, j)-entry of the matrix Λ, which is equal to $NN^{\top}$.

The graph in Figure 3.21(c) is the concurrence graph of the block design in Figure 3.21(a). It has more symmetry than the Levi graph, shown in Figure 3.21(b). However, the block design here cannot be recovered from its concurrence graph.

Likewise, Figure 3.22(c) shows the concurrence graph of the block design in Figure 3.22(a).

The Levi graph always has more vertices than the concurrence graph, so at first sight it seems more complicated. However, consideration of the numbers of edges may change this view.

The number of edges in the Levi graph is always equal to bk. The number of edges in the concurrence graph is equal to $bk(k-1)/2$ if the design is binary, and is somewhat less otherwise. For binary designs, the Levi graph has more edges when $k = 2$, the concurrence graph has more edges when $k \geq 4$, and they both have $3b$ edges when $k = 3$.

Figure 3.23 shows the concurrence graphs for the two block designs with $v = 15$, $b = 7$ and $k = 3$ in Figure 3.7.

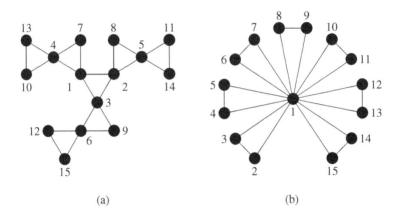

(a)   (b)

Figure 3.23 Concurrence graphs for the two block designs in Figure 3.7

3.3.2 Laplacian matrices

The Laplacian matrix of a general graph is defined in Section 3.2.3.

Concurrence graph

If we apply this definition to the concurrence graph G of a block design Γ we find that the the Laplacian matrix L of the concurrence graph G is a $v \times v$ matrix with (i, j)-entry as follows:

- if $i \neq j$ then $L_{ij} = -(\text{number of edges between } i \text{ and } j) = -\lambda_{ij}$;
- $L_{ii} = \text{valency of } i = \sum_{j \neq i} \lambda_{ij}$.

The off-diagonal entries are the same as those of $-\Lambda$. The diagonal entries make each row sum to zero.

So the graph-theoretic definition of Laplacian matrix gives us exactly the Laplacian matrix L that we defined in Section 3.1.4.

Levi graph

What about the Levi graph? The definition in Section 3.2.3 shows that the Laplacian matrix $\tilde{L}$ of the Levi graph $\tilde{G}$ is a $(v+b) \times (v+b)$ matrix with (i,j)-entry as follows:

- $\tilde{L}_{ii}$ = valency of i

$$= \begin{cases} k & \text{if } i \text{ is a block} \\ \text{replication } r_i \text{ of } i & \text{if } i \text{ is a treatment;} \end{cases}$$

- if $i \neq j$ then $\tilde{L}_{ij} = -(\text{number of edges between } i \text{ and } j)$

$$= \begin{cases} 0 & \text{if } i \text{ and } j \text{ are both treatments} \\ 0 & \text{if } i \text{ and } j \text{ are both blocks} \\ -N_{ij} & \text{if } i \text{ is a treatment and } j \text{ is a block, or vice versa.} \end{cases}$$

This can be written more concisely as

$$\tilde{L} = \begin{bmatrix} R & -N \\ -N^\top & kI_b \end{bmatrix},$$

which is exactly the same as our previous definition of $\tilde{L}$ in Section 3.1.11.

Connectivity

All row-sums of L and of $\tilde{L}$ are zero, so both matrices have 0 as eigenvalue on the appropriate all-1 vector.

In Section 3.1.9 we defined a block design to be *connected* if 0 is a simple eigenvalue of L. This seems a bit arbitrary: why not a simple eigenvalue of $\tilde{L}$? The word *connected* suggests that we are really thinking about a graph. Which one?—the concurrence graph G or the Levi graph $\tilde{G}$? But what the statistician is really interested in is whether or not all simple treatment differences $\tau_i - \tau_j$ can be estimated without bias.

Fortunately, it turns out that these ideas are all equivalent. The following theorem is proved in [5].

Theorem 3.3.5 *The following are equivalent:*

(1) *the value 0 is a simple eigenvalue of L;*
(2) *the concurrence graph G is a connected graph;*
(3) *the Levi graph $\tilde{G}$ is a connected graph;*

(4) *the value 0 is a simple eigenvalue of* $\tilde{L}$;

(5) *the block design* Γ *is connected in the sense that all differences between treatments can be estimated without bias.*

From now on, assume connectivity.

Call the remaining eigenvalues *nontrivial*. They are all nonnegative.

3.3.3 Estimation and variance

We want to estimate all the simple differences $\tau_i - \tau_j$. We want all of these estimators to be unbiased, as discussed in Section 3.1.2, so that they give the correct value on average.

Write V_{ij} for the variance of the best linear unbiased estimator for $\tau_i - \tau_j$. Why does this variance matter? This was partly explained in Sections 3.1.2 and 3.1.10, but now we go a bit further.

When statisticians report the results of an experiment, they report not only the estimates of the treatment differences but also so-called *confidence intervals* for those estimates. What does this mean? If we always present results using a 95% confidence interval, then our interval will contain the true value in 19 cases out of 20. Similarly, if we always present results using a 99% confidence interval, then our interval will contain the true value in 99 cases out of 100, but the cost is that this is a wider interval than the 95% one.

If the random variables have a joint normal distribution and a linear unbiased estimator is used, then the length of the 95% confidence interval for $\tau_i - \tau_j$ is proportional to $\sqrt{V_{ij}}$. The smaller the value of V_{ij}, the smaller is the confidence interval, the closer is the estimate to the true value (on average), and the more likely are we to detect correctly which of τ_i and τ_j is bigger.

This has practical consequences. We can make better decisions about new drugs, about new varieties of wheat, about new engineering materials ... if we make all the V_{ij} small.

We saw in Section 3.1 that

$$V_{ij} = \left(L_{ii}^- + L_{jj}^- - 2L_{ij}^- \right) k\sigma^2 = \left(\tilde{L}_{ii}^- + \tilde{L}_{jj}^- - 2\tilde{L}_{ij}^- \right) \sigma^2. \tag{3.14}$$

3.3.4 Resistance distance

Concurrence graph

We can consider the concurrence graph G as an electrical network, and define the resistance distance R_{ij} between any pair of distinct vertices i and j.

Remark 3.3.6 The resistance distance R_{ij} was written as $R(i, j)$ for an arbitrary graph in Section 3.2. Here we want to use notation specific to the concurrence graph.

Theorem 3.2.32 shows that $R_{ij} = L_{ii}^- + L_{jj}^- - 2L_{ij}^-$. Comparing this with Equation (3.14) gives the following result.

Theorem 3.3.7 *If V_{ij} is the variance of the best linear unbiased estimator of $\tau_i - \tau_j$ and R_{ij} is the resistance distance between vertices i and j in the concurrence graph G, then $V_{ij} = R_{ij} \times k\sigma^2$.*

One advantage of this result is that resistance distances are easy to calculate without matrix inversion if the graph is sparse.

How do we do these calculations? This is how the authors of this chapter do it. If we want to calculate the resistance distance between vertices i and j, we start by assigning voltage $[0]$ at vertex i. Then we send a current x along one of the edges out of i. We are not physicists, so we show the electricity running uphill, and the vertex at the end of that edge gets allocated voltage $[x]$.

We apply one of Kirchhoff's laws at each vertex, and the other of Kirchhoff's laws in each edge.

When we reach vertex j, there are some linear equations to solve, enabling us to give the voltage $[V]$ at vertex j and then calculate the total current I flowing from vertex i to vertex j.

Ohm's law gives $V = IR$, which we use to calculate R_{ij} as V/I.

Example 3.3.8 Figure 3.24 shows the concurrence graph of a block design with $v = 12$, $b = 6$ and $k = 3$. Two vertices are labelled 1 and 2: we want to calculate the resistance distance between them.

Begin by assigning voltage $[0]$ at vertex 1. Send a current x along the edge running down from vertex 1 in the diagram. Ohm's law says that we have to assign voltage $[x]$ to the vertex at the other end of this edge. Then Kirchhoff's current law says that current x flows from that vertex to the only other one to which it is joined, which then has voltage $[2x]$. Hence the current in the direct edge from vertex 1 to that vertex is $2x$.

The pattern in the next triangle to the right is similar. Because the total current leaving the vertex common to these triangles must be equal to the total current into that vertex, we get the values shown.

Now go round in the other direction from vertex 1, starting with a current y in the edge to its left. At this point, we do not know how y is related to x. A similar argument to the one before now assigns multiples of y to the edges and vertices as we go around the diagram clockwise, until we reach a vertex which is joined to vertex 2.

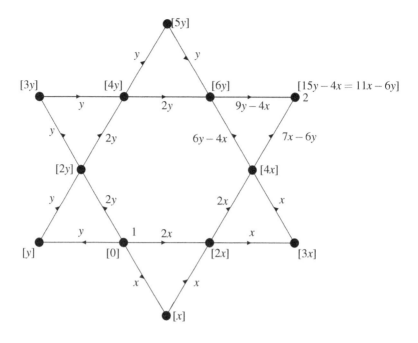

Figure 3.24 Calculating resistance distance in the concurrence graph of a
block design with $v = 12$, $b = 6$ and $k = 3$

Now the current from the vertex with voltage $[4x]$ to the one with voltage
$[6y]$ must be $6y - 4x$. Applying Kirchhoff's current law to the former shows
that the current in the edge from it to vertex 2 is $7x - 6y$. Hence the voltage at
vertex 2 is $11x - 6y$. Likewise, the current flowing into vertex 2 in its other edge
is $9y - 4x$, so the voltage at vertex 2 is $15y - 4x$. Equating these two voltages
shows that $15x = 21y$, so that $5x = 7y$. For the purposes of the calculation, we
can simply put $x = 7$ and $y = 5$. This gives the labelled graph in Figure 3.25.

Figure 3.25 shows that the potential difference between vertices 1 and 2
is $V = 47 - 0 = 47$. The total current flowing from vertex 1 to vertex 2 is $I = 7 + 14 + 5 + 10 = 36$. Hence the effective resistance is $R_{12} = R = V/I = 47/36$.
Theorem 3.3.7 then shows that $V_{12} = 3R_{12}\sigma^2 = (47/12)\sigma^2$.

Remark 3.3.9   Writing out the calculation in Example 3.3.8 in words takes
quite a long time, as does reading those words. We should like to assure the
reader that writing the notation on a diagram of the graph, and doing the relevant calculations, is usually much quicker.

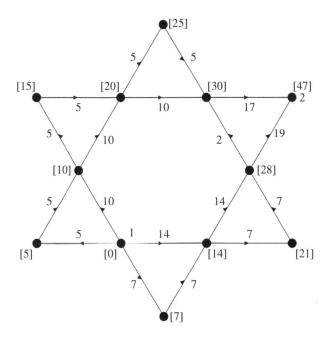

Figure 3.25 One possible set of values for currents and voltages in the graph in Figure 3.24

Levi graph

Likewise, we can consider the Levi graph $\tilde{G}$ as an electrical network, and define the resistance distance $\tilde{R}_{ij}$ between any pair of distinct vertices i and j. We are interested in this chiefly when i and j are treatment vertices. In this case, Equation (3.14) and Theorem 3.2.32 give the following.

Theorem 3.3.10 *If V_{ij} is the variance of the best linear unbiased estimator of $\tau_i - \tau_j$ and $\tilde{R}_{ij}$ is the resistance distance between vertices i and j in the Levi graph $\tilde{G}$, then $V_{ij} = \tilde{R}_{ij} \times \sigma^2$.*

For the block design in Example 3.3.3, we can use the Levi graph in Figure 3.22(b) to calculate the variance V_{46}. There can be no current in any edge with a vertex of degree 1 other than vertex 6. Denote by B the single block containing treatment 6. The Levi graph has just two paths between vertices 4 and B. One of these paths has five edges, the other three. It is easy to see that Kirchhoff's laws are satisfied if we send a current of 3 amps in each edge of the longer path and a current of 5 amps in each edge of the shorter path: see Figure 3.26. This gives a potential of 15 volts at vertex B, and a total incoming current there of 8 amps. Hence there is a current of 8 amps in the edge from vertex B to vertex 6, and so the potential at vertex 6 is $15 + 8$, which is 23. Thus $\tilde{R}_{46} = 23/8$. By Theorem 3.3.10, $V_{46} = (23/8)\sigma^2$.

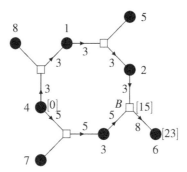

Figure 3.26 Calculating resistance distance in the Levi graph of the block design in Example 3.3.3

Now let us return to the block design in Example 3.3.8. Its Levi graph is shown in Figure 3.27, where B denotes the single block containing the vertex 2. There are two disjoint paths from vertex 1 to vertex B, of lengths 5 and 7. These have resistances 5 and 7 in parallel, so the rule for combining resistances in parallel gives

$$\tilde{R}_{1B} = \frac{1}{\frac{1}{5} + \frac{1}{7}} = \frac{1}{\frac{12}{35}} = \frac{35}{12}.$$

On the other hand, resistances in series are simply added together, and so

$$\tilde{R}_{12} = \tilde{R}_{1B} + \tilde{R}_{B2}.$$

The single path between vertices B and 2 gives $\tilde{R}_{B2} = 1$, and so

$$\tilde{R}_{12} = \tilde{R}_{1B} + 1 = \frac{35}{12} + 1 = \frac{47}{12}.$$

Hence $V_{12} = (47/12)\sigma^2$, which agrees with the calculation using the concurrence graph. As often happens, the calculation using the Levi graph is simpler.

For hand calculation when the graphs are sparse, or for calculations for 'general' graphs with variable v, it may be simpler to use the Levi graph rather than the concurrence graph if $k \geq 3$.

3.3.5 Spanning trees

It turns out that there is a simple relationship between the numbers of spanning trees in the two graphs which interest us.

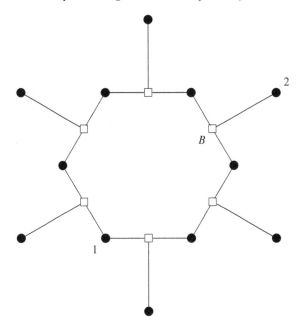

Figure 3.27 Levi graph of the block design in Example 3.3.8

Theorem 3.3.11 *Let G and $\tilde{G}$ be the concurrence graph and Levi graph for a connected incomplete-block design for v treatments in b blocks of size k. Then the number of spanning trees for $\tilde{G}$ is equal to k^{b-v+1} times the number of spanning trees for G.*

Proof Let t and $\tilde{t}$ be the number of spanning trees for G and $\tilde{G}$ respectively. Then the Matrix-Tree Theorem shows that $t = \det L_1$ and $\tilde{t} = \det \tilde{L}_1$, where the subscript 1 denotes the removal of the row and column corresponding to treatment 1.

Extending this subscript notation to the matrices R and N gives $\det L_1 = \det(kR_1 - N_1 N_1^\top)$. Furthermore,

$$
\det \tilde{L}_1 = \det \begin{bmatrix} R_1 & -N_1 \\ -N_1^\top & kI_b \end{bmatrix} = \det \begin{bmatrix} R_1 - k^{-1}(N_1)N_1^\top & -N_1 \\ -N_1^\top + k^{-1}(kI_b)N_1^\top & kI_b \end{bmatrix}
$$

$$
= \det \begin{bmatrix} k^{-1}L_1 & -N_1 \\ 0 & kI_b \end{bmatrix} = k^{-(v-1)} \det L_1 \times k^b,
$$

so $\tilde{t} = \det \tilde{L}_1 = k^{b-v+1} \det L_1 = k^{b-v+1} t.$ □

If $v \geq b+2$ then $\tilde{t} < t$, so a good strategy is to count the number of spanning trees for the Levi graph, then multiply by k^{v-b-1} to obtain the number of spanning trees for the concurrence graph. If $v \leq b$ then do it the other way round.

For example, consider the two graphs in Figure 3.22 for the block design in Example 3.3.3 with $v = 8$, $b = 4$ and $k = 3$. The Levi graph contains four edges to vertices of valency one, so every spanning tree must contain all of those edges. The remaining eight edges form a cycle, so a spanning tree is made by the removal of any one of those. Hence $\tilde{t} = 8$. Theorem 3.3.11 shows that $\tilde{t} = 3^{-3}t$, and so $t = 216$. Counting the number of spanning trees of the concurrence graph directly would be much more complicated.

3.3.6 Measures of optimality

Here we give a very brief introduction to the subject of optimal block designs. For more details, see the book [30], especially Chapters 1–3.

In Section 3.1 we saw that we want all the pairwise variances V_{ij} to be small. If there are more than two treatments, this is a multi-dimensional problem, so there are several different ways of measuring what is best. Here we concentrate on the three measures of optimality which are most often used by statisticians.

Average pairwise variance

The variance of the best linear unbiased estimator of the simple difference $\tau_i - \tau_j$ is

$$V_{ij} = \left(L_{ii}^- + L_{jj}^- - 2L_{ij}^- \right) k\sigma^2 = R_{ij}k\sigma^2.$$

We want all of these values to be small. Now

$$\frac{1}{k\sigma^2} \sum_{1 \leq i < j \leq v} V_{ij} = (v-1)\sum_i L_{ii}^- - \sum_{i \neq j} L_{ij}^- = v\sum_i L_{ii}^- = v\,\mathrm{Trace}(L^-)$$

because the row sums of L^- are zero.

Let the nontrivial eigenvalues of L be $\theta_1, \ldots, \theta_{v-1}$. Then

$$\mathrm{Trace}(L^-) = \frac{1}{\theta_1} + \cdots + \frac{1}{\theta_{v-1}}.$$

Let $\bar{V}$ be the average value of the V_{ij}. Then

$$\bar{V} = \frac{\sum_{1 \leq i < j \leq v} V_{ij}}{v(v-1)/2} = \frac{k\sigma^2}{v(v-1)/2} \times v\left(\frac{1}{\theta_1} + \cdots + \frac{1}{\theta_{v-1}} \right)$$

$$= 2k\sigma^2 \times \frac{1}{\text{harmonic mean of } \theta_1, \ldots, \theta_{v-1}}.$$

In view of Theorem 3.3.7, the foregoing statement is a direct consequence of Theorem 3.2.33.

Definition 3.3.12 (A-optimal)   A block design is *A-optimal* if it minimizes the average $\bar{V}$ of the variances V_{ij} over all block designs which have v treatments in b blocks of size k.

From what we have just shown, this is equivalent to maximizing the harmonic mean of the nontrivial eigenvalues of the Laplacian matrix L over all connected block designs for v treatments in b blocks of size k.

If the design is binary then all diagonal elements of L are equal to $r(k-1)$, and so $\theta_1 + \cdots + \theta_{v-1} = vr(k-1)$. If these nontrivial eigenvalues are all equal then their harmonic mean is

$$\frac{vr(k-1)}{v-1},$$

and so

$$V_{ij} = \bar{V} = 2k\sigma^2 \times \frac{v-1}{vr(k-1)},$$

as we saw in Section 3.1.10 for balanced incomplete-block designs.

If $\theta_1 \ldots, \theta_{v-1}$ are not all equal then their harmonic mean is smaller and so $\bar{V}$ is larger.

Confidence region

When $v > 2$ the generalization of confidence interval is the confidence ellipsoid around the point $(\hat{\tau}_1, \ldots, \hat{\tau}_v)$ in the hyperplane in $\mathbb{R}^v$ with $\sum_i \tau_i = 0$. The volume of this confidence ellipsoid is proportional to

$$\sqrt{\prod_{i=1}^{v-1} \frac{1}{\theta_i}} = (\text{geometric mean of } \theta_1, \ldots, \theta_{v-1})^{-(v-1)/2}$$

$$= \frac{1}{\sqrt{v \times \text{number of spanning trees for } G}},$$

by the Matrix-Tree Theorem.

Definition 3.3.13 (D-optimal)   A block design is *D-optimal* if it minimizes the volume of the confidence ellipsoid for $(\hat{\tau}_1, \ldots, \hat{\tau}_v)$ over all connected block designs for v treatments in b blocks of size k.

From what we have just seen, it is equivalent to say that a block design is D-optimal if it maximizes the geometric mean of the nontrivial eigenvalues of the Laplacian matrix L, or to say that it maximizes the number of spanning trees for the concurrence graph G. Theorem 3.3.11 shows that it is also equivalent to say that it maximizes the number of spanning trees for the Levi graph $\tilde{G}$.

Worst case

Let x be a contrast in $\mathbb{R}^v$. In Section 3.1.10 we showed that the variance of the best linear unbiased estimator of $x^\top \tau$ is $(x^\top L^- x)k\sigma^2$. If we multiply every entry in x by a constant c then this variance is multiplied by c^2; and $x^\top x$ is also multiplied by c^2. The worst case is for contrasts x giving the maximum value of

$$\frac{x^\top L^- x}{x^\top x}.$$

These are precisely the eigenvectors x of L with eigenvalue θ_1, where θ_1 is the smallest nontrivial eigenvalue of L.

Definition 3.3.14 (E-optimal) A block design is *E-optimal* if it maximizes the smallest nontrivial eigenvalue of the Laplacian matrix L over all connected block designs for v treatments in b blocks of size k.

3.3.7 Some optimal designs

Balanced incomplete-block designs

The most well-known theorem about optimal designs was proved in [23, 25].

Theorem 3.3.15 (Kshirsagar, 1958; Kiefer, 1975) *If a balanced incomplete-block design for v treatments in b blocks of size k exists, then it is A-, D- and E-optimal. Moreover, in this case, no non-BIBD is A-, D- or E-optimal.*

Proof Let $T = \text{Trace}(L)$. For any given value of T, the harmonic mean of θ_1, ..., θ_{v-1}, the geometric mean of θ_1, ..., θ_{v-1}, and the minimum of θ_1, ..., θ_{v-1} are all maximized at $T/(v-1)$ when $\theta_1 = \cdots = \theta_{v-1} = T/(v-1)$. This occurs if and only if L is a scalar multiple of $I_v - v^{-1}J_v$.

Since $T = \sum_i(kr_i - \lambda_{ii}) = bk^2 - \sum_i \lambda_{ii}$, the trace is maximized if and only if the design is binary. Among binary designs, the off-diagonal elements of L are equal if and only if the design is balanced. $\square$

Folklore about optimal designs

For many years after it was known that BIBDs are A-, D- and E-optimal, statisticians assumed the following statements, which became a kind of folklore.

- Only binary incomplete-block designs can be optimal.
- Optimal incomplete-block designs must have all their treatment replications as equal as possible.
- Optimal incomplete-block designs must have all their concurrences as equal as possible.

- If an incomplete-block design is optimal on one of the three optimality criteria, then it must be optimal, or close to optimal, on the other two optimality criteria.

So we restricted our search for optimal designs, using these assumptions.

Now we know that all four assumptions are wrong. Here is an example to show that an optimal design may not be binary.

Example 3.3.16 Consider the two designs in Figure 3.8 with $v = 5$, $b = 7$ and $k = 3$. The one in Figure 3.8(a) is binary; the one in Figure 3.8(b) is not. Their Laplacian matrices are given in (3.3) and (3.4) respectively. The former has maximum trace, because the design is binary, while the latter is completely symmetric, and so all of its nontrivial eigenvalues are equal. Figure 3.28 shows their concurrence graphs.

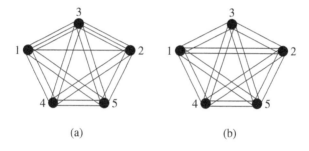

(a)   (b)

Figure 3.28 Concurrence graphs for the two designs in Example 3.3.16

The Laplacian matrix of the design in Figure 3.8(a) has trace 42. Its nontrivial eigenvalues are 13, 10, 10 and 9, with harmonic mean equal to 10.31, geometric mean equal to 10.40 and smallest equal to 9. On the other hand, the nontrivial eigenvalues of the Laplacian matrix of the design in Figure 3.8(b) are all equal to 10, so their harmonic mean, geometric mean and smallest are also all equal to 10. Thus the design in Figure 3.8(a) beats the design in Figure 3.8(b) on the A-criterion and the D-criterion, but is worse on the E-criterion.

In fact, we now know that the design in Figure 3.8(b) is E-optimal for these values of v, b and k. See Section 3.4.3.

Other classes of optimal design

Other general results about classes of optimal designs are fairly limited. Typically either the replications or the nontrivial eigenvalues must all be the same; otherwise, the number of different concurrences, or nontrivial eigenvalues, should be no more than two. Often the class of competing designs has to be restricted.

Here are some of these other results. See [14, 15, 16].

Theorem 3.3.17 (Chêng, 1978)   *Group-divisible designs with two groups in which the between-group concurrence is one more than the within-group concurrence are A-, D- and E-optimal.*

Theorem 3.3.18 (Chêng, 1981)   *Group-divisible designs (with an arbitrary number of groups) in which the between-group concurrence is one more than the within-group concurrence are A-, D- and E-optimal among equireplicate designs whose concurrences differ by at most one.*

Theorem 3.3.19 (Cheng and Bailey, 1991)   *Partially balanced designs with two associate classes, in which the two concurrences differ by 1 and the matrix $(rk)^{-1}L = r^{-1}C$ has an eigenvalue equal to 1, are A-, D- and E-optimal among binary equireplicate designs.*

In particular, square lattice designs are A-, D- and E-optimal among binary equireplicate designs.

Remark 3.3.20   Generalized quadrangles are a special case of partially balanced incomplete-block designs with two associate classes which satisfy these conditions.

Duals

If Γ is an equireplicate incomplete-block design for v treatments in b blocks of size k, with replication r, then its *dual* design Γ' is obtained by interchanging the roles of blocks and treatments, so it has b treatments in v blocks of size r. Even if Γ satisfies the restrictive conditions in any of Theorem 3.3.15 and Theorems 3.3.17–3.3.19, the design Γ' may not. Nevertheless, their optimality properties are linked.

Let G' be the concurrence graph of Γ'. The Levi graphs of the two designs are both equal to $\tilde{G}$. Let t' be the number of spanning trees for G'. Applying Theorem 3.3.11 to Γ' gives $k^{b-v+1}t = \tilde{t} = r^{v-b+1}t'$. Hence t' is maximized if and only if t is maximized. This proves the following.

Theorem 3.3.21   *Among the class of equireplicate block designs for given values of v, b, k and r, a design is D-optimal if and only if its dual design is D-optimal.*

For example, one balanced incomplete-block design for seven treatments in 21 blocks of size three can be made by taking three copies of the design in Figure 3.9(a). The dual design has 21 treatments in seven blocks of size nine.

Some concurrences are equal to three; the rest are equal to one. Nevertheless, this dual design is D-optimal.

There are analogous results for A-optimality and E-optimality, but the proof is messier. We now give the details.

Theorem 3.3.22 *Let Γ be a connected equireplicate incomplete-block design for v treatments replicated r times in b blocks of size k, where $b \geq v$. Let L and L' be the Laplacian matrices of the concurrence graphs G and G' of Γ and Γ'. Then the eigenvalues of L' are the same of those of L, including multiplicity, plus a further $b - v$ eigenvalues all equal to rk.*

Proof Theorem 3.3.5 shows that 0 is a simple eigenvalue of $\tilde{L}$ and hence of L'. The corresponding eigenspace in $\mathbb{R}^b$ consists of the constant vectors.

Equal replication implies that $L = rkI_v - NN^T$. By Theorem 3.2.14, L is nonnegative definite. So is the matrix $NN^\top$. Hence all the eigenvalues of NN^T must be in $[0, rk]$.

Let the vector x in $\mathbb{R}^v$ be an eigenvector of L with eigenvalue θ. Put $z = N^\top x$. Then $N^\top Nz = N^\top NN^\top x = N^\top (rkI_v - L)x = N^\top (rk - \theta)x = (rk - \theta)z$. Hence $L'z = (rkI_b - N^\top N)z = \theta z$. If $z \neq 0$ then θ is an eigenvalue of L'. If $z = 0$ then $Lx = rkx$ and so $\theta = rk$, and conversely.

Let m be the multiplicity of rk as an eigenvalue of L. Then the column space of $N^\top$ has dimension $v - m$, so the row space of N has dimension $v - m$. Its orthogonal complement has dimension $b - v + m$, and all vectors in this are eigenvectors of L' with eigenvalue rk. □

Thus the product of the nontrivial eigenvalues of L' is equal to the product of the nontrivial eigenvalues of L, multiplied by a further $(rk)^{(b-v)}$. It follows that Γ is D-optimal if and only if Γ' is D-optimal, as already shown.

The smallest nontrivial eigenvalue θ_1 of Γ cannot be larger than rk, so it is also the smallest nontrivial eigenvalue of Γ'. Hence Γ is E-optimal if and only if Γ' is E-optimal.

The harmonic mean h of the nontrivial eigenvalues of Γ is given by

$$\frac{v-1}{h} = \frac{1}{\theta_1} + \cdots + \frac{1}{\theta_{v-1}}.$$

The harmonic mean h' of the nontrivial eigenvalues of Γ' is given by

$$\frac{b-1}{h'} = \frac{1}{\theta_1} + \cdots + \frac{1}{\theta_{v-1}} + \frac{b-v}{rk} = \frac{v-1}{h} + \frac{b-v}{rk}.$$

Thus h' is a non-decreasing function of h, and so Γ is A-optimal if and only if Γ' is A-optimal.

3.3.8 Designs with very low replication

Now we concentrate on designs with very low average replication. Our graphical approach gives many results which are counter-intuitive in the sense of disproving some of the folklore in the previous section.

Lowest possible replication

The Levi graph has $v + b$ vertices and bk edges. So, for connectivity, $bk \geq v + b - 1$. The extreme case is $v - 1 = b(k - 1)$. In this case, all connected Levi graphs are trees, and so the D-criterion does not distinguish them.

In a tree, resistance distance is the same as graph distance, so the A-optimal designs have Levi graphs which are stars with a treatment-vertex at the centre: these are just the queen-bee designs.

The queen-bee designs are also are E-optimal: the proof is coming up after the next example.

Example 3.3.23   Consider the two concurrence graphs shown in Figure 3.23 for block designs with $v = 15$, $b = 7$ and $k = 3$. In the graph in Figure 3.23(a), the edges $\{1,3\}$ and $\{2,3\}$ form a cutset of size two whose removal divides the vertices into sets of size five and ten. The Cutset Lemma (Theorem 3.2.17) shows that $\theta_1 \leq 2(5 + 10)/50 = 3/5 < 1$.

On the other hand, for the graph in Figure 3.23(b), we shall now show that the nontrivial Laplacian eigenvalues are 1, 3, and 15, with multiplicities 6, 7 and 1 respectively. Hence $\theta_1 = 1$ for this queen-bee design.

Using the labelling of treatments given in Figure 3.23(b), the top left-hand corner of L is given by

$$
\begin{bmatrix}
14 & -1 & -1 & -1 & -1 & \dots \\
-1 & 2 & -1 & 0 & 0 & \dots \\
-1 & -1 & 2 & 0 & 0 & \dots \\
-1 & 0 & 0 & 2 & -1 & \dots \\
-1 & 0 & 0 & -1 & 2 & \dots \\
& \vdots & & & &
\end{bmatrix} .
$$

Put

$$
x = \begin{bmatrix} 0 \\ 1 \\ -1 \\ 0 \\ 0 \\ 0 \\ \vdots \\ 0 \end{bmatrix}, \quad y = \begin{bmatrix} 0 \\ 1 \\ 1 \\ -1 \\ -1 \\ 0 \\ \vdots \\ 0 \end{bmatrix} \quad \text{and} \quad z = \begin{bmatrix} 14 \\ -1 \\ -1 \\ -1 \\ -1 \\ -1 \\ \vdots \\ -1 \end{bmatrix}.
$$

Then $Lx = 3x$. In fact, this is still true if x is replaced by any contrast $\tau_i - \tau_j$ where i and j are different non-queen treatments in the same block. Moreover, $Ly = y$, and this remains true if y is replaced by any contrast between all non-queen treatments in one block and all non-queen treatments in a different block. Finally, $Lz = 15z$.

Theorem 3.3.24 *Among incomplete-block designs with $v = b(k-1) + 1$, queen-bee designs are E-optimal.*

Proof If the Levi graph is a tree, there must be at least two blocks in which $k-1$ treatments have replication one. The restriction of the Laplacian matrix L to these $2(k-1)$ treatments is

$$
\begin{bmatrix} kI_{k-1} - J_{k-1} & 0 \\ 0 & kI_{k-1} - J_{k-1} \end{bmatrix}.
$$

Let y be the vector with entry $+1$ for those $k-1$ treatments in the first block, entry -1 for those $k-1$ treatments in the other block, and other entries 0. Then

$$
\frac{y^\top Ly}{y^\top y} = \frac{2(k-1)}{2(k-1)} = 1.
$$

Theorem 3.2.16 shows that $\theta_1 \leq 1$. Hence it suffices to show that every queen-bee design has least nontrivial eigenvalue 1.

As above, let y be the vector with entry $+1$ for the non-queen treatments in one block, entry -1 for the non-queen treatments in another block, and other entries 0. The restriction of the Laplacian matrix L to these $2(k-1)$ treatments is as above, and each other row of L has the same entry in all of these $2(k-1)$ positions. Hence $Ly = y$. Thus 1 is an eigenvalue of L with multiplicity at least $b-1$.

Let x be a vector with entries $+1$ and -1 on two non-queen treatments in the same block, and all other entries 0. The restriction of L to these two positions is

$$\begin{bmatrix} k-1 & -1 \\ -1 & k-1 \end{bmatrix}.$$

Each other row of L has the same entries in these two positions, and so $Lx = kx$. Thus k is an eigenvalue of L with multiplicity at least $b(k-2)$.

Let z be the vector with entry $v-1$ for the queen bee and -1 elsewhere. If u is the all-1 vector then $z+u$ has entry v for the queen and 0 elsewhere. Because $Lu = 0$, we have

$$Lz = L(z+u) = v(\text{queen-bee column of } L) = vz.$$

Thus v is an eigenvalue of L with multiplicity at least 1.

The eigenvalue multiplicities so far recorded add up to

$$1 + (b-1) + b(k-2) + 1 = b(k-1) + 1 = v,$$

so there are no further eigenvalues of L. Hence $\theta_1 = 1$ for the queen-bee design. $\qquad\square$

A more detailed investigation in [5] shows that the queen-bee designs are the only E-optimal designs in this case.

Only slightly less extreme

The Levi graph has $v + b$ vertices and bk edges. If it is connected and is not a tree then $bk \geq v + b$. The next case to consider is $v = b(k-1)$. Then every Levi graph has a single cycle.

The number of spanning trees for the Levi graph is equal to the length of the cycle, so the D-optimal designs have a cycle of length $2b$. One example of this is the Levi graph in Figure 3.22(b), where $v = 8$, $b = 4$ and $k = 3$. Another example is in Figure 3.27, where $v = 12$, $b = 6$ and $k = 3$.

Arguments using resistance in the Levi graph show that the A-optimal designs have a Levi graph with a short cycle, and one special treatment in the cycle occurs in every block which is not in the cycle.

Arguments using the Cutset Lemma in the concurrence graph show that the E-optimal designs have similar structure, usually with an even shorter cycle in the Levi graph.

Now we give more details of the A-optimal and E-optimal designs when $v = b(k-1)$. For $2 \leq s \leq b$, denote by $\mathscr{C}(b,k,s)$ the class of designs which can be constructed as follows.

- Construct a design for s treatments in s blocks of size two whose Levi graph is a single cycle.

- If $k > 2$, then insert $k - 2$ extra treatments into each block.
- If $s < b$, then designate one of the original s treatments as a "pseudo-queen".
- Each of the remaining $b - s$ blocks contains the pseudo-queen and $k - 1$ further treatments.

Thus the D-optimal designs are precisely those in $\mathscr{C}(b, k, b)$.

Theorem 3.3.25 *If $v = b(k-1)$ then the A-optimal designs are those in $\mathscr{C}(b, k, s)$, where the value of s is given in Table 3.3.*

k $\backslash$ b	2	3	4	5	6	7	8	9	10	11	12	≥ 13
2	2	3	4	5	6	7	8	4	4	4	3 or 4	3
3	2	3	4	5	6	3	3	3	3	3	2	2
4	2	3	4	5	3	2	2	2	2	2	2	2
5	2	3	4	5	2	2	2	2	2	2	2	2
≥ 6	2	3	4	2	2	2	2	2	2	2	2	2

Table 3.3 *Values of s in Theorem 3.3.25*

This shows that the A-optimal designs become very different from the D-optimal designs as b increases.

Now we construct some non-binary designs when the Levi graph has a single cycle. If $k \geq 3$, then construct a design in the class $\mathscr{C}(b, k, 1)$ as follows.

- Start with a single block of size two containing a single treatment twice.
- Insert $k - 2$ extra treatments into this block.
- If $k > 3$ then designate the treatment which occurs twice in this block as the queen-bee treatment. If $k = 3$ then either treatment in this block may be designated the queen-bee treatment.
- Each of the remaining $b - 1$ blocks contains the queen-bee treatment and $k - 1$ further treatments.

Theorem 3.3.26 *If $v = b(k-1)$ and $b \geq 3$ and $k \geq 3$ then the E-optimal designs are*

- *those in $\mathscr{C}(b, k, b)$ if $3 \leq b \leq 4$;*
- *those in $\mathscr{C}(b, k, 2)$ and those in $\mathscr{C}(b, k, 1)$ if $b \geq 5$.*

Theorem 3.3.27 *If $k = 2$ and $v = b \geq 3$ then the E-optimal designs are*

- *those in $\mathscr{C}(b, 2, b)$ if $b \leq 5$;*
- *those in $\mathscr{C}(b, 2, b)$, those in $\mathscr{C}(b, 2, 3)$ and those in $\mathscr{C}(b, 2, 2)$ if $b = 6$;*
- *those in $\mathscr{C}(b, 2, 3)$ and those in $\mathscr{C}(b, 2, 2)$ if $b \geq 7$.*

The proofs of the above theorems are in [5, 7].

3.3.9 Exercises

Exercise 3.3.1(i) Construct a block design for 18 treatments in the class $\mathscr{C}(6,4,3)$.

(ii) Sketch the Levi graph $\tilde{G}$ of this design.

(iii) Calculate the resistance distance in $\tilde{G}$ between the pseudo-queen-bee treatment and each other treatment (there are three cases).

(iv) Calculate the number of spanning trees in $\tilde{G}$.

(v) Find the number of spanning trees in the concurrence graph G.

Exercise 3.3.2(i) Let $(x_1,\ldots,x_n)$ be a list of n positive real numbers with sum c. Show that, among all such lists, the geometric mean, harmonic mean, and minimal element are all maximized when $x_1 = \cdots = x_n = c/n$.

(ii) Hence show that, if a BIBD with given number of treatments, number of blocks, and block size exists, then it is A-, D- and E-optimal.

3.4 Further topics

We have seen that the values of the various parameters associated with optimality criteria of block designs depend only on the concurrence graph of the design: to find the optimal design we might have to find the graph which maximizes the number of spanning trees, or minimizes the average resistance, or maximizes the smallest Laplacian eigenvalue, or whatever.

This final section will discuss some additional topics. These include:

- Sylvester designs (an interesting class of examples);
- how to recognise the concurrence graph of a block design;
- variance-balanced designs;
- the relation of optimality parameters to other graph invariants such as the Tutte polynomial.

For block designs with block size 2, the design is the same as its concurrence graph (treatments are vertices and blocks are edges). But, for larger block size, there are interesting questions.

In parts of this section, we refer to treatments as *points*, since the ideas are geometric rather than statistical.

3.4.1 Sylvester designs

As conjectured by Euler in 1782 and proved by Tarry in 1900, there is no *affine plane*, or even *pair of orthogonal Latin squares*, of order 6. So there is no square lattice design with more than three replicates for $v = 36$ and $k = 6$. As a substitute, we propose the *Sylvester designs*.

These designs have 36 points and 48 blocks of size 6. Two points are contained in either one or two blocks: the pairs lying in two blocks are the edges of the *Sylvester graph*, a beautiful 36-vertex graph which will be defined below. So the concurrence matrix has 8 on the diagonal, and 1 or 2 off-diagonal, with 2 for edges of the Sylvester graph. Let $A(S)$ is the adjacency matrix of the Sylvester graph. Then the concurrence matrix is $8I + (J - I + A(S))$.

As we have seen, designs with this concurrence matrix will all have the same Laplacian eigenvalues, and so will coincide on the A-, D- and E-criteria.

Conjecture 3.4.1 Sylvester designs are A-, D- and E-optimal among all block designs with 36 points and 48 blocks of size 6.

The outer automorphism of S_6

The symmetric group S_6 has an outer automorphism. This means that S_6 acts in two different ways on sets of six points, say A and B. An element of S_6 which is a transposition on A is a product of three transpositions on B.

The set B can be constructed as the set of 1-factorizations of the complete graph on six vertices (on the vertex set A). The details are given below.

Now we define a graph on the vertex set $A \times B$ (the Cartesian product) by the rule that (a_1, b_1) is joined to (a_2, b_2) if and only if the transposition (a_1, a_2) on A corresponds, by the outer automorphism of S_6, to a product of three transpositions on B, one of which is (b_1, b_2). This is the *Sylvester graph*.

The group S_6 acts on a set A of six points. It also acts on the 15 2-subsets of A (edges of the complete graph, or *duads* in Sylvester's terminology), and on the $15 \cdot 6 \cdot 1/3! = 15$ partitions into three sets of two (1-factors, or Sylvester's *synthemes*). The set B of size 6 is the set of partitions of the duads into five synthemes (1-factorizations, or Sylvester's *synthematic totals*).

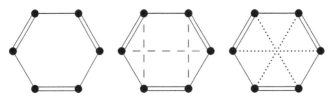

Figure 3.29 Two or three disjoint 1-factors

In Figure 3.29, we have distinguished the synthemes by different styles of line (single, double, dashed and dotted). The first two synthemes in a total must form a 6-cycle, as in the first diagram. The remaining three must use the three long and six short diagonals. There are only two patterns for a syntheme made up of diagonals, shown in dashed and dotted lines. We cannot use the dotted lines since only short diagonals would remain. So the remaining three synthemes each consist of a long diagonal and two perpendicular short diagonals (the dashed lines, and two more obtained by rotating the figure).

This shows that the synthematic total (partition of duads into synthemes) is unique up to isomorphism.

There are 15 choices of the first syntheme, 8 of the second, and $3 \cdot 2 \cdot 1$ of the remaining ones. Since the synthemes can be chosen in any order, the number of synthematic totals is $15 \cdot 8 \cdot 3!/5! = 6$.

Let B be the set of six synthematic totals. Then the group S_6 acts on B, and this action is not isomorphic to the action on the set A of vertices, since the stabilizer of a total doesn't fix a vertex. So there is an outer automorphism (one not induced by conjugation by a group element) mapping the first to the second. Moreover, we can repeat the procedure, and find that the square of this automorphism is inner: if we take the synthematic totals as a new set of vertices of a complete graph, the new synthematic totals correspond in a natural way to the old vertices.

It is remarkable that 6 is the only number n, finite or infinite, for which the symmetric group S_n has an outer automorphism.

The Sylvester graph

The Sylvester graph has an alternative, more combinatorial, definition, as follows: the vertex set is $A \times B$; the pairs (a_1, b_1) and (a_2, b_2) are joined if the duad $\{a_1, a_2\}$ belongs to the unique syntheme which the totals b_1 and b_2 have in common.

The Sylvester graph is distance-transitive, and hence distance-regular (Exercise 3.2.5), with 36 vertices and valency 5. Its diagram is shown in Figure 3.30. The 10 vertices at distance 3 from a given vertex are those in the same row or column of the grid. Its automorphism group is equal to the automorphism group of S_6, and has order 1440.

Its adjacency matrix has eigenvalues 5 (with multiplicity 1), 2 (with multiplicity 16), -1 (multiplicity 10) and -3 (multiplicity 9). From these, the Laplacian eigenvalues of the concurrence matrix are easily computed: the nontrivial ones are 39, 42 and 44.

Its vertices can be regarded as the points of the 6×6 grid $A \times B$. A vertex and its five neighbours lie in distinct rows and columns. The graph contains

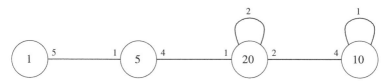

Figure 3.30 The diagram for the Sylvester graph

no 3-cycles or 4-cycles. Any two vertices in different rows and columns lie at distance 1 or 2; if they are not adjacent, they have just one common neighbour.

Starfish

 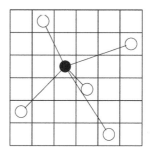

Figure 3.31 A starfish in the Sylvester graph

Definition 3.4.2 (Starfish; galaxy) We define a *starfish* to consist of a vertex and its neighbours; a *galaxy* of starfish is the set of six starfish derived from the vertices in a column of the array. See Figure 3.31.

A Sylvester design

Consider the following 48 sets of size 6:

- the six rows and the six columns of the array;
- the 36 starfish.

It follows from the properties of the graph that two adjacent vertices lie in two blocks (the two starfish defined by these vertices), while any other pair of vertices are contained in a single block.

This design admits the full automorphism group of the graph, which is the automorphism group of S_6 (a group of order 1440).

The design is resolvable, that is, the blocks can be partitioned into eight sets of six, each of which covers all the points. (The resolution classes are the rows, the columns, and the six galaxies of starfish.)

Good designs with lower replication can be obtained by using only rows and columns and some of the galaxies.

More Sylvester designs

As well as this beautiful design, there are others.

Emlyn Williams discovered one using his CycDesign software. It turned out to have the same A-value as the previous one, and a short computation showed that in fact it had the same concurrence matrix. However, the automorphism group of this design is the trivial group.

Another Sylvester design, with 144 automorphisms, was found by Leonard Soicher using semi-Latin squares.

All these designs are resolvable.

For more details, see [6].

Problem 3.4.3   Can the Sylvester designs be classified up to isomorphism?

This would probably be a very big computation!

3.4.2 Sparse versus dense

We have seen that the optimality criteria for block designs tend to agree on designs with dense concurrence graphs, but give very different results in the case where the concurrence graph is sparse.

We have also seen that optimality for block designs tends to agree with desirable characteristics for networks.

Sparse networks occur for the same reason as block designs with low replication, namely resource limitations. So these results are potentially of interest in network theory as well.

BIBDs

Recall that a BIBD for v treatments, with b blocks of size k, has the property that the replication of any treatment is a constant, r, and the concurrence of two treatments is a constant, λ, where

- $bk = vr$;
- $r(k-1) = (v-1)\lambda$.

The concurrence graph of such a design is the λ-*fold complete graph* in which any two vertices are joined by λ edges. Moreover, the design is binary.

For more about BIBDs, we refer to Chapter 16 of Cameron [13], or, for far more detail, to Beth, Jungnickel and Lenz [8].

Steiner triple systems

For $k = 3$ and $\lambda = 1$, such a design is a *Steiner triple system*. The blocks are 3-subsets of the set of points, and two distinct points lie in a unique block.

The two equations for a Steiner triple system assert that

$$2r = v - 1, \qquad 3b = vr,$$

so that $r = (v-1)/2$ and $b = v(v-1)/6$. The condition that these are integers shows that $v \equiv 1$ or $3 \pmod 6$.

The *Fano plane* (Figure 3.10) and the affine plane of order 3 (Figure 3.13) are examples of Steiner triple systems.

In the nineteenth century, Thomas Kirkman showed that this necessary condition is also sufficient for the existence of a Steiner triple system.

Wilson's Theorem

In the early 1970s, Wilson discovered a far-reaching generalization of this theorem. His result has wide applicability; we quote it just for BIBDs.

Suppose that we have a BIBD with given k and λ. Given v, k, λ, the counting equations show that $r = \lambda(v-1)/(k-1)$ and $b = rv/k = \lambda v(v-1)/k(k-1)$. So necessary conditions are that $k-1$ divides $\lambda(v-1)$ and k divides $\lambda v(v-1)$.

Theorem 3.4.4 *If v is sufficiently large (in terms of k and λ), then the above necessary conditions are also sufficient for the existence of a BIBD.*

Of course, this doesn't tell us either how large v has to be, or what to do if the necessary conditions are not satisfied!

3.4.3 Variance-balanced designs

Definition 3.4.5 (Varance-balanced design)   A block design is said to be *variance-balanced* if its Laplacian matrix is totally symmetric, that is, a linear combination of I and the all-1 matrix J. Such a design, if binary, is a BIBD, and hence optimal on all criteria we have discussed; but here we do not assume that the design is binary. For short we write $\text{VB}(v, k, \lambda)$ for a variance-balanced design with given values of these parameters, where λ is the common off-diagonal entry of the concurrence matrix.

The non-binary design with $v = 5$, $k = 3$, and $b = 7$ given in Figure 3.8 is variance-balanced with $\lambda = 2$:

1	1	1	1	2	2	2
1	3	3	4	3	3	4
2	4	5	5	4	5	5

Treatments 1 and 2 concur twice in the first block; any other pair concur in two different blocks.

Variance-balanced designs are not always optimal. Here are two examples of variance-balanced designs with $v = b = 7$ and $k = 6$:

- the design whose blocks are all the 6-subsets of the set of points;
- the design obtained from the Fano plane by doubling each occurrence of a point in a block (so that the first block is the multiset $[1,1,2,2,4,4]$).

The first design, with $\lambda = 5$, is a BIBD, and hence is optimal by Theorem 3.3.15. The second has $\lambda = 4$.

Two questions about variance-balanced designs

Two things we would like to know about variance-balanced designs are

- Given k and λ, for which values of v do $VB(v, k, \lambda)$ designs exist, and what are the possible numbers of blocks of such designs?
- When are variance-balanced designs optimal in some sense?

Morgan and Srivastav [28] have investigated these designs (which they call "completely symmetric").

VB designs with maximum trace

Morgan and Srivastav define two new parameters of a VB design, as follows:

$$r = \left\lfloor \frac{bk}{v} \right\rfloor, \qquad p = bk - vr,$$

so that $bk = vr + p$ and $0 \le p \le v - 1$. Thus, in a BIBD we have $p = 0$. Note that the use of r does not here imply that the design has constant replication!

Definition 3.4.6 (Maximum trace)   Morgan and Srivastav further say that a VB design has *maximum trace* if its parameters satisfy the equation $r(k-1) = (v-1)\lambda$.

In our examples above, $r = \lfloor 7 \cdot 6/7 \rfloor = 6$ and $p = 0$. Since $r(k-1)/(v-1) = 6 \cdot 5/6 = 5$, we see that the first design has maximum trace, but the second does not.

Here is he reason for the term "maximum trace". Since $bk < v(r+1)$, some treatment occurs at most r times on the bk plots. Each occurrence contributes at most $k-1$ edges to the concurrence graph, so the valency of this vertex is at most $r(k-1)$. But the concurrence graph of a VB design is regular, with valency $(v-1)\lambda$; so we have $(v-1)\lambda \leq r(k-1)$, and the trace of the Laplacian matrix (which is $v(v-1)\lambda$) is at most $vr(k-1)$; equality for the trace implies that $(v-1)\lambda = r(k-1)$.

The above argument shows that, in a VB design of maximum trace, any point lies in at least r blocks (counted with multiplicity), with equality if and only if the point occurs at most once in each block. Since $bk = vr + p$, it follows that the number of "bad" points (which occur more than once in some block) is at most p. So if $p = 0$, the design is binary, and is a BIBD or 2-design.

In the examples above with $v = b = 7$ and $k = 6$, we have $r = 6$, $p = 0$, confirming that the first design has maximum trace but the second does not.

Theorem 3.4.7 *A variance-balanced design with maximum trace is E-optimal.*

This was proved by Morgan and Srivastav.

Existence of VB designs of maximum trace

If we have two VB designs on the same set of v points with the same block size k, having parameters λ_1 and λ_2, then the multiset union of the block multi-sets is again VB, with parameter $\lambda_1 + \lambda_2$. The new design is not necessarily of maximum trace; but it is so if one of the VB designs we start with is a BIBD and the other is of maximum trace, or if the sum of their p parameters is less than v.

For example, suppose that $k = 3$. A VB design of maximum trace satisfies $2r = (v-1)\lambda$, so that λ is even or v is odd. Moreover, $\lambda = 1$ is impossible (except for Steiner triple systems), since a non-binary block gives concurrence at least 2. Morgan and Srivastav proved that these necessary conditions are sufficient:

Theorem 3.4.8 *A VB$(v, 3, \lambda)$ design of maximum trace exists provided that $\lambda(v-1)$ is even and $\lambda > 1$.*

Proof A theorem of Hanani [19] shows that a BIBD with $k = 3$ and λ a multiple of 6 exists for all v. Furthermore, he showed that BIBDs with $k = 3$ and other values of λ exist in the following cases:

- for $\lambda \equiv 1$ or $5 \pmod 6$, if $v \equiv 1$ or $3 \pmod 6$;
- for $\lambda \equiv 2$ or $4 \pmod 6$, if $v \equiv 0$ or $1 \pmod 3$;
- for $\lambda \equiv 3 \pmod 6$, if v is odd.

We will construct VB designs for $\lambda = 2$ and $v \equiv 2$ mod 3; they have $p = 1$, so the union of two copies will settle the case $\lambda = 4$. For $\lambda = 5$ or $\lambda = 7$, with v odd, there is a BIBD unless $v \equiv 5 \pmod 6$; in that case we can take a BIBD with $\lambda = 3$ and a VB design with $\lambda = 2$ or $\lambda = 4$.

Here is a construction for VB$(v, 3, 2)$ designs having just one non-binary block. In this case, we must have $v \equiv 2 \pmod 3$.

Suppose first that $v \equiv 2 \pmod 6$. There exist Steiner triple systems of orders $v \pm 1$. Take two such systems, on the point sets $\{1, \ldots, v+1\}$ and $\{1, \ldots, v-1\}$ respectively; let the sets of blocks be $\mathcal{B}_1$ and $\mathcal{B}_2$. Without loss of generality, suppose that the third point of the block B of $\mathcal{B}_1$ containing v and $v+1$ is $v-1$.

Now we take the point set of the new design to be $\{1, \ldots, v\}$. For the blocks, we first remove the block B from $\mathcal{B}_1$; then we replace each occurrence of $v+1$ in any other block with v; the resulting blocks together with the blocks of $\mathcal{B}_2$ and the single non-binary block $[v-1, v-1, v]$ make up the design.

We have to check that $\{v-1, v\}$ lies only in $[v-1, v-1, v]$, while every other pair $\{i, j\}$ lies in two blocks. For the first, note that the only other candidate, namely B, has been removed. For the second, there are two cases:

- $j = v$, $i \neq v-1$: in $\mathcal{B}_1$, there is one block containing i and v, and one containing i and $v+1$ (in which $v+1$ is replaced by v). No block of $\mathcal{B}_2$ can occur.
- $v \notin \{i, j\}$: one block of $\mathcal{B}_1$ and one of $\mathcal{B}_2$ contain $\{i, j\}$, and these two points are unchanged in these blocks.

There is a similar but more elaborate construction when $v \equiv 5 \pmod 6$. In this case, both $v-2$ and $v+2$ are orders of Steiner triple systems.   □

Since there are many non-isomorphic Steiner triple systems, this construction gives rise to many VB designs with $k = 3$.

Example 3.4.9 Consider the case $v = 5$, $k = 3$, $\lambda = 2$. Each block contributes either a triangle or a double edge to the concurrence graph, depending on whether or not it is binary. There are four cases:

- Six triangles and one double edge ($b = 7$): we saw an example.
- Four triangles and four double edges ($b = 8$): take the BIBD consisting of all the 3-subsets of a 4-set and join its four points to the fifth point by four double edges.
- Two triangles and seven double edges ($b = 9$): take a triangle twice and double the seven uncovered edges.
- Ten double edges ($b = 10$): this is a boring design with all its blocks non-binary.

The values of (r,p) in the four cases are $(4,1)$, $(4,4)$, $(5,2)$ and $(6,0)$. So the first two have maximum trace; the others don't.

3.4.4 Recognising a concurrence graph

Given a graph G on v vertices, and an integer k, we would like to know: *Is G the concurrence graph of a block design with block size k?*

If $k = 2$, then the graph is "the same" as the design; the blocks are just edges of the graph.

If the design is binary, then each block contributes a complete graph of size k; so we have to decide whether G is the edge-disjoint union of complete graphs of size k. This is the question which is answered by Wilson's theorem in the case of the λ-fold complete graph. In general, it is necessary that every vertex has valency divisible by $k-1$, and the total number of edges is divisible by $k(k-1)/2$.

What happens in general? We give a condition which extends the above to non-binary blocks.

Weighted cliques

Definition 3.4.10 (Weighted clique) Let $w_1, w_2, \ldots, w_m$ be positive integers, where $m > 1$. A *weighted clique* with weights $w_1, \ldots, w_m$ is a graph on m vertices, in which the ith and jth vertices are joined by $w_i w_j$ edges. Its *weight* is the sum of the weights w_i. (The weights here apply to the *vertices*.)

If all w_i are equal to 1, this is just a complete graph on m vertices, and has weight m.

Theorem 3.4.11 *The graph G is the concurrence graph of a block design with block size k if and only if it is an edge-disjoint union of weighted cliques each with weight k.*

In the design, if a weighted clique with weights $w_1, \ldots, w_m$ corresponds to block j, then the weights are equal to the incidence matrix entries N_{ij} for appropriate values of i.

This generalizes the interpretation of BIBDs as decompositions of the λ-fold complete graph as a disjoint union of ordinary cliques. As we saw, the weighted cliques of weight 3 are a triangle and a double edge.

Decomposition into weighted cliques

Usually the decomposition into weighted cliques, if it exists, is far from unique.

- The Fano plane arises from a decomposition of the 21 edges of K_7 into seven triangles. It is unique up to isomorphism, but there are 30 different ways to make the decomposition (corresponding to the fact that the automorphism group of the Fano plane has index 30 in the symmetric group S_7).
- In our variance-balanced design with $v = 5$, $k = 3$ and $b = 7$, we took the block $[1,1,2]$. However, the block $[1,2,2]$ would have been just as good, and would have given us the same concurrence graph.

3.4.5 Other graph parameters

For the final part of the chapter, we turn to something completely different.

The number of spanning trees of a graph (the D-optimality parameter) also happens to be an evaluation of a famous two-variable polynomial, the *Tutte polynomial*, of the graph. Other evaluations of the Tutte polynomial give lots more information about the graph: number of proper colourings with a given number of colours, number of acyclic or totally cyclic orientations, etc.

We will define the Tutte polynomial and consider how it is related to some of the invariants we have met.

The chromatic polynomial

A *proper colouring* of a graph with a set of q colours is an assignment of colours to the vertices so that adjacent vertices get different colours.

Definition 3.4.12 (Chromatic polynomial) It is well-known that the number of proper colourings of G with a set of q colours is the evaluation at q of a monic polynomial of degree $n = |V(G)|$, known as the *chromatic polynomial* of G.

This is usually proved by "deletion-contraction". It suits our purpose here to give a different proof, using "inclusion-exclusion". The resulting formula is known to some physicists as the *cluster expansion*.

Let S be the the set of all (proper or not) vertex-colourings of G with q colours. For any edge e, let T_e be the set of colourings for which e is improper (has both ends of the same colour); and for $A \subseteq E(G)$ let

$$T_A = \bigcap_{e \in A} T_e,$$

with $T_\emptyset = S$ by convention.

Let $k(A)$ be the number of connected components of the graph with vertex set $V(G)$ and edge set A. A colouring in T_A has the property that all the vertices in a connected component of this graph have the same colour. So $|T_A| = q^{k(A)}$. Then there are $q^{k(A)}$ colourings in which all the edges in A are bad.

So by the Principle of Inclusion and Exclusion, the number of proper colourings is

$$\sum_{A \subseteq E(G)} (-1)^{|A|} q^{k(A)} = P_G(q),$$

where P_G is the chromatic polynomial of G.

Rank

Definition 3.4.13 (Rank)   The *rank* $r(A)$ of a set A of edges of a graph G on n vertices is defined to be the cardinality of the largest acyclic subset of A (that is, the largest edge set contained in A forming a forest). It is easy to see that this is $n - k(A)$, since, in a maximal forest, the number of edges in each connected component is one less than the number of vertices.

Rank has another interpretation. Recall the signed incidence matrix Q of G, as in Figure 3.19. Then $r(A)$ is the rank (in the sense of linear algebra) of the submatrix formed by the rows of the signed incidence matrix indexed by edges in A. The proof is an exercise.

In particular, if $G = (V, E)$ is connected, then $r(E) = n - 1$.

The Tutte polynomial

Definition 3.4.14 (Tutte polynomial)   The *Tutte polynomial* of the graph $G = (V, E)$ is the two-variable polynomial

$$T_G(x, y) = \sum_{A \subseteq E} (x - 1)^{r(E) - r(A)} (y - 1)^{|A| - r(A)}.$$

Many important graph parameters are obtained by plugging in special values for x and y, possibly multiplying by a simple factor.

In particular, putting $x = y = 2$, every term is 1, so that $T_G(2, 2) = 2^{|E|}$.

Other specializations

Assume that G is connected.

Putting $x = 1$, the only nonzero terms are those which have the exponent of $(x - 1)$ equal to 0, that is, $r(A) = r(E)$, so that the graph (V, A) is connected. Similarly, putting $y = 1$, the only nonzero terms are those with $|A| = r(A)$, in other words, the set A contains no cycles.

Hence

- $T_G(1, 2)$ is the number of connected spanning subgraphs of G;
- $T_G(2, 1)$ is the number of spanning forests of G;

- $T_G(1,1)$ is the number of spanning trees of G.

Note that $T_G(1,1)$ is the number associated with D-optimality!

However, other optimality parameters don't appear to be specializations of the Tutte polynomial.

Examples

Neither of the Tutte polynomial and the Laplacian spectrum of G determines the other.

Example 3.4.15 Same Tutte polynomial but different spectra: all trees on n vertices have Tutte polynomial x^{n-1}; but we have seen that they can be very different on A- or E-optimality, and hence on Laplacian spectra.

Same spectrum but different Tutte polynomials: the two strongly regular graphs with the same parameters on 16 vertices have the same Laplacian spectra, but have different Tutte polynomials: see below.

Chromatic polynomial revisited

The formula for the Tutte polynomial looks very similar to the formula we deduced for the chromatic polynomial. Indeed, a little persistence shows that, for a connected graph G on n vertices,

$$P_G(q) = (-1)^{n-1} q T_G(0, -q+1),$$

so, up to a factor q and a sign, the numbers of colourings are values of T_G at integer points on the negative y-axis.

Two strongly regular graphs

Consider the graphs G_1 and G_2 shown in Figure 3.32. On the left is the 4×4 square lattice graph, denoted by $L_2(4)$, in which vertices in the same row or column are joined; and on the right, the Shrikhande graph (which is shown drawn on a torus: nearest neighbours are joined, and opposite edges are identified).

Each graph is *strongly regular* with parameters $(16,6,2,2)$: that is, there are 16 vertices, each vertex has valency 6, and any two vertices have 2 common neighbours, whether or not they are joined.

So the adjacency matrices satisfy $A^2 = 4I + 2J$, and have eigenvalues 6, 2 and -2; the Laplacian eigenvalues are 0, 4 and 8.

Each graph is associated with a square lattice design with three resolution classes. Take a Latin square of order 4, and form a graph whose vertices are the cells, two vertices adjacent if they are *not* in the same row or column and do not contain the same symbol.

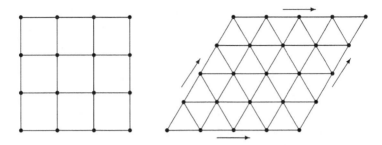

Figure 3.32 Two strongly regular graphs

There are two essentially different Latin squares of order 4: they are the Cayley tables of the Klein group V_4 and the cyclic group C_4. The corresponding graphs are the lattice graph $L_2(4)$ and the Shrikhande graph respectively.

Colouring the graphs

A colouring of the square lattice with four colours is nothing but a *Latin square* of order 4 (see Figure 3.33, where the four symbols – circle, double circle, solid circle and square – form a Latin square). There are 576 Latin squares of order 4, and hence $P_{G_1}(4) = 576$.

However, calculation shows that $P_{G_2}(4) = 240$; so the chromatic polynomials, and hence the Tutte polynomials, are different.

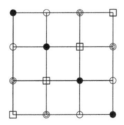

Figure 3.33 A colouring of $L_2(4)$ is a Latin square

Orientations

An orientation of the edges of a graph G is *acyclic* if there are no directed cycles; it is *totally cyclic* if every edge is contained in a directed cycle.

Richard Stanley showed that the number $a(G)$ of acyclic orientations of G is

$$a(G) = |P_G(-1)| = |T_G(0,2)|.$$

It is also known that the number of totally cyclic orientations is $c(G) = |T_G(2,0)|$.

Now recall that the number of spanning trees is $t(G) = T_G(1,1)$.

The Merino–Welsh conjecture

These three numbers are connected by a remarkable conjecture of Merino and Welsh. This mentions the concept of a *bridge*, which is an edge whose removal increases the number of connected components of the graph. For example, the graph on the right in Figure 3.16 contains a bridge.

Conjecture 3.4.16 *If G has no loops or bridges, then $t(G) \leq \max\{a(G), c(G)\}$.*

That is, either the number of acyclic orientations or the number of totally cyclic orientations dominates the number of spanning trees.

The best result so far is by Carsten Thomassen, who showed that this is true for sufficiently sparse graphs (where the number of acyclic orientations wins) and for sufficiently dense graphs (where the number of totally cyclic orientations wins).

Thomassen's Theorem

Theorem 3.4.17 *Let G be a connected graph on n vertices without loops or bridges.*

- *If G has at least $4n$ edges, then $t(G) \leq c(G)$.*
- *If G has at most $16n/15$ edges, then $t(G) \leq a(G)$.*

The first result applies if the average valency is at least 8; the second if it is at most $32/15 = 2.133\ldots$.

Jackson's Theorem

A related theorem of Bill Jackson is intriguing:

Theorem 3.4.18 *Let G be a connected graph without loops or bridges. Then*

$$T_G(1,1)^2 \leq T_G(0,3) \cdot T_G(3,0).$$

Of course, replacing 3 by 2 would give a strengthening of the Merino–Welsh conjecture!

3.4.6 Some open problems

We conclude this chapter with a number of open problems, which could be topics for research projects.

Open problem: a tipping point?

Problem 3.4.19 Given k, is there a number r_0 such that, for designs with block size k and average replication greater than r_0, the different optimality conditions agree?

The number r_0 might also depend on v. However, unpublished work by Robert Johnson and Mark Walters suggests that, for $k = 2$, r_0 might be about 4. This is suggestively similar to Carsten Thomassen's result on the Merino–Welsh conjecture.

If so, what happens for average replication below r_0? There may be further "phase changes".

Open problem: dense simple graphs

For dense simple graphs (those obtained by removing just a few edges from the complete graph), independent studies by Aylin Cakiroglu and Robert Schumacher suggest that, both for optimality and for maximizing the number of acyclic orientations, the best graphs resemble *Turán graphs*: that is, the edges removed should be as close as possible to a disjoint union of complete graphs of the same size.

If we remove vertex-disjoint complete graphs of the same size covering the vertex set, we get a group-divisible design.

Problem 3.4.20 Prove results on the optimality of Turán graphs.

Open problem: adding complete graphs

Aylin Cakiroglu [12] and J. P. Morgan have investigated the following problem. Choose an optimality parameter of block designs which is determined by the concurrence graph of the design (as we have seen for A, D and E). For a nonnegative integer s, and given v, order the simple graphs on v vertices with a fixed number of edges (or the regular simple graphs of prescribed valency) by the rule that $G_1 <_s G_2$ if the union of G_2 with s copies of K_v beats the union of G_1 with s copies of K_v on this parameter.

They showed that this order stabilizes for sufficiently large s. But in cases which could be computed, it stabilizes for $s = 1$ (or at worst for $s = 2$).

Problem 3.4.21 Bound the value of s for which the order stabilizes in terms of v.

One can also make the problem "continuous" by expressing the parameter in terms of s and then allowing s to take real values.

Open problem: variance-balanced designs

Problem 3.4.22   Given k and λ, find necessary and sufficient conditions on v for the existence of a variance-balanced design of maximum trace with these values of v, k, λ.

We saw that this was solved for $k = 3$ and all λ by Morgan and Srivastav.

More generally, there are theorems about decomposing the edge set of a graph into complete graphs of given sizes; find theorems about decomposing the edge set of a graph into weighted k-cliques, with perhaps some restrictions on the cliques (e.g. as few as possible where the weights are not all 1).

Open problems: Finite geometry

In finite geometry one meets many beautiful and symmetrical block designs of various kinds: generalized polygons, near-polygons, Grassmann geometries, ...

Problem 3.4.23   Are these geometries optimal?

Also one meets *geometries of higher rank*, that is, with more than two kinds of objects; they will have various rank 2 geometries as truncations and as residuals (this term is used in a geometric sense, unrelated to its use in design theory). These may be relevant in experimental design, if there are several different kinds of treatment, or of "nuisance factor" to be controlled.

We refer to [11] for an exposition of these geometries.

Problem 3.4.24   What is the relation, if any, between optimality of different truncations or residuals of the same higher-rank geometry?

3.4.7 Exercises

Exercise 3.4.1   Prove that the rank $r(A)$ of a set A of edges of a graph G is equal to the rank of the submatrix of the signed incidence matrix formed by the rows indexed by edges in A.

Exercise 3.4.2(i)   Let A be the adjacency matrix of the 6×6 square lattice graph. Show that

$$A^2 - 2A - 8I = 2J,$$

where J is the all-1 matrix, and hence find the eigenvalues of A.

(ii)   Let B be the adjacency matrix of the Sylvester graph. Show that

$$B^2 + B - 4I = J - A,$$

and hence or otherwise find the eigenvalues of B.

(iii) Hence find the eigenvalues of the concurrence matrix of a Sylvester design.

(iv) Suppose that there were an affine plane of order 6 (a balanced design consisting of 42 blocks of size 6 on 36 points). Take this design and a further set of six pairwise disjoint blocks. Calculate the eigenvalues of the concurrence matrix of such a design, and compare it to a Sylvester design on the A-, D- and E-criteria.

Exercise 3.4.3 Construct a graph on 50 vertices as follows. The vertex set is $\{\alpha, \beta\} \cup A \cup B \cup (A \times B)$, where A and B are 6-sets as in the construction of the Sylvester graph. The edges are:

- $\{\alpha, \beta\}$;
- $\{\alpha, a\}$ for all $a \in A$;
- $\{\beta, b\}$ for all $b \in B$;
- $\{a, (a,b)\}$ and $\{b, (a,b)\}$ for all $a \in A$, $b \in B$;
- all the edges of the Sylvester graph on $A \times B$.

Prove that this graph is strongly regular, with parameters $(50, 7, 0, 1)$. This is the *Hoffman–Singleton graph*: see [20].

Prove further that a regular graph with valency k and diameter 2 has at most $k^2 + 1$ vertices; and that, if equality holds, then the graph is strongly regular with $\lambda = 0$ and $\mu = 1$, and $k = 2, 3, 7$, or possibly 57. (These graphs are the *Moore graphs* of diameter 2; there is a unique graph in each of the first three cases, but the existence of a Moore graph with valency 57 is currently unknown.)

References

[1] R. A. Bailey, *Association Schemes: Designed Experiments, Algebra and Combinatorics*. Cambridge University Press, Cambridge, 2004.

[2] R. A. Bailey, *Design of Comparative Experiments*. Cambridge University Press, Cambridge, 2008.

[3] R. A. Bailey, Variance and concurrence in block designs, and distance in the corresponding graphs, *Michigan Mathematical Journal* **58** (2009), 105–124.

[4] R. A. Bailey and Peter J. Cameron, Combinatorics of optimal designs. In *Surveys in Combinatorics 2009* (eds. S. Huczynska, J. D. Mitchell and C. M. Roney-Dougal), pp. 19–73. London Mathematical Society Lecture Note Series **365**, Cambridge University Press, Cambridge, 2009.

[5] R. A. Bailey and Peter J. Cameron, Using graphs to find the best block designs. In *Topics in Structural Graph Theory* (eds. L. W. Beineke and R. J. Wilson), Encyclopedia of Mathematics and its Applications **147**, pp. 282–317. Cambridge University Press, Cambridge, 2013.

[6] R. A. Bailey, Peter J. Cameron, L. H. Soicher and E. R. Williams, Substitutes for the non-existent square lattice designs for 36 varieties, *Journal of Biological, Agricultural and Environmental Statistics*, **25** (2020), 487–499.

[7] R. A. Bailey and Alia Sajjad, Optimal incomplete-block designs with low replication: A unified approach using graphs, *Journal of Statistical Theory and Practice*, **15**:84 (2021).

[8] T. Beth, D. Jungnickel and H. Lenz, *Design Theory*, 2nd edition, Cambridge University Press, Cambridge, 1999, 2011.

[9] B. Bollobás, *Modern Graph Theory*. Springer, New York, 1998.

[10] Bose memorial session, *Sankhyā*, Series B **54** (1992), i–viii.

[11] F. Buekenhout and A. M. Cohen, *Diagram Geometry*. Springer, Berlin, 2013.

[12] S. A. Cakiroglu, Optimal regular graph designs, *Statistics and Computing* **28** (2018), 103–112.

[13] P. J. Cameron, *Combinatorics: Topics, Techniques, Algorithms*. Cambridge University Press, Cambridge, 1996.

[14] C.-S. Chêng, Optimality of certain asymmetrical experimental designs, *Annals of Statistics* **6** (1978), 1239–1261.

[15] C.-S. Chêng, Maximizing the total number of spanning trees in a graph: two related problems in graph theory and optimum design theory, *Journal of Combinatorial Theory Series B* **31** (1981), 240–248.

[16] C.-S. Cheng and R. A. Bailey, Optimality of some two-associate-class partially balanced incomplete-block designs, *Annals of Statistics* **19** (1991), 1667–1671.

[17] R. Christensen, *Plane Answers to Complex Questions: The Theory of Linear Models*. Springer-Verlag, New York, 1987.

[18] B. R. Clarke, *Linear Models: The Theory and Application of Analysis of Variance*. John Wiley & Sons, Hoboken, 2008.

[19] H. Hanani, The existence and construction of balanced incomplete block designs, *Annals of Mathematical Statistics* **32** (1961), 361–386.

[20] A. J. Hoffman and R. R. Singleton, On Moore graphs with diameters 2 and 3, *IBM J. Research Development* **4** (1960), 497–504.

[21] Derek Holton and John Sheehan, *The Petersen Graph*, Australian Mathematical Society Lecture Series **7**, Cambridge University Press, Cambridge, 1993.

[22] M. Kagan and B. Mata, A physics perspective on the resistance distance for graphs, *Mathematics in Computer Science* **13** (2019), 105–115.

[23] J. Kiefer, Construction and optimality of generalized Youden designs. In *A Survey of Statistical Design and Linear Models* (ed. J. N. Srivastava), pp. 333–353. North-Holland, Amsterdam (1975).

[24] M. Krivelevich and B. Sudakov, Sparse pseudo-random graphs are Hamiltonian, *Journal of Graph Theory* **42** (2003), 17–33.

[25] A. M. Kshirsagar, A note on incomplete block designs, *Annals of Mathematical Statistics* **29** (1958), 907–910.

[26] A. D. Mednykh and I. A. Mednykh, On the structure of the Jacobian group of circulant graphs. (Russian) *Dokl. Akad. Nauk* **469** (2016), no. 5, 539–543; translation in *Dokl. Math.* **94** (2016), no. 1, 445–449.

[27] B. Mohar, Some applications of Laplace eigenvalues of graphs. In *Graph Symmetry: Algebraic Methods and Applications* (eds. G. Hahn and G. Sabidussi), pp. 225–275. Springer, Dordrecht, 1997.

[28] J. P. Morgan and S. K. Srivastav, The completely symmetric designs with blocksize three, *Journal of Statistical Planning and Inference* **106** (2002), 21–30.

[29] D. J. Saville and G. R. Wood, *Statistical Methods: A Geometric Primer*. Springer-Verlag, New York, 1996.

[30] K. R. Shah and B. K. Sinha, *Theory of Optimal Designs*, Lecture Notes in Statistics **54**, Springer-Verlag, New York, 1989.

4

Symbolic dynamics and the stable algebra of matrices

Mike Boyle and Scott Schmieding

Abstract

We give an introduction to a topic in the "stable algebra of matrices", as related to certain problems in symbolic dynamics. We introduce enough symbolic dynamics to explain these connections, but the algebra is of independent interest and can be followed with little attention to the symbolic dynamics. This "stable algebra of matrices" involves the study of properties and relations of square matrices over a semiring $\mathscr{S}$ which are invariant under two fundamental equivalence relations: shift equivalence and strong shift equivalence. When $\mathscr{S}$ is a field, these relations are the same, and matrices over $\mathscr{S}$ are shift equivalent if and only if the nonnilpotent parts of their canonical forms are similar. We give a detailed account of these relations over other rings and semirings, especially $\mathbb{Z}$, $\mathbb{Z}_+$ and $\mathbb{R}_+$. When $\mathscr{S}$ is a ring, this involves module theory and algebraic K-theory. We discuss in detail and contrast the problems of characterizing the possible spectra, and the possible nonzero spectra, of nonnegative real matrices. We also review key features of the automorphism group of a shift of finite type; the recently introduced stabilized automorphism group; and the work of Kim, Roush and Wagoner giving counterexamples to Williams' shift equivalence conjecture.

Keywords: shift equivalence, shift of finite type, strong shift equivalence, integer matrix, nonnegative matrix, nonzero spectrum, automorphism group, algebraic K-theory

Mathematics Subject Classification: 06-01, 15B36, 15B48, 19-01, 37B10

4.1 Overview

The bulk of this work is devoted to an exposition of algebraic properties and relations of square matrices which are invariant under similarity and certain "stabilizing" relations, with connections to symbolic dynamics, linear algebra and algebraic K-theory. For matrices over a field, this amounts simply to neglecting the nilpotent part of their action; for matrices over other rings, even $\mathbb{Z}$, the analogous stabilization is more subtle. For matrices over $\mathbb{Z}_+$, it becomes very subtle indeed. The algebra of these relations arises naturally in problems of symbolic dynamics. However, quite apart from symbolic dynamics, we believe this algebra is an important topic in the theory of matrices. We aim to give a presentation useful for both students and experts.

The slice of the "stable algebra of matrices" we study involves matrix relations and properties invariant under two fundamental equivalence relations: shift equivalence and strong shift equivalence. We begin with definitions of these relations. Let $\mathscr{C}$ be a category, with $\mathscr{M}$ its class of morphisms and $\mathscr{E}$ its class of endomorphisms (morphisms with domain = codomain). Elements A, B of $\mathscr{E}$ are *elementary strong shift equivalent* if there are morphisms U, V in $\mathscr{M}$ such that $A = UV$ and $B = VU$. Here, UV is the composition[1] [2] of the morphisms U, V. In general, U and V do *not* have to be endomorphisms, although the equations force A and B to be endomorphisms[3]. We refer to (U, V) as an elementary strong shift equivalence from A to B, or between A and B. We use ESSE as a multipurpose abbreviation whose meaning should be clear in context. The relation ESSE is only defined for endomorphisms, but depends on $\mathscr{M}$. We may refer to ESSE in $\mathscr{C}$, or in $\mathscr{M}$.

We say endomorphisms A, B are *similar* (isomorphic as endomorphisms) if there is an invertible morphism[4] U such that $B = U^{-1}AU$. Now suppose that A and B are automorphisms (invertible endomorphisms) in a category $\mathscr{C}$, with an ESSE $A = UV$ and $B = VU$. Then U and V must be invertible[5], with $B = U^{-1}AU$. So, for automorphisms ESSE is similarity. We see that ESSE for endomorphisms is some kind of endomorphism generalization of similarity.

---

[1] Unless otherwise specified, we adopt the convention that morphisms are arrows from a domain object on the left to a codomain object on the right. So, for the composition UV of morphisms U, V to be defined, the codomain of U must equal the domain of V.

[2] Note, when the morphisms U, V are functions, our domain/codomain convention means that we are reading composition from left to right: for an input x, the output $(UV)(x)$ is $V(U(x))$. The chosen convention is irrelevant (and safely ignored) until Section 4.8, as explained there.

[3] E.g., domain(A) = domain(U) = codomain(V) = codomain(A).

[4] A morphism A from object D to object C is invertible if there are morphisms U, V such that $UA = 1_{\text{codom}(A)}$ and $AV = 1_{\text{dom}(A)}$. In this case, U must equal V, as $U = U(AV) = (UA)V = V$; U is the inverse of A, denoted A^{-1}.

[5] E.g., U is invertible because $U(VA^{-1}) = 1_{\text{dom}(U)}$ and $(B^{-1}V)U = 1_{\text{codom}(U)}$.

We will be primarily interested in the case that $\mathscr{C}$ is a category whose morphisms are $\mathscr{M}(\mathscr{S})$, the finite matrices over a semiring[6] $\mathscr{S}$, with composition of morphisms given by matrix multiplication, so $\mathscr{E}$ is the set of square matrices over $\mathscr{S}$. The objects of $\mathscr{C}$ are the sets $\mathscr{S}^n$ ($\mathscr{S}^n$ is the set of n-tuples with entries in $\mathscr{S}$). With our default domain convention for morphisms, we take $\mathscr{S}^n$ to be a set of row vectors. If $\mathscr{S}$ is a ring $\mathscr{R}$, then by matrix multiplication an $m \times n$ matrix over $\mathscr{R}$ defines an $\mathscr{R}$-module homomorphism from $\mathscr{S}^m$ to $\mathscr{S}^n$, and $\mathscr{C}$ can (of course) be identified with the category of $\mathscr{R}$-module homomorphisms of free, finitely generated left $\mathscr{R}$-modules. In this category, ESSE is not an equivalence relation.

Returning to a general category $\mathscr{C}$, we define strong shift equivalence to be the equivalence relation which is the transitive closure of ESSE. That is, endomorphisms A, B are *strong shift equivalent* if there there are endomorphisms $A_0, \ldots, A_\ell$ and morphisms U_i, V_i, such that $A = A_0$, $B = A_\ell$, and for $1 \leq i \leq \ell$ we have $A_{i-1} = U_i V_i$, $A_i = V_i U_i$. We refer to the finite sequence $(U_i, V_i), 1 \leq i \leq \ell$, as a strong shift equivalence from A to B. The relation SSE, like ESSE, is only defined for endomorphisms, and depends on $\mathscr{M}$. We use SSE as a multipurpose abbreviation. For a semiring $\mathscr{S}$, SSE-$\mathscr{S}$ denotes SSE in $\mathscr{M} = \mathscr{M}(\mathscr{S})$, as described above. SE (shift equivalence, see below) and SSE arise in various settings within and outside symbolic dynamics [7].

The simplicity of the SSE definition is utterly deceptive. For example, SSE-$\mathbb{Z}_+$ was introduced by Williams to classify shifts of finite type up to topological conjugacy. A half century later, we do not know if the relation SSE-$\mathbb{Z}_+$ is even decidable.

To study strong shift equivalence, one is naturally led (by Williams) to the more tractable relation of shift equivalence. Given an SSE $(U_i, V_i), 1 \leq i \leq \ell$ from $A = A_0$ to $B = A_\ell$ in a category $\mathscr{C}$, let $U = U_1 \cdots U_\ell$ and $V = V_\ell \cdots V_1$. Then we have

$$A^\ell = UV, \qquad B^\ell = VU,$$
$$AU = UB, \qquad VA = BV.$$

So, endomorphisms A, B are defined to be *shift equivalent* in $\mathscr{C}$ if there are morphisms U, V and a positive integer ℓ satisfying the four displayed equations.

---

[6] A matrix "over" a set $\mathscr{S}$ is simply a matrix all of whose entries are in $\mathscr{S}$. By a semiring $\mathscr{S}$, we mean a set with addition and multiplication satisfying all the ring axioms except possibly existence of additive inverses. By definition, we require a semiring (in particular, a ring) to contain a multiplicative unit.

[7] E.g., SE arises in ergodic theory [90] and Conley index theory [100]. For SSE-$\mathbb{Z}_+$ in knot theory, see [50, 136]. For SSE related to the nonzero spectrum (and other invariants) of primitive real matrices, see [23, 25]. For a connection of SSE to category theory, see [65]. For SE and SSE in other symbolic dynamics settings, see Section 4.3.9.

One can check that SE is an equivalence relation on the endomorphisms in $\mathscr{C}$. Clearly SSE implies SE. The converse holds for $\mathscr{M}(\mathscr{S})$ for many rings $\mathscr{S}$, but fails even for some integral domains, and fails for the semiring $\mathscr{S} = \mathbb{Z}_+$.

Let $\mathscr{R}$ be a ring. Then, SSE-$\mathscr{R}$ and SE-$\mathscr{R}$ can be described by other generating relations. We say square matrices A, B in $\mathscr{M}(\mathscr{R})$ are similar over $\mathscr{R}$ (SIM-$\mathscr{R}$) if there exists a matrix U invertible over R such that $B = U^{-1}AU$.[8] Given a square matrix A, we define a nilpotent extension of A to be a square matrix with a block form $\left(\begin{smallmatrix} A & X \\ 0 & N \end{smallmatrix}\right)$ or $\left(\begin{smallmatrix} A & 0 \\ X & N \end{smallmatrix}\right)$, with N a nilpotent matrix. A zero extension of A is a nilpotent extension in which N is a zero matrix. Then, the following hold (by Theorem 4.7.16).

(1) SSE-$\mathscr{R}$ is generated by similarity and zero extensions[9].
(2) SE-$\mathscr{R}$ is generated by similarity and nilpotent extensions.

So, we study invariants of similarity which persist under zero or nilpotent extensions. We will see that this "stable" viewpoint can be useful even for studying properties of real nonnegative matrices.

Now, we turn to the organization of the eight sections which follow.

In Section 4.2, we introduce shifts of finite type, the most fundamental symbolic dynamical systems. A square nonnilpotent matrix A over $\mathbb{Z}_+$ defines a shift of finite type (SFT), σ_A; we will see how dynamical properties and relations correspond to "stable algebra" invariants of the matrix A. By no means do we give a full introduction to SFTs. We say enough to explain how the stable algebra invariants arise in symbolic dynamics: to inform those interested in the dynamics, and provide motivation and a concrete model system for those interested in the algebra.

In Section 4.3, after brief general remarks about strong shift equivalence, we give an extensive discussion of shift equivalence, including several example cases. For a general ring $\mathscr{R}$, SE-$\mathscr{R}$ is characterized by isomorphism of certain associated $\mathscr{R}[t]$-modules.

The theory of SE and SSE can be entirely recast in terms of polynomial matrices; we do this in Section 4.4. This is essential for later K-theory connections. One key is the presentation of a directed graph by a polynomial matrix, a construction of interest for anyone using directed graphs.

A classical problem of linear algebra, the nonnegative inverse eigenvalue

---

[8] A matrix is invertible over $\mathscr{R}$ if it has an inverse matrix all of whose entries are in $\mathscr{R}$. If a ring $\mathscr{R}$ has the invariant basis property (IBP), then an invertible matrix over $\mathscr{R}$ must be square. If a ring $\mathscr{R}$ is a commutative ring, or an integral group ring, then it has the IBP [150]. For a ring without the IBP, "similar" square matrices can have different sizes.

[9] That is, SSE-$\mathscr{R}$ is the smallest equivalence relation on square matrices over $\mathscr{R}$ containing SIM-$\mathscr{R}$ and closed under taking nilpotent extensions.

problem (NIEP), asks which multisets of complex numbers can be the spectrum of a nonnegative matrix. A stable version of the NIEP asks which multisets of nonzero complex numbers can be the nonzero part of the spectum of a nonnegative matrix. In contrast to the notoriously difficult NIEP, the stable version has a transparent solution (not a transparent proof). We review results and conjectures related to this stable approach in Section 4.5, including new commentary on a realization theorem of Tom Laffey.

The simple definition of SSE takes us to the higher mathematics of algebraic K-theory. In Section 4.6, we give an introduction to enough algebraic K-theory to make the later statements understandable, with a little context[10].

Let $\mathscr{R}$ be a ring. SE-$\mathscr{R}$ is a topic in the theory of $\mathscr{R}$-modules; the refinement of SE-$\mathscr{R}$ by SSE-$\mathscr{R}$ is understood (incompletely, to date) with algebraic K-theory. In Section 4.7, we give the algebraic K-theoretic characterization of the refinement of SE-$\mathscr{R}$ by SSE-$\mathscr{R}$. Here the class group of the category of nilpotent endomorphisms over $\mathscr{R}$ plays a key role.

In Section 4.8, we give an overview of results on the automorphism group of a shift of finite type: its actions and representations, and related problems. We also (briefly) introduce the mapping class group and the stabilized automorphism group of a shift of finite type. We review two crucial representations (dimension and SGCC) of the automorphism group which become key ingredients to the work of Kim and Roush/Wagoner giving counterexamples to the Williams conjecture that SE-$\mathbb{Z}_+$ implies SSE-$\mathbb{Z}_+$.

In Section 4.9, we give an overview of the counterexample work showing SE-$\mathbb{Z}_+$ does not imply SSE-$\mathbb{Z}_+$, even for primitive matrices. This work takes place within the machinery of Wagoner's CW complexes for SSE, which give a different, homotopy-based approach to SSE. The counterexample invariant of Kim and Roush is a relative sign-gyration number. Wagoner later formulated a different counterexample invariant, with values in a certain group coming out of algebraic K-theory.

Our exposition is an elaboration of the course we gave at the 2019 Yichang G2D2 school, accessible to graduate students, but of broader interest. After an introductory section, the next four sections are by Boyle; the last four are by Schmieding. Each section after the first has the form of a lecture, together with a Notes section including further details, proof and remarks.

---

[10] A reader attracted by this glimpse can look into various algebraic K-theory introductions, with their different scopes, prerequisites and publication dates (e.g., in order of publication, [7, 99, 137, 123, 150]). The short exposition [156] gives a lovely perspective on K-theory as a "theory of assembly".

The appendices are intended to enhance the value of this work both as exposition and reference.

Fundamental problems remain open in the theory of strong shift equivalence, and the related theory of shifts of finite type and their automorphism groups. We hope that our exposition may encourage some contribution to their solution.

Acknowledgements. For feedback and corrections on our manuscript, we thank Rosemary Bailey, Peter Cameron, Tullio Ceccherini-Silberstein, Sompong Chuysurichay, Ricardo Gomez, Emmanuel Jeandel, Charles Johnson, Johan Kopra, Wolfgang Krieger, Michael Maller, Alexander Mednykh, Akihiro Munemasa, Michael Shub, Yaokun Wu and Yinfeng Zhu. We thank especially Yaokun Wu, without whose vision and organization this work would not exist.

4.2 Basics

In this first section, we review the fundamentals of shifts of finite type and the algebraic invariants of the matrices which present them.

4.2.1 Topological dynamics

By a topological dynamical system (or system), we will mean here a homeomorphism of a compact metric space to itself. This is one setting for studying how points can/must/typically behave over time, that is, how points move under iteration of the homeomorphism. A system/homeomorphism $S : X \to X$ is often written as a pair (X, S). Formally: we have a category, in which

- an object is a topological dynamical system,
- a morphism $\phi : (X, S) \to (Y, T)$ is a continuous map $\phi : X \to Y$ such that $\phi T = S\phi$, i.e. the following diagram commutes.

$$
\begin{array}{ccc}
X & \xrightarrow{\ S\ } & X \\
\phi \downarrow & & \downarrow \phi \\
Y & \xrightarrow{\ T\ } & Y
\end{array}
$$

The morphism is a topological conjugacy (an isomorphism in our category) if ϕ is a homeomorphism.

Now, let $\phi : X \to Y$ be a topological conjugacy. We can think of ϕ as follows: ϕ renames points without changing the mathematical structure of the system.

- Because ϕ is a homeomorphism, it gives new names to points in essentially the same topological space.
- Because $\phi T = S\phi$, the renamed points move as they did with their original names. E.g., with $y = \phi x$,

$$\cdots \xrightarrow{S} x \xrightarrow{S} Sx \xrightarrow{S} S^2 x \xrightarrow{S} \cdots$$

$$\cdots \xrightarrow{T} y \xrightarrow{T} Ty \xrightarrow{T} T^2 y \xrightarrow{T} \cdots$$

As ϕ respects the mathematical structure under consideration, we see that the systems (X, S) and (Y, T) are essentially the same, just described in a different language. (Perhaps, points of X are described in English, and points of Y are described in Chinese, and ϕ gives a translation.)

We are interested in "dynamical" properties/invariants of a topological dynamical system – those which are respected by topological conjugacy. For example, suppose $\phi : (X, S) \to (Y, T)$ is a topological conjugacy, and x is a fixed point of S: i.e., $S(x) = x$. Then $\phi(x)$ is a fixed point of T. The proof is trivial: $T(\phi(x)) = \phi(Sx) = \phi(x)$. (We almost don't need a proof: however named, a fixed point is a fixed point.) So the cardinality of the fixed point set, $\mathrm{card}(\mathrm{Fix}(S))$, is a dynamical invariant.

Notation 4.2.1   For $k \in \mathbb{N}$, S^k is S iterated k times. E.g., $S^2 : x \mapsto S(S(x))$.

If $\phi : (X, S) \to (Y, T)$ is a topological conjugacy, then ϕ is also is a topological conjugacy $(X, S^k) \to (Y, T^k)$, for all k (because $\phi T = S\phi \implies \phi T^k = S^k \phi$). So, the sequence $(|\mathrm{card}(\mathrm{Fix}(S^k))|)_{k \in \mathbb{N}}$ is a dynamical invariant of the system (X, S).

4.2.2 Symbolic dynamics

Let $\mathscr{A}$ be a finite set. Then $\mathscr{A}^{\mathbb{Z}}$ is the set of functions from $\mathbb{Z}$ to $\mathscr{A}$. We write an element x of $\mathscr{A}^{\mathbb{Z}}$ as a doubly infinite sequence, $x = \ldots x_{-1} x_0 x_1 \ldots$, with each x_k an element of $\mathscr{A}$. (The bisequence defines the function $k \mapsto x_k$.) Often $\mathscr{A}$ is called the alphabet, and its elements are called symbols.

Let $\mathscr{A}$ have the discrete topology and let $\mathscr{A}^{\mathbb{Z}}$ have the product topology. Then $\mathscr{A}^{\mathbb{Z}}$ is a *compact metrizable space* (Remark 4.2.27). For one metric compatible with the topology, given $x \neq y$ set $\mathrm{dist}(x, y) = 1/(M+1)$, where $M = \min\{|k| : x_k \neq y_k\}$. Points x, y are close when they have the same central word, $x_{-M} \ldots x_M = y_{-M} \ldots y_M$, for large M.

The *shift map* $\sigma : \mathscr{A}^{\mathbb{Z}} \to \mathscr{A}^{\mathbb{Z}}$ is defined by $(\sigma x)_n = x_{n+1}$. (This is the "left shift": visually, a symbol in box $n+1$ moves left into box n.) The shift map $\sigma : \mathscr{A}^{\mathbb{Z}} \to \mathscr{A}^{\mathbb{Z}}$ is easily checked to be a homeomorphism. The system $(\mathscr{A}^{\mathbb{Z}}, \sigma)$ is called the full shift on n symbols, if $n = |\mathscr{A}|$. One notation: a dot over a symbol indicates it occurs in the zero coordinate. Then

$$\sigma : x \mapsto \sigma(x)$$

$$\sigma : \ldots x_{-2} x_{-1} \overset{\bullet}{x_0} x_1 x_2 \ldots \mapsto \ldots x_{-2} x_{-1} x_0 \overset{\bullet}{x_1} x_2 \ldots \ .$$

A *subshift* is a subsystem $(X, \sigma|_X)$ of some full shift $(\mathscr{A}^{\mathbb{Z}}, \sigma)$ (i.e. X is a closed subset of $\mathscr{A}^{\mathbb{Z}}$, and $\sigma(X) = X$). For notational simplicity, we generally write $(X, \sigma|_X)$ as (X, σ). Among the subshifts, the subshifts of finite type (also called *shifts of finite type*, or *SFTs*) are a fundamental class, with varied applications (Remark 4.2.26).

Definition 4.2.2 (Subshift of finite type) By definition, a subshift (X, σ) is a subshift of finite type (SFT) if there is a finite set $\mathscr{F}$ of words on the alphabet $\mathscr{A}$ of X such that X is the subset of points x in $\mathscr{A}^{\mathbb{Z}}$ such that no subword $x_m \cdots x_n$ is in $\mathscr{F}$.

4.2.3 Edge SFTs

We are interested in relating dynamical properties and relations of SFTs to matrix algebra. A key to this is given by a particular class of SFTs, the edge SFTs, which are presented by matrices.

Notation 4.2.3 For us, always, "*graph*" means "directed graph". Given an ordering of the vertices, $v_1, \ldots v_n$, the *adjacency matrix* A of the graph is defined by setting $A(i, j)$ to be the number of edges from vertex v_i to vertex v_j. For simplicity we often just refer to vertices $1, \ldots, n$.

Definition 4.2.4 (Edge SFT) Given a square matrix A over $\mathbb{Z}_+$, we let Γ_A denote a graph with adjacency matrix A. Let $\mathscr{E}$ be the set of edges of Γ_A. X_A is the set of doubly infinite sequences $x = \ldots x_{-2} x_{-1} x_0 x_1 x_2 \ldots$ such that each x_n is in $\mathscr{E}$, and for all n the terminal vertex of x_n equals the inital vertex of x_{n+1}. (So the points in X_A correspond to doubly infinite walks through Γ_A.) The system (X_A, σ) is the edge shift, or edge SFT, defined by A (Remark 4.2.28). We may also use the notation σ_A to denote the map $\sigma : X_A \to X_A$.

Example 4.2.5 (X_A, σ) is the full shift on the two symbols a, b. The graph has a single vertex, denoted as 1.

$$A = (2), \qquad \Gamma_A \quad = \quad a \,\overset{\frown}{\underset{\smile}{\bigcirc}}\, b \quad = \quad a \,\overset{\curvearrowright}{\bigcirc} \cdot \overset{\curvearrowleft}{\bigcirc}\, b$$

Example 4.2.6 The edge set $\mathscr{E}$ is $\{a,b,c,d\}$, and the vertex set is $\{1,2\}$:

$$A = \begin{pmatrix} 1 & 2 \\ 1 & 0 \end{pmatrix}, \qquad \Gamma_A \quad = \quad a\,\raisebox{-1ex}{\Large\textcircled{\scriptsize 1}}\,\overset{\displaystyle c}{\underset{\displaystyle d}{\rightrightarrows}}\,b\,\raisebox{-1ex}{\Large\textcircled{\scriptsize 2}} \quad = \quad a\,\bigcirc\cdot\,\overset{\displaystyle c}{\underset{\displaystyle d}{\rightrightarrows}}\,b\,\cdot$$

Here *...aabdc...* can occur in a point of X_A, but not *...bc...* .

4.2.4 The continuous shift-commuting maps

Notation 4.2.7 For a subshift (X,σ) and $n \in \mathbb{N}$, $\mathscr{W}_n(X)$ denotes the set of X-words of length n:

$$\mathscr{W}_n(X) = \{x_0 \ldots x_{n-1} : x \in X\}$$
$$= \{x_{i+n} \ldots x_{i+n-1} : x \in X\}, \quad \text{for every } i \in \mathbb{Z}.$$

Block codes

Suppose that (X,σ) and (Y,σ) are subshifts. Suppose $\Phi : \mathscr{W}_N(X) \to \mathscr{W}_1(Y)$, and j,k are integers, with $j + N - 1 = k$. Then, for $x \in X$, we can define a bisequence $y = \phi(x)$ by the rule $y_n = \Phi(x_{n+j} \ldots x_{n+k})$, for all n.

For example, with $N = 4$, $j = -1$ and $k = 2$:

$$\cdots\ x_{-1}x_0x_1x_2\ \cdots\ x_{n-1}x_nx_{n+1}x_{n+2}\ \cdots$$
$$\downarrow \qquad\qquad\qquad \downarrow$$
$$\cdots\qquad y_0\qquad \cdots\qquad y_n\qquad\qquad \cdots$$

where $y_0 = \Phi(x_{-1}x_0x_1x_2)$ and $y_n = \Phi(x_{n-1}x_nx_{n+1}x_{n+2})$. The point y is defined by "sliding" the rule Φ along x. For some rules Φ, the image of ϕ is contained in the subshift (Y,σ).

Definition 4.2.8 (Block code) The rule Φ above is called a *block code*. The map ϕ defined by j and Φ is called a sliding block code (or a block code, or just a code). The map ϕ has range n if Φ above can be chosen with $(j,k) = (-n,n)$ (i.e., $x_{-n} \cdots x_n$ determines $(\phi x)_0$).

The following result is fundamental for symbolic dynamics, though it is easy to prove (Theorem 4.2.29).

Theorem 4.2.9 (Curtis–Hedlund–Lyndon) *Suppose that* (X,σ) *and* (Y,σ) *are subshifts, and* $\phi : X \to Y$. *Then the following are equivalent:*

(1) ϕ is continuous and $\sigma\phi = \phi\sigma$.
(2) ϕ is a block code.

The CHL theorem tells us that the morphisms between subshifts are given by block codes.

We'll define some examples by stating the rule $(\phi x)_0 = \Phi(x_i \ldots x_{i+N-1})$.

Example 4.2.10 The shift map σ and its powers are sliding block codes. E.g., $(\sigma x)_0 = x_1$, $(\sigma^2 x)_0 = x_2$ and $(\sigma^{-1} x)_0 = x_{-1}$.

Example 4.2.11 Let (X, σ) be the full shift on the two symbols $0, 1$. Define $\phi : (X, \sigma) \to (X, \sigma)$ by $(\phi x)_0 = x_0 + x_1$ (mod 2). For example, if we have $x = \ldots 11\overset{\bullet}{0}1110001\ldots$, then $\phi(x) = \ldots 011\overset{\bullet}{0}01001 \ldots$.

Higher block presentations

Given a subshift (X, σ) and $k \in \mathbb{N}$, we define $X^{[k]}$ to be the image of X under the block code $\phi : x \mapsto y$, where, for each n, the symbol y_n is $x_n \ldots x_{n+k-1}$, the X-word (block) of length k beginning at x_n. We might put parentheses around this word for visual clarity. For example, with $k = 2$,

$$x = \ldots x_{-1} \overset{\bullet}{x_0} x_1 x_2 x_3 x_4 \ldots$$

$$\phi(x) = \ldots (x_{-1} x_0) \overset{\bullet}{(x_0 x_1)} (x_1 x_2)(x_2 x_3)(x_3 x_4) \ldots$$

For each k, the map $\phi : X \to X^{[k]}$ is easily checked to be a topological conjugacy, $(X, \sigma) \to (X^{[k]}, \sigma)$.

Definition 4.2.12 The subshift $(X^{[k]}, \sigma)$ is the k-block presentation of (X, σ).

Proposition 4.2.13 *For a subshift (X, σ), the following are equivalent.*

(1) (X, σ) is a shift of finite type.
(2) (X, σ) is topologically conjugate to an edge SFT.

We leave the (not difficult) proof of Proposition 4.2.13 to Remark 4.2.39.

By Proposition 4.2.13, in order to relate the dynamical relations and properties of arbitrary SFTs to matrix algebra, it suffices to relate the dynamical relations and properties of edge SFTs to their defining matrices. So, we will be concerned from here almost exclusively with SFTs which are edge SFTs. For example, to classify SFTs up to topological conjugacy, it suffices to determine when matrices A, B over $\mathbb{Z}_+$ define edge SFTs σ_A, σ_B which are topologically conjugate.

4.2.5 Powers of an edge SFT

Proposition 4.2.14 *Let n be a positive integer. Then the nth power system (X_A, σ^n) is topologically conjugate to the edge SFT (X_{A^n}, σ) defined by A^n.*

The proposition holds because, in a graph with adjacency matrix A, the number of paths of length n from vertex i to vertex j is $A^n(i,j)$ (Proposition 4.2.30). E.g. let $n = 2$, and let $\mathcal{V}$ be the vertex set of Γ_A. Let $\mathcal{G}$ be the graph with vertex set $\mathcal{V}$ for which an edge from i to j is a two-edge path (ab) from i to j in $\mathcal{G}$. Since A^2 is an adjacency matrix for $\mathcal{G}$, we may take for (X_{A^2}, σ) the edge SFT on edge paths in $\mathcal{G}$. We have (Remark 4.2.32) a topological conjugacy $\phi : (X_A, \sigma^2) \to (X_{A^2}, \sigma)$, defined by $(\phi x)_n = x_{2n} x_{2n+1}$, for $n \in \mathbb{Z}$:

$$
\begin{array}{ccc}
\ldots x_{-2}x_{-1}\overset{\bullet}{x_0}x_1x_2x_3\ldots & \xrightarrow{\ \sigma^2\ } & \ldots x_{-2}x_{-1}x_0x_1\overset{\bullet}{x_2}x_3\ldots \\
\phi \downarrow & & \phi \downarrow \\
\ldots (x_{-2}x_{-1})(\overset{\bullet}{x_0}x_1)(x_2x_3)\ldots & \xrightarrow{\ \sigma\ } & \ldots (x_{-2}x_{-1})(x_0x_1)(\overset{\bullet}{x_2}x_3)\ldots
\end{array}
$$

The inverse system (X_A, σ^{-1}) is conjugate to $(X_{A^\top}, \sigma)$, the edge SFT defined by the transpose of A (Remark 4.2.31).

4.2.6 Periodic points and nonzero spectrum

Given a subshift (X, σ), let $\text{Fix}(\sigma^k) = \{x \in X : \sigma^k x = x\}$. We can regard the sequence $(|\text{Fix}(\sigma^k)|)_{k \in \mathbb{N}}$ as the *periodic data* of the system (Remark 4.2.33). For an edge SFT (X_A, σ_A), we will derive from A a complete invariant for the periodic data.

Periodic data $\leftrightarrow$ trace sequence of A

The point x is a fixed point for σ iff $x = \ldots a\overset{\bullet}{a}aaa\ldots$ for some edge a whose terminal vertex is the same as its initial vertex. The number of edges from vertex i to vertex i is $A(i,i)$. So, in X_A,

$$|\text{Fix}(\sigma)| = \sum_i A(i,i) = \text{tr}(A) \, .$$

Likewise, a length k path whose initial and terminal vertices are equal gives a fixed point of σ^k, and

$$|\text{Fix}(\sigma)^k| = \text{trace}(A^k).$$

Thus $(|\text{Fix}(\sigma^k)|)_{k \in \mathbb{N}} = (\text{tr}(A^k))_{k \in \mathbb{N}}.$

Trace sequence of A $\leftrightarrow$ $\det(I - tA)$

There is a standard equation in the indeterminate t (Proposition 4.2.34):

$$\frac{1}{\det(I - tA)} = \exp \sum_{k=1}^{\infty} \frac{1}{n} \operatorname{tr}(A^k) t^k .$$

From this, one sees the trace sequence and $\det(I - tA)$ determine each other (Corollary 4.2.36). (This mutual determination holds for a matrix over any torsion-free commutative ring (Remark 4.2.37)).

$\det(I - tA)$ $\leftrightarrow$ nonzero spectrum of A

Definition 4.2.15 (Nonzero spectrum) If a matrix A has characteristic polynomial $t^m \prod_{i=1}^{k} (t - \lambda_i)$, with the λ_i nonzero, then the *nonzero spectrum* of A is $(\lambda_1, \ldots, \lambda_k)$. Here – by abuse of notation (Remark 4.5.37) – the m-tuple is used as notation for a multiset: the multiplicity of entries of $(\lambda_1, \ldots, \lambda_k)$ matters, but not their order. For example, $(2, 1, 1)$ and $(1, 2, 1)$ denote the same nonzero spectrum, but $(2, 1)$ is different.

If A has nonzero spectrum $\Lambda = (\lambda_1, \ldots, \lambda_k)$, then

$$\det(I - tA) = \prod_{i=1}^{k} (1 - \lambda_i t) .$$

For example,

$$A = \begin{pmatrix} 3 & 0 & 0 & 0 \\ 0 & 3 & 0 & 0 \\ 0 & 0 & 5 & 0 \\ 0 & 0 & 0 & 0 \end{pmatrix} , \qquad I - tA = \begin{pmatrix} 1 - 3t & 0 & 0 & 0 \\ 0 & 1 - 3t & 0 & 0 \\ 0 & 0 & 1 - 5t & 0 \\ 0 & 0 & 0 & 1 \end{pmatrix}$$

$$\Lambda = (3, 3, 5) , \qquad \det(I - tA) = (1 - 3t)^2 (1 - 5t) .$$

The nonzero spectrum and the polynomial $\det(I - tA)$ determine each other.

4.2.7 Classification of SFTs

Problem 4.2.16 (Classification Problem) Given square matrices A, B over $\mathbb{Z}_+$, determine whether they present SFTs which are topologically conjugate.

There are trivial ways to produce infinitely many distinct matrices which define the same SFT. E.g.,

$$(2) , \begin{pmatrix} 2 & 0 \\ 0 & 0 \end{pmatrix} , \begin{pmatrix} 2 & 0 \\ 1 & 0 \end{pmatrix} , \begin{pmatrix} 2 & 1 \\ 0 & 0 \end{pmatrix} , \begin{pmatrix} 2 & 0 & 0 \\ 1 & 0 & 0 \\ 1 & 0 & 0 \end{pmatrix} , \begin{pmatrix} 2 & 1 & 1 \\ 0 & 0 & 1 \\ 0 & 0 & 0 \end{pmatrix} , \ldots$$

Every SFT (X_A, σ) equals one which is defined by a matrix which is *non-degenerate* (has no zero row and no zero column) (Remark 4.2.38). We can avoid the trivial problem by considering only nondegenerate matrices. Still, in the nontrivial case (the case that X_A contains infinitely many points), there are nondegenerate matrices of unbounded size which define SFTs topologically conjugate to (X_A, σ) (Remark 4.2.41).

4.2.8 Strong shift equivalence of matrices, classification of SFTs

Definition 4.2.17 (Semiring) A *semiring* is a set with operations addition and multiplication satisfying all the ring axioms, except that an element is not required to have an additive inverse. In this chapter, the semiring is always assumed to contain a multiplicative identity, 1.

Below, $\mathscr{S}$ is a subset of a semiring (Remark 4.2.42) containing 0 and 1. For $\mathscr{S}$ a subset of $\mathbb{R}$, $\mathscr{S}_+$ denotes $\mathscr{S} \cap \{x \in \mathbb{R} : x \geq 0\}$. We are especially interested in $\mathscr{S} = \mathbb{Z}, \mathbb{Z}_+, \mathbb{R}, \mathbb{R}_+$.

Let A and B be square matrices over $\mathscr{S}$ (not necessarily of the same size).

Definition 4.2.18 (Elementary strong shift equivalent, ESSE) A and B are *elementary strong shift equivalent* over $\mathscr{S}$ (ESSE-$\mathscr{S}$) if there exist matrices R, S over $\mathscr{S}$ such that $A = RS$ and $B = SR$.

Note, if a matrix R is $m \times n$, and S is a matrix such that RS and SR are well defined, then S must be $n \times m$, and the matrices RS and SR must be square.

Definition 4.2.19 (Strong shift equvalent, SSE) A and B are *strong shift equivalent* over $\mathscr{S}$ (SSE-$\mathscr{S}$) if there are matrices $A = A_0, A_1, \ldots, A_\ell = B$ over $\mathscr{S}$ such that A_i and A_{i+1} are ESSE-$\mathscr{S}$, $0 \leq i < \ell$.

The number ℓ above is called the lag of the strong shift equivalence.

The relation ESSE-$\mathscr{S}$ is reflexive and symmetric. Easy examples show that ESSE-$\mathscr{S}$ is not transitive (Remark 4.2.47); but SSE-$\mathscr{S}$, the transitive closure of ESSE-$\mathscr{S}$, is an equivalence relation. Williams' introduction of strong shift equivalence in [153] – the foundation for all later work on the classification of shifts of finite type – is explained by the following theorem.

Theorem 4.2.20 (Williams 1973) *(Remark 4.2.43) Suppose that A and B are square matrices over $\mathbb{Z}_+$. Then the following are equivalent:*

(1) *A and B are SSE-$\mathbb{Z}_+$.*
(2) *The SFTs defined by A and B are topologically conjugate.*

Proof The difficult implication (2) $\implies$ (1) follows from the Decomposition Theorem (Remark 4.2.44). We will prove the easy direction, (1) $\implies$ (2).

It suffices to consider an ESSE over $\mathbb{Z}_+$, $A = RS, B = SR$. Define a square matrix M with block form $\begin{pmatrix} 0 & R \\ S & 0 \end{pmatrix}$, and edge SFT (X_M, σ). Then

$$M^2 = \begin{pmatrix} RS & 0 \\ 0 & SR \end{pmatrix} = \begin{pmatrix} A & 0 \\ 0 & B \end{pmatrix}.$$

The system (X_M, σ^2) is a disjoint union of two subsystems, $(X_1, \sigma^2|X_1)$ and $(X_2, \sigma^2|X_2)$. The shift map $\sigma : X_1 \to X_2$ gives a topological conjugacy between these subsystems.

For all i, j, we have $A(i, j) = \sum_k R(i, k)S(k, j)$. Therefore, we may choose a bijection $\alpha : a \mapsto rs$ from the set of Γ_A edges to the set of R, S paths in Γ_M (an R, S path is an R edge followed by an S edge) which respects the initial and the terminal vertex. Similarly we choose a bijection $\beta : b \mapsto sr$ from Γ_B edges to S, R paths in Γ_M. We define a conjugacy $\phi_\alpha : (X_A, \sigma) \to (X_1, \sigma^2|X_1)$, $\phi_\alpha : \ldots x_{-1}x_0x_1 \cdots \mapsto \ldots (r_{-1}s_{-1})(r_0s_0)(r_1s_1) \ldots$, by replacing each x_n with $\alpha(x_n)$. We define a conjugacy $\phi_\beta : (X_B, \sigma) \to (X_2, \sigma^2|X_2)$ in the same way.

We now have a conjugacy $c(R, S) : X_A \to X_B$ as the composition, $c(R, S) = \phi_\alpha \sigma \phi_\beta^{-1}$,

$$c(R, S) : \ldots x_{-1}x_0x_1 \ldots \mapsto \ldots (r_{-1}s_{-1})(r_0s_0)(r_1s_1) \ldots$$
$$\mapsto \ldots (s_{-1}r_0)(s_0r_1)(s_1r_2) \ldots$$
$$\mapsto \ldots y_{-1}y_0y_1 \ldots \ . \qquad \square$$

The technical statements of the next remark are not needed at all before Sections 4.8 and 4.9.

Remark 4.2.21 Let (R, S) be an ESSE-$\mathbb{Z}_+$, with $A = RS$ and $B = SR$. Let $c(R, S)$ be a topological conjugacy from (X_A, σ) to (X_B, σ) defined as in the proof above. The conjugacy $c(R, S)$ is uniquely determined by (R, S) when all entries of A and B are in $\{0, 1\}$ (as the bijections α, β are then unique). But, in general, the conjugacy depends on the choice of those bijections. With appropriate choice of those bijections, we have the following:

(1) $c(R, S)c(S, R) = \sigma_A$, the shift map on X_A .
(2) $(c(R, S))^{-1} = c(S, R)\sigma_A^{-1} = \sigma_B^{-1}c(S, R)$.
(3) $c(I, A) = \text{Id}$, and $c(A, I) = \sigma_A$.

Also: with $c(R, S)$, x_0x_1 determines y_0; with $(c(R, S))^{-1}$, $y_{-1}y_0$ determines x_0.

4.2.9 Shift equivalence

Despite the seeming simplicity of its definition, SSE over $\mathbb{Z}_+$ is a very difficult relation to understand fully. Consequently, Williams introduced shift equivalence.

Definition 4.2.22 (Shift equivalent, SE)   Let A, B be square matrices over a semiring $\mathscr{S}$. Then A, B are shift equivalent over $\mathscr{S}$ (SE-$\mathscr{S}$) if there exist matrices R, S over $\mathscr{S}$ and a positive integer ℓ such that the following hold:

$$A^\ell = RS , \quad B^\ell = SR , \quad AR = RB , \quad SA = BS .$$

Here, (R, S) is a shift equivalence of lag ℓ from A to B.

The next proposition is an easy exercise (Proposition 4.2.45).

Proposition 4.2.23   *Let $\mathscr{S}$ be a semiring.*

(1) *SE over $\mathscr{S}$ is an equivalence relation.*
(2) *SSE over $\mathscr{S}$ implies SE over $\mathscr{S}$.*

4.2.10 Williams' shift equivalence conjecture

Conjecture 4.2.24   (Williams, 1974) [153] Suppose that A, B are two square matrices which are SE-$\mathbb{Z}_+$. Then they are SSE-$\mathbb{Z}_+$.

Despite the seeming complexity of its definition, shift equivalence is much easier to understand than strong shift equivalence, as we shall see. A positive solution to Williams' Conjecture would have been a very satisfactory solution to the classification problem for SFTs. Alas ... there are counterexamples to the conjecture, due to Kim and Roush (building on work of Wagoner, and Kim–Roush–Wagoner). The first Kim–Roush counterexample was in 1992. We recall now a definition fundamental for the theory of nonnegative matrices (as we will review in Section 4.5).

Definition 4.2.25 (Primitive matrix)   A primitive matrix is a square matrix such that every entry is a nonnegative real number and for some positive integer k, every entry of A^k is positive.

By far the most important case of Williams' conjecture is the case that the matrices A, B are primitive. An edge SFT defined from a nondegenerate matrix A is *mixing*[11] if and only if A is primitive. The mixing SFTs play a role among SFTs very much analogous to the role played by primitive matrices in the theory of nonnegative matrices.

---

[11]   See [93] for the definition of the dynamical property "mixing", which we do not need.

The Kim–Roush counterexample for primitive matrices came in 1999.

Over twenty years later, we have no new theorem or counterexample for primitive matrices over $\mathbb{Z}_+$. The Kim–Roush counterexamples require quite special constructions (reviewed in Section 4.9). The proof method can work only in special SE-$\mathbb{Z}$ classes, and can never show that there is an infinite family of primitive matrices which are SE-$\mathbb{Z}_+$ but are pairwise not SSE-$\mathbb{Z}_+$ (see Section 4.9.6).

The gap between SE-$\mathbb{Z}_+$ and SSE-$\mathbb{Z}_+$?

How big is the gap between SE-$\mathbb{Z}_+$ and SSE-$\mathbb{Z}_+$? We really don't know.

Suppose that A is ANY square matrix over $\mathbb{Z}_+$ such that A is primitive (for some n, every entry of A^n is positive), and $A \neq (1)$. (The case $A = (1)$ is trivial.) As we approach a half century following Williams' conjecture, we cannot verify or rule out either of the following statements.

(1) There is an algorithm which takes as input any square matrix B over $\mathbb{Z}_+$ and decides whether A and B are SSE-$\mathbb{Z}_+$.

(2) There are infinitely many matrices which are SE-$\mathbb{Z}_+$ to A and which are pairwise not SSE-$\mathbb{Z}_+$.

Regarding the first item above: we do not know upper bounds on the lag of a possible SSE or the sizes of the matrices in its chain of ESSEs. (See Remark 4.2.47–Example 4.2.49 for more on lag issues.) Also, for example, the "1×1 case", in which A in Conjecture 4.2.24 is assumed to be 1×1, is completely open. This is called the "Little Shift Equivalence Conjecture" in [16, Problem 3].) We will see (Remark 4.3.47) that a square matrix over $\mathbb{Z}_+$ is SE-$\mathbb{Z}_+$ to $(k) \iff$ its nonzero spectrum is (k). But, for every positive integer $k > 1$, we do not know whether a matrix SE over $\mathbb{Z}_+$ to (k) must be SSE over $\mathbb{Z}_+$ to (k) (Remark 4.2.50). Remarkably, even for two 2×2 matrices over $\mathbb{Z}_+$, we do not know whether SE-$\mathbb{Z}_+$ implies SSE-$\mathbb{Z}_+$ (although, here there are significant partial results, e.g. [4, 5, 36, 154]). Nevertheless … perhaps the situation is not hopeless.

(1) If A is a matrix over $\mathbb{R}_+$ with $\det(I - tA) = 1 - \lambda t$, then A is SSE over $\mathbb{R}_+$ to (λ). (Over $\mathbb{R}_+$, the "1×1 case" is solved!)

Despite limited progress, we think the proof framework for this result of Kim and Roush is promising for proving that SE-$\mathbb{R}_+$ implies SSE-$\mathbb{R}_+$ for positive matrices (Remark 4.2.51).

(2) In recent years we have (at last) gained a much better (but not complete) understanding of strong shift equivalence over a ring, as discussed in Section 4.7. This gives more motivation for investigation, and new ideas to explore.

4.2.11 Notes

This subsection contains various remarks, proofs and comments referenced in earlier parts of Section 4.2.

Remark 4.2.26 By way of Markov partitions, shifts of finite type are a fundamental tool in the theory of smooth dynamical systems. SFTs have applications to coding theory, general topological dynamics, ergodic theory, C^*-algebras, cellular automata and geometric group theory. Our focused introduction to the stable algebra related to SFTs will avoid all of this. For a comprehensive introduction to the theory of shifts of finite type and some related topics, including a supplement reviewing recent developments, see the 2021 edition [93] of the classic 1995 text [92] of Lind and Marcus. This crystal-clear book is intended to be widely accessible; a math graduate student can easily read it without guidance. The 2021 edition includes a long supplement reviewing recent developments. The lucid 1998 book [82] of Kitchens is also valuable, providing additional depth on various topics and developing the basic theory of countable state Markov shifts (a topic not covered by Lind and Marcus).

Remark 4.2.27 Let the finite set $\mathscr{A}$ have the discrete topology. Then compactness of $\mathscr{A}^{\mathbb{Z}}$ follows from a diagonal argument from the chosen metric, or from Tychonoff's theorem (the product topology is the same as the topology coming from the chosen metric).

Suppose that X is a closed nonempty subset of $\mathscr{A}^{\mathbb{Z}}$. A "cylinder set" is a set C in X of the following form: there is a point $x \in X$, and $i \leq j$ in $\mathbb{Z}$, such that $C = \{y \in X : y_n = x_n \text{ if } i \leq n \leq j\}$. The cylinder sets form a basis for the topology on X.

The cylinder sets are closed open. A subset of X is closed open if and only if it is the union of finitely many cyinders. By definition, a metric space is zero dimensional if there is a base for the topology consisting of closed open sets. Therefore X is zero dimensional.

Remark 4.2.28 To be careful, we'll be a little pedantic.

Two different but isomorphic graphs define different but isomorphic SFTs. The topological conjugacy of SFTs in this case is rather trivial. If the graph isomorphism gives a map on edges $e \mapsto \bar{e}$, then the topological conjugacy ϕ is defined by $(\phi x)_n = \overline{x_n}$, for all n.

In the other direction, given just the matrix A, a graph $\mathscr{G}$ with adjacency matrix A is only defined up to graph isomorphism. If A is $n \times n$, then there is an ordering of the vertices, $v_1, v_2, \ldots, v_n$, such that $A(i, j)$ is the number of edges from v_i to v_j. For simplicity, we often just regard the vertex set as $\{1, 2, \ldots, n\}$, with $v_i = i$.

Theorem 4.2.29 (Curtis–Hedlund–Lyndon)   *Suppose that (X, σ) and (Y, σ) are subshifts, and $\phi : X \to Y$. Then the following are equivalent:*

(1) *ϕ is continuous and shift-commuting.*
(2) *There are integers j, k with $j \leq k$, such that for $N = k - j + 1$ there is a function $\Phi : \mathcal{W}_N(X) \to \mathcal{W}_1(Y)$, such that for all n in $\mathbb{Z}$ and x in X, $(\phi x)_n = \Phi(x_{n+j} \ldots x_{n+k})$.*

Proof   (1) $\implies$ (2) Suppose ϕ is continuous, hence uniformly continuous, on X. There is an $\varepsilon > 0$ such that for y, y' in Y, $y_0 \neq (y')_0 \implies \mathrm{dist}(y_0, (y')_0) > \varepsilon$. By the uniform continuity, there is $m \in \mathbb{N}$ such that for x, w in X,

$$x_{-m} \ldots x_m = w_{-m} \ldots w_m \implies (\phi x)_0 = (\phi w)_0 \,.$$

This gives a rule $\Phi : \mathcal{W}_{2m+1}(X) \to \mathcal{W}_1(Y)$ such that for all x in X, $(\phi x)_0 = \Phi(x_{-m} \ldots x_m)$. Because ϕ is shift commuting, we then get for all n that

$$\Phi(x_{n-m} \ldots x_{n+m}) = \Phi((\sigma^n x)_{-m} \ldots (\sigma^n x)_m) = (\phi(\sigma^n x))_0 = (\sigma^n (\phi x))_0 = (\phi x)_n \,.$$

We leave the proof of (2) $\implies$ (1) as an exercise. $\square$

There are other ways to state the CHL theorem. (We didn't copy the original statement in [62].)

Proposition 4.2.30   *Let a graph have adjacency matrix A. Then the number of paths of length n from vertex i to vertex j is $A^n(i, j)$.*

Proof   A length 2 path from i to j is, for some vertex k, an edge from i to k followed by an edge from k to j. The number of such paths is $\sum_k A(i, k) A(k, j) = A^2(i, j)$. The claim for paths of length n follows by induction, considering paths of length $n - 1$ followed by path of length 1. $\square$

Remark 4.2.31   Suppose that A is a square matrix over $\mathbb{Z}_+$, with transpose $A^\top$. From a graph $\mathcal{G}$ with adjacency matrix A, let $\mathcal{G}^{\mathrm{reversed}}$ be the graph with the same vertex set as G, and edges with the same names but with reversed direction (an edge e from i to j in $\mathcal{G}$ becomes an edge e from j to i in $\mathcal{G}^{\mathrm{reversed}}$. Then $A^\top$ is an adjacency matrix for $\mathcal{G}^{\mathrm{reversed}}$.

Now, there is a topological conjugacy $\phi : (X_A, \sigma^{-1}) \to (X_{A^\top}, \sigma)$, defined by the rule $(\phi x)_n = x_{-n}$, for $n \in \mathbb{Z}$.

Remark 4.2.32   The topological conjugacy $\phi : (X_A, \sigma^2) \to (X_{A^2}, \sigma)$ is not a block code. This does not contradict the CHL theorem, because (X_A, σ^2) is not a subshift.

Remark 4.2.33 Formally, the "periodic data" for a system (X,S) is the isomorphism class of the system $(\text{Per}(S),S)$, with the periodic points, $\text{Per}(S)$, given the discrete topology (i.e., ignore topology). (Here "system" relaxes our terminology in this chapter that the domain must be compact.)

A complete invariant for the periodic data is one such that two systems agree on the invariant if and only if they have the same periodic data.

Now, one complete invariant of the periodic data of a system is simply the function which assigns to n the cardinality of the set of points of least period n. (A point has least period n if its orbit is finite with cardinality n.) For a subshift (X,σ), there is a finite number q_n of points of least period n, and the sequence (q_n) is a complete invariant of the periodic data. Let $\tau_n = |\text{Fix}(\sigma^n)|$. The sequence (q_n) determines the sequence $(\tau_n)_{n=1}^{\infty}$. For our systems, each τ_n is a nonnegative integer, and in this case the converse holds: the sequence (τ_n) determines the sequence (q_n). E.g., $q_1 = \tau_1$, $q_2 = \tau_2 - \tau_1$, $\dots$, $q_6 = \tau_6 - \tau_3 - \tau_2 + \tau_1$, $\dots$. (The formal device for producing a systematic formula for this inclusion-exclusion pattern is Möbius inversion.) So, "we may regard" (τ_n) as the periodic data in the sense that it is a complete invariant for the periodic data.

Proposition 4.2.34 *Suppose that A is a matrix with entries in $\mathbb{C}$. Then*

$$\frac{1}{\det(I - tA)} = \exp \sum_{n=1}^{\infty} \frac{1}{n} \text{tr}(A^n) t^n . \tag{4.1}$$

Proof Recall that $-\log(1-x) = x + \frac{x^2}{2} + \frac{x^3}{3} + \cdots$. Let $(\lambda_1,\dots,\lambda_n)$ be the nonzero spectrum of A. Then

$$\exp\left(\sum_{n=1}^{\infty} \frac{1}{n} \text{tr}(A^n) t^n\right) = \exp\left(\sum_{n=1}^{\infty} \frac{1}{n} \left(\sum_i \lambda_i^n\right) t^n\right)$$

$$= \exp\left(\sum_i \left(\sum_{n=1}^{\infty} \frac{1}{n} (\lambda_i t)^n\right)\right) = \prod_i \exp\left(\sum_{n=1}^{\infty} \frac{(\lambda_i t)^n}{n}\right)$$

$$= \prod_i \exp\left(-\log(1 - \lambda_i t)\right) = \prod_i \frac{1}{(1 - \lambda_i t)} = \frac{1}{\det(I - tA)} .$$

$\square$

(The last proposition remains true as an equation in formal power series if $\mathbb{C}$ is replaced by a torsion-free commutative ring $\mathcal{R}$. In this case, $\mathbb{N}$ is a multiplicative subset of $\mathcal{R}$ containing no zero divisor, and all the power series coefficients make sense in the localization $\mathcal{R}[\mathbb{N}^{-1}]$.)

Remark 4.2.35 (The zeta function) Suppose (X,S) is a dynamical system such that for all n in $\mathbb{N}$, $|\mathrm{Fix}(S^n)| < \infty$. Then the (Artin–Mazur) zeta function of the system is defined to be

$$\zeta(t) = \exp\left(\sum_{n=1}^{\infty} \frac{1}{n}|\mathrm{Fix}(S^n)|t^n\right).$$

This is defined at least as a formal power series; it is defined as an analytic function inside the radius of convergence. The zeta function (where it is defined) is the premier complete invariant of the periodic data. For an edge SFT defined from a matrix A, we see $\zeta(t) = 1/\det(I - tA)$.

Corollary 4.2.36 *Suppose that A is a square matrix over $\mathbb{C}$. Then $\det(I - tA)$ and the sequence $(\mathrm{tr}(A^n))$ determine each other.*

Proof The nontrivial implication, the fact that the trace sequence determines $\det(I - tA)$, follows from Proposition 4.2.34. The proposition also is easily used to prove the reverse implication; but we may also simply notice that $\det(I - tA)$ determines the nonzero spectrum $(\lambda_1, \ldots, \lambda_n)$ of A, which determines $\mathrm{tr}(A^k) = \sum_i (\lambda_i)^k$. □

Remark 4.2.37 For any $n \times n$ matrix A over any commutative ring, the polynomial $\det(I - tA)$ determines the trace sequence $(\mathrm{tr}(A^k))_{k=1}^{\infty}$; if the ring is torsion-free, then conversely $(\mathrm{tr}(A^k))_{k=1}^{\infty}$ must determine $\det(I - tA)$. To see this, let us write $\det(I - tA)$ as $1 - f(t) = 1 - f_1 t - f_2 t^2 \cdots - f_n t^n$, and let τ_k denote $\mathrm{tr}(A^k)$. Then the claimed determinations are easily proved by induction from Newton's identities,[12] valid over any commutative ring:

$$\tau_k = kf_k + \sum_{i=1}^{k-1} f_i \tau_{k-i}, \quad \text{if } 1 \le k \le N,$$

$$= \sum_{i=1}^{n} f_i \tau_{k-i}, \quad \text{if } k > n.$$

To see the torsion-free assumption is not extraneous, let $\mathscr{R}$ be the ring $\mathbb{Z}_2 \times \mathbb{Z}_2$, and consider the matrices

$$A = ((0,1)), \qquad B = \begin{pmatrix} (0,0) & (1,1) \\ (1,0) & (0,1) \end{pmatrix}.$$

Here, $\det(I - tA) = 1 - t(0,1) \ne 1 - t(0,1) - t^2(1,0) = \det(I - tB)$, but $\mathrm{tr}(A^n) = \mathrm{tr}(B^n) = (0,1)$ for every positive integer n.

One of the ways to prove Newton's identities is to take the derivative of the log of both sides of (4.1), and equate coefficients in the resulting equation

---

[12] As $\det(I - tA)$ is the reversed characteristic polynomial, Newton's identities can alternatively be (and usually are) stated in terms of coefficients of the characteristic polynomial.

of power series. This makes sense at the level of formal power series when the ring is torsion free, in particular for a polynomial ring $\mathbb{Z}[\{x_{ij}\}]$, where $\{x_{ij} : 1 \leq i, j \leq n\}$ is a set of n^2 commuting variables. Then, given A over any commutative ring $\mathscr{R}$, using the ring homomorphism $\mathbb{Z}[\{x_{ij}\}] \to \mathscr{R}$ induced by $x_{ij} \mapsto A(i,j)$, from the Newton identities over $\mathbb{Z}[\{x_{ij}\}]$ we obtain the Newton identities for A.

Remark 4.2.38   A matrix is *degenerate* if it has a zero row or a zero column. The *nondegenerate core* of a square matrix is the largest principal submatrix C which is nondegenerate. If row i or column i of A is zero, then remove row i and column i. Continue until a nondegenerate matrix C is reached. This matrix is the nondegenerate core of A.

When the matrices have all entries in $\mathbb{Z}_+$, $X_C = X_A$, because if an edge occurs as x_n for some point of X_A, then the edge must be followed and preceded by arbitrarily long paths in Γ_A.

Remark 4.2.39   Proposition 4.2.13 states that every SFT is topologically conjugate to an edge SFT. Because a subshift is topologically conjugate to each of its higher block presentations, in order to prove Proposition 4.2.13 it suffices to prove the next result. As defined in notation 4.2.7, we use $\mathscr{W}_k(X)$ to denote the set of X-words of length k.

Proposition 4.2.40   *Suppose that (X, σ) is a subshift of finite type on alphabet $\mathscr{A}$. Let $\mathscr{F}$ be a finite set of words such that X equals the set of points x on alphabet $\mathscr{A}$ such that no word $\mathscr{F}$ occurs in x. Suppose that $N > 1$ and that $N \geq \max\{length of W : W \in \mathscr{F}\}$.*

Then the N-block presentation of (X, σ) is an edge SFT, with edge set $\mathscr{W}_N(X)$.

Proof   We define a directed graph G. The vertex set is $\mathscr{W}_{N-1}(X)$. The edge set is $\mathscr{W}_N(X)$. An edge $W_1 \dots W_N$ is an edge from vertex $W_1 \dots W_{N-1}$ to vertex $W_2 \dots W_N$. Clearly, the edge SFT (X_A, σ) defined from G contains the N-block presentation $(X^{[N]}, \sigma)$. Conversely, the condition on $\mathscr{F}$ implies that every point of (X_A, σ) is a point in $X^{[N]}$.   □

Remark 4.2.41   For an edge SFT X_A, let $A^{[k]}$ be a transition matrix for its k-block presentation. Note, $A^{[k]}$ is not degenerate. For example, let $A = A^{[1]} = (2)$, with edge set $\mathscr{E}_1 = \{a, b\}$. The vertex sets $\mathscr{W}_2$ and $\mathscr{W}_3$ for 2 and 3 block presentations are $\{a, b\}$ and $\{aa, ab, ba, bb\}$. With the lexicographic orderings on these sets (ordered as written), we get the corresponding adjacency matrices

$$A^{[2]} = \begin{pmatrix} 1 & 1 \\ 1 & 1 \end{pmatrix}, \qquad A^{[3]} = \begin{pmatrix} 1 & 1 & 0 & 0 \\ 0 & 0 & 1 & 1 \\ 1 & 1 & 0 & 0 \\ 0 & 0 & 1 & 1 \end{pmatrix}.$$

In general, if X_A is infinite, then the size of $A^{[k]}$ must go to infinity with k.

Remark 4.2.42 The use of SSE over semirings goes beyond the study of SSE over the positive set of an ordered ring. SSE over the Boolean semiring $\{0, 1\}$, in which $1 + 1 = 1$, ends up being quite relevant to some constructions over $\mathbb{R}_+$ [25], and to relating topological conjugacy and flow equivalence of SFTs [15]. The Boolean semiring cannot be embedded in a ring, as $1 + 1 = 1$ would then force $1 = 0$.

Remark 4.2.43 For simplicity, we take some liberties with the statement of Proposition 4.2.40. "Edge SFTs" don't appear in Williams' paper; he used a more abstract approach to associate SFTs to matrices over $\mathbb{Z}_+$.

Williams' 1973 paper [153] contained a "proof" (erroneous) of his conjecture. The 1974 conjecture appeared in the erratum. One of the most important papers in symbolic dynamics also included perhaps its most famous mistake.

Remark 4.2.44 We say a little about the Decomposition Theorem, even though we won't have space to explain it well, because it is a very important feature of SSE. Lind and Marcus give a nice presentation of the Decomposition Theorem [93].

The Decomposition Theorem tells us that when there is a conjugacy of edge SFTs $\phi : (X_A, \sigma) \to (X_B, \sigma)$, there is another matrix C, an SSE-$\mathbb{Z}_+$ from C to A given by a string of column amalgamations, and an SSE -$\mathbb{Z}_+$ from C to B given by a string of row amalgamations, such that the associated conjugacies $\alpha : (X_C, \sigma) \to (X_A, \sigma)$ and $\beta : (X_C, \sigma) \to (X_A, \sigma)$ give $\phi = \alpha^{-1}\beta$.

For x in X_C: $(\alpha x)_0$ and $(\beta x)_0$ depend only on x_0.

A column amalgamation $C \to D$ is an ESSE $C = RS$, $D = SR$, such that S is a zero-one matrix with each column containing exactly one nonzero entry. For example,

$$C = \begin{pmatrix} 1 & 1 & 5 \\ 2 & 2 & 3 \\ 1 & 1 & 2 \end{pmatrix} = \begin{pmatrix} 1 & 5 \\ 2 & 3 \\ 1 & 2 \end{pmatrix} \begin{pmatrix} 1 & 1 & 0 \\ 0 & 0 & 1 \end{pmatrix} = RS\,,$$

$$D = \begin{pmatrix} 3 & 8 \\ 1 & 2 \end{pmatrix} = \begin{pmatrix} 1 & 1 & 0 \\ 0 & 0 & 1 \end{pmatrix} \begin{pmatrix} 1 & 5 \\ 2 & 3 \\ 1 & 2 \end{pmatrix} = SR\,.$$

Row amalgamations are correspondingly given by amalgamating rows rather than columns.

The Decomposition Theorem, or a relative, is a tool for the characterization of nonzero spectra of primitive real matrices [23]; for Parry's cohomological characterization of SSE-$\mathbb{Z}_+ G$ [32]; and for studying SSE over dense subrings of $\mathbb{R}$ [25].

Proposition 4.2.45 *Let $\mathscr{S}$ be a semiring.*

(1) *SE over $\mathscr{S}$ is indeed an equivalence relation.*
(2) *SSE over $\mathscr{S}$ implies SE over $\mathscr{S}$.*

Proof (1) If (R_1, S_1) is a shift equivalence of lag ℓ_1 from A to B, and (R_2, S_2) is a shift equivance of lag ℓ_2 from B to C, then $(R_1 R_2, S_2 S_1)$ satisfies the equations to be a shift equivalence of lag $\ell_1 + \ell_2$ from A to C. (For example, $R_1 R_2 S_2 S_1 = R_1 B^{\ell_1} S_1 = R_1 S_1 A^{\ell_1} = A^{\ell_2} A^{\ell_1} = A^{\ell_1 + \ell_2}$.)

(2) Suppose that we are given a lag ℓ SSE from A to B:
$A = A_0, A_1, \ldots, A_\ell = B; \qquad A_i = R_i S_i$ and $A_{i+1} = S_i R_i, \quad$ for $0 \le i < \ell$.
Set $R = R_1 R_2 \ldots R_\ell, \quad S = S_\ell \ldots S_2 S_1$.
Then (R, S) is a shift equivalence of lag ℓ from A to B. $\quad\square \qquad\qquad\qquad \square$

Next we state one of the interesting partial results on Williams' conjecture, which we will use later.

Theorem 4.2.46 *(K. Baker [4]) Suppose that A, B are positive 2×2 integral matrices with nonnegative determinant which are similar over the integers. Then A, B are strong shift equivalent over $\mathbb{Z}_+$.*

Remark 4.2.47 *(Nilpotence and lag)* Let $A = RS, B = SR$ be an ESSE over a semiring $\mathscr{S}$. Suppose that $m \ge 2$ is the smallest positive integer such that $A^m = 0$. Then B is also nilpotent (because $B^{m+1} = SA^m R = 0$), but $B^{m-2} \ne 0$ (because $B^{m-2} = 0$ would force $A^{m-1} = RB^{m-2}S = 0$). Thus if ℓ is the lag of an SSE-$\mathscr{S}$ from A to a zero matrix, then $\ell \ge m - 1$. For example, there is a lag 2 SSE-$\mathbb{R}$ from $\begin{pmatrix} 0 & 1 & 0 \\ 0 & 0 & 1 \\ 0 & 0 & 0 \end{pmatrix}$ to (0), but there is no ESSE-$\mathbb{R}$ from $\begin{pmatrix} 0 & 1 & 0 \\ 0 & 0 & 1 \\ 0 & 0 & 0 \end{pmatrix}$ to (0).

For an example involving primitive matrices, consider the matrix $A = (2)$ and its 3-block presentation matrix $B = A^{[3]}$ in Remark 4.2.41. There is a lag 2 SSE-$\mathbb{Z}_+$ between (2) and B. But there cannot be an ESSE-$\mathbb{R}$ of B and (2): if $RS = (2)$ and $SR = B$, then R and S have rank 1, so SR has rank at most 1, contradicting B having rank 2.

The next example (extracted from Norbert Riedel's paper [120], which has more) shows that the lag of an SSE-$\mathbb{Z}_+$ is not just a matter of nilpotence.

Example 4.2.48 (*Bad lag at size 2 from geometry.*) For each positive integer k, set $A_k = \left(\begin{smallmatrix} k & 2 \\ 1 & k \end{smallmatrix}\right)$ and $B_k = \left(\begin{smallmatrix} k-1 & 1 \\ 1 & k+1 \end{smallmatrix}\right)$. For each k, the matrices A_k, B_k are SSE over $\mathbb{Z}_+$. However, the minimum lag of an SE-$\mathbb{Z}_+$ between A_k, B_k (and therefore the minimum lag of an SSE-$\mathbb{Z}_+$ between A_k, B_k) goes to infinity as $k \to \infty$.

Proof sketch First, note that A_k and B_k have the same nonzero spectrum ($k + \sqrt{2}, k - \sqrt{2}$), and $\mathbb{Z}[k + \sqrt{2}] = \mathbb{Z}[\sqrt{2}]$, which is the ring of algebraic integers in $\mathbb{Q}[\sqrt{2}]$. This ring is well known to have class number 1. By Theorem 4.3.26, A_k and B_k are similar over $\mathbb{Z}$. Then, by Theorem 4.2.46, A_k and B_k are SSE over $\mathbb{Z}_+$. By induction one checks that for each n, there are polynomials $P_1^{(n)}, P_2^{(n)}$ with positive integral coefficients such that $\deg(P_1^{(n)}) = \deg(P_2^{(n)}) + 1$ and, for all k, n, we have

$$(A_k)^n = \begin{pmatrix} P_1^{(n)}(k) & 2P_2^{(n)}(k) \\ P_2^{(n)}(k) & P_1^{(n)}(k) \end{pmatrix}.$$

Now suppose that R, S are matrices over $\mathbb{Z}_+$ and $\ell \in \mathbb{N}$ such that $AR = RB$, $SA = BS$, $RS = A^\ell$. The first two equations force R, S to have the forms

$$R = \begin{pmatrix} b - a & a + b \\ a & b \end{pmatrix}; \qquad a, b, b - a \in \mathbb{Z}_+$$

$$S = \begin{pmatrix} b - a & 2a - b \\ a & b \end{pmatrix}; \qquad a, b, b - a, 2a - b \in \mathbb{Z}_+$$

and from this one can check that RS has the form

$$RS = \begin{pmatrix} a & 2b \\ b & a \end{pmatrix}, \qquad a, b, 2b - a \in \mathbb{Z}_+ .$$

For fixed n, $\lim_k 2P_2^{(n)}(k)/P_1^{(n)}(k) = \infty$. Thus given $\ell_0 \in \mathbb{N}$, for all sufficiently large k we have for $n \leq \ell_0$ that $P_1^{(n)}(k) > 2P_2^{(n)}(k)$. Thus, for such k, the lag of an SE-$\mathbb{Z}_+$ between A_k and B_k is greater than ℓ_0. $\qquad\square$

It is worth noting that Riedel's argument showing the smallest lag of an SE-$\mathbb{Z}_+$ goes to infinity with k works just as well with $\mathbb{Q}_+$ or $\mathbb{R}_+$ in place of $\mathbb{Z}_+$: bad lags can happen for "geometric" reasons, without nilpotence or arithmetic issues. On the other hand, bad lags can happen for strictly arithmetic reasons, as the next example shows.

Example 4.2.49 (*Bad lag at size 2 from arithmetic.*) Given $\ell \in \mathbb{N}$, there are 2×2 positive integral matrices A, B such that (i) A, B are SE-$\mathbb{Z}_+$, with minimum lag at least ℓ, and (ii) A, B are SE-$\mathbb{Q}_+$ with lag 2.[13]

Proof sketch We list steps to check. Given a prime q, and positive integer x, set $A_x = \begin{pmatrix} q & x \\ 0 & 1 \end{pmatrix}$.

Step 1. Suppose that (R, S) gives an SE-$\mathbb{Z}$ from A_x to A_y: $A_x R = R A_y$, etc. Then (perhaps after replacing R, S with $-R, -S$) R has the form $\begin{pmatrix} \pm q^k & z \\ 0 & 1 \end{pmatrix}$, where k is a nonnegative integer. It follows that $\pm x \equiv q^k y \mod (q-1)$.

Step 2. Suppose that there is a smallest positive integer k such that $q^k x \equiv \pm y$ mod $(q-1)$. Then A_x, A_y are SE-$\mathbb{Z}$, but any such shift equivalence has lag at least k.

Step 3. Choose p prime such that $p - 1 > 2(2\ell + 5)$. Then choose q prime such that p divides $q - 1$ (this is possible by Dirichlet's theorem [98]). Because $p - 1 \geq 2\ell + 5$, by the Pigeonhole Principle we may choose j a positive integer such that $1 \leq j \leq 2\ell + 5$ and also for $1 \leq k \leq \ell + 2$ we have $j \not\equiv \pm q^k$ mod p. Define $x = (q-1)/p$ and $y = j(q-1)/p$. Then A_x and A_y are SE-$\mathbb{Z}$ with minimum lag at least $\ell + 2$. Also, $0 < x < y < (1/2)q$ and $y < qx$.

Step 4. For $z \in \{x, y\}$, define the positive integral matrix

$$M_z = \begin{pmatrix} 1 & 0 \\ 1 & 1 \end{pmatrix} \begin{pmatrix} q & z \\ 0 & 1 \end{pmatrix} \begin{pmatrix} 1 & 0 \\ -1 & 1 \end{pmatrix} = \begin{pmatrix} q-z & z \\ q-z-1 & 1+z \end{pmatrix}.$$

This SIM-$\mathbb{Z}$ gives a lag 1 SE-$\mathbb{Z}$ between A_z and M_z. If follows that there can be no SE-$\mathbb{Z}$ from M_x to M_y with lag smaller than ℓ.

Step 5. It remains to produce the lag 2 SE-$\mathbb{Q}_+$ between M_x and M_y. For the eigenvalues q and 1, M_z has right eigenvectors $v = (1, 1)^{\text{tr}}$ and $w_z = (-z, q - z - 1)^{\text{tr}}$. Let U be the 2×2 matrix such that $Uv = v$ and $Uw_x = w_y$. Then $R = M_y U$, $S = M_x U^{-1}$ gives a lag 2 SE-$\mathbb{Q}$ between M_x and M_y. It remains to

[13] In Example 4.2.49, we don't know any obstruction to the existence of an example for which condition (ii) is replaced by "A, B are ESSE-$\mathbb{Q}_+$ and SSE-$\mathbb{Z}_+$".

check R, S are nonnegative. We have

$$
R = M_y U = \begin{pmatrix} q - y & y \\ q - y - q & y + 1 \end{pmatrix} \frac{1}{q-1} \begin{pmatrix} q - x - 1 + y & x - y \\ -x + y & q + x - y - 1 \end{pmatrix}
$$
$$
= \frac{1}{q-1} \begin{pmatrix} q^2 - q(x+1) - xy & qx - y \\ q^2 - q(x+2) - xy + 1 & q(x+1) - (y+1) \end{pmatrix}.
$$

From the last sentence of Step 3, we see the entries of $M_y U$ are positive. The matrix S is obtained from R by interchanging the roles of x and y, and S is likewise positive. □

Remark 4.2.50 By the way, here is an example due to Jonathan Ashley ("Ashley's eight by eight", from [82, Example 2.2.7]) of a primitive matrix A SE-$\mathbb{Z}$ to (2), but not known to be SSE-$\mathbb{Z}_+$ to (2). A is the 8×8 matrix which is the sum of the permutation matrices for the permutations (12345678) and (8)(1)(263754).

Remark 4.2.51 For more on the problem of SSE over $\mathbb{R}$, focused on the case of positive matrices, see [25]. (Kim and Roush proved that primitive matrices over $\mathbb{R}_+$ are SSE-$\mathbb{R}_+$ to positive matrices. So, the case of SSE-$\mathbb{R}_+$ of positive matrices handles the primitive positive trace case.) The method here, due to Kim and Roush, is to derive from a path of similar positive matrices an SSE-$\mathbb{R}_+$ between the endpoints. Kim and Roush were able to reduce this to considering positive matrices of equal size, similar over $\mathbb{R}$; and in the "1×1 case", to produce such a path.

However, even when both A and B are 2×2 positive real matrices, the problem of when they are SSE-$\mathbb{R}_+$ is open. It is embarassing that we are not more clever.

Remark 4.2.52 To understand when SE-$\mathbb{Z}_+$ matrices A, B are SSE-$\mathbb{Z}_+$, it is best to focus on the fundamental case that A and B are primitive. (Then consider irreducible matrices, then general matrices, modulo a solution of the primitive case.) For primitive matrices over $\mathbb{Z}_+$, SE-$\mathbb{Z}_+$ is equivalent to SSE-$\mathbb{Z}$ (Proposition 4.3.4). For primitive matrices over $\mathbb{Z}_+$, it has been important to study a reformulation of the problem: when does SSE-$\mathbb{Z}$ imply SSE-$\mathbb{Z}_+$? This formulation was essential for the Wagoner complex setting for the Kim–Roush counterexamples [76] to Williams' conjecture, and for some arguments for a general subring R of $\mathbb{R}$ (see [25]). For some subrings R of $\mathbb{R}$, SE-R_+ does not even imply SSE-$\mathbb{R}$, as we will see.

4.3 Shift equivalence and strong shift equivalence over a ring

In this section, we present basic facts about shift equivalence and strong shift equivalence over rings, with various example classes.

4.3.1 SE-$\mathbb{Z}_+$: dynamical meaning and reduction to SE-$\mathbb{Z}$

First we give the dynamical meaning of SE-$\mathbb{Z}_+$.

Definition 4.3.1 (Eventually conjugate) Homeomorphisms S and T are eventually conjugate if S^n, T^n are conjugate for all but finitely many positive integers n.

Theorem 4.3.2 *Let A, B be square matrices over $\mathbb{Z}_+$. The following are equivalent (Proposition 4.3.42).*

(1) *A, B are shift equivalent over $\mathbb{Z}_+$.*
(2) *The SFTs (X_A, σ), (X_B, σ) are eventually conjugate.*

Next we consider how SE-$\mathbb{Z}$ and SE-$\mathbb{Z}_+$ are related. Recall Definition 4.2.25: a primitive matrix is a square nonnegative real matrix such that some power is positive.

Example 4.3.3 The matrices (1) and $\begin{pmatrix} 1 & 1 \\ 1 & 0 \end{pmatrix}$ are primitive.

The matrices $\begin{pmatrix} 1 & 1 \\ 0 & 0 \end{pmatrix}$, $\begin{pmatrix} 1 & 1 \\ 0 & 1 \end{pmatrix}$ and $\begin{pmatrix} 0 & 1 \\ 1 & 0 \end{pmatrix}$ are not primitive.

Proposition 4.3.4 *(Proposition 4.3.48) Suppose that two primitive matrices over a subring $\mathscr{R}$ of the reals are SE over $\mathscr{R}$. Then they are SE over $\mathscr{R}_+$. (Recall that $\mathscr{R}_+ = \mathscr{R} \cap \{x \in \mathbb{R} : x \geq 0\}$.)*

For primitive matrices, the classification up to SE-$\mathbb{Z}_+$ reduces to the tractable problem of classifying up to SE-$\mathbb{Z}$. The Proposition becomes false if the hypothesis of primitivity is removed (Remark 4.3.49).

4.3.2 Strong shift equivalence of matrices over a ring

Let $\mathscr{R}$ be a ring. Recall, $\mathrm{GL}(n, \mathscr{R})$ is the group of $n \times n$ matrices invertible over $\mathscr{R}$; $U \in \mathrm{GL}(n, \mathscr{R})$ if there is a matrix V over $\mathscr{R}$ with $UV = VU = I$. This matrix V is denoted U^{-1}. If $\mathscr{R}$ is commutative, then $U \in \mathrm{GL}(n, \mathscr{R})$ if and only if $\det U$ is a unit in $\mathscr{R}$.

Square matrices A, B are similar over $\mathscr{R}$ (SIM-$\mathscr{R}$) if there exists U in $\mathrm{GL}(n, \mathscr{R})$ such that $B = U^{-1}AU$.

Our viewpoint: SE and SSE of matrices over a ring $\mathscr{R}$ are *stable versions of similarity* of matrices over $\mathscr{R}$.

By a "stable version of similarity" we mean an equivalence relation on square matrices which coarsens the relation of similiarity, and is obtained by allowing some kind of neglect of the nilpotent part of the matrix multiplication[14]. (This will be less vague soon.)

Proposition 4.3.5 (Maller–Shub) *SSE over a ring $\mathscr{R}$ is the equivalence relation on square matrices over $\mathscr{R}$ generated by the following relations on square matrices A, B over $\mathscr{R}$.*

(1) *(Similarity over $\mathscr{R}$) For some n, A and B are $n \times n$ and there is a matrix U in $\mathrm{GL}(n, \mathscr{R})$ such that $A = U^{-1}BU$.*

(2) *(Zero extension) There exists a matrix X over $\mathscr{R}$ such that in block form,*
$$B = \begin{pmatrix} A & X \\ 0 & 0 \end{pmatrix} \text{ or } B = \begin{pmatrix} A & 0 \\ X & 0 \end{pmatrix} .$$

Proof A similarity or a zero extension produces an ESSE:

If $A = U^{-1}BU$, then $A = (U^{-1}B)U$ and $B = U(U^{-1}B)$.

If $B = \begin{pmatrix} A & X \\ 0 & 0 \end{pmatrix}$, then $B = \begin{pmatrix} I \\ 0 \end{pmatrix} (A \quad X)$ and $A = (A \quad X) \begin{pmatrix} I \\ 0 \end{pmatrix}$.

If $B = \begin{pmatrix} A & 0 \\ X & 0 \end{pmatrix}$, then $B = \begin{pmatrix} A \\ X \end{pmatrix} (I \quad 0)$ and $A = (I \quad 0) \begin{pmatrix} A \\ X \end{pmatrix}$.

Conversely, given $A = RS$ and $B = SR$, Maller and Shub constructed in [96] a similarity of zero extensions[15]:

$$\begin{pmatrix} I & 0 \\ S & I \end{pmatrix} \begin{pmatrix} A & R \\ 0 & 0 \end{pmatrix} = \begin{pmatrix} 0 & R \\ 0 & B \end{pmatrix} \begin{pmatrix} I & 0 \\ S & I \end{pmatrix} . \qquad \square$$

SSE-$\mathscr{R}$ coarsens SIM-$\mathscr{R}$ by allowing "zero extensions". The analogue of Proposition 4.3.5 for SE, Theorem 4.7.16, will replace zero extensions with nilpotent extensions.

The relation SSE-$\mathscr{R}$ can be very subtle indeed, as we will see. Fortunately, if $\mathscr{R}$ is $\mathbb{Z}$, or a field, then SSE-$\mathscr{R}$ = SE-$\mathscr{R}$.

---

[14] The term "stable" has had diverse use. We think of "stable algebra of matrices" as a large subject in which we consider one meaningful topic.

[15] The paper [96] did not consider general rings, or state Proposition 4.3.5 explicitly even for $\mathbb{Z}$. However, Boyle heard Maller, in a talk in the 1980s, state and prove the content of Proposition 4.3.5 in full generality.

4.3.3 SE, SSE and $\det(I - tA)$

Let $\mathscr{R}$ be a commutative ring, and A a square matrix over $\mathscr{R}$. As explained in Remark 4.2.37, the polynomial $\det(I - tA)$ determines the trace sequence $(\operatorname{tr}(A^n))_{n=1}^{\infty}$, and that sequence determines $\det(I - tA)$ if $\mathscr{R}$ is torsion-free.

If A and B are SSE over $\mathscr{R}$, then one easily sees $(\operatorname{tr}(A^n))_{n=1}^{\infty} = (\operatorname{tr}(B^n))_{n=1}^{\infty}$, simply because $\operatorname{tr}(RS) = \operatorname{tr}(SR)$. To see that in addition $\det(I - tA) = \det(I - tB)$, apply the Maller-Shub characterization Proposition 4.3.5.

If there is a lag ℓ shift equivalence over $\mathscr{R}$ between A and B, then $(\operatorname{tr}(A^k))_{k=\ell}^{\infty} = (\operatorname{tr}(B^k))_{k=\ell}^{\infty}$. We shall see below that if $\mathscr{R}$ is an integral domain, then $\det(I - tA)$ is also an invariant of SE-$\mathscr{R}$ (because it is an invariant of shift equivalence over the field of fractions of $\mathscr{R}$).

But in some cases, the trace of a matrix need not be an invariant of SE-$\mathscr{R}$. Suppose that $\mathscr{R}$ is a ring with a nilpotent element a (i.e., $a \neq 0$ and $a^k = 0$ for some positive integer k). For example, let $\mathscr{R} = \mathbb{Z}[t]/(t^2)$ and $a = t$. Consider the 1×1 matrices $A = (a)$ and $B = (0)$. Then A and B are SE-$\mathscr{R}$ but $\operatorname{tr}(A) \neq \operatorname{tr}(B)$. This by the way gives an easy example of a ring $\mathscr{R}$ for which SE-$\mathscr{R}$ and SSE-$\mathscr{R}$ are not the same relation.[16]

If a commutative ring $\mathscr{R}$ has no nilpotent element, then $\det(I - tA)$ will be an invariant of the SE-$\mathscr{R}$ class of A (Remark 4.3.43).

4.3.4 Shift equivalence over a ring $\mathscr{R}$

We consider shift equivalence over a ring $\mathscr{R}$, by cases.

$\mathscr{R}$ is a field

Suppose that A is a square matrix over $\mathscr{R}$. There is an invertible U over $\mathscr{R}$ such that $U^{-1}AU$ has the form $\begin{pmatrix} A' & 0 \\ 0 & N \end{pmatrix}$, where A' is invertible and N is triangular with zero diagonal. (For $\mathscr{R} = \mathbb{C}$, use the Jordan form.)

Example 4.3.6 $\mathscr{R} = \mathbb{R}, \quad U^{-1}AU = \begin{pmatrix} \sqrt{2} & 0 & 0 & 0 \\ 0 & 0 & 1 & 0 \\ 0 & 0 & 0 & 1 \\ 0 & 0 & 0 & 0 \end{pmatrix}, \quad A' = (\sqrt{2}).$

A' (as a vector space endomorphism) is isomorphic to the restriction of A to the largest invariant subspace on which A acts invertibly. Abusing notation, we call A' **the nonsingular part** of A (keeping in mind that A' is only well defined up to similarity over the field $\mathscr{R}$).

---

[16] Somehow this easy example was missed for many years, perhaps because the rings arising in symbolic dynamics are generally without nilpotents.

Proposition 4.3.7 *A square matrix A over a field $\mathscr{R}$ is SSE over $\mathscr{R}$ to its nonsingular part, A'.*

Proof A' reaches $U^{-1}AU$ by a string of zero extensions. $\qquad\square$

Exercise 4.3.8 Suppose that $\det A = 0$, and $U^{-1}AU = \begin{pmatrix} A' & 0 \\ 0 & N \end{pmatrix}$, and ℓ is the smallest positive integer such that $N^\ell = 0$. Then the smallest lag of an SSE over $\mathscr{R}$ from A to A' is ℓ (Remark 4.2.47).

From Proposition 4.3.7, square matrices A,B are SSE-$\mathscr{R}$ if and only if their nonsingular parts A',B' are SSE-$\mathscr{R}$. Likewise for SE-$\mathscr{R}$.

Proposition 4.3.9 *Suppose that A,B are square nonsingular matrices over the field $\mathscr{R}$. The following are equivalent.*

(1) *A and B have the same size and are similar over $\mathscr{R}$.*

(2) *A and B are SE-$\mathscr{R}$.*

(3) *A and B are SSE-$\mathscr{R}$.*

Proof $(1) \implies (3) \implies (2)$: Clear.

$(2) \implies (1)$: Let (R,S) be a lag ℓ SE over $\mathscr{R}$ from A to B:

$$A^\ell = RS\,, \quad B^\ell = SR\,, \quad AR = RB\,, \quad BS = SA\,.$$

Suppose that A is $m \times m$ and B is $n \times n$. Then R is $m \times n$. Hence $m = n$, because

$$m \;=\; \operatorname{rank}(RS) \;\leq\; \operatorname{rank}(R) \;\leq\; \min\{m,n\} \;\leq n$$

and likewise $n \leq m$. Now $\det(A^\ell) = (\det R)(\det S)$, hence $\det R \neq 0$. Then $AR = RB$ gives $B = R^{-1}AR$. $\qquad\square$

Corollary 4.3.10 *Suppose that matrices A,B are SE over a field $\mathscr{R}$. Then $\det(I - tA) = \det(I - tB)$.*

Proof By Proposition 4.3.9, the nonsingular parts A',B' of A,B are similar over $\mathscr{R}$. Therefore they have the same spectrum, which is the nonzero spectrum of A and B. Therefore $\det(I - tA) = \det(I - tB)$. $\qquad\square$

When matrices A,B have entries in a field $\mathscr{R}$ contained in $\mathbb{C}$, similarity over $\mathscr{R}$ is equivalent to similarity over $\mathbb{C}$, and the Jordan form of the nonsingular part is a complete invariant for similarity of A,B over $\mathscr{R}$.

$\mathcal{R}$ **is a principal ideal domain**

The principal ideal domain of greatest interest to us is $\mathcal{R} = \mathbb{Z}$, the integers. The PID case is like the field case, but with more arithmetic structure. In place of the Jordan form, we use a classical fact. Recall, an *upper triangular* matrix is a square matrix with only zero entries below the diagonal (i.e., $i > j \implies A(i,j) = 0$). A lower triangular matrix is a square matrix with only zero entries above the diagonal. A matrix is triangular if it is upper or lower triangular. Block triangular matrices are defined similarly, for block structures on square matrices which use the same index sets for rows and columns.

Theorem 4.3.11 (PID Block Triangular Form) *[109] Suppose that $\mathcal{R}$ is a principal ideal domain (e.g., $\mathcal{R} = \mathbb{Z}$ or a field). Suppose that A is a square matrix over $\mathcal{R}$ and $p_1, \dots, p_k$ are monic irreducible polynomials with coefficients in $\mathcal{R}$ such that the characteristic polynomial of A is $\chi_A(t) = \prod_i p_i(t)$.*

Then A is similar over $\mathcal{R}$ to a block triangular matrix, with diagonal blocks A_i, $1 \le i \le k$, such that p_i is the characteristic polynomial of A_i.

Example 4.3.12 Suppose that A is 6×6 and

$$\chi_A(t) = (t-3)(t^2+5)(t+1)(t)(t).$$

Then there is some U in $\mathrm{GL}(6, \mathbb{Z})$ such that $U^{-1}AU$ has the form

$$\begin{pmatrix} 3 & * & * & * & * & * \\ 0 & a & b & * & * & * \\ 0 & c & d & * & * & * \\ 0 & 0 & 0 & -1 & * & * \\ 0 & 0 & 0 & 0 & 0 & * \\ 0 & 0 & 0 & 0 & 0 & 0 \end{pmatrix}$$

in which $\begin{pmatrix} a & b \\ c & d \end{pmatrix}$ has characteristic polynomial $t^2 + 5$.

Corollary 4.3.13 (Corollary of PID Block Triangular Form) *Let A ba a square matrix over the PID $\mathcal{R}$. Then A is similar over $\mathcal{R}$ to a matrix with block form $\begin{pmatrix} A' & X \\ 0 & N \end{pmatrix}$, where $\det(A') \ne 0$ and N is upper triangular with zero diagonal.*

As in the field case, we call A' the *nonsingular part* of A. (Note that A' is defined up to similarity over $\mathcal{R}$).

Corollary 4.3.14 (Nonsingularity) *For $\mathcal{R}$ a principal ideal domain, any non-nilpotent square matrix over $\mathcal{R}$ is SSE-$\mathcal{R}$ to its nonsingular part (hence, SE-$\mathcal{R}$ to its nonsingular part).*

Proof A' reaches $U^{-1}AU$ by a string of zero extensions.   □

Corollary 4.3.13 can easily fail even for a Dedekind domain, such as the algebraic integers in a number field [24].

$\mathscr{R}$ is $\mathbb{Z}$

Exercise 4.3.15 (Remark 4.3.47) Suppose that A is square over $\mathbb{Z}_+$ and that $\det(I - tA) = 1 - nt$, where n is a positive integer. Then A is SE over $\mathbb{Z}_+$ to the 1×1 matrix (n).

The classification of matrices over $\mathbb{Z}$ up to SE-$\mathbb{Z}$ reduces to the classification of nonsingular matrices over $\mathbb{Z}$ up to SE-$\mathbb{Z}$. If A, B are SE-$\mathbb{Z}$, then A, B are SE-$\mathbb{R}$, so their nonsingular parts are similar over $\mathbb{R}$; in particular A, B have the same nonzero spectrum.

It is NOT true that a square matrix over $\mathbb{Z}_+$ must be SE-$\mathbb{Z}_+$ to a nonsingular matrix.

Exercise 4.3.16 (Exercise 4.3.46) The primitive matrix $A = \begin{pmatrix} 1 & 0 & 0 & 1 \\ 0 & 1 & 0 & 1 \\ 0 & 1 & 1 & 0 \\ 1 & 0 & 1 & 0 \end{pmatrix}$ has nonzero spectrum $(2, 1)$. If A is SE-$\mathbb{Z}_+$ to a nonsingular matrix B, then B must be a primitive 2×2 matrix with nonzero spectrum $(2, 1)$. Prove that no such B exists.

Proposition 4.3.17 *Suppose that A, B are nonsingular matrices over $\mathbb{Z}$ with $|\det(A)| = 1$. The following are equivalent.*

(1) *A, B are SE-$\mathbb{Z}$.*
(2) *A, B are SIM-$\mathbb{Z}$.*

Proof (2) $\Longrightarrow$ (1): Clear.
 (1) $\Longrightarrow$ (2): An SE (R, S) over $\mathbb{Z}$ from A to B is also an SE (R, S) over the field $\mathbb{Q}$. Therefore A, B have the same size, $n \times n$. Now $A^\ell = RS$ forces $\det A$ to divide $\det A^\ell$, so $|\det R| = 1$. This implies $R \in \mathrm{GL}(n, \mathbb{Z})$. Then $AR = RB$ gives $B = R^{-1}AR$.   □

Example 4.3.18 (Nonsingular A, B which are SIM-$\mathbb{Q}$, but not SE-$\mathbb{Z}$.) Let $A = \begin{pmatrix} 3 & 4 \\ 1 & 1 \end{pmatrix}$ and $B = \begin{pmatrix} 3 & 2 \\ 2 & 1 \end{pmatrix}$. A and B have the same characteristic polynomial, $p(t) = t^2 - 4t - 1$; as p has no repeated root, A and B are similar over $\mathbb{Q}$.

Now suppose A, B are SE-$\mathbb{Z}$. Because $\det A = -1$, they are SIM-$\mathbb{Z}$: there is R in $\mathrm{GL}(2, \mathbb{Z})$ such that $B = R^{-1}AR$. Therefore $(B - I) = R^{-1}(A - I)R$. This is a contradiction, because $B - I = \begin{pmatrix} 2 & 2 \\ 2 & 0 \end{pmatrix}$ is zero mod 2, but $A - I = \begin{pmatrix} 2 & 4 \\ 1 & 0 \end{pmatrix}$ is not.

What else? Here is a quick overview.

Theorem 4.3.19   *Let p be a monic polynomial in $\mathbb{Z}[t]$ with no zero root. Let $\mathcal{M}(p)$ be the set of matrices over $\mathbb{Z}$ with characteristic polynomial p.*
If p has no repeated root, then the following hold.

(1) *All matrices in $\mathcal{M}(p)$ are SIM-$\mathbb{Q}$ (and therefore SE-$\mathbb{Q}$).*
(2) *$\mathcal{M}(p)$ is the union of finitely many SIM-$\mathbb{Z}$ classes (hence finitely many SE-$\mathbb{Z}$ classes).*
(3) *It is can happen (depending on p) that in $\mathcal{M}(p)$, SIM-$\mathbb{Z}$ properly refines SE-$\mathbb{Z}$.*

If p has a repeated root, then $\mathcal{M}(p)$ contains infinitely many SE-$\mathbb{Z}$ classes, but only finitely many SE-$\mathbb{Q}$ classes.

Example 4.3.20   (Easily checked.) For $n \in \mathbb{N}$, the matrices $\begin{pmatrix} 1 & n \\ 0 & 1 \end{pmatrix}$ are similar over $\mathbb{Q}$, but pairwise not similar over $\mathbb{Z}$.

Lastly, we report on some decidablity issues for shift equivalence over $\mathbb{Z}$.

Theorem 4.3.21   *Suppose that A, B are square matrices over $\mathbb{Z}$.*

(1) *(Grunewald [55]; see also [56].) There is an algorithm to decide whether A, B are SIM-$\mathbb{Z}$. The general algorithm is not practical.*
(2) *(Kim and Roush) [70]) There is an algorithm to decide whether A, B are SE-$\mathbb{Z}$. The general algorithm is not practical.*

4.3.5 SIM-$\mathbb{Z}$ and SE-$\mathbb{Z}$: some example classes

The proof of the next result, from [24], is an exercise.

Theorem 4.3.22   *Suppose that a, b are integers and $a > |b| > 0$. Let $\mathcal{M}$ be the set of 2×2 matrices over $\mathbb{Z}$ with eigenvalues a, b. Then the following hold.*

(1) *Every matrix in $\mathcal{M}$ is SIM-$\mathbb{Z}$ to a triangular matrix $M_x = \begin{pmatrix} a & x \\ 0 & b \end{pmatrix}$.*
(2) *M_x and M_y are SIM-$\mathbb{Z}$ if and only if $x = \pm y \bmod (a - b)$.*
(3) *M_x and M_y are SE-$\mathbb{Z}$ if and only if $x \sim y$, where $\sim$ is the equivalence relation generated by $x \sim y$ if $x = \pm qy \bmod (a - b)$ for a prime q dividing a or b.*

Example 4.3.23   Suppose that $a = 6$, $b = 1$. Then $\mathcal{M}$ is the union of three SIM-$\mathbb{Z}$ classes and two SE-$\mathbb{Z}$ classes.

Example 4.3.24   Suppose that $a = 6$, $b = 2$. Then $\mathcal{M}$ is the union of two SIM-$\mathbb{Z}$ classes and one SE-$\mathbb{Z}$ class.

Exercise 4.3.25 (Proposition 4.3.50) Use Theorem 4.3.22 to prove the follow-ing: the matrix $\begin{pmatrix} 256 & 7 \\ 0 & 1 \end{pmatrix}$ is not SE-$\mathbb{Z}$ to its transpose. Then show $\begin{pmatrix} 256 & 7 \\ 0 & 1 \end{pmatrix}$ is SE-$\mathbb{Z}$ to a primitive matrix, which cannot be SE-$\mathbb{Z}$ to its transpose.

The next theorem states a result relating a matrix similarity problem to alge-braic number theory, and the analogous result for shift equivalence. The simi-larity result is a special case of a theorem of Latimer and MacDuffee [89]; Olga Taussky provided a simple proof in this special case, which generalizes nicely to the SE-$\mathbb{Z}$ situation (Theorem 4.3.51). In the next theorem, for $\mathscr{R} = \mathbb{Z}[\lambda]$ or $\mathscr{R} = \mathbb{Z}[1/\lambda]$, $\mathscr{R}$-ideals I, I' are equivalent if they are equivalent as $\mathscr{R}$-modules, which in this case means there is a nonzero c in $\mathbb{Q}[\lambda]$ such that $cI = I'$. By an ideal class of $\mathscr{R}$ we mean an equivalence class of nonzero $\mathscr{R}$-ideals.

Theorem 4.3.26 *Suppose that p is monic irreducible in $\mathbb{Z}[t]$, and $p(\lambda) = 0$, where $0 \neq \lambda \in \mathbb{C}$. Let $\mathscr{M}$ be the set of matrices over $\mathbb{Z}$ with characteristic polynomial p. There are bijections:*

(1) $\mathscr{M}/(\text{SIM-}\mathbb{Z}) \leftrightarrow$ *Ideal classes of $\mathbb{Z}[\lambda]$ [89, 139]*

(2) $\mathscr{M}/(\text{SE-}\mathbb{Z}) \leftrightarrow$ *Ideal classes of $\mathbb{Z}[1/\lambda]$ [28].*

We will see in Propositon 4.3.52 that, if λ is a nonzero algebraic integer, then $\mathbb{Z}[\lambda]$ has finite class number.

The number theory connection is useful. For example, it follows from the above-mentioned Proposition that $\mathscr{M}$ in Theorem 4.3.26 contains only finitely many SIM-$\mathbb{Z}$ classes ([109, Theorem III.14]). In the case that $\mathbb{Z}[\lambda]$ is a full ring of quadratic integers, one can often simply look up the class number of $\mathbb{Z}[\lambda]$ in a table.

4.3.6 SE-$\mathbb{Z}$ via direct limits

Let A be an $n \times n$ matrix over $\mathbb{Z}$. We choose to let A act on row vectors. From the action $A : \mathbb{Z}^n \to \mathbb{Z}^n$ one can form the direct limit group, on which there is a group automorphism $\hat{A} : G_A \to G_A$ induced by A.

We will take a very concrete presentation, $\hat{A} : G_A \to G_A$, for the induced automorphism of the direct limit group (Remark 4.3.53).

The eventual image V_A

Define rational vector spaces $W_k = \{vA^k : v \in \mathbb{Q}^n\}$ and $V_A = \bigcap_{k \in \mathbb{N}} W_k$. Then

$$\mathbb{Q}^n \supset W_1 \supset W_2 \supset W_3 \supset \cdots$$
$$\dim(W_{k+1}) = \dim(W_k) \implies W_{k+1} = W_k$$
$$V_A = W_n.$$

V_A is the "eventual image" of A as an endomorphism of the rational vector space $\mathbb{Q}^n$. V_A is the largest invariant subspace of $\mathbb{Q}^n$ on which A acts as a vector space isomorphism.

The pair $(G_A, \hat{A})$

G_A is the subset of V_A eventually mapped by A into the integer lattice: $G_A := \{v \in V_A : \exists k \in \mathbb{N}, vA^k \in \mathbb{Z}^n\}$. The automorphism $\hat{A}$ of G_A is defined by restriction, $\hat{A} : v \mapsto vA$.

Example 4.3.27   Suppose that $|\det A| = 1$. Then $V_A = \mathbb{Q}^n$, $G_A = \mathbb{Z}^n$.

Example 4.3.28 $A = (2)$. Then $V_A = \mathbb{Q}$, and G_A is the group of dyadic rationals: $G_A = \mathbb{Z}[1/2] = \{n/2^k : n \in \mathbb{Z}, k \in \mathbb{Z}_+\}$.

Example 4.3.29   Similarly, for a positive integer k, if $A = (k)$ then $G_A = \mathbb{Z}[1/k]$. For positive integers k and m, the following are equivalent:

- $\mathbb{Z}[1/k] = \mathbb{Z}[1/m]$;
- k and m are divisible by the same primes;
- $\mathbb{Z}[1/k]$ and $\mathbb{Z}[1/m]$ are isomorphic groups.

Example 4.3.30 $A = \begin{pmatrix} 2 & 0 \\ 0 & 5 \end{pmatrix}$. $G_A = \{(x,y) : x \in \mathbb{Z}[1/2], y \in \mathbb{Z}[1/5]\}$.

Example 4.3.31 $B = \begin{pmatrix} 2 & 1 \\ 0 & 5 \end{pmatrix}$. The groups G_B and G_A are not isomorphic. (G_B is not the sum of a 2-divisible subgroup and a 5-divisible subgroup (Remark 4.3.54).)

Definition 4.3.32   Two pairs $(G_A, \hat{A}), (G_B, \hat{B})$ are isomorphic if there is a group isomorphism $\phi : G_A \to G_B$ such that $\phi \hat{B} = \hat{A} \phi$. (In other words, $\hat{A}$ and $\hat{B}$ are isomorphic, in the category of group automorphisms; or, equivalently, in the category of group endomorphisms.)

Proposition 4.3.33   *Let A, B be square matrices over $\mathbb{Z}$. The following are equivalent (Proposition 4.3.55).*

(1) *A and B are SE-$\mathbb{Z}$.*

(2) *There is an isomorphism of direct limit pairs* $(G_A, \hat{A})$ *and* $(G_B, \hat{B})$.

There is a natural way to make G_A above into an ordered group (in an important class of ordered groups, the *dimension groups* (Remark 4.3.56)). Then, by analogy with Proposition 4.3.33, there an ordered group characterization of SE-$\mathbb{Z}_+$ (Remark 4.3.58).

Example 4.3.34 Let $A = (2)$ and $B = \begin{pmatrix} 1 & 1 \\ 1 & 1 \end{pmatrix}$. These matrices are SE-$\mathbb{Z}$ (and even ESSE-$\mathbb{Z}_+$).

- $V_A = \mathbb{Q}$ and $G_A = \mathbb{Z}[1/2]$.
- $V_B = \{(x,x) : x \in \mathbb{Q}\}$, the eigenline corresponding to the eigenvalue 2, and $G_B = \{(x,x) : x \in \mathbb{Z}[1/2]\}$.
- $\phi : x \mapsto (x,x)$ defines a group isomorphism $G_A \to G_B$ such that $\phi \hat{B} = \hat{A} \phi$.

4.3.7 SE-$\mathbb{Z}$ via polynomials

It will be important for us to put everything we've done with shift equivalence into a polynomial setting. (To our knowledge, the polynomial-shift equivalece connection was first explicitly pointed out by Wagoner (Remark 4.3.57).) We use $\mathbb{Z}[t]$, the ring of polynomials in one variable with integer coefficients.

Let A be an $n \times n$ matrix over $\mathbb{Z}$. Recall

- $V_A = \bigcap_{k \in \mathbb{N}} W_k = \bigcap_{k \in \mathbb{N}} \{ vA^k : v \in \mathbb{Q}^n \}$,
- $G_A := \{ v \in V_A : \exists k \in \mathbb{N}, vA^k \in \mathbb{Z}^n \}$,
- $\hat{A}$ is the automorphism of G_A given by $\hat{A} : x \mapsto xA$.

We regard the direct limit group G_A as a $\mathbb{Z}[t]$-module, by letting t act as $(\hat{A})^{-1}$. (This choice of t action will match $\mathrm{cok}(I - tA)$ below.) Isomorphism of the pairs $(G_A, \hat{A}), (G_B, \hat{B})$ is equivalent to isomorphism of G_A, G_B as $\mathbb{Z}[t]$-modules. So, we sometimes simply refer to a pair $(G_A, \hat{A})$ as a $\mathbb{Z}[t]$-module. To summarize, we have the following.

Proposition 4.3.35 *Suppose A, B are square matrices over $\mathbb{Z}$. The following are equivalent.*

(1) *A, B are SE over $\mathbb{Z}$.*
(2) *$(G_A, \hat{A})$ and $(G_B, \hat{B})$ are isomorphic $\mathbb{Z}[t]$-modules.*

Next we get another presentation of these $\mathbb{Z}[t]$-modules.

4.3.8 Cokernel of $(I - tA)$, a $\mathbb{Z}[t]$-module

Let A be an $n \times n$ over $\mathbb{Z}$, and let I be the $n \times n$ identity matrix. View $\mathbb{Z}[t]^n$ as a $\mathbb{Z}[t]$-module: for v in $\mathbb{Z}[t]^n$ and $c \in \mathbb{Z}[t]$, the action of c is to send v to cv, where $cv = c(v_1, \ldots, v_n) = (cv_1, \ldots, cv_n)$. The map $(I - tA) : \mathscr{R}^n \to \mathscr{R}^n$, by $v \mapsto v(I - tA)$, is a $\mathbb{Z}[t]$-module homomorphism, as $(cv)(I - tA) = c(v(I - tA))$.

Now define the cokernel $\mathrm{cok}(I - tA) := \mathbb{Z}[t]^n / \mathrm{Image}(I - tA)$, where Image $(I - tA) = \{v(I - tA) \in \mathbb{Z}[t]^n : v \in \mathbb{Z}[t]^n\}$. An element of $\mathrm{cok}(I - tA)$ is a coset, $v + \mathrm{Image}(I - tA)$, denoted $[v]$. $\mathrm{cok}(I - tA)$ is a $\mathbb{Z}[t]$-module, with $c : [v] \mapsto [cv]$.

Reminder: we use row vectors to define the module.

Proposition 4.3.36 *Let A be a square matrix over $\mathbb{Z}$. Then the $\mathbb{Z}[t]$-modules $\mathrm{cok}(I - tA)$ and $(G_A, \hat{A})$ are isomorphic.*

Proof Define $\phi : G_A \to \mathrm{cok}(I - tA)$ by $x \mapsto [x(tA)^k]$, where k (depending on x) is any nonnegative integer large enough that $xA^k \in \mathbb{Z}^n$. For a proof, check that this ϕ is a well-defined isomorphism of $\mathbb{Z}[t]$-modules. $\qquad\square$

Let us see how this works out in a concrete example.

Example 4.3.37 $A = (2)$. Here $G_A = \mathbb{Z}[1/2] = \bigcup_{k \geq 0} (\frac{1}{2})^k \mathbb{Z}$ and $\mathbb{Z}[t] = \bigcup_{k \geq 0} t^k \mathbb{Z}$. The isomorphism $\phi : G_A \to \mathrm{cok}(I - tA)$ is defined by

$$\phi : \mathbb{Z}[1/2] \to \mathbb{Z}[t]/(1 - 2t)\mathbb{Z}[t]$$
$$(1/2)^k n \mapsto [t^k n], \qquad \text{for } n \text{ in } \mathbb{Z}, \ k \in \mathbb{Z}_+ .$$

The isomorphism ϕ takes $(1/2)^k \mathbb{Z}$ to $[t^k \mathbb{Z}]$. The cokernel relation mimics the G_A relation $(1/2)^k n = (1/2)^{k+1}(2n)$. In more detail, to check that ϕ in this example is a $\mathbb{Z}[t]$-module isomorphism, check the following (some details are provided).

- ϕ is well defined.
  Because $[x] \in \mathrm{cok}(1 - 2t)$, we have $[x] = [2tx]$, so

$$(1/2)^k n \ \mapsto \ [t^k n]$$
$$(1/2)^{k+1}(2n) \ \mapsto \ [t^{k+1}(2n)] = [(t^k n)(2t)] = [t^k n] .$$

- ϕ is a group homomorphism.
- ϕ is a $\mathbb{Z}[t]$-module homomorphism :

$$t\phi((1/2)^k n) \ = \ t[t^k n] = [t^{k+1} n] ,$$
$$\phi(t((1/2)^k n)) \ = \ \phi(\hat{A}^{-1}((1/2)^k n)) = \phi((1/2)((1/2)^k n)) = [t^{k+1} n] .$$

- ϕ is surjective.

- ϕ is injective.

  Given $\phi((1/2)^k n) = [t^k n] = [0]$, there exists p in $\mathbb{Z}[t]$ such that $t^k n = (1 - 2t)p$. This forces $n = 0$. (Otherwise $p \neq 0$, and then $(1 - t)p = p - tp$ with nonzero coefficients at different powers of t, contradicting $t^k n = (1 - 2t)p$.)

Corollary 4.3.38 *For square matrices A, B over $\mathbb{Z}$, the following are equivalent.*

(1) *The matrices A, B are SE-$\mathbb{Z}$.*
(2) $cok(I - tA), cok(I - tB)$ *are isomorphic $\mathbb{Z}[t]$ modules.*

Remark 4.3.39 Consider now $\mathbb{Z}[t, t^{-1}]$, the ring of Laurent polynomials in one variable. Given the $\mathbb{Z}[t]$-module G_A, with t acting as $\hat{A}^{-1}$, there is a unique way to extend the $\mathbb{Z}[t]$-module action on G_A to a $\mathbb{Z}[t, t^{-1}]$-module action (t^{-1} must act as $\hat{A}$). A map $\phi : G_A \to G_B$ is a $\mathbb{Z}[t]$-module isomorphism if and only if it is a $\mathbb{Z}[t, t^{-1}]$-module isomorphism.

Consequently, SE-$\mathbb{Z}$ can be (and has been) characterized using $\mathbb{Z}[t, t^{-1}]$-modules above in place of the $\mathbb{Z}[t]$-modules.

4.3.9 Other rings for other systems

We have looked at SFTs presented by matrices over $\mathbb{Z}_+$, and considered algebraic invariants in terms of these matrices (e.g. SE-$\mathbb{Z}$, SSE-$\mathbb{Z}$). There are cases (Remark 4.3.59, Remark 4.3.60) of SFTs with additional structure, or SFT-related systems, for which there is very much the same kind of theory, but with $\mathbb{Z}$ replaced by an integral group ring $\mathbb{Z}G$, and $\mathbb{Z}_+$ replaced by $\mathbb{Z}_+G$. We will say a little about one case, to indicate the pattern, and help motivate our interest in SSE over more general rings.

Let G be a finite group, and let $\mathbb{Z}_+G = \{\sum_{g \in G} n_g g : n_g \in \mathbb{Z}_+\}$, the "positive" semiring in $\mathbb{Z}G$. By a G-SFT we mean an SFT together with a free, continuous shift-commuting G-action. A square matrix over $\mathbb{Z}_+G$ can be used to define an SFT T_A with such a G-action. Two G-SFTs are isomorphic if there is a topological conjugacy between them intertwining the G-actions. Every G-SFT is isomorphic to some G-SFT T_A.

Remark 4.3.40 We list below some correspondences (Remark 4.3.60).

(1) SSE-$\mathbb{Z}_+G$ of matrices is equivalent to conjugacy of their G-SFTs.
(2) If n is a positive integer, then $(T_A)^n$ and T_{A^n} are conjugate G-SFTs.
(3) SE-$\mathbb{Z}_+G$ of matrices is equivalent to eventual conjugacy of their G-SFTs. [30, Prop. B.11].
(4) If G is abelian, then the polynomial $\det(I - tA)$ encodes the periodic data.

(5) If A is a square nondegenerate matrix over $\mathbb{Z}_+G$, then the SFT T_A is mixing if and only if A is G-primitive [30, Prop. B.8].

(6) G-primitive matrices are SE-$\mathbb{Z}G$ if and only if they are SE-$\mathbb{Z}_+G$ [30, Prop. B.12].

We add comments for some items in Remark 4.3.40.

(2) The G-action for $(T_A)^n$ above is the G-action given for T_{A^n}.

(4) The determinant is defined for commutative rings, and $\mathbb{Z}G$ is commutative if and only if the group G is abelian. Above, the polynomial $\det(I - tA)$ has coefficients in the ring $\mathbb{Z}G$. For abelian G, by definition two G-SFTs have the same "periodic data" if there is a shift-commuting – not necessarily continuous – bijection between their periodic points which respects the G-action.

(5), (6) By definition, a G-primitive matrix is a square matrix A over $\mathbb{Z}_+G$ such that for some positive integer k, every entry of A^k has the form $\sum_{g \in G} n_g g$ with every n_g a positive integer.

We note one feature of the $\mathbb{Z}$ situation which does NOT translate to $\mathbb{Z}G$. Recall, SE-$\mathbb{Z} \implies$ SSE-$\mathbb{Z}$. In contrast, for many G, the relationship of SE-$\mathbb{Z}G$ and SSE-$\mathbb{Z}G$ is highly nontrivial, as we will see.

4.3.10 The module-theoretic formulation of SE over a ring

It is basically an observation that arguments for SE-$\mathbb{Z}$ adapt to prove the statements collected below in Theorem 4.3.41. We will outline how this goes. We spell out a few details in the Notes (Remark 4.3.61).

Let $\mathscr{R}$ be a (not necessarily commutative) ring. By 'module', we will mean left module. For example, $\mathscr{R}^n$ is an $\mathscr{R}$-module, with an element r acting from the left by $v \mapsto rv$.

Let A be an $n \times n$ matrix over $\mathscr{R}$. The rule $v \mapsto vA$ defines a map $\mathscr{R}^n \to \mathscr{R}^n$. This is an $\mathscr{R}$-module endomorphism, because $(rv)A = r(vA)$. Considering $\mathscr{R}^n$ as an additive group, let G be the direct limit group defined as in (Remark 4.3.53) by the action of A. We will let $\widehat{A}$ denote the group automorphism of G defined, in the notation of (Remark 4.3.53), by $[(v,m)] \mapsto [(vA,m)]$.

Because A is an $\mathscr{R}$-module endomorphism, there is an induced $\mathscr{R}$-module structure on the direct limit group $(r[(v,m)] = [(rv,m)])$, with respect to which $\widehat{A}$ is an $\mathscr{R}$-module automorphism. G_A becomes an $\mathscr{R}[t,t^{-1}]$-module by having t^{-1} act as $\widehat{A}$ and t act as $\widehat{A}^{-1}$. Call this the direct limit $\mathscr{R}[t,t^{-1}]$-module of A. By restriction of action, it becomes the direct limit $\mathscr{R}[t]$-module of A.

The $n \times n$ matrix $I - tA$ acts on the $\mathscr{R}[t]$-module $(\mathscr{R}[t])^n$ by $v \mapsto v(I - tA)$. The cokernel of this map, $\text{cok}(I - tA)$, is an $\mathscr{R}[t]$-module. The action of t on

$\text{cok}(I - tA)$ has an inverse ($[v] \mapsto [vA]$), so we may also consider $\text{cok}(I - tA)$ as an $\mathscr{R}[t, t^{-1}]$-module.

Theorem 4.3.41 *Let $\mathscr{R}$ be a ring, and A a square matrix over $\mathscr{R}$. Then $\text{cok}(I - tA)$ and the direct limit module of A are isomorphic, as $\mathscr{R}[t]$-modules and as $\mathscr{R}[t, t^{-1}]$-modules. For square matrices A, B over $\mathscr{R}$, the following are equivalent:*

(1) *A and B are SE-$\mathscr{R}$.*
(2) *The direct limit $\mathscr{R}[t]$-modules of A and B are isomorphic.*
(3) *The $\mathscr{R}[t]$-modules $\text{cok}(I - tA)$ and $\text{cok}(I - tB)$ are isomorphic.*
(4) *The direct limit $\mathscr{R}[t, t^{-1}]$-modules of A and B are isomorphic.*
(5) *The $\mathscr{R}[t, t^{-1}]$-modules $\text{cok}(I - tA)$ and $\text{cok}(I - tB)$ are isomorphic.*

4.3.11 Notes

This subsection contains various remarks, proofs and comments referenced in earlier parts of Section 4.3.

Proposition 4.3.42 *For square matrices A, B over $\mathbb{Z}_+$, the following are equivalent.*

(1) *A and B are SE-$\mathbb{Z}_+$*
(2) *A^k and B^k are SSE-$\mathbb{Z}_+$, for all but finitely many k.*
  (So the SFTs defined by A and B are eventually conjugate.)
(3) *A^k and B^k are SE-$\mathbb{Z}_+$, for all but finitely many k.*

Proof
(1) $\implies$ (2) Suppose that matrices R, S give a lag ℓ SE-$\mathbb{Z}_+$ from A to B. Because $AR = RB$ and $SA = BS$, we have for k in $\mathbb{Z}_+$ that

$$(A^k R)(S) = A^k(RS) = A^{k+\ell}$$
$$(S)(A^k R) = S(RB^k) = (SR)B^k = B^{k+\ell} \,.$$

(2) $\implies$ (3) This is trivial.
(3) $\implies$ (1) This argument, due to Kim and Roush, is not so trivial; see [93]. SE-$\mathbb{Z}$ of A^k and B^k does not always imply SE-$\mathbb{Z}$ of A and B, because there are different choices of kth roots of eigenvalues. For example, consider $A = (3)$, $B = (-3)$ and $k = 2$. The very rough idea of the Kim–Roush argument is that when k is a prime very large (with respect to every number field generated by the eigenvalues), then the implication does reverse. $\square$

Remark 4.3.43 If A and B are SE over a ring $\mathscr{R}$, then by Theorem 4.7.10 there is a nilpotent matrix N over $\mathscr{R}$ such that B is SSE over $\mathscr{R}$ to $\left(\begin{smallmatrix} A & 0 \\ 0 & N \end{smallmatrix}\right)$. For $\mathscr{R}$ commutative, it follows that $\det(I - tA)$ fails to be an invariant of SE-$\mathscr{R}$ if and only if there is a nilpotent matrix N over $\mathscr{R}$ such that $\det(I - tN) \neq 1$. We check next that this requires $\mathscr{R}$ to contain a nilpotent element.

Proposition 4.3.44 *Suppose that N is a nilpotent matrix over a commutative ring $\mathscr{R}$ and $\det(I - tN) \neq 1$. Then $\mathscr{R}$ contains a nilpotent element.*

Proof Let $\det(I - tN) = 1 + \sum_{i=1}^{k} c_i t^i$, with $c_k \neq 0$. Suppose that N is $n \times n$, and take m in $\mathbb{N}$ such that $N^m = 0$. Then the polynomial $\det((I - tN)^m)$ has degree at most $n(m-1)$. For any r,

$$\det(I - tN)^r = (\det(I - tN))^r = (1 + c_1 t + \cdots + c_k t^k)^r.$$

This polynomial equals $(c_k)^r t^{kr}$ together with terms of lower degree. Therefore, if $r > n(m-1)$, we must have $(c_k)^r = 0$. $\qquad\square$

Remark 4.3.45 By the way, it can happen that matrices A, B shift equivalent over a commutative ring $\mathscr{R}$ have $\operatorname{tr}(A^n) = \operatorname{tr}(B^n)$ for all n while $\det(I - tA) \neq \det(I - tB)$. For example, let $\mathscr{R}$ be $\mathbb{Z} \cup \{a\}$, with $a^2 = 2a = 0$. Then set $A = \left(\begin{smallmatrix} 0 & 1 \\ a & 0 \end{smallmatrix}\right)$ and $B = (0)$.

Exercise 4.3.46 The primitive matrix $A = \left(\begin{smallmatrix} 1 & 0 & 0 & 1 \\ 0 & 1 & 0 & 1 \\ 0 & 1 & 1 & 0 \\ 1 & 0 & 1 & 0 \end{smallmatrix}\right)$ has nonzero spectrum $(2, 1)$. Prove that A is not SE-$\mathbb{Z}_+$ to a nonsingular matrix.

Proof Such a matrix B would be 2×2 primitive with diagonal entries $1, 2$ (a diagonal entry 3 would force B to have spectral radius greater than 2). But, then B has spectral radius at least as large as the spectral radius of $\left(\begin{smallmatrix} 1 & 1 \\ 1 & 2 \end{smallmatrix}\right)$, which is greater than 2. (A more informative obstruction, due to Handelman, shows that A is not SE-$\mathbb{Z}_+$ to a matrix of size less than 4 [24, Cor. 5.3].) $\qquad\square$

In the proof above, we used the following corollary of Theorem 4.5.6: for nonnegative square matrices C, B, with $C \leq B$ and $C \neq B$ and B primitive, the spectral radius of B is stricty greater than that of C.

Remark 4.3.47 (Proof of Exercise 4.3.15)

Here we use some basic theory of nonnegative matrices reviewed in Section 4.5. There is a permutation matrix P such that $P^{-1}AP$ is block triangular with each diagonal block either (0) or an irreducible matrix. Because the nonzero spectrum is a singleton (n), only one of these blocks is not zero, and this block B must be primitive. There is an SSE-$\mathbb{Z}_+$ by zero extensions from A to B. Now there is an SE-$\mathbb{Z}$ from B to (n). Because B is primitive, this implies there is an SE-$\mathbb{Z}_+$ from B to (n).

Proposition 4.3.48 *Suppose that two primitive matrices over a subring $\mathscr{R}$ of the reals are SE over $\mathscr{R}$. Then they are SE over $\mathscr{R}_+$.*

Proof See [93] for a proof. With matrices R, S giving a lag ℓ SE over $\mathscr{R}$ from A to B, the basic idea is to use linear algebra and the Perron Theorem (Section 4.5) to show (possibly after replacing (R, S) with $(-R, -S)$) that for large n, the matrices $A^n R$ and SA^n will be positive. Then the pair RA^n and $A^n S$ implements an SE over $\mathscr{R}_+$ with lag $\ell + 2n$. □

Remark 4.3.49 An example of myself and Kaplansky, recorded in [10], shows that two irreducible nonnegative matrices can be SE-$\mathbb{Z}$ but not SE-$\mathbb{Z}_+$. The example corrects [116, Remark 4, Sec.5] and shows that [36, Lemma 4.1] should be stated for primitive rather than irreducible matrices (the proof is fine for the primitive case).

Proposition 4.3.50 *The matrix* $\begin{pmatrix} 256 & 7 \\ 0 & 1 \end{pmatrix}$ *is not SE-$\mathbb{Z}$ to its transpose.*

Proof First, suppose a, b are integers such that $a > |b| > 0$. Let M_x denote the matrix $\begin{pmatrix} a & x \\ 0 & b \end{pmatrix}$. Now suppose x, y are integers such that $xy = 1 \mod (a - b)$. Then the matrices $(M_x)^{\mathrm{tr}}$ and M_y are SIM-$\mathbb{Z}$:

$$\begin{pmatrix} a & 0 \\ x & b \end{pmatrix} \begin{pmatrix} a-b & y \\ x & (1-xy)/(b-a) \end{pmatrix} = \begin{pmatrix} a-b & y \\ x & (1-xy)/(b-a) \end{pmatrix} \begin{pmatrix} a & y \\ 0 & b \end{pmatrix}.$$

Thus M_x and $(M_x)^\top$ are SE-$\mathbb{Z}$ if and only if M_x and M_y are SE-$\mathbb{Z}$. Fix $a = 256 = 2^8, b = 1$. Theorem 4.3.22 implies that M_x and M_y are SE-$\mathbb{Z}$ if and only if there are integers j, m such that $2^m x = \pm 2^j y \mod 255$. Because 2 is a unit in $\mathbb{Z}/255\mathbb{Z}$, and $xy = 1 \mod 255$, this holds if and only if there is a nonnegative integer n such that $x^2 = \pm 2^n \mod 255$. Because 2 and -2 are not squares mod 5, they are not squares mod 255. Because $2^8 = 1 \mod 255$, the only squares mod 255 in $\{\pm 2^n : n \geq 0\}$ are 1, 4, 16 and 64. The square 49 is not on this list. Therefore the matrix $\begin{pmatrix} 256 & 7 \\ 0 & 1 \end{pmatrix}$ and its transpose are not SE-$\mathbb{Z}$. □

The following fact from [24] facilitates constructions of primitive matrices realizing the algebraic invariants above: any 2×2 matrix over $\mathbb{Z}$ with integer eigenvalues a, b with $a > |b|$ is SE-$\mathbb{Z}$ to a primitive matrix. In our example,

$$\begin{pmatrix} 1 & 0 \\ 1 & 1 \end{pmatrix} \begin{pmatrix} 256 & 7 \\ 0 & 1 \end{pmatrix} \begin{pmatrix} 1 & 0 \\ -1 & 1 \end{pmatrix} = \begin{pmatrix} 249 & 7 \\ 248 & 8 \end{pmatrix} := B.$$

It is an easy exercise to show that when matrices A, B are shift equivalent, if one of A, B is shift equivalent to its transpose then so is the other. Consequently, the matrix B displayed above cannot be SE-$\mathbb{Z}$ to its transpose.

We have not seen the method of Proposition 4.3.50 used to distinguish the SE-$\mathbb{Z}$ classes of a primitive matrix and its transpose, but examples of such were produced long ago. The matrix $A = \left(\begin{smallmatrix} 19 & 5 \\ 4 & 1 \end{smallmatrix} \right)$ is an early example, due to Köllmer, of a primitive matrix not SIM-$\mathbb{Z}$ (hence not SE-$\mathbb{Z}$, as $|\det(A)| = 1$) to its transpose (for an elementary proof, see [114, Ch.V, Sec.4]). The connection of $\mathrm{SL}(2, \mathbb{Z})$ to continued fractions leads to a computable characterization of SIM-$\mathbb{Z}$ for 2×2 unimodular matrices, exploited by Cuntz and Krieger as another method to produce 2×2 primitive integer matrices not SE-$\mathbb{Z}$ to their transposes (see [36, Corollary 2.2]).[17] Lind and Marcus use another connection to $\mathbb{Z}[\lambda]$ ideal classes to give an example of a primitive integral matrix not SE-$\mathbb{Z}$ to its transpose, [93, Example 12.3.2]. There are much earlier papers which give many cases in which a square integer matrix and its transpose must correspond to inverse ideal classes of an associated ring (see [140] and its connections in the literature), and these ideal classes may differ. However, this still leaves the issue of realizing the algebraic invariants in primitive matrices.

Next, we restate Theorem 4.3.26 and sketch the proof coming from Taussky's work [139].

Theorem 4.3.51 *Suppose that p is monic irreducible in $\mathbb{Z}[t]$, and $p(\lambda) = 0$, with $0 \neq \lambda \in \mathbb{C}$. Let $\mathscr{M}$ be the set of matrices over $\mathbb{Z}$ with characteristic polynomial p. Then there are bijections*

$$\mathscr{M}/(\mathrm{SIM} - \mathbb{Z}) \to \text{Ideal classes of } \mathbb{Z}[\lambda], \qquad \text{and}$$

$$\mathscr{M}/(\mathrm{SE} - \mathbb{Z}) \to \text{Ideal classes of } \mathbb{Z}[1/\lambda].$$

Proof If A is in $\mathscr{M}$, then A has a right eigenvector r_A for λ. The eigenvector can be chosen with entries in the field $\mathbb{Q}[\lambda]$ (solve $(\lambda I - A)r = 0$ using Gaussian elimination). Then, after multiplying r_A by a suitable element of $\mathbb{Z}$ to clear denominators, we may assume the entries of r_A are in $\mathbb{Z}[\lambda]$. Let $I(r_A)$ be the ideal of the ring $\mathbb{Z}[\lambda]$ generated by the entries of r_A. Let $\mathscr{I}(r_A)$ be the ideal class of $\mathbb{Z}[\lambda]$ which contains $I(r_A)$. Now it is routine to check that the map $A \mapsto \mathscr{I}(r_A)$ is well defined and induces the first bijection.

For the second bijection, just repeat this Taussky argument, with the ring $\mathbb{Z}[1/\lambda]$ in place of $\mathbb{Z}[\lambda]$, and say $\mathscr{J}(r_A)$ denoting the $\mathbb{Z}[1/\lambda]$ ideal generated by the entries of r_A. The rule $\mathscr{I}(r_A) \mapsto \mathscr{J}(r_A)$ induces a surjective map from the set of ideal classes of $\mathbb{Z}[\lambda]$ to those of $\mathbb{Z}[1/\lambda]$, which corresponds to the lumping of SIM-$\mathbb{Z}$ classes to SE-$\mathbb{Z}$ classes.

There is more detail and comment on this in [28]. □

---

[17] For a dimension group viewpoint, read [36, Theorem 2.1]) as: the SFTs defined by irreducible unimodular 2×2 matrices over $\mathbb{Z}_+$ are topologically conjugate if and only if they have isomorphic dimension groups and equal entropy.

It is important to note above that the ring $\mathbb{Z}[\lambda]$ is not in general equal to $\mathcal{O}_\lambda$, the full ring of algebraic integers in $\mathbb{Q}[\lambda]$. When $\mathbb{Z}[\lambda]$ is a proper subset of $\mathcal{O}_\lambda$, its class number will strictly exceed that of $\mathcal{O}_\lambda$ (in this case, a principal $\mathbb{Z}[\lambda]$ ideal cannot be an $\mathcal{O}_\lambda$ ideal).

Proposition 4.3.52 *Suppose that λ is a nonzero algebraic integer. Then the class number of $\mathbb{Z}[\lambda]$ is finite.*

Proof Let n be the the dimension of $\mathbb{Q}[\lambda]$ as a rational vector space. As free abelian groups, $\mathcal{O}_\lambda$ and $\mathbb{Z}[\lambda]$ (and all of their nonzero ideals) have rank n. For R equal to $\mathbb{Z}[\lambda]$ or $\mathcal{O}_\lambda$, the following are equivalent conditions on R-ideals I, I'.

- I, I' are equivalent as R-ideals.
- there is a nonzero $c \in \mathbb{Q}[\lambda]$ such that $cI = I'$.
- I, I' are isomorphic as R-modules.

Because the class number of $\mathcal{O}_\lambda$ is finite, there is a finite set $\mathscr{J}$ of $\mathcal{O}_\lambda$ ideals such that every nonzero $\mathcal{O}_\lambda$ ideal is equivalent to an element of $\mathscr{J}$. Let N be a positive integer such that $N\mathcal{O}_\lambda \subset \mathbb{Z}[\lambda]$.

Now suppose I is a $\mathbb{Z}[\lambda]$ ideal, with $\{\gamma_1, \ldots, \gamma_n\}$ a $\mathbb{Z}$-basis of I. Set $J = \{\sum_{i=1}^n r_i \gamma_i : r_i \in \mathcal{O}_\lambda, 1 \le i \le n\}$. There is a nonzero c in $\mathbb{Q}[\lambda]$ such that $cJ \in \mathscr{J}$. The $\mathbb{Z}[\lambda]$ modules I, cI are isomorphic. We have $NJ \subset I \subset J$, and therefore $|J/I| \le n^N$. There are only finitely many abelian subgroups of J with index at most n^N in J. It follows that there are only finitely many possibilities for cI as a $\mathbb{Z}[\lambda]$ module, and this finishes the proof. □

Proposition 4.3.52 is a (very) special case of the Jordan–Zassenhaus theorem (see [118]).

Remark 4.3.53 We'll recall the general notion of direct limit of a group endomorphism, and see in the $\mathbb{Z}$ case that our concrete presentation really is isomorphic to the general vesion. The concrete version has its merits, but the general version is essential.

For a group endomorphism $\phi : G \to G$, take the union of the disjoint sets (G, n), $n \in \mathbb{Z}_+$. Define an equivalence relation on $\bigcup_{n \in \mathbb{Z}_+} (G, n)$: $(g, m) \sim (h, n)$ if there exist j, k in $\mathbb{Z}_+$ such that $(\phi^j(g), j + m) = \phi^k(h), n + k)$. Define $\varinjlim_\phi G$ to be the quotient set $\left(\bigcup_{n \in \mathbb{Z}_+} (G, n)\right) / \sim$. The operation on $\varinjlim_\phi G$ given by $[(g, m)] + [(h, n)] = [(\phi^n(g) + \phi^m(h), m + n)]$ is well defined and makes $\varinjlim_\phi G$ a group. The endomorphism ϕ induces a group automorphism $\widehat{\phi}$ given by $\widehat{\phi} : [(g, n)] \mapsto [(\phi(g), n)]$. The inverse of $\widehat{\phi}$ is defined by $[(g, n)] \mapsto [(g, n + 1)]$.

In our case, A acts on $\mathbb{Z}^n$ as $x \mapsto xA$, and we may define a map $\psi : G_A \to \varinjlim_\phi G$ by $x \mapsto [(xA^m, m)]$ where $m = m(x)$ is sufficiently large that $xA^m \in \mathbb{Z}^n$.

One can check that ψ is a well defined group automorphism, with $\psi \circ A = \hat{A} \circ \psi$.

Remark 4.3.54 If $A = \left(\begin{smallmatrix} 2 & 0 \\ 0 & 5 \end{smallmatrix}\right)$ and $B = \left(\begin{smallmatrix} 2 & 1 \\ 0 & 5 \end{smallmatrix}\right)$, we will show that G_A is the sum of a 2-divisible group and a 5-divisible group, but G_B is not.

For $M = A$ or B, and $\lambda = 2$ or 5, let $H_{M,\lambda} = \{v \in G_M : \lambda^{-k}v \in G_M, \text{for all } k \in \mathbb{N}\}$. An isomorphism $G_B \to G_A$ must send $H_{B,\lambda}$ to $H_{A,\lambda}$, for $\lambda = 2, 5$. For $M = A$ or $M = B$, because the eigenvalues $2, 5$ are relatively prime, we can check $H_{M,\lambda} = G_M \cap \{v \in \mathbb{Q}^2 : vM = \lambda v\}$. Clearly $G_A = H_{A,2} \oplus H_{A,5}$. In contrast, $G_B \neq H_{B,2} \oplus H_{B,5}$. For example, $(1,0) \in G_B$, and $(1,0)$ is uniquely a sum of vectors on the two eigenlines, $(1,0) = (1/3)(3,-1) + (1/3)(0,1)$. But $(1/3)(0,1) \notin H_{B,5}$.

Proposition 4.3.55 *Let A, B be square matrices over $\mathbb{Z}$. Then the following are equivalent.*

(1) A and B are SE-$\mathbb{Z}$.
(2) There is an isomorphism of direct limit pairs $(G_A, \hat{A})$ and $(G_B, \hat{B})$.

Proof We will give a proof with the general direct limit definition in Remark 4.3.53, rather than using the more concrete version of the group involving eventual images. The general proof is easier.

(1) $\implies$ (2) Suppose that R, S gives the lag ℓ shift equivalence: $A^\ell = RS$, etc. First note that the rule $[(x,n)] \to [(xR,n)]$ gives a well defined map $\phi : G_A \to G_B$, because $[(xAR, n+1)] = [(xRB, n+1)] = [(xR,n)]$. Check ϕ is a group homomorphism. Similarly, define $\psi : G_B \to G_A$ by $[(y,m)] \mapsto [(yS, m+\ell)]$.

Now, $\psi(\phi([(x,n)])) = \psi([(xR,n)]) = [(xRS, n+\ell)] = [(xA^\ell, n+\ell)] = [(x,n)]$. Similarly, $\phi(\psi([(y,n)])) = [(y,n)]$. Thus, the homomorphism ϕ is an isomorphism, with $\phi^{-1} = \psi$. Finally, $\hat{A}\phi = \phi\hat{B}$, because $\phi(\hat{A}([(x,n)])) = \phi([(xA,n)]) = [(xAR,n)] = [(xRB,n)] = \hat{B}([(xR,n)]) = \hat{B}(\phi([(x,n)]))$.

(2) $\implies$ (1) Suppose that $\phi : G_A \to G_B$ gives the isomorphism of pairs. Check that there must be $N > 0$ and a matrix R such that $\phi : [(x,0)] \to [(xR,N)]$. After postcomposing with the automorphism $[(y,N)] \mapsto [(y,0)]$, we may suppose that $N = 0$. There must be a matrix S and $\ell > 0$ such that the inverse map is $[(y,0)] \mapsto [(yS, \ell)]$.

Now, $[(v,0)] = [(vA^\ell, \ell)] = [(vRS, \ell)]$ for every v. So, if v is a standard basis vector, then for all large k, $vA^{\ell+k} = vRSA^k$. Thus for all large k, $RSA^k = A^{k+\ell}$. Similarly, for all large k we have $SRB^k = B^{k+\ell}$, $(AR)B^k = (RB)B^k$ and $(BS)A^k = (SA)A^k$. Thus for all large k we get a shift equivalence with lag $\ell + 2k$,

$$B(SA^k) = (SA^k)A, \qquad (RB^k)(SA^k) = R(SA^k)A^k = A^{\ell+2k}$$
$$A(RB^k) = (RB^k)B, \qquad (SA^k)(RB^k) = S(A^kB)B^k = B^{\ell+2k}. \qquad \square$$

Remark 4.3.56 (Dimension groups) The dimension groups are an important class of ordered groups arising from functional analysis [53], with important applications in C^*-algebras [42] and topological dynamics [52, 51, 41]. We consider only countable groups. As a group, a dimension group is a direct limit of the form

$$\mathbb{Z}^{n_1} \xrightarrow{A_1} \mathbb{Z}^{n_2} \xrightarrow{A_2} \mathbb{Z}^{n_3} \xrightarrow{A_3} \cdots$$

for which nonnegative integral matrices A_n defined the bonding homomorphisms. For v in $\mathbb{Z}^{n_k}$ (we use row vectors), the element $[(v, n_k)]$ of the group is in the positive set if $vA^{n_1}A^{n_2}\cdots A^{n_j} \in \mathbb{Z}_+$ for some (hence for every large) nonnegative integer j. Every torsion free countable abelian group is isomorphic as an unordered group to a dimension group. Effros, Handelman and Shen have given an elegant and important abstract characterization of the ordered groups which are isomorphic to dimension groups [43].

The dimension groups were introduced to the theory of SFTs (where they play a fundamental role) by Wolfgang Krieger [85], in 1980.

Remark 4.3.57 We (and others) learned the contents of Propositions 2.7.1, 2.8.1, Corollary 2.8.3 and Remark 2.8.4 from Wagoner in person, at conferences or at MSRI, by some time in the 1980s or early 1990s. Early references in print are perhaps somewhat scattered and implicit. Wagoner's module viewpoint is evident (if not very quotable) in a 1987 paper [142, pp. 92,120]. There is an explicit statement of the correspondence of $\mathbb{Z}[t, t^{-1}]$-module class and SE-$\mathbb{Z}$ class in my 1993 review [14, Sec. 5.4]. The content of Proposition 2.7.1 is contained in the standard 1995 Lind–Marcus text [92] (see more on this in Section 4.8); however, there is no explicit mention in [92] of the $\mathbb{Z}[t]$ modules. Ordered $\mathbb{Z}_+[t, t^{-1}]$-module versions of SE-$\mathbb{Z}_+$ are given in [12, Lec. III; Secs. 2.2, 3.1, 3.2] and [33, Sec. 5].

Mischaikow and Weibel make good use of the $\mathbb{Z}[t]$-module version of shift equivalence in their study [100] of the (homological) Conley index. Interesting parts of this go beyond the contents of our Section 4.3.

Remark 4.3.58 For a square matrix A over $\mathbb{Z}_+$, the group G_A above becomes an ordered group, (G_A, G_A^+), by defining the positive set $G_A^+ = \{x \in G_A : \exists k \in \mathbb{N}, xA^k \geq 0\}$. The ordered group (G_A, G_A^+) is a dimension group (set every bonding map A_n equal to A). Now $(G_A, G_A^+, \hat{A})$ is an ordered $\mathbb{Z}[t]$ module (the action of t takes G_A^+ to G_A^+), and is sometimes called a dimension module. (Sometimes the unordered group G_A is referred to as a dimension group. We have tried to avoid this.)

For A, B over $\mathbb{Z}_+$, SE-$\mathbb{Z}_+$ of A, B is equivalent to existence of an isomorphism $G_A \to G_B$ which intertwines $\hat{A}$ and $\hat{B}$ and sends G_A^+ onto G_B^+. For more on this, see Lind and Marcus [93].

Remark 4.3.59 Parry and Tuncel made the first beyond-$\mathbb{Z}$ connection of this sort in [115], as they studied conjugacies of SFTs taking one Markov measure to another. The matrices they considered are not taken explicitly from a group ring, but the connection to an integral group ring of a finitely generated free abelian group emerges in [97].

Remark 4.3.60 It was Bill Parry who introduced the presentation of G-SFTs by matrices over $\mathbb{Z}_+G$, and the conjugacy/SSE-$\mathbb{Z}_+G$ correspondence. Parry never published a proof (although one can see the ideas emerging from the earlier paper with Tuncel, [115]). For an exposition with proofs, see [32] and [30, Appendices A,B]. The items (2), (4) the list in Remark 4.3.40 are not proved explicitly in [32], but they should not be difficult to verify following the exposition of [32]. For further development of relations between the $\mathbb{Z}_+G$ matrices and their G-SFTs, see [20, Appendix]. The exposition in [32] includes Parry's connection between SSE-$\mathbb{Z}_+G$ and cohomology of functions [32, Theorem 2.7.1], which is the heart of the matter. When G is not abelian, one needs to be careful about left vs. right actions; [30, Appendix A] explains this, and corrects a left/right error in the presentation in [32].

Remark 4.3.61 We spell out some details of the outline given in Section 4.3.10.

By the action $v \mapsto rv$ of $\mathcal{R}$ on $\mathcal{R}^n$, we mean $(v_1,\ldots,v_n) \mapsto (rv_1,\ldots,rv_n)$, with rv_i given by multiplication in $\mathcal{R}$.

With notation as in Remark 4.3.53, an element of the direct limit group $G = G(A)$ has the form $[(v,m)]$, with $v \in \mathcal{R}^n$ and m an integer. In G, $[(v,m)] = [(vA,m+1)]$. Given $r \in \mathcal{R}$, for the rule $r[(v,m)] = [(rv,m)]$ to give a well-defined map on G we need $[(rv,m)] = [(rvA,m+1)]$. This holds because

$$[(rv,m)] = [((rv)A,m+1)] = [(r(vA),m+1)].$$

Then $\hat{A}$ is an $\mathcal{R}$-module homomorphism, because $r\hat{A} : [(v,m)] \mapsto [(rvA,m)] = \hat{A}r$. Similarly, $r\hat{A}^{-1} : [(v,m)] \mapsto [(rv,m+1)] = \hat{A}^{-1}r$.

By definition, $\mathrm{cok}(I-tA) = (\mathcal{R}[t])^n / \{v(I-tA) : v \in (\mathcal{R}[t])^n\}$. This cokernel is an $\mathcal{R}[t]$-module: an element p in $\mathcal{R}[t]$ acts on $\mathrm{cok}(I-tA)$ by the rule that $[v] \mapsto [pv]$. The rule is well defined on $\mathrm{cok}(I-tA)$ because

$$\{pv(I-tA) : v \in \mathcal{R}^n\} \subset \{v(I-tA) : v \in \mathcal{R}^n\}.$$

Then $[v] \mapsto [vA]$ defines an $\mathcal{R}[t]$-module endomorphism (f, say) of $\mathrm{cok}(I-tA)$ which is an inverse to the action of t, because $f(t[v]) = f([tv]) = [tvA] = [tvA + v(I-tA)] = [v]$.

The claimed equivalences of the items in Theorem 4.3.41 are proved just as for $\mathscr{R} = \mathbb{Z}$, with the observations that the isomorphisms constructed (e.g., for Prop. 4.3.36) give module isomorphisms as required.

4.4 Polynomial matrices

We will define SFTs, and the algebraic and classification structures around them, using polynomial matrices. This is essential for the K-theory connections to come.

4.4.1 Background

Before we move on to the polynomial matrices, we review background on flow equivalence and vertex SFTs. Later, this will be context for the polynomial approach.

Flow equivalence of SFTs

Two homeomorphisms are *flow equivalent* if there is a homeomorphism between their mapping tori which takes orbits onto orbits preserving the direction of the suspension flow (Remark 4.4.22). Roughly speaking: two homeomorphisms are flow equivalent if their suspension flows move in the same way, but at different speeds. If SFTs are topologically conjugate, then they are flow equivalent, but the converse is not true.

An $n \times n$ matrix C over $\mathbb{Z}$ defines a map $\mathbb{Z}^n \to \mathbb{Z}^n$, $v \mapsto vC$, with $\text{Image}(C) = \{vC : v \in \mathbb{Z}^n\}$, and cokernel group $\text{cok}_{\mathbb{Z}}(C) = \mathbb{Z}^n/\text{Image}(C)$.

Theorem 4.4.1 *If SFTs defined by $\mathbb{Z}_+$ matrices A, B are flow equivalent, then*

(1) $\det(I - A) = \det(I - B)$, *and*

(2) $\text{cok}_{\mathbb{Z}}(I - A)$ *and* $\text{cok}_{\mathbb{Z}}(I - B)$ *are isomorphic abelian groups.*

Above, (1) is due to Bill Parry and Dennis Sullivan [113]; (2) is due to Rufus Bowen and John Franks [9]. The group $\text{cok}_{\mathbb{Z}}(I - A)$ is called the Bowen–Franks group of the SFT defined by $\mathbb{Z}_+$-matrix A. The group $\text{cok}_{\mathbb{Z}}(I - A)$ determines $|\det(I - A)|$, except for the sign of $\det(I - A)$ in the case $\det(I - A) \neq 0$ (Remark 4.4.23).

When A is irreducible and A is not a permutation matrix, the converse of the theorem holds (John Franks, [48]). So, in this case the Bowen–Franks group determines the flow equivalence class, up to knowing the sign of $\det(I - A)$.

Vertex SFTs

Once upon a time, before edge SFTs, SFTs were presented only by matrices with entries in $\{0,1\}$. Such a matrix can be viewed as the adjacency matrix of a graph without parallel edges (i.e., for each vertex pair (i,j), there is at most one edge from i to j). We can then define a "vertex SFT" as we defined edge SFT, but using bisequences of vertices rather than bisequences of edges to describe infinite walks through the graph.

A vertex SFT is quite natural, especially if one starts from subshifts. A subshift (X,σ) is a "topological Markov shift" if whenever points x, y satisfy $x_0 = y_0$, the bisequence $z = \ldots x_{-3}x_{-2}x_{-1}x_0 y_1 y_2 y_3 \ldots$ is also a point in X. (That is, the past of x and the future of y can be glued together at their common present to form a point. This is a topological analogue of the independence property of a Markov measure.) One can check that a topological Markov shift is the same object as a vertex SFT, with the alphabet of the subshift being the vertex set (Remark 4.4.24).

Remark 4.4.2 Defining SFTs (as edge SFTs) with matrices over $\mathbb{Z}_+$ has some significant advantages over defining SFTs (as vertex SFTs) with matrices over $\{0,1\}$, as follows.

- *Functoriality.* Recall, $(X_A, (\sigma_A)^n)$ is conjugate to the edge SFT defined by A^n, whereas A^n cannot define a vertex SFT if A^n has an entry greater than 1.
- *Conciseness.* E.g., an edge SFT defined by the perfectly transparent 2×2 matrix $A = \begin{pmatrix} 1 & 4 \\ 4 & 15 \end{pmatrix}$ has a (rather large) alphabet of 24 symbols; as a vertex SFT, it would be defined by a 24×24 zero-one matrix. And while A^n is 2×2 for all n, the size of the matrix presenting the vertex SFT $(X_A, (\sigma_A)^n)$ goes to infinity as $n \to \infty$.
- *Proof techniques.* Defining the SFTs directly with matrices over $\mathbb{Z}_+$ allows other proof techniques (Remark 4.4.25).

We shall see that some advantages of defining SFTs with $\mathbb{Z}_+$ rather than $\{0,1\}$ matrices are repeated, as we compare defining SFTs with polynomial rather than $\mathbb{Z}_+$ matrices.

4.4.2 Presenting SFTs with polynomial matrices

The *length* of a path $e_1 \ldots e_n$ of n edges in a graph is n. (We also think of n as the time taken at unit speed to traverse the path.) An $n \times n$ matrix A with polynomial entries in $t\mathbb{Z}_+[t]$ presents a graph Γ_A as follows.

- $\{1, \ldots, n\}$ is a subset of the vertex set of Γ_A.

- For each monomial entry t^k of $A(i,j)$, there is a distinct path of k edges from vertex i to vertex j. We call such a path an *elementary path* in Γ_A. (E.g. if $A(i,j) = 2t^3$, then from i to j there are two elementary paths of length 3.)
- There are no other edges, and distinct elementary paths do not intersect at intermediate vertices.

Above, the vertex set $\{1,\dots,n\}$ is a *rome* (Remark 4.4.26) for the graph Γ_A: every sufficiently long path hits the rome. ("All roads lead to Rome ...")

Example 4.4.3 Below, the rome vertex set is $\{1,2\}$; the additional vertices are unnamed black dots; and there are five elementary paths in Γ_A.

$$A = \begin{pmatrix} 2t & t^2 + t^3 \\ t^2 & 0 \end{pmatrix}, \qquad \Gamma_A = \quad$$

Given A over $t\mathbb{Z}_+[t]$, let $A^\sharp$ be the adjacency matrix for the graph Γ_A. In Example 4.4.3, $A^\sharp$ would be 6×6. (The vertex set of the graph is the rome, together with $k-1$ additional vertices for each monomial t^k.) We can think of A as being a way to present the edge SFT defined by the matrix $A^\sharp$.

Conciseness. Obviously, we can present many SFTs (and, various interesting families of SFTs) much more concisely with polynomial matrices than with matrices over $\mathbb{Z}_+$. For example, a theorem of D. Perrin shows that any number which can be the entropy of an SFT is the entropy of an SFT defined by a 2×2 matrix over $t\mathbb{Z}_+[t]$. (Remark 4.4.27)

Definition 4.4.4 (Elementary matrix) An elementary matrix is a square matrix equal to the identity except in at most a single offdiagonal entry.

The polynomial presentation offers more than conciseness. To see this, we need a little preparation. I_k denotes the $k \times k$ identity matrix.

Definition 4.4.5 (Stabilized elementary equivalence) Suppose that $\mathscr{R}$ is a ring. *Stabilized elementary equivalence* is the equivalence relation $\sim$ on square matrices C over $\mathscr{R}$ generated by the following two relations.

(1) $C \sim C \oplus I_k$, for $k \in \mathbb{N}$. (E.g., (2) $\sim \begin{pmatrix} 2 & 0 \\ 0 & 1 \end{pmatrix}$.)
(2) $C \sim D$ if there is an elementary matrix E such that $D = CE$ or $D = EC$.

Above, condition (1) is the "stabilized" part. A stabilized elementary equivalence from C to D is a finite sequence of the elementary matrix moves, taking C to D.

Given $C \sim D$, for either type of relation, we have

(1) $\det C = \det D$, if $\mathscr{R}$ is commutative, and
(2) the $\mathscr{R}$-modules $\mathrm{cok}(C)$, $\mathrm{cok}(D)$ are isomorphic (Proposition 4.4.28).

When working in a stable setting, we often say just "elementary equivalence" instead of "stabilized elementary equivalence".

4.4.3 Algebraic invariants in the polynomial setting

Example 4.4.6 If all nonzero entries of A have degree one, then the relation of A and $A^\sharp$ is obvious: for example,

$$A = \begin{pmatrix} t & 2t \\ t & 0 \end{pmatrix} = t\begin{pmatrix} 1 & 2 \\ 1 & 0 \end{pmatrix} = tA^\sharp, \qquad \Gamma_A =$$

Here, $I - A$ equals $I - tA^\sharp$. It follows, of course, that

$$\det(I - A) = \det(I - tA^\sharp), \quad \text{and}$$
$$\mathrm{cok}_{\mathbb{Z}[t]}(I - A) \cong \mathrm{cok}_{\mathbb{Z}[t]}(I - tA^\sharp).$$

These two statements hold for general A over $t\mathbb{Z}_+[t]$, for the following reason.

Proposition 4.4.7 *There is a stabilized elementary equivalence over the ring* $\mathbb{Z}[t]$ *from* $I - A$ *to* $I - tA^\sharp$.

Next we show the essential ideas of the proof of the proposition. Given an $n \times n$ matrix A over $t\mathbb{Z}_+[t]$, let $\mathscr{H}_A$ be the $n \times n$ labeled graph in which a monomial t^k of $A(i,j)$ gives rise to an edge from i to k labeled t^k.

Example 4.4.8

$$A = \begin{pmatrix} 2t & t+t^4 \\ t^2 & 0 \end{pmatrix}, \qquad \mathscr{H}_A =$$

Note: the graph Γ_A with adjacency matrix $A^\sharp$ is obtained from $\mathscr{H}_A$ by replacing each path labeled t^k with a path of length k. The graph $\mathscr{H}_{tA^\sharp}$ is the graph Γ_A with each edge labeled by t.

We can decompose the graph move $\mathcal{H}_A \to \mathcal{H}_{tA^\sharp}$ into steps, $\mathcal{H}_0 \to \mathcal{H}_1 \to \cdots \to \mathcal{H}_4$, with one vertex added at each step. The labeled graph $\mathcal{H}_{i+1}$ is obtained from $\mathcal{H}_i$ by replacing some edge labeled t^k with a path of two edges: an edge labeled t followed by an edge labeled t^{k-1}. There will be matrices A_i over $t\mathbb{Z}_+[t]$ such that $\mathcal{H}_i = \mathcal{H}_{A_i}$, with $A_0 = A$ and $A_4 = tA^\sharp$. Here is the data for the step $\mathcal{H}_0 \to \mathcal{H}_1$:

$$A = A_0 = \begin{pmatrix} 2t & t+t^4 \\ t^2 & 0 \end{pmatrix}, \qquad \mathcal{H}_A = \mathcal{H}_0 =$$

$$B = A_1 = \begin{pmatrix} 2t & t & t \\ t^2 & 0 & 0 \\ 0 & t^3 & 0 \end{pmatrix}, \qquad \mathcal{H}_{A_1} = \mathcal{H}_1 =$$

Let us see how the move $A \to A_1$ in the example above is accomplished at the matrix level, by a stabilized elementary equivalence over the ring $\mathbb{Z}[t]$.

First, define the matrix $A \oplus 0 = \begin{pmatrix} 2t & t+t^4 & 0 \\ t^2 & 0 & 0 \\ 0 & 0 & 0 \end{pmatrix}$. The move $A \to A \oplus 0$ is the same as the elementary stabilization move $(I - A) \to (I - A) \oplus 1$. Then multiply $(I - A) \oplus 1$ by elementary matrices to get $(I - A_1)$. This is a small computation:

$$(I - B) = \begin{pmatrix} 1 & 0 & 0 \\ 0 & 1 & 0 \\ 0 & 0 & 1 \end{pmatrix} - \begin{pmatrix} 2t & t & t \\ t^2 & 0 & 0 \\ 0 & t^3 & 0 \end{pmatrix}$$

$$(I - B)E_1 = \begin{pmatrix} 1 - 2t & -t & -t \\ -t^2 & 1 & 0 \\ 0 & -t^3 & 1 \end{pmatrix} \begin{pmatrix} 1 & 0 & 0 \\ 0 & 1 & 0 \\ 0 & t^3 & 1 \end{pmatrix}$$

$$= \begin{pmatrix} 1 - 2t & -t - t^4 & -t \\ -t^2 & 1 & 0 \\ 0 & 0 & 1 \end{pmatrix} = I - \begin{pmatrix} 2t & t + t^4 & t \\ t^2 & 0 & 0 \\ 0 & 0 & 0 \end{pmatrix} := I - C$$

$$E_2(I - C) = \begin{pmatrix} 1 & 0 & t \\ 0 & 1 & 0 \\ 0 & 0 & 1 \end{pmatrix} \begin{pmatrix} 1 - 2t & -t - t^4 & -t \\ -t^2 & 1 & 0 \\ 0 & 0 & 1 \end{pmatrix}$$

$$= \begin{pmatrix} 1 - 2t & -t - t^4 & 0 \\ -t^2 & 1 & 0 \\ 0 & 0 & 1 \end{pmatrix}$$

$$= \begin{pmatrix} 1 & 0 & 0 \\ 0 & 1 & 0 \\ 0 & 0 & 1 \end{pmatrix} - \begin{pmatrix} 2t & t + t^4 & 0 \\ t^2 & 0 & 0 \\ 0 & 0 & 0 \end{pmatrix} = (I - A) \oplus 1 .$$

The example computation above contains the ideas of the general proof that there is a stabilized elementary equivalence from $(I - A)$ to $(I - tA^\sharp)$.

Corollary 4.4.9 *Let A be a square matrix over $t\mathbb{Z}_+[t]$, with $A^\sharp$ the adjacency matrix of Γ_A. Then*

(1) $\det(I - A) = \det(I - tA^\sharp)$.
(2) *The $\mathbb{Z}[t]$-modules $\mathrm{cok}(I - A)$, $\mathrm{cok}(I - tA^\sharp)$ are isomorphic.*

Proof   The claim follows because the matrices $(I - A)$, $(I - tA^\sharp)$ are connected by a sequence of the two relations $\sim$ generating stabilized elementary equivalence. □

Thus algebraic data of the polynomial matrix $(I - A)$ captures

(1) the nonzero spectrum (by $\det(I - A)$), and
(2) the SE-$\mathbb{Z}$ class of $A^\sharp$ (given by the isomorphism class of the $\mathbb{Z}[t]$-module $\mathrm{cok}(I - A)$).

4.4.4 Polynomial matrices: from elementary equivalence to conjugate SFTs

For square matrices A, B over $\mathbb{Z}_+$, the SFTs $(X_A, \sigma), (X_B, \sigma)$ are topologically conjugate if and only if A, B are SSE-$\mathbb{Z}_+$. We will find a relation on polynomial matrices corresponding to topological conjugacy of the SFTs they define.

Notation 4.4.10 With $i \neq j$, let $E_{ij}(x)$ denote the elementary matrix with (i, j) entry defined to be x, and other entries matching the identity. The size of the square matrix $E_{ij}(x)$ is suppressed from the notation (but evident in context).

E.g., $E_{12}(t^2)$ could denote $\begin{pmatrix} 1 & t^2 \\ 0 & 1 \end{pmatrix}$ or $\begin{pmatrix} 1 & t^2 & 0 \\ 0 & 1 & 0 \\ 0 & 0 & 1 \end{pmatrix}$.

There is now a very pleasant surprise.

Theorem 4.4.11 *Suppose that A, B are square matrices over $t\mathbb{Z}_+[t]$, with $E(I - A) = (I - B)$ or $(I - A)E = (I - B)$, where $E = E_{ij}(t^k)$.*

Then A, B define topologically conjugate SFTs (i.e., $B^\sharp$ and $A^\sharp$ define topologically conjugate edge SFTs).

Remark 4.4.12 Suppose that $E = E_{ij}(t^k)$, A is square with entries in $t\mathbb{Z}_+[t]$, and $(I - B) = E(I - A)$ or $(I - B) = (I - A)E$. Then one easily checks (it will be obvious from the next example) that the following are equivalent:

(1) The entries of B are in $t\mathbb{Z}_+[t]$.

(2) $A(i, j) - t^k \in t\mathbb{Z}_+[t]$.

Proof ideas for Theorem 4.4.11 The ideas of the proof of Theorem 4.4.11 should be clear from the next example.

Example 4.4.13 Suppose A is matrix over $t\mathbb{Z}_+[t]$, $A = \begin{pmatrix} a & b + t^3 & c \\ d & e & f \\ g & h & i \end{pmatrix}$, with $b \in t\mathbb{Z}_+[t]$ (i.e., not only $A(1, 2)$, but also $A(1, 2) - t^3$, is in $t\mathbb{Z}_+[t]$). Now multiply $I - A$ from the left by the elementary matrix $E = E_{12}(t^3)$,

$$E(I-A) = \begin{pmatrix} 1 & t^3 & 0 \\ 0 & 1 & 0 \\ 0 & 0 & 1 \end{pmatrix} \begin{pmatrix} 1-a & -b-t^3 & -c \\ -d & 1-e & -f \\ -g & -h & 1-i \end{pmatrix}$$

$$= \begin{pmatrix} 1-a+\mathbf{t^3}(-\mathbf{d}) & -b-t^3+\mathbf{t^3}(1-\mathbf{e}) & -c+\mathbf{t^3}(-\mathbf{f}) \\ -d & 1-e & -f \\ -g & -h & 1-i \end{pmatrix}$$

$$= \begin{pmatrix} 1 & 0 & 0 \\ 0 & 1 & 0 \\ 0 & 0 & 1 \end{pmatrix} - \begin{pmatrix} a+t^3d & b+t^3e & c+t^3f \\ d & e & f \\ g & h & i \end{pmatrix}.$$

We then define a matrix B over $t\mathbb{Z}_+[t]$ by setting $I - B = E(I - A)$, so,

$$A = \begin{pmatrix} a & b+t^3 & c \\ d & e & f \\ g & h & i \end{pmatrix}, \quad B = \begin{pmatrix} a+t^3d & b+t^3e & c+t^3f \\ d & e & f \\ g & h & i \end{pmatrix}.$$

Producing Γ_A *from* Γ_B. Suppose that $\tau = \tau_1 \tau_2 \tau_3$ is the elementary path in Γ_A from vertex 1 to vertex 2 corresponding to the term t^3 above. Let $|p|$ denote the length (number of edges) in a graph path p. We obtain Γ_B from Γ_A as follows.

(1) Remove the elementary path τ from Γ_A;
(2) For each elementary path v of Γ_A beginning at vertex 2, put in an elementary path $\tilde{v}$ beginning at vertex 1, such that
   (i) $|\tilde{v}| = |\tau| + |v| = 3 + |v|$,
   (ii) the terminal vertices of v and $\tilde{v}$ agree.

For example,

with $|\tilde{v}| = |\tau| + |v|$.

Defining the conjugacy $\phi : X_{A^\sharp} \to X_{B^\sharp}$. Wherever the elementary path τ occurs in a point x of X_A, it must be followed by an elementary path v. Now define $\phi(x)$ be replacing each path τv with the elementary path $\tilde{v}$:

- If $x_{k+1} \ldots x_{k+|\tau v|} = \tau v$, with v an elementary path in Γ_A,
  then $(\phi x)_{k+1} \ldots (\phi x)_{k+|\tau v|} = \tilde{v}$.

- Otherwise, $(\phi x)_n = x_n$.

If we look at a succession of elementary paths τ and v_i, the code looks like:

$$\ldots\ v_{-1}\ \ \tau v_1\ v_2\ \tau v_3\ v_4\ \tau v_5\ v_6\ v_7\ \ldots$$

$$\phi \downarrow$$

$$\ldots\ v_{-1}\ \ \tilde{v}_1\ v_2\ \tilde{v}_3\ v_4\ \tilde{v}_5\ v_6\ v_7\ \ldots$$

This map ϕ is well defined because an elementary path v following τ has no edge in common with τ (because the initial and terminal vertices of τ are different) (Remark 4.4.29). Given that ϕ is well defined, it is straightforward to check that ϕ defines a topological conjugacy $(X_{A^\sharp}, \sigma) \to (X_{B^\sharp}, \sigma)$.

Above, we considered $E(I - A) = (I - B)$. Suppose that instead we define a matrix C by $(I - A)E = (I - C)$. No surprise: the matrix C also defines an SFT conjugate to that defined by A. In this case, instead of a conjugacy $X_A \to X_B$ based on $\tau v \mapsto \tilde{v}$ as above, we have a conjugacy $X_A \to X_C$ based on $v\tau \mapsto \tilde{v}$, where v is an elementary path in Γ_A with *terminal* vertex 1. $\qquad\square$

If C is a square matrix, then $C \oplus 1$ is the square matrix with block form $\begin{pmatrix} C & 0 \\ 0 & 1 \end{pmatrix}$.

Definition 4.4.14 (Positive equivalence) (Remark 4.4.30) Suppose that $\mathscr{P}$ is a subset of a ring $\mathscr{R}$. Let $\mathscr{S}$ be a set of square matrices over $\mathscr{R}$ which is "1-stabilized", in the sense that

$$C \in \mathscr{S} \implies (C \oplus 1) \in \mathscr{S}.$$

Positive equivalence of matrices in $\mathscr{S}$ (with respect to $\mathscr{P}$) is the equivalence relation on $\mathscr{S}$ generated by the following relations (where C, D must *both* be in $\mathscr{S}$):

(1) $C \sim C \oplus 1$.
(2) $EC = D$ or $CE = D$, where $E = E_{ij}(r)$, with $i \neq j$ and $r \in \mathscr{P}$.

If $\mathscr{P}$ is not specified, then by default we assume that $\mathscr{P} = \mathscr{R}$.

For $I - A$ in $\mathscr{S}$, the requirement that $\mathscr{S}$ is closed under the move $(I - A) \to (I - A) \oplus 1$ is equivalent to the requirement that the set $\{A : I - A \in \mathscr{S}\}$ is closed under the move $A \to A \oplus 0$.

Now suppose that $\mathscr{R} = \mathbb{Z}[t]$. We let $\mathscr{M}(X)$ denote the set of matrices with entries in a set X, and put $I - \mathscr{M}(X) = \{I - A : A \in \mathscr{M}(X)\}$. Suppose that $\{A, B\} \subset \mathscr{M}(\mathbb{Z}_+[t])$; $E = E_{ij}(f)$, with $f \in \mathbb{Z}[t]$; and $(I - B) = E(I - A)$. Writing $f = c - d$ with $\{c, d\} \subset \mathbb{Z}_+[t]$, we see that $A_{ij} - c \in \mathbb{Z}_+[t]$, and then that

$E_{ij}(c)(I-A) = (I-C)$ with $C \in \mathcal{M}(\mathbb{Z}_+[t])$. This equivalence is a composition of equivalences of the form $E_{ij}(t^k)(I-A_r) = (I-A_{r+1})$, with $\{A_r, A_{r+1}\} \subset \mathcal{M}(\mathbb{Z}_+[t])$ and $k \geq 0$. Considering likewise $(I-C) = E_{ij}(d)$, we see that the positive equivalence of $I-A$ and $I-B$ with respect to $\mathscr{P} = \mathscr{R}$ gives rise to a positive equivalence with respect to $\mathscr{P} = \{t^k : k \geq 0\}$. The same holds if $(I-A)E = (I-B)$. To summarize, matrices in $I - \mathcal{M}(\mathbb{Z}_+[t])$ are positive equivalent with respect to $\mathscr{P} = \mathscr{R}$ if and only if they are positive equivalent with respect to $\mathscr{P} = \{t^k : k \geq 0\}$. Given $\{A,B\} \subset \mathcal{M}(t\mathbb{Z}_+[t])$, the possibility that $k = 0$ can be excluded, and the positive equivalence for $\mathscr{P} = \mathscr{R}$ gives rise to a positive equivalence over $\mathscr{P} = \{t^k : k > 0\}$.

It then follows from Theorem 4.4.11 that positive equivalent matrices in $I - \mathcal{M}(t\mathbb{Z}_+[t])$ define topologically conjugate SFTs. Also, if $I - A \in I - (t\mathbb{Z}_+[t])$, then so is $(I-A) \oplus 1$, and A and $A \oplus 1$ define conjugate SFTs. Consequently we have the following.

Theorem 4.4.15 *Suppose that matrices $(I-A)$ and $(I-B)$ are positive equivalent in $I - \mathcal{M}(t\mathbb{Z}_+[t])$. Then A, B define topologically conjugate SFTs.*

The converse of Theorem 4.4.15 is "true up to a technicality" (Remark 4.4.31). For a true converse, we expand the collection of matrices allowed to present SFTs, from $\mathcal{M}(t\mathbb{Z}_+[t])$ to a slightly larger class, NZC. (On first exposure, it is fine to pretend that NZC $= \mathcal{M}(t\mathbb{Z}_+[t])$. But we will give statements for NZC, just to tell the truth.)

4.4.5 Classification of SFTs by positive equivalence in $I -$ NZC

For a matrix M over $\mathbb{Z}[t]$, let M_0 be M evaluated at $t = 0$.

Definition 4.4.16 (NZC condition)   Let NZC be the set of square matrices A over $\mathbb{Z}_+[t]$ such that A_0 is nilpotent.

Example 4.4.17 A and B are in NZC; C and D are not:

$$A = \begin{pmatrix} t^3+t & 3t^5 \\ t & 3t^5 \end{pmatrix}, \quad B = \begin{pmatrix} t^3 & 1 \\ t & 3t^5 \end{pmatrix}, \quad C = (1), \quad D = \begin{pmatrix} t^3 & 5t^2+2 \\ 1+t^7 & 3t^5 \end{pmatrix},$$

$$A_0 = \begin{pmatrix} 0 & 0 \\ 0 & 0 \end{pmatrix}, \quad B_0 = \begin{pmatrix} 0 & 1 \\ 0 & 0 \end{pmatrix}, \quad C_0 = (1), \quad D_0 = \begin{pmatrix} 0 & 2 \\ 1 & 0 \end{pmatrix}.$$

If A is in $\mathcal{M}(t\mathbb{Z}_+[t])$, then $A_0 = 0$. A matrix in NZC can have some entries with nonzero constant term, but not too many.

Why the term NZC? Here is the heuristic.

If, for example, $A(i,j) = t^4$, then in the graph with adjacency matrix $A^\sharp$, there is an elementary path, from i to j, of 4 edges. We consider this a path taking

4 units of time to traverse. The time to traverse concatenations of elementary paths is the sum of the times for its elementary paths. A nonzero term 1 in A_0 is considered as $1 = t^0$, giving a path taking zero time to traverse. "NZC" then refers to "No Zero Cycles", where a zero cycle is a cycle in the graph with adjacency matrix $A^\sharp$) taking zero time to traverse.

In the case NZC, one can make good sense of this heuristic, and everything works (Remark 4.4.32). But for a matrix A over $\mathbb{Z}_+[t]$ with zero cycles, we can't make sense of how A defines an SFT (let alone how multiplication by elementary matrices might induce topological conjugacies).

We do get a classification statement parallelling the SSE-$\mathbb{Z}_+$ setup of Williams.

Theorem 4.4.18 *For matrices A, B in NZC, the following are equivalent.*

(1) *$(I - A)$ and $(I - B)$ are positive equivalent in $I -$ NZC.*

(2) *A and B define topologically conjugate SFTs.*

Proof (1) $\implies$ (2) We saw this for positive equivalence in $I - \mathcal{M}(t\mathbb{Z}_+[t])$. It works similarly for matrices in $I -$ NZC [33, 15]. Note that, for matrices A and B from $\mathcal{M}(\mathbb{Z}_+[t])$, and E a basic elementary matrix with $(I - B) = E(I - A)$ or $(I - B) = (I - A)E$, we have $A \in$ NZC $\iff B \in$ NZC. One can check this by considering a correspondence of cycle paths, similar to the correspondence of paths in Example (4.4.13). Alternatively, one can use that for $A \in \mathcal{M}(\mathbb{Z}_+[t])$, we have $A \in$ NZC $\iff (\det(I - A))|_{t=0} = 1$.

(2) $\implies$ (1): $(I - A)$ is positive equivalent in $I -$ NZC to the matrix $I - tA^\sharp$, likewise $(I - B)$. So it suffices to get the positive equivalence for matrices $(I - tA^\sharp)$, $(I - tB^\sharp)$, assuming that the edge SFTS for $A^\sharp, B^\sharp$ are conjugate, that is, the $\mathbb{Z}_+$ matrices $A^\sharp, B^\sharp$ are SSE over $\mathbb{Z}_+$. It suffices to show the positive equivalence given an elementary SSE, $A^\sharp = RS, B^\sharp = SR$. For this, define matrices $A_0, A_1, \ldots, A_4$ in NZC : defined in block form as follows:

$$\begin{pmatrix} tRS & 0 \\ 0 & 0 \end{pmatrix}, \begin{pmatrix} tRS & 0 \\ tS & 0 \end{pmatrix}, \begin{pmatrix} 0 & R \\ tS & 0 \end{pmatrix}, \begin{pmatrix} 0 & 0 \\ tS & tSR \end{pmatrix}, \begin{pmatrix} 0 & 0 \\ 0 & tSR \end{pmatrix}.$$

(Notice that A_2 is in NZC, but is not in $\mathcal{M}(t\mathbb{Z}_+[t])$.)

The following Polynomial Strong Shift Equivalence Equations (PSSE Equations), taken from [33], give a positive equivalence in $I -$ NZC between $(I - A_i)$

and $(I - A_{i+1})$, for $0 \leq i < 4$.

$$\begin{pmatrix} I - tRS & 0 \\ -tS & I \end{pmatrix} \begin{pmatrix} I & 0 \\ tS & I \end{pmatrix} = \begin{pmatrix} I - tRS & 0 \\ 0 & I \end{pmatrix} \qquad \text{is} \quad (I - A_1)E_1 = (I - A_0),$$

$$\begin{pmatrix} I & R \\ 0 & I \end{pmatrix} \begin{pmatrix} I & -R \\ -tS & I \end{pmatrix} = \begin{pmatrix} I - tRS & 0 \\ -tS & I \end{pmatrix} \qquad \text{is} \quad E_2(I - A_2) = (I - A_1),$$

$$\begin{pmatrix} I & -R \\ -tS & I \end{pmatrix} \begin{pmatrix} I & R \\ 0 & I \end{pmatrix} = \begin{pmatrix} I & 0 \\ -tS & I - tSR \end{pmatrix} \qquad \text{is} \quad (I - A_2)E_3 = (I - A_3),$$

$$\begin{pmatrix} I & 0 \\ tS & I \end{pmatrix} \begin{pmatrix} I & 0 \\ -tS & I - tSR \end{pmatrix} = \begin{pmatrix} I & 0 \\ 0 & I - tSR \end{pmatrix} \qquad \text{is} \quad E_4(I - A_3) = (I - A_4).$$

One can check that each of the four equivalences given by the PSSE equations is a composition of basic positive equivalences in NZC. That finishes the proof. $\square$

Tools for construction. One way to construct a conjugacy between SFTs defined by matrices A, B over $\mathbb{Z}_+$ is to find an SSE over $\mathbb{Z}_+$ from A to B. The polynomial matrix setting gives another way: find a chain of elementary positive equivalences from $I - tA$ to $I - tB$. This is not a strict advantage; it is an alternative tool. There are results for which the only known proof uses this tool (Remark 4.4.33).

4.4.6 Functoriality: flow equivalence in the polynomial setting

We will consider one satisfying feature of presenting SFTs by matrices in NZC (or, just in $\mathcal{M}(t\mathbb{Z}_+[t])$) (Remark 4.4.34). With the $\mathbb{Z}_+$ matrix presentation, the algebraic invariant for conjugacy, SE-$\mathbb{Z}$, does not have an obvious natural relationship to algebraic invariants for flow equivalence (e.g. Bowen–Franks group, $\det(I - A)$). In the polynomial setting, we do see that natural relationship.

Let $\mathcal{M}$ be the set of matrices $I - A$ with A in NZC. We say that the matrices $(I - A)$ and $(I - B)$ are related by changing positive powers, and write this as $(I - A) \sim_+ (I - B)$, if they become equal after changing positive powers of t to other positive powers. For example,

$$\begin{pmatrix} 1 - t^2 - t^5 & -t - t^3 \\ -t^2 & 1 \end{pmatrix} \sim_+ \begin{pmatrix} 1 - t^2 - t^3 & -t^4 - t^5 \\ -t^7 & 1 \end{pmatrix} \sim_+ \begin{pmatrix} 1 - 2t & -2t \\ -t & 1 \end{pmatrix}.$$

The next result is one version for SFTs of the Parry–Sullivan characterization of flow equivalence of subshifts.

Theorem 4.4.19 *Suppose that A, B are matrices in NZC. The following are equivalent.*

(1) A, B define flow equivalent SFTs.
(2) $(I - A)$ and $(I - B)$ are equivalent, under the equivalence relation generated by (i) positive equivalence in $I - $NZC and (ii) $\sim_+$.

We do not give a proof for this theorem here, see (Remark 4.4.35). But, it is intuitive: flow equivalence arises from conjugacy and time changes, and the time changes are addressed by the $\sim_+$ relation.

Given a matrix $A = A(t)$ in $\mathcal{M}(t\mathbb{Z}_+[t])$, or in NZC, let $A(1)$ be the matrix defined entrywise by the (augmentation) homomorphism $\mathbb{Z}[t] \to \mathbb{Z}$ which sends t to 1. For example,

$$A = A(t) = (3t) , \quad B = B(t) = (t^2 + 2t^3) , \quad A(1) = (3) = B(1) .$$

From Theorem 4.4.19, one can check for SFTs defined by A, B from NZC:

(1) Flow equivalent SFTs defined by $A(t), B(t)$ from NZC produce isomorphic groups $\text{cok}_\mathbb{Z}(I - A(1)), \text{cok}_\mathbb{Z}(I - B(1))$.
(2) The group $\text{cok}_\mathbb{Z}(I - A(1))$ is the *Bowen–Franks group* of the SFT defined by A. (Remark 4.4.23)

We sometimes use notation $\text{cok}_\mathscr{R}$ to emphasize that a cokernel is an $\mathscr{R}$-module. (A $\mathbb{Z}$-module is just an abelian group.)

Recall, for $A = A(t)$ in NZC, the isomorphism class of $\text{cok}_{\mathbb{Z}[t]}(I - A(t))$ is the SE-$\mathbb{Z}$ class of the SFT. There is a functor, induced by $t \mapsto 1$:

$$\mathbb{Z}[t]\text{-modules} \to \mathbb{Z}[1]\text{-modules} = \mathbb{Z}\text{-modules} = \text{abelian groups}$$
$$\text{cok}_{\mathbb{Z}[t]}(I - A(t)) \mapsto \text{cok}_\mathbb{Z}(I - A(1)) .$$

So, this functor gives a presentation of

$$\text{SE-}\mathbb{Z} \text{ class} \to \text{Bowen–Franks group} .$$

This shows us how algebraic invariants of flow equivalence and topological conjugacy are naturally related in the polynomial setting.

Example 4.4.20 Let $A = (3t)$ and $B = (t^2 + 2t^3)$. The $\mathbb{Z}[t]$-modules $\text{cok}(I - A)$ and $\text{cok}(I - B)$ are not isomorphic. (For example, $\det(I - A) \neq \det(I - B)$.) However they do define SFTs which are flow equivalent, with Bowen–Franks group

$$\text{cok}(I - A(1)) = \text{cok}(I - B(1)) = \text{cok}(-2) = \mathbb{Z}/(-2)\mathbb{Z} = \mathbb{Z}/2\mathbb{Z} .$$

There is a useful analog of positive equivalence for constructing maps which give a flow equivalence, using multiplications by elementary matrices over $\mathbb{Z}$ rather than $\mathbb{Z}[t]$ (Remark 4.4.36). Also, the passage from SSE-$\mathbb{Z}_+$ of matrices

A, B to positive equivalence of matrices $(I - tA), (I - tB)$ works with an integral group ring $\mathbb{Z}_+G$ in place of $\mathbb{Z}_+$, as noted in [33, 15].

Remark 4.4.21 (Category theory)   For an approach to the classification of SFTs (and flow equivalence) through category theory, see the substantive recent paper [65] of Jeandel. In Jeandel's work, again matrices over $\mathbb{Z}_+$ and $\mathbb{Z}_+[t]$ play roles related to strong shift equivalence and flow equivalence. For an earlier approach to flow equivalence through category theory, see the paper [34] of Costa and Steinberg.

4.4.7 Notes

This subsection contains various remarks, proofs and comments referenced in earlier parts of Section 4.4.

Remark 4.4.22 (Flow equivalence background)   It takes more space than we will spend to give a reasonably understandable introduction to flow equivalence; see for example [18, 19] for background and definitions for flow equivalence of subshifts. However, the description to come of the Parry–Sullivan theorem [113] for SFTs will be quite adequate for our purposes, as a description of what flow equivalence is equivalent to.

Unexpectedly, tools developed for flow equivalence of SFTs turned out to be quite useful for certain classification problems in C^*-algebras (see for example [44, 119, 122] and their references).

Remark 4.4.23   For a square matrix C over $\mathbb{Z}$, one can check that the group $\text{cok}_{\mathbb{Z}}(C)$ is infinite when $\det C = 0$, and $|\text{cok}_{\mathbb{Z}}(C)| = |\det C|$ when $\det C \neq 0$. The groups arising as $\text{cok}_{\mathbb{Z}}(C)$ are the finitely generated abelian groups. The group $\text{cok}_{\mathbb{Z}}(C)$ may be determined algorithmically by computing the Smith normal form of C.

We refer to "the" Bowen–Franks group class associated to an SFT. Formally, the group depends on the presentation; really, we are talking about "the" group up to isomorphism. Americans of a certain age may remember Bill Clinton being mocked for a reply, "It depends on what you mean by the word 'is'."

In math, we really do need to keep track.

Remark 4.4.24   Topological Markov shifts were defined as "intrinsic Markov chains" by Bill Parry in the 1964 paper [112]. Parry's paper has the independence of past and future conditioned on the present, and this being presented by a zero-one transition matrix, essentially as a vertex shift.

But before Parry, there was Claude Shannon's astonishing monograph [134] in the 1940s, which launched information theory. Shannon already was looking at something we could understand as a Markov shift, with half of the variational principle proved in Parry's paper. Shannon even used polynomials to present those Markov shifts, just as we describe.

A zero-one matrix can be used to define an edge SFT or a vertex SFT. Yes, they are topologically conjugate SFTs. (The two block presentation of the vertex SFT is the edge SFT.)

Remark 4.4.25 As a postdoc, one of us heard a talk of John Franks on his classification of irreducible SFTs up to flow equivalence. Edge SFTs were a bit new; he announced for the suspicious that, for his proofs, zero-one matrices just weren't enough.

Remark 4.4.26 The "rome" term was introduced in the paper [8], which also gave a proof that $\det(I - A) = \det(I - tA^\sharp)$.

Remark 4.4.27 The entropy of an SFT defined by a matrix B over $\mathbb{Z}_+$ is the log of the spectral radius λ of B. If $\lambda > 1$ is the spectral radius of a primitive matrix over $\mathbb{Z}$, Perrin constructs a 2×2 matrix A over $t\mathbb{Z}_+[t]$ such that $A^\sharp$ is primitive with spectral radius λ [117]. (The condition that $A^\sharp$ is primitive is a significant part of the result.)

Proposition 4.4.28 *Suppose that U, C, V are matrices over a ring $\mathcal{R}$; U and V are invertible over $\mathcal{R}$; and $D = UCV$. Then $\mathrm{cok}_\mathcal{R} C$ and $\mathrm{cok}_\mathcal{R} D$ are isomorphic as $\mathcal{R}$-modules.*

Proof We consider C, D acting by matrix multiplication on row vectors; of course, the same fact holds for the action on column vectors. Corresponding to the action being on row vectors, we are considering left $\mathcal{R}$-modules (c in $\mathcal{R}$ sends v to cv), so that matrix multiplication gives an $\mathcal{R}$-module homomorphism (for example, $(cv)D = c(vD)$).

Let C be $j \times k$, and let D be $m \times n$. Then

$$\mathrm{cok}_\mathcal{R} C = \mathcal{R}^k / \mathrm{Image}(C) = \mathcal{R}^k / \{vC : v \in \mathcal{R}^j\},$$
$$\mathrm{cok}_\mathcal{R} D = \mathcal{R}^n / \mathrm{Image}(D) = \mathcal{R}^n / \{vD : v \in \mathcal{R}^m\}.$$

Define an $\mathcal{R}$-module isomorphism $\phi : \mathcal{R}^k \to \mathcal{R}^n$ by $\phi : w \mapsto wV$. (For most rings of interest, necessarily $j = m$ and $k = n$.) To show that ϕ induces the isomorphism $\mathrm{cok}_\mathcal{R} C \to \mathrm{cok}_\mathcal{R} D$, it suffices to show $\phi : \mathrm{Image}(C) \to \mathrm{Image}(D)$ and $\phi^{-1} : \mathrm{Image}(D) \to \mathrm{Image}(C)$. For $xC \in \mathrm{Image}(C)$,

$$\phi(xC) = xCV = (xU^{-1})(UCV) = (xU^{-1})D \in \mathrm{Image}(D).$$

For $yD \in \text{Image}(D)$,

$$\phi^{-1}(yD) = yDV^{-1} = y(UCV)V^{-1} = yUC \in \text{Image}(C). \qquad \square$$

Recall, for $A = A(t)$ in NZC, the isomorphism class of $\text{cok}_{\mathbb{Z}[t]}(I - A(t))$ determines the SE-$\mathbb{Z}$ class of the SFT, and conversely.

Remark 4.4.29 If the initial and terminal vertices of τ were the same, then we could apply the ϕ "rule" to a point $x = \dots \tau\tau\overset{\bullet}{\tau}\tau\tau \dots$ (with τ beginning at x_0) in contradictory ways, according to the two groupings

$$\dots (\tau\tau)(\tau\tau)(\overset{\bullet}{\tau}\tau)(\tau\tau)(\tau\tau)(\tau\tau)(\tau\tau)\dots$$

$$\dots (\tau\tau)(\tau\tau)(\tau\overset{\bullet}{\tau})(\tau\tau)(\tau\tau)(\tau\tau)(\tau\tau)\dots .$$

Remark 4.4.30 The move to polynomial algebraic invariants was pushed by Wagoner, who wanted to exploit analogies between SFT invariants and algebraic K-theory. Positive equivalence was born in the Kim–Roush–Wagoner papers [80, 81] as a tool, and taken further in [33] (see also [15]). The framework developed from considering conjugacy of SFTs via positive equivalence is called "Positive K-theory" (or, Nonnegative K-theory). This reflects the heuristic connection to algebraic K-theory. We will see that the connection is more than heuristic.

The term "positive equivalence" arises from its genesis in our application. We defined positive equivalence rather generally; there is nothing a priori about $\mathcal{M}$ which must involve positivity. Also, if $E_{ij}(-t^k)(I - A) = (I - B)$, then $E_{ij}(t^k)(I - B) = (I - A)$ – and so multiplications by elementary matrices $E_{ij}(-t^k)(I-A)$ are allowed. If U is a product of elementary matrices over $\mathcal{R}[t]$, such that $U(I-A) = (I-B)$, with A, B in I − NZC, it need not be the case $I - A$ and $I - B$ are positive equivalent. Each elementary step must be from a matrix in $\mathcal{M}$ to a matrix in $\mathcal{M}$.

Remark 4.4.31 If A, B in $\mathcal{M}(t\mathbb{Z}_+[t])$ have polynomial entries with all coefficients in $\{0,1\}$, and define topologically conjugate SFTs, then one can show that $I - A$ and $I - B$ are positive equivalent in $I - \mathcal{M}(t\mathbb{Z}_+[t])$. But, in general, the converse of the theorem is not true; for example, the matrices $(1 - 2t)$ and $\begin{pmatrix} 1-t & -t \\ -t & 1-t \end{pmatrix}$ define SFTs which are conjugate; but, there is not a string of elementary positive equivalences of square matrices over $t\mathbb{Z}_+[t]$, from one to the other. To see this, check the following claim: if $E = E_{ij}(t^k)$ with $k \geq 0$, and A, B are positive equivalent in $\mathcal{M}(t\mathbb{Z}_+[t])$, and $A(i,i) = 2t + \sum_{k \geq 2} a_k t^k$, then $B(i,i) = 2t + \sum_{k \geq 2} b_k t^k$.

Remark 4.4.32 We can expand NZC further, and consider matrices A over

$\mathbb{Z}[t, t^{-1}]$ with no cycles taking zero time or negative time, and make good sense of their presenting SFTs, and positive equivalence of these matrices $I - A$ as classifying SFTs. This isn't necessary for classification of SFTs, but might be convenient for some construction.

Remark 4.4.33 For example, constructions of SFTs and topological conjugacies between them, using polynomial matrices and basic positive equivalences, were the proof method for the result in [80, 81] of Kim, Roush and Wagoner (a result quite important for SFTs). The hardest step was a construction of brutal complication. But without their proof, we would have no proof at all.

Remark 4.4.34 Edge SFTs are related in a simple and transparent way to their defining matrices over $\mathbb{Z}_+$. When using a matrix A in NZC, or even just in $\mathcal{M}(t\mathbb{Z}_+[t])$, to define an SFT; we did it by way of the edge SFT defined from $A^{\sharp}$. The relationship between A and $A^{\sharp}$ is not very tight. There is some freedom about what matrix $A^{\sharp}$ is produced. That can be eliminated by precise choices, but these in generality become complicated and rather artificial.

So, for A in NZC, one would like to have a presentation of an SFT more simply and transparently related to A, and with an elementary positive equivalence presented transparently. There is such a presentation: the "path SFT" presented by A (see [15]).

For a square matrix A over $\mathbb{Z}_+$, and a positive integer n, the systems (X_A, σ^n) and (X_{A^n}, σ) are topologically conjugate. For a polynomial matrix A, A^n generally does not define an SFT conjugate to the nth power system of the SFT defined by A. But, in the path SFT presentation, we recover a natural way to pass to powers of the SFT, which works equally well for negative powers (no passage to transposes needed).

Caveat. We considered three matrix presentations of SFTs: by matrices over $\{0,1\}$, $\mathbb{Z}_+$ and $t\mathbb{Z}_+[t]$. The polynomial presentations have the greatest scope. But we certainly still need edge SFTs: usually the most convenient choice, sometimes the only choice, as for Wagoner's SSE-$\mathbb{Z}_+$ complex.

We also need vertex SFTs. Every topological Markov shift in the sense of Parry (also known as a 1-step shift of finite type) is a vertex SFT, up to naming of symbols. But, not every topological Markov shift is equal to an edge SFT up to naming of symbols. For an example, consider the vertex SFT with adjacency matrix $\left(\begin{smallmatrix} 1 & 1 \\ 1 & 0 \end{smallmatrix}\right)$. This vertex SFT cannot be an edge SFT after renaming symbols as edges in some directed graph, because a nondegenerate adjacency matrix for a graph with exactly two edges is either (2), $\left(\begin{smallmatrix} 1 & 0 \\ 0 & 1 \end{smallmatrix}\right)$ or $\left(\begin{smallmatrix} 0 & 1 \\ 1 & 0 \end{smallmatrix}\right)$.

Remark 4.4.35 See [15] for a proof of this version of the Parry–Sullivan result (Theorem 4.4.19) [113]. For a careful discussion of flow equivalence for subshifts, and related issues, see [18, 19], which includes references and a detailed proof of the Parry–Sullivan result.

Remark 4.4.36   For this version of positive equivalence, see the paper [15] and papers citing it.

4.5 Inverse problems for nonnegative matrices

In this section, we study certain inverse spectral problems, and related problems, for nonnegative matrices. We are especially interested in inverse problems which involve the realization of "stable algebra" invariants, such as the nonzero spectrum.

4.5.1 The NIEP

Definition 4.5.1 (Nonnegative matrix)   A real matrix is nonnegative if every entry is in $\mathbb{R}_+$. A real matrix is positive if every entry is positive.

We recall some definitions. If the real matrix A has characteristic polynomial $\chi_A(t) = \prod_{i=1}^{n}(t - \lambda_i)$, then the *spectrum* of A is $(\lambda_1, \ldots, \lambda_n)$. We refer to the spectrum as an n-tuple by abuse of notation (Remark 4.5.37): the ordering of the λ_i does not matter but the multiplicity does matter. The λ_i are in $\mathbb{C}$. Similarly, if $\chi_A(t) = t^j \prod_{i=1}^{k}(t - \lambda_i)$, with the λ_i nonzero, then the *nonzero spectrum* of A is $(\lambda_1, \ldots, \lambda_k)$.

Problem 4.5.2   *The NIEP* (nonnegative inverse eigenvalue problem): What can be the spectrum of an $n \times n$ nonnegative matrix A over $\mathbb{R}$?

Work on the NIEP goes back to (at least) the following result.

Theorem 4.5.3 (Suleĭmanova 1949)   *[138] Suppose that* $\Lambda = (\lambda_1, \ldots, \lambda_n)$ *is a list of real numbers;* $\sum_i \lambda_i > 0$; *and* $i > 1 \implies \lambda_i < 0$. *Then* Λ *is the spectrum of a nonnegative matrix.*

In fact, under the assumptions of Suleĭmanova's theorem, the companion matrix of the polynomial $\prod_i(t - \lambda_i)$ is nonnegative (Remark 4.5.38).

There is a huge and active literature on the NIEP; see the survey [68] for an overview and extensive bibliography. Despite a rich variety of interesting results, a complete solution is not known at size n if $n > 4$.

Theorem 4.5.4 (Johnson–Loewy–London Inequalities) *(Theorem 4.5.39)*
Suppose that A is an $n \times n$ nonnegative matrix. Then for all k, m in $\mathbb{N}$,

$$\text{tr}(A^{mk}) \geq \frac{\left(\text{tr}(A^m)\right)^k}{n^{k-1}}.$$

The JLL inequalities, proved independently by Johnson and by Loewy and London, give a quantitative version of an easy compactness result: for $n \times n$ nonnegative matrices A with $\text{trace}(A) \geq \tau > 0$, there is a positive lower bound to $\text{trace}(A^k)$ which depends only on τ, n, k. We will use the JLL inequalities later.

4.5.2 Stable variants of the NIEP

Throughout this section, $\mathscr{R}$ denotes a subring of $\mathbb{R}$.

Problem 4.5.5 (Inverse problem for nonzero spectrum) What can be the nonzero spectrum of a nonnegative matrix A over $\mathscr{R}$? What can be the nonzero spectrum of an irreducible or primitive matrix over $\mathscr{R}$?

The case $\mathscr{R} = \mathbb{Z}$ asks, what are the possible periodic data for shifts of finite type? This is the connection to "stable algebra" for symbolic dynamics (and the original impetus for the paper [23]). Later, we will also consider the realization in nonnegative matrices of more refined stable algebra structure.

To begin we review relevant parts of the Perron–Frobenius theory of nonnegative matrices (Remark 4.5.40). This will let us reduce the different flavors of Problem 4.5.5 to the primitive case.

4.5.3 Primitive matrices

Recall Definition 4.2.25: a primitive matrix is a square nonnegative matrix A such that for some positive integer k, A^k is positive. (Then, A^n is positive for all $n \geq k$.) The next theorem is the heart of the theory of nonnegative matrices (Remark 4.5.41). Recall, the *spectral radius* of a square matrix with real (or complex) entries is the maximum of the moduli of the eigenvalues (the radius of the smallest circle in $\mathbb{C}$ with center 0 which contains the spectrum).

Theorem 4.5.6 (Perron) *Suppose that the matrix A is primitive, with spectral radius λ. Then the following hold.*

(1) λ *is a simple root of the characteristic polynomial χ_A.*
(2) *If ν is another root of χ_A, then $|\nu| < \lambda$.*

(3) *There are left and right eigenvectors ℓ, r of A for λ which have all entries positive.*
(4) *The only nonnegative eigenvectors of A are the eigenvectors for the spectral radius.*

Example 4.5.7   We list three nonprimitive nonnegative matrices for which a conclusion of the Perron theorem fails.

$$A = \begin{pmatrix} 0 & -1 \\ 1 & 0 \end{pmatrix}, \qquad B = \begin{pmatrix} 1 & 0 \\ 0 & 1 \end{pmatrix}, \qquad C = \begin{pmatrix} 0 & 1 \\ 1 & 0 \end{pmatrix}.$$

Then A has spectrum $(i, -i)$; the spectral radius of A is 1, but 1 is not an eigenvalue of A. B has spectrum $(1, 1)$; the spectral radius of B is 1, and 1 is a repeated root of χ_B. C has spectrum $(1, -1)$; the spectral radius 1 is an eigenvalue, but $1 = |-1|$.

Example 4.5.8   The matrix $A = \begin{pmatrix} 0 & 3 \\ 4 & 1 \end{pmatrix}$ is primitive with spectrum $(4, -3)$. There is a positive left eigenvector for eigenvalue 4, but not for 3:

$$(1,1)\begin{pmatrix} 0 & 3 \\ 4 & 1 \end{pmatrix} = 4(1,1) \quad \text{and} \quad (-4,3)\begin{pmatrix} 0 & 3 \\ 4 & 1 \end{pmatrix} = -3(-4,3).$$

4.5.4 Irreducible matrices

Definition 4.5.9 (Irreducible matrix)   An irreducible matrix is an $n \times n$ nonnegative matrix A such that

$$\{i, j\} \subset \{1, \ldots, n\} \quad \Longrightarrow \quad \exists k > 0 \text{ such that } A^k(i, j) > 0.$$

Every primitive matrix is irreducible.

Example 4.5.10

$$A = \begin{pmatrix} 1 & 1 & 1 \\ 1 & 1 & 1 \\ 0 & 0 & 0 \end{pmatrix} \quad B = \begin{pmatrix} 1 & 1 \\ 0 & 1 \end{pmatrix} \quad C = \begin{pmatrix} 0 & 1 & 0 \\ 0 & 0 & 1 \\ 1 & 0 & 0 \end{pmatrix} \quad D = \begin{pmatrix} 0 & 0 & 1 \\ 0 & 0 & 1 \\ 1 & 1 & 0 \end{pmatrix}.$$

For all $n \in \mathbb{N}$, we see sign patterns:

$$A^n = \begin{pmatrix} + & + & + \\ + & + & + \\ 0 & 0 & 0 \end{pmatrix} \quad B^n = \begin{pmatrix} + & + \\ 0 & + \end{pmatrix} \quad C^{3n} = \begin{pmatrix} + & 0 & 0 \\ 0 & + & 0 \\ 0 & 0 & + \end{pmatrix}$$

$$D^{2n} = \begin{pmatrix} + & + & 0 \\ + & + & 0 \\ 0 & 0 & + \end{pmatrix}.$$

A and B are not irreducible. C and D are irreducible, but not primitive.

Block permutation structure

If $n > 1$, then an $n \times n$ cyclic-permutation matrix is irreducible but not primitive. This is representative of the general irreducible case.

Theorem 4.5.11   *For a square nonnegative matrix A, the following are equivalent.*

(1) *A is irreducible.*

(2) *There is a permutation matrix Q and a positive integer p such that $Q^{-1}AQ$ has the block structure of a cyclic permutation,*

$$
Q^{-1}AQ = \begin{pmatrix}
0 & A_1 & 0 & 0 & \cdots & 0 \\
0 & 0 & A_2 & 0 & \cdots & 0 \\
 & & \cdots & & & \\
0 & 0 & 0 & 0 & \cdots & A_{p-1} \\
A_p & 0 & 0 & 0 & \cdots & 0
\end{pmatrix}
$$

such that each of the cyclic products $D_1 = A_1 A_2 \cdots A_p$, $D_2 = A_2 A_3 \cdots A_1$, ... , $D_p = A_p A_1 \cdots A_{p-1}$ is a primitive matrix.

The integer p above is called the period of the irreducible matrix A. (If $p = 1$, then A is primitive.) For A above, A^p is block diagonal, with diagonal blocks $D_1, \ldots, D_p$.

From the block permutation structure, one can show the following (in which D could be any of the matrices D_i above).

Theorem 4.5.12 (Irreducible to primitive reduction)   *Suppose that A is an irreducible matrix with period p. Then there is a primitive matrix D such that $\det(I - tA) = \det(I - t^p D)$.*

It is not hard to check that the converse of this theorem is also true (Proposition 4.5.42).

Reduction in terms of nonzero spectrum

Recall here that the matrix A has nonzero spectrum $(\lambda_1, \ldots, \lambda_k)$ if and only if $\det(I - tA) = \prod_{i=1}^{k}(1 - \lambda_i t)$. The statement $\det(I - tA) = \det(I - t^p D)$ has an equivalent description (Proposition 4.5.43): if Λ is the nonzero spectrum of D, then $\Lambda^{1/p}$ is the nonzero spectrum of A. Here, $\Lambda^{1/p}$ is defined by replacing each entry of Λ with the list of its pth roots in $\mathbb{C}$. If Λ has k entries, then $\Lambda^{1/p}$ has pk entries.

Example 4.5.13   Suppose that $\det(I - tD) = (1 - 8t)(1 - 7t)^2$, and also that

$\det(I - tA) = \det(I - t^3 D)$. Let $\xi = e^{2\pi i/3}$. The nonzero spectrum of D is $\Lambda = (8,7,7)$. The nonzero spectrum of A is

$$\Lambda^{1/3} = \left(\ 2, \xi 2, \xi^2 2,\quad 7^{1/3}, \xi 7^{1/3}, \xi^2 7^{1/3},\quad 7^{1/3}, \xi 7^{1/3}, \xi^2 7^{1/3}\ \right).$$

Multiplicity of zero in the spectrum

Apart from one exception: if a nonzero spectrum is realized by an irreducible matrix over $\mathscr{R}$ of size $n \times n$, then it can also be realized at any larger size, by an irreducible matrix over $\mathscr{R}$ of the same period.

The one exception: if $\mathscr{R} = \mathbb{Z}$, then an irreducible matrix with spectral radius 1 can only be a cyclic permutation matrix.

Also: if $n \times n$ is the smallest size primitive matrix realizing the nonzero spectrum Λ, then $pn \times pn$ is the smallest size irreducible matrix realizing $\Lambda^{1/p}$.

Conclusion. Knowing the possible spectra of irreducible matrices over a subring $\mathscr{R}$ of $\mathbb{R}$ reduces to knowing the possible nonzero spectra of primitive matrices over $\mathscr{R}$, and the smallest dimension in which they can be realized.

4.5.5 Nonnegative matrices

Exercise 4.5.14 (Remark 4.5.40) Suppose that A is a square nonnegative matrix. Then there is a permutation matrix P such $P^{-1}AP$ is block triangular, such that each diagonal block is either irreducible or (0).

For A nonnegative as above, let A_i be the ith diagonal block, with characteristic polynomial p_i. Then the characteristic polynomial of A is $\chi_A(t) = \prod_i p_i(t)$, and the nonzero spectrum is given by $\det(I - tA) = \prod_i \det(I - tA_i)$.

So, the spectrum of a nonnegative matrix is an arbitrary disjoint union of spectra of irreducible matrices, together with an arbitrary repetition of 0.

There are constructions and constraints which work best at the level of nonnegative matrices (e.g., JLL). Still, one approach to the NIEP is to focus on the primitive case (which gives the irreducible case, and then the general case). Obstructions might be more simply formulated in this case. Moreover, in real applications, a nonnegative matrix must often be irreducible or primitive. (For symbolic dynamics: definitely.) A realization statement for nonnegative matrices does not give a realization statement for irreducible or primitive matrices. So, we focus on primitive matrices. But even in this restricted case, no satisfactory general characterization is known or conjectured.

Conclusion. We will focus on the nonzero spectrum of primitive matrices. And here, at last, we find simplicity.

4.5.6 The Spectral Conjecture

Let $\Lambda = (\lambda_1, \ldots, \lambda_k)$ be a k-tuple of nonzero complex numbers. We will give three simple conditons Λ must satisfy to be the nonzero spectrum of a primitive matrix over $\mathscr{R}$.

Definition 4.5.15 (Perron value)   For a tuple $\Lambda = (\lambda_1, \ldots, \lambda_k)$ of complex numbers:

- λ_i is a *Perron value* for Λ if λ_i is a positive real number and $\lambda_i > |\lambda_j|$ for $i \neq j$.
- $\text{tr}(\Lambda) = \sum_{i=1}^{k} \lambda_i$.
- $\Lambda^n = \left((\lambda_1)^n, \ldots, (\lambda_k)^n \right)$,   if $n \in \mathbb{N}$.

Proposition 4.5.16 (Necessary conditions)   *Suppose that $\Lambda = (\lambda_1, \ldots, \lambda_k)$ is the nonzero spectrum of a primitive matrix over a subring $\mathscr{R}$ of $\mathbb{R}$. Then the following hold.*

(1) Perron Condition:
   Λ *has a Perron value.*
(2) Coefficients Condition (Remark 4.5.44):
   The polynomial $p(t) = \prod_{i=1}^{k} (t - \lambda_i)$ has all its coefficients in $\mathscr{R}$.
(3) Trace Condition:
   If $\mathscr{R} \neq \mathbb{Z}$, then for all positive integers n, ℓ:
   (i) $\text{tr}(\Lambda^n) \geq 0$, *and*
   (ii) $\text{tr}(\Lambda^n) > 0 \implies \text{tr}(\Lambda^{n\ell}) > 0$.
(4) *If $\mathscr{R} = \mathbb{Z}$, then for all positive integers n, $\text{tr}_n(\Lambda) \geq 0$.*

(We define $\text{tr}_n(\Lambda)$, the nth net trace of Λ, below.)

Proof (1) By the Perron theorem, Λ has a Perron value.
(2) The characteristic polynomial of a matrix over a ring has coefficients in the ring. For some $m \geq 0$, the characteristic polynomial of A is $t^m p(t)$. So, p has coefficients in the ring.
(3)(i) $\text{tr}(\Lambda) = \text{tr}(A)$, and $\text{tr}(\Lambda^n) = \text{tr}(A^n)$. The trace of a nonnegative matrix is nonnegative.
  (ii) Suppose that $\text{tr}(\Lambda^n) > 0$. Then $\text{tr}(A^n) > 0$ and $A^n \geq 0$. Therefore $\text{tr}(A^n)^\ell > 0$. But, $\text{tr}(A^n)^\ell = \text{tr}(A^{n\ell}) = \text{tr}(\Lambda^{n\ell})$.
(4) Suppose that $\mathscr{R} = \mathbb{Z}$. Conditions (i) and (ii) hold, but a stronger condition holds.

   Consider A as the adjacency matrix of a graph. A loop is a path with the same minimal and terminal vertex. The number of loops of length n is trace(A^n).

A loop is minimal if it is not a concatenation of copies of a shorter loop. So, for example,

$$\text{number of minimal loops of length } 1 = \text{trace}(A)$$
$$\text{number of minimal loops of length } 2 = \text{trace}(A^2) - \text{trace}(A) \, .$$

For example, let $\Lambda = (2, i, -i, i, -i, 1)$. Then $\text{tr}(\Lambda^2) - \text{tr}(\Lambda) = 1 - 3 = -2 < 0$. This Λ cannot be the nonzero spectrum of a matrix over $\mathbb{Z}_+$, even though Λ satisfies conditions (1), (2) and (3).

The number of minimal loops of length n, $\text{tr}_n(\Lambda)$, can be expressed as a function of the traces of powers of Λ using Möbius inversion:

$$\text{tr}_n(\Lambda) := \sum_{d|n} \mu(n/d) \, \text{tr}(\Lambda^d) \, ,$$

where μ is the Möbius function,

$$\mu : \mathbb{N} \to \{-1, 0, 1\}$$
$$: n \mapsto 0 \quad \text{if } n \text{ is not squarefree}$$
$$: n \mapsto (-1)^e \quad \text{if } n \text{ is the product of } e \text{ distinct primes.} \qquad \Box$$

Conjecture 4.5.17 (Spectral Conjecture, Boyle–Handelman 1991 [23])   Let $\mathscr{R}$ be a subring of $\mathbb{R}$. Suppose that $\Lambda = (\lambda_1, \ldots, \lambda_k)$ is an k-tuple of complex numbers. Then Λ is the nonzero spectrum of some primitive matrix over $\mathscr{R}$ if and only the above conditions (1)–(4) hold.

Example 4.5.18   (Unbounded realization size) Suppose that $\mathscr{R} = \mathbb{R}$. Given $0 < \varepsilon < (1/2)$, set

$$\Lambda_\varepsilon = \left(1 \, , \, i\sqrt{(1-\varepsilon)/2} \, , \, -i\sqrt{(1-\varepsilon)/2} \right) .$$

This Λ_ε satisfies the conditions of the Spectral Conjecture.

But, if a nonnegative $n \times n$ matrix A has nonzero spectrum Λ_ε, then

$$\text{tr}(\Lambda_\varepsilon^2) \geq \frac{(\text{tr}\,\Lambda_\varepsilon)^2}{n} \, , \quad \text{by the JLL inequality, and therefore}$$

$$\varepsilon \geq \frac{1^2}{n} = 1/n \, .$$

So, as ε goes to zero, the size of A must go to infinity.

Definition 4.5.19 (Eventually positive matrix)   A matrix A is *eventually positive* (EP) if for all large $k > 0$, A^k is positive.

Theorem 4.5.20 (Handelman)   *(Remark 4.5.46) Suppose that A is a square matrix over $\mathscr{R}$ whose spectrum has a Perron value.*

(1) *If $\mathscr{R} \neq \mathbb{Z}$, then A is similar over $\mathscr{R}$ to an EP matrix [57].*
(2) *If $\mathscr{R} = \mathbb{Z}$, then A is SSE over $\mathscr{R}$ to an EP matrix [58].*

In particular, the Spectral Conjecture would be true if we were allowed to replace Λ with Λ^k, k large. With $\mathscr{R} \neq \mathbb{Z}$, and Λ an n-tuple, we could even realize Λ^k with a positive matrix which is $n \times n$.

Let's consider existing results on the Spectral Conjecture.

4.5.7 Boyle–Handelman Theorem

Theorem 4.5.21 (Boyle–Handelman [23]) *The Spectral Conjecture is true if $\mathscr{R} = \mathbb{R}$.*

We refer to this as the *BH Theorem*.

Remark 4.5.22 The focus on nonzero spectra in [23] grew out of symbolic dynamics, as indicated by Section 4.2.6. However, in his 1981 paper [66], Charles Johnson had already called attention to the potential impact of adding zeros to a candidate spectrum of a nonnegative matrix (Remark 4.5.45).

Remark 4.5.23 The problem of determining the possible nonzero spectra of primitive symmetric matrices is quite different. If a k-tuple is the nonzero spectrum of a nonnegative symmetric matrix, then it is achieved by a matrix whose size is bounded above by a function of k [67]. Adding more zeros to the spectrum doesn't help.

The Boyle–Handelman Theorem is a corollary of a stronger result.

Theorem 4.5.24 (Subtuple Theorem [23]) *Suppose that Λ satisfies the conditions of the Spectral Conjecture, and a subtuple of Λ containing the Perron value of Λ is the nonzero spectrum of a primitive matrix over $\mathscr{R}$. (For example, this holds if the Perron value is in $\mathscr{R}$.) Then Λ is the nonzero spectrum of a primitive matrix over $\mathscr{R}$.*

The proof of the Subtuple Theorem uses ideas from symbolic dynamics. The proof is constructive, in the sense that one could make it a formal algorithm. But the construction is very complicated, and uses matrices of enormous size. It has no practical value as a general algorithm.

Theorem 4.5.25 (Boyle–Handelman–Kim–Roush [23]) *Suppose that Λ satisfies the conditions of the Spectral Conjecture, $\mathrm{tr}(\Lambda) > 0$ and $\mathscr{R} \neq \mathbb{Z}$. Then Λ is the nonzero spectrum of a primitive matrix over $\mathscr{R}$.*

Proof We will outline the proof.

(1) By the BH Theorem, there is a primitive matrix A over $\mathbb{R}$ with nonzero spectrum Λ.

(2) Given A primitive with positive trace, a theorem of Kim and Roush produces a positive matrix which is SSE-$\mathbb{R}_+$ to A (hence, has the same nonzero spectrum as A).

(3) There are matrices U, C over $\mathscr{R}$ such that $U^{-1}CU = A$ and $\det U = 1$.

(4) U is a product of elementary matrices over $\mathbb{R}$, equal to I except in a single off diagonal entry. By density of $\mathscr{R}$ in $\mathbb{R}$, these can be perturbed to elementary matrices over $\mathscr{R}$. Thus U can be perturbed to a matrix V over $\mathscr{R}$ with determinant 1.

(5) Because $U^{-1}CU$ is positive, if V is close enough to U then $V^{-1}CV$ is positive. $\qquad\qquad\qquad\qquad\qquad\qquad\qquad\qquad\qquad\qquad\qquad\qquad\square$

Remark 4.5.26 Suppose that $\mathscr{R} \neq \mathbb{Z}$. It would be very satisfying to see the Spectral Conjecture proved in the remaining case, $\mathrm{tr}(\Lambda) = 0$, by some analogous perturbation argument. We have no idea how to do this, or if it can be done.

4.5.8 The Kim–Ormes–Roush Theorem

Theorem 4.5.27 (Kim–Ormes–Roush)   *[69] For $\mathscr{R} = \mathbb{Z}$, the Spectral Conjecture is true.*

Let us note an immediate corollary.

Corollary 4.5.28   *For $\mathscr{R} = \mathbb{Q}$, the Spectral Conjecture is true.*

Remark 4.5.29 Polynomial matrices and formal power series play a fundamental role in the proof. The Kim–Ormes–Roush Theorem gives us a complete understanding of the possible periodic data for SFTs. The proof, though quite complicated, is much more tractable than the proof of the BH Theorem. The use of power series leads to an interesting analytical approach to the NIEP [87].

4.5.9 Status of the Spectral Conjecture

The conjecture is true for $\mathbb{R}$, $\mathbb{Q}$ and $\mathbb{Z}$; in the positive trace case; under the Subtuple Theorem assumption; and in other special cases. It is very hard to doubt the conjecture.

One expects the case $\mathscr{R} = \mathbb{Z}$ to be the hardest case. Perhaps it is feasible to prove the Spectral Conjecture by adapting the Kim–Ormes–Roush proof.

4.5.10 Laffey's Theorem

Laffey [86] gave a constructive version of the Boyle–Handelman Theorem in the case that $\mathscr{R} = \mathbb{R}$ and the candidate spectrum Λ, satisfying the necessary conditions of the Spectral Conjecture for $\mathbb{R}$, also satisfies $\mathrm{tr}(\Lambda^k) > 0$ for all $k \geq 2$.

The primitive matrix which Laffey constructs to realize Λ has a rather classical form, and there is a comprehensible formula giving an upper bound on the size of the smallest N given by the construction. From here, we give some remarks on Laffey's theorem.

The Coefficients Condition In the case of $\mathbb{R}$, the Coefficients Condition of the Spectral Conjecture follows automatically from the Trace Conditions (Remark 4.5.44).

Laffey's upper bound Laffey gave an upper bound on the smallest size N of a primitive matrix A realizing a given $\Lambda = (\lambda_1, \ldots, \lambda_n)$. If A is primitive with nonzero spectrum Λ and $c > 0$, then cA is primitive with nonzero spectrum $c\Lambda = (c\lambda_1, \ldots, c\lambda_n)$. So, to consider an upper bound N, for simplicity we consider just the special case that the Perron value of Λ is $\lambda_1 = 1$.

Laffey's explicit, computable formula giving an upper bound for N is rather complicated. But, using the Perron value $\lambda_1 = 1$, and considering only the nontrivial case $n \geq 2$, it can be shown that Laffey's bound implies

$$N \leq \kappa_n \left(\frac{1}{MG} \right)^n \tag{4.2}$$

where κ_n depends only on n,

$$G = 1 - \max\{|\lambda_i| : 2 \leq i \leq n\} ,$$
$$M = \min\{\mathrm{tr}(\Lambda)^n : n \geq 2 \} .$$

The numbers κ_n obtained from the estimate grow very rapidly; for example, $\kappa_n \geq n^n$.

This bound is certainly nonoptimal! For example, suppose $0 < \varepsilon < 1$. The nonzero spectrum $(1, -1 + \varepsilon)$ is realized by the primitive matrix $\begin{pmatrix} 0 & 1 \\ 1 - \varepsilon & \varepsilon \end{pmatrix}$. But here, $\varepsilon = G$, and as ε goes to zero the upper bound in (Section 4.2) goes to infinity.

Nevertheless: this is a transparent and meaningful bound. The bound involves only n, M and G. The spectral gap G appears repeatly in the use of primitive matrices (and more generally), e.g. for convergence rates. Also, neither of the terms $1/M$ and $1/G$ can simply be deleted, as we note next.

The tracial floor term $1/M$ If Laffey's formula for an upper bound on N were replaced by a formula of the form $N \le f(G,n)$, then even at $n = 3$ the formula could not give a correct bound, on acount of the JLL Inequalities (Proposition 4.5.47).

The spectral gap term $(1/G)$ If Laffey's formula for an upper bound on N were replaced by a formula of the form $N \le f(M,n)$, then even at $n = 4$, the formula could not give a correct bound (Proposition 4.5.48).

Example 4.5.30   Let $\Lambda = (1.1, \xi, \overline{\xi})$, where $\xi = \exp(\pi i/10)$ and $\overline{\xi}$ is its complex conjugate. Laffey stated that there is a 128×128 primitive matrix realizing this nonzero spectrum.

The matrix form The primitive matrix with nonzero spectrum Λ has (for sufficiently large k) the banded form

$$
\begin{pmatrix}
x_1 & 1 & 0 & 0 & \cdots & 0 & 0 & 0 \\
x_2 & x_1 & 2 & 0 & \cdots & 0 & 0 & 0 \\
x_3 & x_2 & x_1 & 3 & \cdots & 0 & 0 & 0 \\
x_4 & x_3 & x_2 & x_1 & \cdots & 0 & 0 & 0 \\
\cdots & \cdots & \cdots & \cdots & \cdots & \cdots & \cdots & \cdots \\
x_{k-2} & x_{k-3} & x_{k-4} & \cdots & \cdots & x_1 & k-2 & 0 \\
x_{k-1} & x_{k-2} & x_{k-3} & \cdots & \cdots & x_2 & x_1 & k-1 \\
x_k & x_{k-1} & x_{k-2} & \cdots & \cdots & x_3 & x_2 & x_1
\end{pmatrix}.
$$

For the relation of the matrix entries to Λ, see Laffey's paper [86].

Limits of the argument A lot of the complication of the BH proof involves complications of $\mathrm{tr}(\Lambda^n) = 0$ for a variety of sets of n. These general difficulties are not addressed in Laffey's result. Laffey's argument also proves the Spectral Conjecture over any subfield $\mathscr{R}$ of $\mathbb{R}$, under the restriction $\mathrm{tr}(\Lambda^k) > 0$ for $k > 1$. But it does not work for all $\mathscr{R}$. The argument uses division by integers in $\mathscr{R}$.

4.5.11 The Generalized Spectral Conjectures

The NIEP refines to an even harder question: what can be the Jordan form of a square nonnegative matrix over $\mathbb{R}$? We refer to [68, Section 9] for a discussion. A rather sobering example of Laffey and Meehan [88] shows that $(3+t, 3-t, -2, -2, -2)$ is the spectrum of a 5×5 nonnegative matrix if $t > (16\sqrt{6})^{1/2} - 39 \approx 0.437\ldots$, but it is the spectrum of a diagonalizable nonnegative matrix if and only if $t \ge 1$.

Suppose that A is a nonnilpotent square matrix over $\mathbb{R}$. The *nonsingular part* of A is a nonsingular matrix A' over $\mathbb{R}$ such that A is similar to the direct sum of A' and a nilpotent matrix. (A' is only defined up to similarity over $\mathbb{R}$.) Analogously to the Spectral Conjecture (Conjecture (4.5.17)), we have the following.

Conjecture 4.5.31 (Boyle–Handelman)   If B is a square real matrix satisfying the necessary conditions of the Spectral Conjecture, then B is the nonsingular part of some primitive matrix over $\mathbb{R}$.

Let A, B be square matrices over $\mathbb{R}$, with nonsingular parts A', B'. Recall, the following are equivalent:

(1) A' and B' are SIM-$\mathbb{R}$ (similar over $\mathbb{R}$).
(2) A and B are SE-$\mathbb{R}$ (shift equivalent over $\mathbb{R}$).
(3) A and B are SSE-$\mathbb{R}$ (strong shift equivalent over $\mathbb{R}$).

So, the conjecture above is a special case of each of the following conjectures (Remark 4.5.49).

Conjecture 4.5.32 ((Weak) Generalized Spectral Conjecture, Boyle–Handelman 1991)   Suppose that A is a square matrix over a subring $\mathscr{R}$ of $\mathbb{R}$, and the nonzero spectrum of A satisfies the necessary conditions of the Spectral Conjecture.
   Then A is SE-$\mathscr{R}$ to a primitive matrix.

Conjecture 4.5.33 ((Strong) Generalized Spectral Conjecture, Boyle–Handelman 1993)   Suppose that A is a square matrix over a subring $\mathscr{R}$ of $\mathbb{R}$, and the nonzero spectrum of A satisfies the necessary conditions of the Spectral Conjecture.
   Then A is SSE-$\mathscr{R}$ to a primitive matrix.

The Strong GSC is the strongest viable conjecture we know which reflects the idea that the only obstruction to expressing stable algebra in a primitive matrix is the nonzero spectrum obstruction.
   In the next result, a "nontrivial unit" is a unit in the ring not equal to ± 1. (The assumption of a nontrivial unit is probably an artifact of the proof.)

Theorem 4.5.34 *[24, Theorem 3.3] Let $\mathscr{R}$ be a unital subring of $\mathbb{R}$. Suppose that either $\mathscr{R} = \mathbb{Z}$ or $\mathscr{R}$ is a Dedekind domain with a nontrivial unit. Let A be a square matrix with entries from $\mathscr{R}$ whose nonzero spectrum Λ satisfies the necessary conditions of the Spectral Conjecture and consists of elements of $\mathscr{R}$. Then A is algebraically shift equivalent[18] over $\mathscr{R}$ to a primitive matrix.*

---

[18] "Algebraically shift equivalent over $\mathscr{R}$" was the notation in [24] for what we are calling

The following corollary is immediate.

Corollary 4.5.35 *Conjecture 4.5.31 is true under the additional assumption that the spectrum of B is real.*

For example, the corollary covers the case that B in Conjecture 4.5.31 is a diagonal matrix. (For example, if B is diagonal with a Laffey–Meehan spectrum $(3+t, 3-t, -2, -2, -2)$, for any $t > 0$). On the other hand, the following (embarassing) open problem indicates how little we know.

Problem 4.5.36 Suppose that A is a 2×2 matrix over $\mathbb{Z}$ with irrational eigenvalues satisfying the conditions of the Spectral Conjecture. Prove that A is SE-$\mathbb{Z}$ to a primitive matrix.

When the Generalized Spectral Conjectures were made, it was not known whether SE-$\mathscr{R}$ implied SSE-$\mathscr{R}$ for every ring $\mathscr{R}$. We now know that there are many rings over which SSE properly refines SE [31], including some subrings of $\mathbb{R}$. So, the weak and strong conjectures are not *a priori* equivalent. [31] explains this, and corrects a left/right error in the presentation.

So, the weak and strong conjectures are not *a priori* equivalent. Nevertheless, it can be proved, for every subring $\mathscr{R}$ of $\mathbb{R}$, that if any matrix in a given SE-$\mathscr{R}$ class is primitive, then every matrix in that SE-$\mathscr{R}$ class is SSE-$\mathscr{R}$ to a primitive matrix [29]. So, we now know that the weak and strong conjectures are equivalent.

4.5.12 Notes

This subsection contains various remarks, proofs and comments referenced in earlier parts of Section 4.5.

Remark 4.5.37 "Abuse of notation" is a use of notation to mean something it does not literally represent, for simplicity. For example, describing the spectrum correctly as a multiset (set with multiplicities) seems to divert more mental energy than one uses to be aware that an n-tuple is not literally a multiset.

Remark 4.5.38 The companion matrix characterization for Suleĭmanova's theorem is attributed in [86] to Shmuel Friedland.

Theorem 4.5.39 *(JLL Inequalities) Let A be an $n \times n$ nonnegative matrix.*

SE-$\mathscr{R}$, shift equivalence over the ring $\mathscr{R}$. Also, [24, Prop.2.4] established that SE and SSE are equivalent over a Dedekind domain, so the conclusion could have been stated for strong shift equivalence.

Then for all k, m in $\mathbb{N}$:

$$\text{trace}(A^{mk}) \geq \frac{\left(\text{trace}(A^m)\right)^k}{n^{k-1}} .$$

Expressed at $m = 1$ in terms of the spectrum $(\lambda_1, \ldots, \lambda_n)$, the inequality becomes

$$\sum_{i=1}^{n} (\lambda_i)^k \geq \frac{\left(\sum_{i=1}^{t} \lambda_i\right)^k}{n^{k-1}} . \qquad (4.3)$$

This result was proved independently by Loewy and London [94], and by Johnson [66]. (Johnson's explicit statement was only for $m = 1$, the essential case.) The proof of this insightful result is not difficult.

(1) If B is an $n \times n$ nonnegative matrix, and $k \in \mathbb{N}$, then $\text{tr}(B^k) \geq \sum_{i=1}^{n} \left(B(i,i)\right)^k$. (Because: the $B(i,i)^k$ are some of the terms contributing to $\text{tr}(B^k)$, and the other terms are nonnegative.)
(2) Now suppose $\tau = \text{tr}(B) > 0$, and solve the problem: if $x_1, \ldots, x_n$ are non-negative numbers with positive sum τ, what is the minimum possible for the sum $s_k = \sum_{i=1}^{n} (x_i)^k$?

You can check (with Lagrange multipliers, say, or Hölder's inequality) that the minimum is achieved at $(x_1, \ldots, x_n) = (\tau/n, \tau/n, \ldots, \tau/n)$. (Intuitively, there is no other candidate, because there is a minimum and the minimum is not achieved at $(x_1, \ldots, x_n) = (\tau, 0, \ldots, 0)$.) Then, for $B = A^m$,

$$\text{tr}(B^k) \geq s_k \geq \sum_{i=1}^{n} (\tau/n)^k = n(\tau/n)^k = \tau^k/n^{k-1} ,$$

$$\text{tr}(A^{mk}) = \text{tr}((A^m)^k) \geq (\text{tr}(A^m))^k/n^{k-1} . \qquad \square$$

Remark 4.5.40 There are a number of excellent works available on the Perron–Frobenius theory of nonnegative matrices; Seneta's classic book [133] includes one introduction. The short exposition [17], appealing to an argument of Michael Brin, covers the heart of the theory (statements in this chapter), but not all parts of it.

Remark 4.5.41 Briefly: why is the Perron theorem so important?

Suppose that A is primitive with spectral radius λ. Let ℓ, r be positive left, right eigenvectors for λ, such that $\ell r = (1)$. The Perron theorem implies that, for many purposes, for large n, A^n is very well approximated by the rank one positive matrix $\lambda^n r\ell$.

What is "very well approximated"? Let μ be the second highest eigenvalue modulus. There is a matrix B with spectral radius μ such that $A = (\lambda r\ell) + B$, with $(\lambda r\ell)B = 0 = B(\lambda r\ell)$, so $A^n = (\lambda^n r\ell) + B^n$. Entries of B^n cannot grow at an exponential rate greater than μ^n; but every entry of A^n grows at the exponentially greater rate λ^n.

Proposition 4.5.42 *Suppose that D is a primitive matrix over a subring $\mathscr{R}$ of $\mathbb{R}$, and p is a positive integer. Then there is an irreducible matrix A over $\mathscr{R}$ with period p such that $\det(I - tA) = \det(I - t^p D)$.*

Proof We give a proof for $p = 4$ (which should make the general case obvious). Define $A = \begin{pmatrix} 0 & D & 0 & 0 \\ 0 & 0 & I & 0 \\ 0 & 0 & 0 & I \\ I & 0 & 0 & 0 \end{pmatrix}$. We compute a product

$$(I - tA)U = \begin{pmatrix} I & -tD & 0 & 0 \\ 0 & I & -tI & 0 \\ 0 & 0 & I & -tI \\ -tI & 0 & 0 & I \end{pmatrix} \begin{pmatrix} I & 0 & 0 & 0 \\ t^3 I & I & 0 & 0 \\ t^2 I & 0 & I & 0 \\ tI & 0 & 0 & I \end{pmatrix} \cdot = \begin{pmatrix} I - t^4 D & -tD & 0 & 0 \\ 0 & I & -tI & 0 \\ 0 & 0 & I & -tI \\ 0 & 0 & 0 & I \end{pmatrix}$$

Noting $\det U = 1$, we see $\det(I - tA) = \det\big((I - tA)U\big) = \det(I - t^4 D)$. $\quad\square$

Proposition 4.5.43 *Suppose that A, D are square real matrices, $\det(I - tA) = \det(I - t^p D)$, and the nonzero spectrum of D is $\Lambda = (\lambda_1, \ldots, \lambda_k)$.*
Then the nonzero spectrum of A is $\Lambda^{1/p}$.

Proof Because the nonzero spectrum of D is $\Lambda = (\lambda_1, \ldots, \lambda_k)$, we can write $\det(I - tD) = \prod_{i=1}^k (1 - \lambda_i t)$. Therefore, $\det(I - t^p D) = \prod_{i=1}^k (1 - \lambda_i t^p)$. Given λ_i, let $\mu_{i1}, \ldots, \mu_{ip}$ be a list of its pth roots in $\mathbb{C}$. Then,

$$(1 - \lambda_i t^p) = \prod_{j=1}^p (1 - \mu_{ij} t).$$

Thus, $\det(I - tA) = \prod_{i=1}^k \prod_{i=1}^p (1 - \mu_{ij} t)$, and it follows that the nonzero spectrum of A is $\Lambda^{1/p}$. $\quad\square$

Remark 4.5.44 The Coefficients Condition of the Spectral Conjecture holds if the ring $\mathscr{R}$ contains $\mathbb{Q}$ and if $\operatorname{tr}(\Lambda^n) \in \mathscr{R}$ for all positive integers n. For a self contained proof of this, consider the companion matrix C to the polynomial

$$\prod_{i=1}^k (t - \lambda_i) = t^k - c_1 t^{k-1} - c_2 t^{k-2} - \cdots .$$

Clearly $c_1 \in \mathscr{R}$ iff $\operatorname{tr}(\Lambda) \in \mathscr{R}$. Now suppose $c_1, \ldots, c_{j-1}$ are in $\mathscr{R}$, and $j \leq k$. From this assumption and the form of C, we have that jc_j equals an element of $\mathscr{R}$ plus $\operatorname{tr}(\Lambda^j)$.

In particular, the Coefficients Condition is redundant if $\mathscr{R} = \mathbb{R}$, because $\operatorname{tr}(\Lambda^n) \geq 0$ implies $\operatorname{tr}(\Lambda^n) \in \mathbb{R}$.

Remark 4.5.45 In [66, Section 4], Johnson wrote the following. "Suppose the set of numbers $\{\lambda_1, \ldots, \lambda_m\}$ is not the spectrum of an $m \times m$ nonnegative matrix. Is it possible to "save" this set by appending $n - m > 0$ zeros, that is, might $\{\lambda_1, \ldots, \lambda_m, 0, \ldots, 0\}$ be the spectrum of a nonnegative matrix?" Johnson also (among the many results in [66]) gave an example of such a "save" with $m = 4$ (minimum possible); established the JLL inequality (4.3); and noted that it "is more likely to be satisfied as zeroes are added to the proposed spectrum."

Remark 4.5.46 Given a square matrix A over $\mathscr{R} \neq \mathbb{Z}$ with a Perron value λ, Handelman finds a matrix U invertible over $\mathscr{R}$ such that $U^{-1}AU$ has positive left and right eigenvectors for λ. This matrix $U^{-1}AU$ must be eventually positive. He also exhibits an obstruction to this in the case $\mathscr{R} = \mathbb{Z}$: if ℓ, r are left, right integral eigenvectors for λ, then the minimum inner product $\ell \cdot r$ does not improve with similarity, and if it is smaller than the size of A then it is impossible to find U invertible over $\mathbb{Z}$ such that $U^{-1}AU$ has the positive left, right eigenvectors. But, if needed, Handelman produces an SSE-$\mathbb{Z}$ to a larger matrix (of smallest size possible) for which he produces the desired U.

Proposition 4.5.47 *If Laffey's formula for an upper bound on N were replaced by a formula of the form $N \leq f(G, n)$, then even for $n = 3$ the formula could not give a correct bound.*

Proof To show this, it suffices to exhibit a family $\{\Lambda_\varepsilon : 0 < \varepsilon < 1/2\}$ of 3-tuple nonzero spectra of primitive matrices, with spectral gaps bounded away from zero, which cannot be realized by matrices of bounded size.

Set $\Lambda_\varepsilon = (1, i\sqrt{(1-\varepsilon)/2}, -i\sqrt{(1-\varepsilon)/2})$. Each Λ_ε satisfies the conditions of the Spectral Conjecture for $\mathbb{R}$, with spectral gap greater than $1/2$. But, if Λ_ε is the nonzero spectrum of an $N \times N$ matrix, we have already seen from the JLL inequalities that $N \geq 1/\varepsilon$. □

Proposition 4.5.48 *If Laffey's formula for an upper bound on N were replaced by a formula of the form $N \leq f(M, n)$, then even for $n = 4$ the formula could not give a correct bound.*

Proof It suffices to find a family $\{\Lambda_\varepsilon : 0 < \varepsilon < \varepsilon_0\}$ of 4-tuple nonzero spectra of primitive matrices, with

$$\inf_\varepsilon \inf\{\operatorname{tr}((\Lambda_\varepsilon)^k) : k \in \mathbb{N}\} > 0\,,$$

such that the Λ_ε cannot be the nonzero spectra of matrices of bounded size.

Let $\Lambda_\varepsilon = (1, 1 - \varepsilon, 0.9i, -0.9i)$, with $0 < \varepsilon < \varepsilon_0 = 0.0001$ (to avoid computation). Each Λ_ε is the nonzero spectrum of a primitive matrix over $\mathbb{R}$. For $\Lambda =$

$(1, 1, 0.9i, -0.9i)$, for $n \in \mathbb{N}$, $\mathrm{tr}(\Lambda^{2n}) = 2$ and $\mathrm{tr}(\Lambda^{2n+1}) = 2 - 2(0.9)^n$ and therefore $\mathrm{tr}(\Lambda^n) \geq 2 - 2(0.9) = .2$. With ε_0 small enough, likewise $\inf_\varepsilon \inf\{\mathrm{tr}((\Lambda_\varepsilon)^k) : k \in \mathbb{N}\} > 0$.

Suppose for some positive integer K, for each Λ_ε there is a nonnegative matrix A_ε of size $K \times K$ with nonzero spectrum Λ_ε. Then by compactness, there is a subsequence of the sequence $(A_{1/n})$ which converges to a nonnegative matrix A. The spectrum is a continuous function of the matrix entries, so A has nonzero spectrum $\Lambda = (1, 1, 0.9i, -0.9i)$. By the Perron-Frobenius spectral constraints, A cannot be irreducible, and Λ is the union of nonzero spectra of irreducible matrices, (1) and $(1, 0.9i, -0.9i)$. But $1^2 + (0.9i)^2 + (-0.9i)^2 = -1.8 < 0$, a contradiction.   □

Remark 4.5.49 The Weak Generalized Spectral Conjecture was stated in the 1991 publication [23]. The Strong Generalized Spectral Conjecture was stated in the 1993 publication [14]. Although Handelman was not a coauthor of the latter paper, the Strong conjecture was a conjecture by both of us.

4.6 A brief introduction to algebraic K-theory

Shift equivalence and strong shift equivalence are relations on sets of matrices over a semiring. Algebraic K-theory offers many tools for such a setting[19], so it is natural to suspect algebraic K-theory might be useful for studying the relations of shift and strong shift equivalence. This suspicion is correct, and we will present two cases where this happens:

(1) For a general ring $\mathscr{R}$, the refinement of SE-$\mathscr{R}$ by SSE-$\mathscr{R}$.
(2) Wagoner's obstruction map detecting a difference between SE-$\mathbb{Z}_+$ and SSE-$\mathbb{Z}_+$

The first is a purely algebraic problem, motivated by applications to symbolic dynamics, and to topics in algebra. The second, Wagoner's obstruction map, is concerned with an "order" problem, and is one of two known methods to produce counterexamples to Williams' conjecture (discussed in Section 4.2).

Sections 4.6 and 4.7 will focus on addressing the first item above. Section 4.8 will discuss automorphisms of shifts of finite type, an important topic in its own right. Section 4.8 is also used partly to prepare for Section 4.9, which addresses the second item above.

---

[19] At the beginning of the book *Algebraic K-theory and Its Applications* [123], the author Jonathan Rosenberg writes "Algebraic K-theory is the branch of algebra dealing with linear algebra over a ring".

To begin, we introduce some necessary background from algebraic K-theory, relevant for Section 4.7.

4.6.1 K_1 of a ring $\mathscr{R}$

The *abelianization* of a group G, denoted G_{ab}, is the quotient $G/[G,G]$ of G by its commutator subgroup $[G,G]$ (the subgroup generated by all commutators $[g,h] = g^{-1}h^{-1}gh$ for $g,h \in G$).

Given a ring $\mathscr{R}$, consider the group $\mathrm{GL}(n,\mathscr{R})$ of invertible $n \times n$ matrices over $\mathscr{R}$. If one wishes to understand the structure of this group, a natural question one may ask is: what is the abelianization of $\mathrm{GL}(n,\mathscr{R})$ (the quotient of this group G by its commutator subgroup $[G,G]$)? While the answer may be fairly complicated depending on n and $\mathscr{R}$, Whitehead [151] in 1950 made a beautiful observation: by stabilizing, the commutator subgroup becomes more accessible.

To describe Whitehead's result, first let us say that by stabilizing, we mean the following.

Definition 4.6.1 For any n, there is a group homomorphism

$$\mathrm{GL}(n,\mathscr{R}) \hookrightarrow \mathrm{GL}(n+1,\mathscr{R})$$

$$A \mapsto \begin{pmatrix} A & 0 \\ 0 & 1 \end{pmatrix}$$

and we define

$$\mathrm{GL}(\mathscr{R}) = \varinjlim \mathrm{GL}(n,\mathscr{R}).$$

The group $\mathrm{GL}(\mathscr{R})$ is often called the *stabilized general linear group* (over the ring $\mathscr{R}$).

An important collection of invertible matrices are the *elementary matrices*. A matrix $E \in \mathrm{GL}(n,\mathscr{R})$ is an elementary matrix if E agrees with the identity except in at most one off-diagonal entry. The following observation may be familiar from linear algebra: if E is an $n \times n$ elementary matrix and B is any $n \times n$ matrix, then

(1) EB is obtained from B by an elementary row operation (adding a multiple of one row of B to another row of B);
(2) BE is obtained from B by an elementary column operation (adding a multiple of one column of B to another column of B).

We define $\mathrm{El}(n,\mathscr{R})$ to be the subgroup of $\mathrm{GL}(n,\mathscr{R})$ generated by $n \times n$ elementary matrices.

Like $\mathrm{GL}(\mathscr{R})$, we can also stabilize the elementary subgroups. The homomorphisms in Definition 4.6.1 map $\mathrm{El}(n,\mathscr{R})$ to $\mathrm{El}(n+1,\mathscr{R})$, and we define

$$\mathrm{El}(\mathscr{R}) = \varinjlim \mathrm{El}(n,\mathscr{R}).$$

If $X \in \mathrm{El}(\mathscr{R})$, then X can be written as a product of elementary matrices

$$X = \prod_{i=1}^{k} E_i.$$

It follows that, for any matrix $A \in \mathrm{GL}(\mathscr{R})$, XA is obtained from A by performing a sequence of row operations, and AX is obtained from A by performing a sequence of column operations.

Note that when we write AX and XA, A and X may be of different sizes. However, the process of stabilization allows us replace A with $A \oplus I$ or X with $X \oplus I$ as necessary to carry out the multiplication.

The group $\mathrm{El}(\mathscr{R})$ turns out to be the key to analyzing the abelianization of $\mathrm{GL}(\mathscr{R})$.

Theorem 4.6.2 (Whitehead)   *For any ring $\mathscr{R}$, $[\mathrm{GL}(\mathscr{R}),\mathrm{GL}(\mathscr{R})] = \mathrm{El}(\mathscr{R})$.*

A proof of this can be found in a number of places; for example, see [150, Chapter III]. To see why the commutator $[\mathrm{GL}(\mathscr{R}),\mathrm{GL}(\mathscr{R})]$ is contained in $\mathrm{El}(\mathscr{R})$, one can check that if $A \in \mathrm{GL}(n,\mathscr{R})$, then

$$\begin{pmatrix} A & 0 \\ 0 & A^{-1} \end{pmatrix} = \begin{pmatrix} I & A \\ 0 & I \end{pmatrix} \begin{pmatrix} I & 0 \\ -A^{-1} & I \end{pmatrix} \begin{pmatrix} I & A \\ 0 & I \end{pmatrix} \begin{pmatrix} 0 & -I \\ I & 0 \end{pmatrix}$$

and that the last matrix in the above lies in $\mathrm{El}(\mathscr{R})$, so that $\begin{pmatrix} A & 0 \\ 0 & A^{-1} \end{pmatrix}$ is always in $\mathrm{El}(\mathscr{R})$. Now observe that we have

$$\begin{pmatrix} ABA^{-1}B^{-1} & 0 \\ 0 & I \end{pmatrix} = \begin{pmatrix} A & 0 \\ 0 & A^{-1} \end{pmatrix} \begin{pmatrix} B & 0 \\ 0 & B^{-1} \end{pmatrix} \begin{pmatrix} (BA)^{-1} & 0 \\ 0 & BA \end{pmatrix}$$

so any commutator lies in $\mathrm{El}(\mathscr{R})$.

Definition 4.6.3 (K_1 of a ring)   For a ring $\mathscr{R}$, the first algebraic K-group (of $\mathscr{R}$) is defined by

$$K_1(\mathscr{R}) = \mathrm{GL}(\mathscr{R})_{\mathrm{ab}} = \mathrm{GL}(\mathscr{R})/\mathrm{El}(\mathscr{R}).$$

We use $[A]$ to refer to the class of a matrix A in $K_1(\mathscr{R})$.

The second equality in the above definition is precisely Whitehead's theorem. Note:

(1) $K_1(\mathscr{R})$ is always an abelian group. (We will write it additively.)

(2) As noted before, multiplying a matrix A by an elementary matrix from the left (resp. right) corresponds to performing an elementary row (resp. column) operation on A. Thus the group $K_1(\mathscr{R})$ coincides with equivalence classes of (stabilized) invertible matrices over $\mathscr{R}$, where two matrices are equivalent if one can be obtained from the other by a sequence of elementary row and column operations.

(3) The group operation in $K_1(\mathscr{R})$ is, by definition,

$$[A] + [B] = [AB]$$

where again the product AB is defined because we have stabilized. However, the group operation is equivalently defined by

$$[A] + [B] = \left[\begin{pmatrix} A & 0 \\ 0 & B \end{pmatrix} \right].$$

To see this, as we noted before, for any $A \in \mathrm{GL}(\mathscr{R})$, the matrix $\begin{pmatrix} A & 0 \\ 0 & A^{-1} \end{pmatrix}$ is in $\mathrm{El}(\mathscr{R})$. Since we have stabilized, we may assume that A and B are the same size, and

$$[AB] = \left[\begin{pmatrix} A & 0 \\ 0 & I \end{pmatrix} \begin{pmatrix} B & 0 \\ 0 & I \end{pmatrix} \right] \left[\begin{pmatrix} B^{-1} & 0 \\ 0 & B \end{pmatrix} \right]$$

$$= \left[\begin{pmatrix} A & 0 \\ 0 & I \end{pmatrix} \begin{pmatrix} B & 0 \\ 0 & I \end{pmatrix} \begin{pmatrix} B^{-1} & 0 \\ 0 & B \end{pmatrix} \right] = \left[\begin{pmatrix} A & 0 \\ 0 & B \end{pmatrix} \right].$$

Historically, one of Whitehead's main motivations was to define what is now called *Whitehead torsion*. If $f \colon X \to Y$ is a homotopy equivalence between two finite CW complexes, Whitehead showed how to define a certain torsion class $\tau(f)$ in $K_1(\mathbb{Z}\pi_1(X))$. He showed that f is a simple homotopy equivalence (one obtained through some finite sequence of elementary moves) if and only if $\tau(f) = 0$. For more on this, see [123, Section 2.4], to which we refer for the topological concepts mentioned here.

What about computing $K_1(\mathscr{R})$? In general this is a difficult problem, but there are many cases where the answer is accessible, and we give some examples shortly.

Before discussing these examples, suppose now that $\mathscr{R}$ is commutative. Then there is a determinant homomorphism

$$\det \colon K_1(\mathscr{R}) \to \mathscr{R}^{\times}$$
$$\det([A]) = \det(A),$$

where $\mathscr{R}^{\times}$ is the multiplicative group of units of $\mathscr{R}$. The kernel of the determinant map is denoted by

$$SK_1(\mathscr{R}) = \ker \det.$$

Since the determinant map is surjective and right split (by identifying $\mathscr{R}^{\times}$ with $GL(1, \mathscr{R})$), we get an exact sequence of abelian groups

$$0 \to SK_1(\mathscr{R}) \longrightarrow K_1(\mathscr{R}) \xrightarrow{\det} \mathscr{R}^{\times} \to 0$$

and

$$K_1(\mathscr{R}) \cong SK_1(\mathscr{R}) \oplus \mathscr{R}^{\times}.$$

The determinant map turns out to be very useful in actually computing $K_1(\mathscr{R})$; often, it is actually an isomorphism.

Here are a few examples of K_1 for some rings.

(1) When $\mathscr{R}$ is a field, or even a Euclidean domain, the group $SK_1(\mathscr{R})$ is trivial, and $K_1(\mathscr{R}) \cong \mathscr{R}^{\times}$. When $\mathscr{R}$ is a field, this is just the classical fact that, over a field, any invertible matrix A can be row and column reduced to the matrix $(\det(A)) \oplus I$. When $\mathscr{R}$ is a Euclidean domain, $SK_1(\mathscr{R}) = 0$ as well (see [150, Ex. 1.3.5]). Thus for example

$$K_1(\mathbb{Z}) \cong \mathbb{Z}/2\mathbb{Z} = \{1, -1\}$$

where we've identified $\{1, -1\}$ with the group of units in $\mathbb{Z}$.

(2) If $\mathscr{R}$ is an integrally closed subring of a finite extension of $\mathbb{Q}$, then $SK_1(\mathscr{R}) = 0$ (this is a deep theorem of Bass, Milnor, and Serre; see [6, 4.3]).

(3) When G is an abelian group, the integral group ring $\mathbb{Z}G$ is commutative, so $SK_1(\mathbb{Z}G)$ is defined. There are finite abelian groups G for which $SK_1(\mathbb{Z}G) \neq 0$; for example, if $H = \mathbb{Z}/4\mathbb{Z} \times \mathbb{Z}/2\mathbb{Z} \times \mathbb{Z}/2\mathbb{Z}$, then $SK_1(\mathbb{Z}H) \cong \mathbb{Z}/2\mathbb{Z}$ [110, Example 5.1]). In general, the calculation of $SK_1(\mathbb{Z}G)$ is very nontrivial (see [110]).

This last example is especially important in topology (see [123, Section 4] for a brief discussion of this), and in addition, has applications to symbolic dynamics; see [30].

4.6.2 $NK_1(\mathscr{R})$

We introduce now a certain algebraic K-group called $NK_1(\mathscr{R})$. This group will play a key role for us later, when we discuss strong shift equivalence and shift equivalence over a ring $\mathscr{R}$.

Any homomorphism of rings $f \colon \mathcal{R} \to \mathcal{S}$ induces, for each n, a homomorphism of groups $\mathrm{GL}(n, \mathcal{R}) \to \mathrm{GL}(n, \mathcal{S})$ and hence a group homomorphism $\mathrm{GL}(\mathcal{R}) \to \mathrm{GL}(\mathcal{S})$. The homomorphism f then induces a group homomorphism

$$f_* \colon K_1(\mathcal{R}) \to K_1(\mathcal{S}).$$

In fact, the assignment $\mathcal{R} \to K_1(\mathcal{R})$ defines a functor from the category of rings to the category of abelian groups. For any ring $\mathcal{R}$, we may consider the ring of polynomials $\mathcal{R}[t]$ over $\mathcal{R}$, and there is a ring homomorphism

$$ev_0 \colon \mathcal{R}[t] \to \mathcal{R}$$
$$p(t) \mapsto p(0).$$

This induces a homomorphism

$$(ev_0)_* \colon K_1(\mathcal{R}[t]) \to K_1(\mathcal{R})$$

and the kernel of this map is denoted by

$$NK_1(\mathcal{R}) = \ker \left(K_1(\mathcal{R}[t]) \xrightarrow{(ev_0)_*} K_1(\mathcal{R}) \right).$$

Thus by definition, $NK_1(\mathcal{R})$ is a subgroup of $K_1(\mathcal{R}[t])$. In particular, it is always an abelian group.

The group $NK_1(\mathcal{R})$ is important in algebraic K-theory. It appears (among other places) in the Fundamental Theorem of Algebraic K-theory, relating the K-groups of $\mathcal{R}[t]$ and $\mathcal{R}[t, t^{-1}]$ to the K-groups of $\mathcal{R}$ (see [150, Chapter III]).

Here are a few facts about $NK_1(\mathcal{R})$:

(1) If $\mathcal{R}$ is a Noetherian regular ring (see [123, Chapter 3] for definitions), then $NK_1(\mathcal{R}) = 0$. In particular, if $\mathcal{R}$ is a field, a PID, or a Dedekind domain, then $NK_1(\mathcal{R}) = 0$ (see [150, III.3.8]).

(2) A theorem of Farrell [46] shows that if $NK_1(\mathcal{R}) \neq 0$, then it is not finitely generated as an abelian group.

Thus, to summarize the above two items: $NK_1(\mathcal{R})$ very often vanishes, but when it does not vanish, it is large (as an abelian group).

There are rings $\mathcal{R}$ for which $NK_1(\mathcal{R}) \neq 0$. For an easy example, take any commutative ring $\mathcal{R}$, and let $\mathcal{S} = \mathcal{R}[s]/(s^2)$. Then $NK_1(\mathcal{S}) \neq 0$. Indeed, over the ring $\mathcal{S}[t]$, the 1×1 matrix $(1 + st)$ is invertible, and hence we can consider its class $[(1 + st)] \in K_1(\mathcal{S}[t])$. Clearly $[(1 + st)]$ lies in $NK_1(\mathcal{S})$, and the class $[1 + st]$ is nontrivial in $K_1(\mathcal{S}[t])$ since $\det(1 + st) \neq 1$.

Here are some more interesting examples:

Example 4.6.4 (1) $NK_1(\mathbb{Q}[t^2,t^3,z,z^{-1}]) \neq 0$ (see [130] for details on this calculation). This is a nontrivial fact: since the ring $\mathbb{Q}[t^2,t^3,z,z^{-1}]$ is reduced (has no nontrivial nilpotent elements), we have $NK_1(\mathscr{R}) \subset SK_1(\mathscr{R}[t])$ (see Exercise 4.6.6 below), and often it is not easy to determine whether SK_1 vanishes[20].

(2) There are finite groups G for which $NK_1(\mathbb{Z}G) \neq 0$; for example, for $G = \mathbb{Z}/4\mathbb{Z}$, $NK_1(\mathbb{Z}[\mathbb{Z}/4\mathbb{Z}]) \neq 0$ (details for this particular G can be found in [148]).

See [29] for an application of the example (1) above. The example (2) above of integral group rings of finite groups is relevant for applications to symbolic dynamics (see [30]). In general, the calculation of $NK_1(\mathbb{Z}G)$ for G a finite group is complicated, and not fully known (see e.g. [59], [148]).

The following is a very useful tool for studying $NK_1(\mathscr{R})$. The result is often referred to as Higman's trick: see [63].

Theorem 4.6.5 (Higman) *Let $\mathscr{R}$ be a ring and let A be a matrix in $\mathrm{GL}(\mathscr{R}[t])$ such that $[A] \in NK_1(\mathscr{R})$. Then there exists a nilpotent matrix N over $\mathscr{R}$ such that $[A] = [I - tN]$ in $NK_1(\mathscr{R})$.*

Sketch of proof Use the fact that we are in the stabilized setting to kill off powers of t from A using elementary operations, arriving at a matrix of the form $A_0 + A_1 t$. Since $[A] \in NK_1(\mathscr{R})$, $[A_0] = 0 \in K_1(\mathscr{R})$, so $[A] = [I + B_1 t]$ for some B_1 over $\mathscr{R}$. Since the matrix $I + B_1 t$ is invertible over $\mathscr{R}[t]$, B_1 must be nilpotent. □

A more detailed proof of Theorem 4.6.5 may be found in [150, III.3.5.1].

Exercise 4.6.6 (Remark 4.6.15) Suppose that $\mathscr{R}$ is a commutative ring which is reduced, that is, has no nontrivial nilpotent elements. Prove that $NK_1(\mathscr{R}) \subset SK_1(\mathscr{R}[t])$.

Exercise 4.6.7 (Remark 4.6.16) Let $\mathscr{R}$ be a principal ideal domain. Prove that $NK_1(\mathscr{R}) = 0$.

4.6.3 $\mathrm{Nil}_0(\mathscr{R})$

Higman's trick suggests there is a connection between the group $NK_1(\mathscr{R})$ and the structure of nilpotent matrices over the ring $\mathscr{R}$. This is indeed the case, and we describe this relationship quite explicitly in this subsection (Remark

---

[20] To be convinced of the difficulties in determining whether SK_1 vanishes, see the introduction of Oliver's very thorough book [110].

4.6.19). To begin, we first define another group coming from algebraic K-theory, the *class group of the category of nilpotent endomorphisms over* $\mathscr{R}$. That's quite a long name, and we usually just call it "nil zero (of $\mathscr{R}$)", since it is denoted by $\mathrm{Nil}_0(\mathscr{R})$.

Definition 4.6.8 (Nil$_0$ of a ring) Let $\mathscr{R}$ be a ring. Define $\mathrm{Nil}_0(\mathscr{R})$ to be the free abelian group on the set of generators

$$\{[N] \mid N \text{ is a nilpotent matrix over } \mathscr{R}\}$$

together with the following relations:

(1) $[N_1] = [N_2]$ if $N_1 = P^{-1}N_2P$ for some $P \in \mathrm{GL}(\mathscr{R})$;

(2) $[N_1] + [N_2] = \left[\begin{pmatrix} N_1 & B \\ 0 & N_2 \end{pmatrix}\right]$ for any matrix B over $\mathscr{R}$;

(3) $[0] = 0$.

Where does the group $\mathrm{Nil}_0(\mathscr{R})$ come from? First let us recall some definitions. Consider the category **Nil$\mathscr{R}$** whose objects are pairs (P, f) where P is a finitely generated projective $\mathscr{R}$-module (a direct summand of a free module) and f is a nilpotent endomorphism of P, and where a morphism from (P, f) to (Q, g) is given by an $\mathscr{R}$-module homomorphism $\alpha \colon P \to Q$ for which the square below commutes.

$$\begin{array}{ccc} P & \xrightarrow{f} & P \\ \alpha \downarrow & & \downarrow \alpha \\ Q & \xrightarrow{g} & Q \end{array}$$

The category **Nil$\mathscr{R}$** has a notion of exact sequence by defining

$$(P_1, f_1) \to (P_2, f_2) \to (P_3, f_3)$$

to be exact if the corresponding sequence of $\mathscr{R}$-modules

$$P_1 \to P_2 \to P_3$$

is exact, i.e. Image$(P_1 \to P_2) = \ker(P_2 \to P_3)$ (see (Remark 4.6.20) for how **Nil$\mathscr{R}$** with this notion of exact sequence fits into a more general setting). Given this, define $K_0(\mathbf{Nil}\mathscr{R})$ to be the free abelian group on isomorphism classes of objects (P, f) in **Nil$\mathscr{R}$**, together with the relation:

$$[(P_1, f_1)] + [(P_3, f_3)] = [(P_2, f_2)]$$

whenever $0 \to (P_1, f_1) \to (P_2, f_2) \to (P_3, f_3) \to 0$ is exact.

Let **Proj**$\mathscr{R}$ denote the category of finitely generated projective $\mathscr{R}$-modules and consider the standard notion of an exact sequence in **Proj**$\mathscr{R}$. We can likewise define the group $K_0(\mathbf{Proj}\mathscr{R})$ to be the free abelian group on isomorphism classes of objects in **Proj**$\mathscr{R}$ with the similar relations:

$$[P_1] + [P_3] = [P_2]$$

whenever $0 \to P_1 \to P_2 \to P_3 \to 0$ is exact in **Proj**$\mathscr{R}$.

These relations are equivalent to the set of relations

$$[P_1] + [P_2] = [P_1 \oplus P_2], \qquad P_1, P_2 \text{ in } \mathbf{Proj}\mathscr{R}$$

since any exact sequence of projective $\mathscr{R}$-modules splits. Thus $K_0(\mathbf{Proj}\mathscr{R})$ is isomorphic to the group completion of the abelian monoid of isomorphism classes of finitely generated projective $\mathscr{R}$-modules under direct sum, which is often given as the definition of the group $K_0(\mathscr{R})$.

There is a functor **Nil**$\mathscr{R} \to \mathbf{Proj}\mathscr{R}$ given by $(P, f) \mapsto P$, and this functor respects exact sequences, so there is an induced map on the level of the K_0 groups defined above

$$K_0(\mathbf{Nil}\mathscr{R}) \to K_0(\mathbf{Proj}\mathscr{R}).$$

The kernel of this map is isomorphic to $\mathrm{Nil}_0(\mathscr{R})$ (details of this isomorphism can be found in [150, Chapter II]).

The following formalizes the connection between $NK_1(\mathscr{R})$ and nilpotent matrices over $\mathscr{R}$.

Theorem 4.6.9 *The map*

$$\Psi \colon \mathrm{Nil}_0(\mathscr{R}) \to NK_1(\mathscr{R})$$
$$\Psi \colon [N] \mapsto [I - tN] \tag{4.4}$$

is an isomorphism of abelian groups.

Exercise 4.6.10   (Remark 4.6.17) Show the map Ψ defined in (4.4) is a well-defined group homomorphism.

Towards showing Ψ is an isomorphism, given Higman's theorem 4.6.5 above, one obvious thing to try is to define an inverse map

$$NK_1(\mathscr{R}) \to \mathrm{Nil}_0(\mathscr{R})$$
$$[I - tN] \mapsto [N]. \tag{4.5}$$

This in fact works: this map turns out to be well-defined, and is an inverse to the map Ψ. This is classically done, in algebraic K-theory, using a fair amount of machinery and long exact sequences coming from localization results (for example, see [150, III.3.5.3]). Later we will see there is an alternative, more elementary, proof using strong shift equivalence theory.

We will make frequent use of the isomorphism (4.4) above in later sections.

Exercise 4.6.11   (Remark 4.6.18) Consider an upper triangular matrix N over $\mathscr{R}$ with zero diagonal. Then $I - tN$ lies in $\mathrm{El}(\mathscr{R}[t])$, and hence $[I - tN] = 0$ in $NK_1(\mathscr{R})$. Using the relations defining $\mathrm{Nil}_0(\mathscr{R})$, show that the class of such an N must be zero in $\mathrm{Nil}_0(\mathscr{R})$.

The isomorphism $NK_1(\mathscr{R}) \cong \mathrm{Nil}_0(\mathscr{R})$ is only one instance of a larger phenomenon, which, loosely speaking, relates the K-theory of polynomial rings $\mathscr{R}[t]$ (in fact, certain localizations of them) to the K-theory of endomorphisms over the ring $\mathscr{R}$ (Remark 4.6.21). The strong shift equivalence theory also fits nicely into this framework, and we'll describe this in a little more detail later.

4.6.4 K_2 of a ring $\mathscr{R}$

This short subsection gives a definition and a few very basic properties of the group K_2 of a ring, motivated by its appearance later in Section 4.9. For a more thorough introduction to K_2, see either [99] or [150, III. Sec. 5].

Roughly speaking, $K_2(\mathscr{R})$ measures the existence of "extra relations" among elementary matrices over $\mathscr{R}$. We make this more formal below, but the idea is that elementary matrices always satisfy a certain collection of relations which do not depend on the ring. The group $K_2(\mathscr{R})$ is a way to detect additional relations coming from the ring.

Let $\mathscr{R}$ be a ring. Given $n \geq 1$ and $1 \leq i \neq j \leq n$, let $E_{i,j}(r)$ denote the matrix which has r in the i, j entry, and agrees with the identity matrix everywhere else. Recall the group $\mathrm{El}(n, \mathscr{R})$ of $n \times n$ elementary matrices over $\mathscr{R}$ is generated by matrices $E_{i,j}(r)$, $i \neq j$. It is straightforward to check that $\mathrm{El}(n, \mathscr{R})$ always satisfies certain relations: for any $r, s \in \mathscr{R}$, we have

(1) $E_{i,j}(r)E_{i,j}(s) = E_{i,j}(r+s)$.

(2) $[E_{i,j}(r), E_{k,l}(s)] = \begin{cases} I & \text{if } i \neq l \text{ and } j \neq k \\ E_{i,l}(rs) & \text{if } i \neq l \text{ and } j = k \\ E_{k,j}(-sr) & \text{if } j \neq k \text{ and } i = l. \end{cases}$

The key here is that these relations are satisfied by $\mathrm{El}(n, \mathscr{R})$ for *every* ring. This perhaps motivates defining the following group:

Definition 4.6.12 (Steinberg group of a ring)   Let $\mathscr{R}$ be a ring and $n \geq 3$. The nth Steinberg group $\mathrm{St}(n, \mathscr{R})$ has generators $x_{i,j}(r)$, where $1 \leq i \neq j \leq n$ and $r \in \mathscr{R}$, and relations:

(1) $x_{i,j}(r) x_{i,j}(s) = x_{i,j}(r+s)$.

$$(2) \quad [x_{i,j}(r), x_{k,l}(s)] = \begin{cases} 1 & \text{if } i \neq l \text{ and } j \neq k \\ x_{i,l}(rs) & \text{if } i \neq l \text{ and } j = k \\ x_{k,j}(-sr) & \text{if } j \neq k \text{ and } i = l. \end{cases}$$

The map

$$x_{i,j}(r) \mapsto E_{i,j}(r)$$

defines a surjective group homomorphism

$$\theta_n \colon \mathrm{St}(n, \mathscr{R}) \to \mathrm{El}(n, \mathscr{R}).$$

The relations for $\mathrm{St}(n, \mathscr{R})$ and $\mathrm{St}(n+1, \mathscr{R})$ imply there is a well-defined group homomorphism

$$\mathrm{St}(n, \mathscr{R}) \to \mathrm{St}(n+1, \mathscr{R})$$
$$x_{ij}(r) \mapsto x_{ij}(r)$$

and we define

$$\mathrm{St}(\mathscr{R}) = \varinjlim \mathrm{St}(n, \mathscr{R})$$

and assemble the θ_n's to get a group homomorphism

$$\theta \colon \mathrm{St}(\mathscr{R}) \to \mathrm{El}(\mathscr{R}).$$

Finally, we define

$$K_2(\mathscr{R}) = \ker \theta.$$

It turns out the sequence

$$K_2(\mathscr{R}) \to \mathrm{St}(\mathscr{R}) \to \mathrm{El}(\mathscr{R})$$

is the universal central extension of the group $\mathrm{El}(\mathscr{R})$. The group $K_2(\mathscr{R})$ is precisely the center of $\mathrm{St}(\mathscr{R})$, and so is always abelian. Furthermore, the assignment $\mathscr{R} \to K_2(\mathscr{R})$ is functorial; see [150, III, Sec.5] for more details on this.

An observation we make use of later is the following. An expression of the form

$$\prod_{i=1}^{k} E_i = 1,$$

where E_i are elementary matrices, can be used to produce an element of $K_2(\mathcal{R})$: lift each E_i to some x_i in $\mathrm{St}(\mathcal{R})$ and consider

$$x = \prod_i^k x_i \in \mathrm{St}(\mathcal{R}).$$

Then $x \in K_2(\mathcal{R})$, although in general this element may depend on the choice of lifts.

Example 4.6.13 Let $\mathcal{R} = \mathbb{Z}$, and consider

$$E = E_{1,2}(1)E_{2,1}(-1)E_{1,2}(1) = \begin{pmatrix} 0 & 1 \\ -1 & 0 \end{pmatrix}.$$

One can check directly that $E^4 = I$, so we can consider the element of $K_2(\mathbb{Z})$

$$x = \left(x_{1,2}(1)x_{2,1}(-1)x_{1,2}(1) \right)^4.$$

Milnor in [99, Sec. 10] proves that x is nontrivial in $K_2(\mathbb{Z})$, $x^2 = 1$, and x is actually the only nontrivial element of $K_2(\mathbb{Z})$. Thus we have

$$K_2(\mathbb{Z}) \cong \mathbb{Z}/2\mathbb{Z}.$$

It turns out (see [150, Chapter V]) that $K_2(\mathbb{Z}[t]) \cong K_2(\mathbb{Z})$. Given $m \geq 1$, there is a split surjection $K_2(\mathbb{Z}[t]/(t^m)) \to K_2(\mathbb{Z})$ (that is, there is a subgroup of the left which maps bijectively to the right), and we can define the group $K_2(\mathbb{Z}[t]/(t^m), (t))$ to be the kernel of this split surjection. In [141], van der Kallen proved that $K_2(\mathbb{Z}[t]/(t^2), (t)) \cong \mathbb{Z}/2\mathbb{Z}$, a fact which will prove to be useful later in Section 4.9. More generally, the following was proved by Geller and Roberts.

Theorem 4.6.14 ([121, Section 7]) *For any $m \geq 2$, the group $K_2(\mathbb{Z}[t]/(t^m), (t))$ is isomorphic to $\bigoplus_{k=2}^m \mathbb{Z}/k\mathbb{Z}$.*

4.6.5 Notes

These notes contain some remarks, proofs, and solutions of exercises for Section 4.6.

Remark 4.6.15 (Solution to Exercise 4.6.6)

If $\mathcal{R}$ is commutative and reduced, the only units in $\mathcal{R}[t]$ are degree zero. Thus for a nilpotent matrix N over $\mathcal{R}$, since $I - tN$ is invertible, we have $\det(I - tN) = 1$. So, together with Higman's trick (Theorem 4.6.5), we have $NK_1(\mathcal{R}) \subset SK_1(\mathcal{R}[t])$.

Remark 4.6.16   (Solution to Exercise 4.6.7)

By Higman's trick (Theorem 4.6.5), it suffices to show that if N is a nilpotent matrix over $\mathscr{R}$ then $[I - tN] = 0$ in $K_1(\mathscr{R}[t])$. Given N nilpotent, by Theorem 4.3.11 from Section 4.3, there exists some $P \in GL(\mathscr{R})$ such that $P^{-1}NP$ is upper triangular with zero diagonal. Then

$$[I - tN] = [P^{-1}(I - tN)P] = [I - t(P^{-1}NP)].$$

in $K_1(\mathscr{R}[t])$. Since $P^{-1}NP$ is upper triangular with zero diagonal, $I - t(P^{-1}NP)$ lies in $El(\mathscr{R}[t])$, and hence $[I - t(P^{-1}NP)] = 0$ in $K_1(\mathscr{R}[t])$.

Remark 4.6.17   (Solution to Exercise 4.6.10)

Since $[I - tN_1] + [I - tN_2] = [I - t(N_1 \oplus N_2)] = [(I - tN_1) \oplus (I - tN_2)]$ in $K_1(\mathscr{R}[t])$, Ψ respects the group operations. To see it is well-defined it suffices to check Ψ on the relations for $\text{Nil}_0(\mathscr{R})$. For the first relation of $\text{Nil}_0(\mathscr{R})$, if N is a nilpotent matrix over $\mathscr{R}$ and $P \in GL(\mathscr{R})$ then

$$[I - tN] = [P^{-1}(I - tN)P] = [I - t(P^{-1}NP)]$$

in $NK_1(\mathscr{R})$. For the second relation, suppose N_1, N_2 are nilpotent matrices and B is some matrix over $\mathscr{R}$ and consider

$$\begin{pmatrix} I - tN_1 & -tB \\ 0 & I - tN_2 \end{pmatrix}.$$

Since $I - tN_2$ is invertible over $\mathscr{R}[t]$, we can consider the block matrix in $El(\mathscr{R}[t])$ given by

$$E = \begin{pmatrix} I & tB(I - tN_2)^{-1} \\ 0 & I \end{pmatrix}.$$

Then

$$E \begin{pmatrix} I - tN_1 & -tB \\ 0 & I - tN_2 \end{pmatrix} = \begin{pmatrix} I - tN_1 & 0 \\ 0 & I - tN_2 \end{pmatrix}$$

so the second relation is preserved by Ψ. The third relation is obvious.

Remark 4.6.18   (Solution to Exercise 4.6.11)

If N is size one or two then this is immediate from relation (2) in the definition of $\text{Nil}_0(\mathscr{R})$. Now if N is upper triangular of size $n \geq 2$ with zero diagonal, then there is some matrix B such that

$$N = \begin{pmatrix} N_1 & B \\ 0 & 0 \end{pmatrix}$$

where N_1 is upper triangular of size $n - 1$ with zero diagonal. Now use relation (2) of $\text{Nil}_0(\mathscr{R})$ and induction.

Remark 4.6.19 For a more abstract viewpoint, the connection between NK_1 and the class group $\mathrm{Nil}_0(\mathscr{R})$ of nilpotent endomorphisms over $\mathscr{R}$ essentially comes from the localization sequence in algebraic K-theory, together with identifying the category of $\mathscr{R}[t]$-modules of projective dimension less than or equal to 1 which are t-torsion (i.e. are annihilated by t^k for some k) with the category of pairs (P, f) where P is a finitely generated projective $\mathscr{R}$-module and f is a nilpotent endomorphism of P; see [150, Chapter III] for more on this viewpoint.

Remark 4.6.20 The category **Nil**$\mathscr{R}$ equipped with the notion of exact sequence as defined here is a particular case of the more general concept, introduced by Quillen, of an *exact category*, a category equipped with some notion of exact sequences which satisfy some conditions. Such a category has enough structure to define K-groups of the category; our definition of $K_0(\mathbf{Nil}\mathscr{R})$ coincides with K_0 of the exact category **Nil**$\mathscr{R}$. See [150, II Sec. 7] for details regarding this viewpoint.

Remark 4.6.21 One may also define a class group for endomorphisms over a ring $\mathscr{R}$. Define **End**$\mathscr{R}$ to be the category whose objects are pairs (P, f) where P is a finitely generated projective $\mathscr{R}$-module and $f \colon P \to P$ is an endomorphism, and a morphism $(P, f) \to (Q, g)$ is given by an $\mathscr{R}$-module homomorphism $h \colon P \to Q$ such that $hf = gh$. Analogous to **Nil**$\mathscr{R}$, we call a sequence

$$(P_1, f_1) \to (P_2, f_2) \to (P_3, f_3)$$

in **End**$\mathscr{R}$ exact if the associated sequence of $\mathscr{R}$-modules

$$P_1 \to P_2 \to P_3$$

is exact. Then $K_0(\mathbf{End}\mathscr{R})$ is defined to be the free abelian group on isomorphism classes of objects (P, f) in **End**$\mathscr{R}$ together with the relations

$$[(P_1, f_1)] + [(P_3, f_3)] = [(P_2, f_2)]$$

whenever

$$0 \to (P_1, f_1) \to (P_2, f_2) \to (P_3, f_3) \to 0 \tag{4.6}$$

is exact in **End**$\mathscr{R}$.

There is a forgetful functor **End**$\mathscr{R} \to$ **Proj**$\mathscr{R}$ given by $(P, f) \mapsto P$ and an induced group homomorphism on the level of K_0

$$K_0(\mathbf{End}\mathscr{R}) \to K_0(\mathbf{Proj}\mathscr{R})$$

$$[(P, f)] \mapsto [P]. \tag{4.7}$$

Now define $\mathrm{End}_0(\mathscr{R})$ to be the kernel of this homomorphism. The group

$\mathrm{End}_0(\mathscr{R})$ has a presentation analogous to the one given in Definition 4.6.8: $\mathrm{End}_0(\mathscr{R})$ is generated by

$$\{[A] \mid A \text{ is a square matrix over } \mathscr{R}\}$$

subject to the relations

(1) $[A_1] = [A_2]$ if $A_1 = P^{-1}A_2P$ for some $P \in \mathrm{GL}(\mathscr{R})$;

(2) $[A_1] + [A_2] = \left[\begin{pmatrix} A_1 & B \\ 0 & A_2 \end{pmatrix}\right]$ for any matrix B over $\mathscr{R}$;

(3) $[0] = 0$.

There is an equivalence relation on square matrices over $\mathscr{R}$ defined by $A \sim_{end} B$ if $[A] = [B]$ in $\mathrm{End}_0(\mathscr{R})$. A natural question is how this relation compares to the relations of strong shift equivalence and shift equivalence over $\mathscr{R}$. In fact, this is settled in the commutative case by the following theorem of Almkvist (which was also proved, and greatly generalized, by Grayson in [54]). In the theorem, for $\mathscr{R}$ commutative we let $\tilde{\mathscr{R}}$ denote the multiplicative subgroup of $1 + t\mathscr{R}[[t]]$ given by

$$\tilde{\mathscr{R}} = \left\{ \frac{p(t)}{q(t)} \mid p(t), q(t) \in \mathscr{R}[t] \text{ and } p(0) = q(0) = 1 \right\}.$$

Theorem 4.6.22 ([1, 2])   *Let $\mathscr{R}$ be a commutative ring. The map*

$$\begin{aligned} \mathrm{End}_0(\mathscr{R}) &\to \tilde{\mathscr{R}} \\ [A] &\mapsto \det(I - tA) \end{aligned} \tag{4.8}$$

is an isomorphism.

There is an extension of Theorem 4.6.22 to general (i.e. not necessarily commutative) rings due to Sheiham [135].

As a consequence of the theorem, if $\mathscr{R}$ is an integral domain then the relation $\sim_{end}$ is coarser than shift equivalence over $\mathscr{R}$. For example, when $\mathscr{R} = \mathbb{Z}$ and A over $\mathbb{Z}_+$ presents a shift of finite type (X_A, σ_A), knowing the class $[A]$ in $\mathrm{End}_0(\mathbb{Z})$ is the same as knowing the zeta function $\zeta_{\sigma_A}(t)$. Also (see Section 4.3.3), $\det(I - tA)$ is an invariant of SSE-$\mathscr{R}$ for any commutative ring $\mathscr{R}$, but there are commutative rings for which the trace is not an invariant of shift equivalence, and for such a ring $\mathscr{R}$, SE-$\mathscr{R}$ does not refine $\sim_{end}$.

For a symbolic system presented by a matrix A over a noncommutative ring (for example, the integral group ring $\mathbb{Z}G$ where G is nonabelian), Theorem 4.6.22 suggests the class $[A]$ in $\mathrm{End}_0(\mathscr{R})$ can serve as an analogue of the zeta function of the symbolic system presented by A.

4.7 The algebraic K-theoretic characterization of the refinement of strong shift equivalence over a ring by shift equivalence

Let $\mathscr{R}$ be a semiring. Recall that square matrices A, B are *elementary strong shift equivalent over* $\mathscr{R}$ (ESSE-$\mathscr{R}$ for short, denoted $A \stackrel{\text{esse-}\mathscr{R}}{\approx} B$) if there exist matrices R, S over $\mathscr{R}$ such that

$$A = RS, \quad B = SR.$$

Recall also from Section 4.3 the following two equivalence relations defined on the collection of square matrices over $\mathscr{R}$:

(1) Square matrices A and B are *strong shift equivalent over* $\mathscr{R}$ (SSE-$\mathscr{R}$ for short, denoted $A \stackrel{\text{sse-}\mathscr{R}}{\approx} B$) if there exists a chain of elementary strong shift equivalences over $\mathscr{R}$ from A to B:

$$A = A_0 \stackrel{\text{esse-}\mathscr{R}}{\approx} A_1 \stackrel{\text{esse-}\mathscr{R}}{\approx} \cdots \stackrel{\text{esse-}\mathscr{R}}{\approx} A_{n-1} \stackrel{\text{esse-}\mathscr{R}}{\approx} A_n = B.$$

(2) Square matrices A and B are *shift equivalent over* $\mathscr{R}$ (SE-$\mathscr{R}$ for short, denoted $A \stackrel{\text{se-}\mathscr{R}}{\approx} B$) if there exists matrices R, S over $\mathscr{R}$ and a number $l \in \mathbb{N}$ such that

$$A^l = RS, \quad B^l = SR$$
$$AR = RB, \quad BS = SA.$$

The group $\mathscr{R}^n$ is a (left) $\mathscr{R}$-module, by the obvious definition $r : x \mapsto rx$. For an $n \times n$ square matrix A over $\mathscr{R}$ there is an $\mathscr{R}$-module endomorphism $\mathscr{R}^n \to \mathscr{R}^n$ given by $x \mapsto xA$ We can form the direct limit $\mathscr{R}$-module

$$G_A = \varinjlim \{\mathscr{R}^n, x \mapsto xA\}.$$

This was introduced in the case $\mathscr{R} = \mathbb{Z}$ in Section 4.3.6. The $\mathscr{R}$-module G_A becomes an $\mathscr{R}[t, t^{-1}]$-module by defining $x \cdot t^{-1} = xA$. The following result was a part of Theorem 4.3.41.

Proposition 4.7.1 *For square matrices A, B over $\mathscr{R}$, we have*

$A \stackrel{\text{se-}\mathscr{R}}{\approx} B$ if and only if G_A and G_B are isomorphic as $\mathscr{R}[t, t^{-1}]$-modules.

This proposition shows that shift equivalence over a ring $\mathscr{R}$ has a nice classical algebraic interpretation.

4.7.1 Comparing shift equivalence and strong shift equivalence
over a ring

Recall that for any semiring $\mathscr{R}$ and square matrices A, B over $\mathscr{R}$,

$$A \xrightarrow{\text{sse-}\mathscr{R}} B \Longrightarrow A \xrightarrow{\text{se-}\mathscr{R}} B.$$

Sections 4.2 and 4.3 discussed various aspects of both shift equivalence and strong shift equivalence, especially in the central cases of $\mathscr{R} = \mathbb{Z}_+$ and $\mathscr{R} = \mathbb{Z}$. Recall Conjecture 4.2.24 from Section 4.2:

Conjecture 4.7.2 (Williams' shift equivalence conjecture, 1974)   If A and B are square matrices over $\mathbb{Z}_+$ which are shift equivalent over $\mathbb{Z}_+$, then A and B are strong shift equivalent over $\mathbb{Z}_+$.

There are counterexamples to Williams' conjecture; we discuss some of these in Section 4.9. We can generalize the conjecture in the obvious way to arbitrary semirings, and rephrase as a more general problem:

Problem 4.7.3 (General Williams problem)   Suppose that $\mathscr{R}$ is a semiring, and A, B are square matrices over $\mathscr{R}$. If A and B are shift equivalent over $\mathscr{R}_+$, must A and B be strong shift equivalent over $\mathscr{R}_+$?

Williams' original shift equivalence conjecture concerns the case $\mathscr{R} = \mathbb{Z}_+$, and is most immediately linked to shifts of finite type, through its relation to topological conjugacy (as discussed in Section 4.2). It turns out, even in the case $\mathscr{R} = \mathbb{Z}_+$ the answer to Williams problem is 'not always'. We will talk more about this later, but for now let us consider the following picture, which outlines how the general Williams problem can be approached:

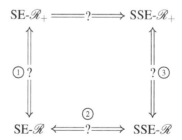

Looking at the picture above, Williams' problem concerns the top arrow. The picture describes how the problem can be broken down into a few parts: an 'algebra' part (②) in the picture), and two 'order' parts (① and ③) in the picture). In key cases, the answer to ① is yes for a fundamental subclass of matrices over $\mathscr{R}_+$. Recall, for $\mathscr{R} \subset \mathbb{R}$, a matrix A is primitive if there exists k

such that A^k has all positive entries. Then as shown in Proposition 4.3.48, we have:

Theorem 4.7.4 *Suppose that $\mathscr{R} \subset \mathbb{R}$ and $\mathscr{R}_+ = \mathscr{R} \cap \mathbb{R}_+$. If A and B are primitive matrices over $\mathscr{R}$, then A and B are shift equivalent over $\mathscr{R}$ if and only if they are shift equivalent over $\mathscr{R}_+$.*

This result says that, when the ring is a subring of $\mathbb{R}$, we can reduce the question of SE-$\mathscr{R}_+$ of primitive matrices to the purely algebraic question of SE-$\mathscr{R}$.

Part ② is the main topic of this and the next section. Part ③, in the case $\mathscr{R} = \mathbb{Z}_+$, contains the remaining core of Williams' problem, and will be discussed in Section 4.9.

4.7.2 The Algebraic Shift Equivalence Problem

We consider now ②, which we can restate as:

Problem 4.7.5 (Algebraic Shift Equivalence Problem, [146])   Let $\mathscr{R}$ be a ring and A, B be square matrices over $\mathscr{R}$. If A and B are shift equivalent over $\mathscr{R}$, must A and B be strong shift equivalent over $\mathscr{R}$?

Williams gave an argument in [152, Lemma 4.6] (which needed an additional step, later given in [154]) showing that, when $\mathscr{R} = \mathbb{Z}$, the answer to Problem 4.7.5 is yes. Effros also gave a similar argument, in an unpublished work, in the case $\mathscr{R} = \mathbb{Z}$, and it was observed in [24] that both arguments work in the case $\mathscr{R}$ is a principal ideal domain. It was then shown by Boyle and Handelman [24] that the answer to Problem 4.7.5 is also yes when $\mathscr{R}$ is a Dedekind domain. The Boyle–Handelman paper [24] was published in 1993, and after that point no further progress was made; in fact, it was still not known whether the answer to Problem 4.7.5 might be yes for *every* ring. Now, from recent work [31], we know the answer to Problem 4.7.5 is not always yes, and we have a pretty satisfactory characterization (Corollary 4.7.11) of the rings $\mathscr{R}$ for which the relations SE-$\mathscr{R}$ and SSE-$\mathscr{R}$ are the same. It turns out to depend on some K-theoretic properties of the ring $\mathscr{R}$ in question, and we'll spend the remainder of the section discussing how this works.

In short, the answer to Problem 4.7.5 turns out to depend on the group $NK_1(\mathscr{R})$. Before getting into the precise statements, recall from Proposition 4.3.5 in Section 4.3 that SSE-$\mathscr{R}$ is the relation generated by similarity and extensions by zero. Since the direct limit module associated to a nilpotent matrix is clearly trivial, it is reasonable to suspect that determining the strong shift equivalence classes of nilpotent matrices is connected to determining which nilpotent matrices over the ring can be obtained from the zero matrix (up to similarity) by extensions by zero. In fact this is the case, and the question of which nilpotent matrices over the ring can be obtained from the zero matrix (up to similarity) by extensions by zero turns out to be governed by $\text{Nil}_0(\mathscr{R})$.

Fix now a ring $\mathscr{R}$. For a matrix A over $\mathscr{R}$, we let $[A]_{sse}$, $[A]_{se}$ denote the strong shift equivalence (respectively shift equivalence) class of A over $\mathscr{R}$ (we suppress the $\mathscr{R}$ in the notation, as it is cumbersome). We define the following sets

$$SSE(\mathscr{R}) = \{[A]_{sse} \mid A \text{ is a square matrix over } \mathscr{R}\}$$
$$SE(\mathscr{R}) = \{[A]_{se} \mid A \text{ is a square matrix over } \mathscr{R}\}.$$

Since matrices which are strong shift equivalent over $\mathscr{R}$ must be shift equivalent over $\mathscr{R}$, there is a well-defined map of sets

$$\pi \colon SSE(\mathscr{R}) \to SE(\mathscr{R})$$
$$\pi \colon [A]_{sse} \mapsto [A]_{se}. \tag{4.9}$$

Problem 4.7.5 is equivalent to determining whether π is injective. We'll discuss when this happens, and in fact, we will do much more: we will describe the fiber over a class $[A]_{se}$ in terms of some K-theoretic data involving $NK_1(\mathscr{R})$.

4.7.3 Strong shift equivalence and elementary equivalence

From here on, we identify a square matrix M over $\mathscr{R}[t]$ with its class in the stabilization of matrices given by

$$M_n(\mathscr{R}[t]) \hookrightarrow M_{n+1}(\mathscr{R}[t])$$
$$M \mapsto \begin{pmatrix} M & 0 \\ 0 & 1 \end{pmatrix}.$$

Definition 4.7.6 (Elementary equivalence)   Let $\mathscr{R}$ be a ring. We say square matrices M, N over $\mathscr{R}[t]$ are *elementary equivalent over* $\mathscr{R}[t]$, denoted $M \overset{\text{El-}\mathscr{R}[t]}{\sim} N$, if there exist $E, F \in \text{El}(\mathscr{R}[t])$ such that $EMF = N$.

Note that, as when we first met the definition of K_1 of a ring, matrices M, N over $\mathscr{R}[t]$ are elementary equivalent over $\mathscr{R}[t]$ if and only if they (after stabilizing!) can be transformed into each other through a sequence of elementary row and column operations.

Remark 4.7.7   Given square matrices M, N over $\mathscr{R}[t]$, it may be tempting to ask why we don't just define $M \overset{\text{El-}\mathscr{R}[t]}{\sim} N$ if and only if $[M] = [N]$ in $K_1(\mathscr{R}[t])$, but this doesn't make sense: since M, N may not be invertible over $\mathscr{R}[t]$, we can't consider their class in $K_1(\mathscr{R}[t])$.

The following is one of the key results for studying strong shift equivalence over a ring $\mathscr{R}$.

Theorem 4.7.8 ([31, Theorem 7.2]) *Let $\mathscr{R}$ be a ring. For any square matrices A, B over $\mathscr{R}$, we have*

$$A \stackrel{sse-\mathscr{R}}{\sim} B \quad \text{if and only if} \quad I - tA \stackrel{\text{El-}\mathscr{R}[t]}{\sim} I - tB.$$

To see how this fits into the endomorphism $\leftrightarrow$ polynomial philosophy, consider a finitely generated free $\mathscr{R}$-module P and an endomorphism $f : P \to P$ (one may allow more generally P to be finitely generated projective; see (Remark 4.7.17)). The endomorphism f gives P the structure of an $\mathscr{R}[t]$-module with t acting by f, and the similarity class of f (over $\mathscr{R}$) corresponds to the isomorphism class of the $\mathscr{R}[t]$-module. The direct limit $\mathscr{R}[t, t^{-1}]$-module $\mathscr{M}_f = \varinjlim\{P, v \mapsto f(v)\}$ is isomorphic as an $\mathscr{R}[t, t^{-1}]$-module to $P \otimes_{\mathscr{R}[t]} \mathscr{R}[t, t^{-1}]$, and it follows that passing from the similarity class of f to the shift equivalence class of f is, in the polynomial world, the same as 'localizing at t' (note that, apart from here, our convention for the direct limit modules is that t^{-1} acts by f). On the endomorphism side, the strong shift equivalence relation lies between the similarity relation and the shift equivalence relation, and Theorem 4.7.8 tells us the meaning of the strong shift equivalence relation in the polynomial world.

We can summarize the above in Table 4.1, where A_f denotes a matrix over $\mathscr{R}$ representing $f : P \to P$ in a chosen basis for P as a free $\mathscr{R}$-module (Remark 4.7.18).

In the chart, GL-$\mathscr{R}[t]$-equivalence of (stabilized) matrices C and D means there exists $U, V \in \mathrm{GL}(\mathscr{R}[t])$ such that $UCV = D$. The ?? entry indicates that we do not have a good intrinsic interpretation of SSE-$\mathscr{R}$ at the $\mathscr{R}[t]$-module level.

Theorem 4.7.8 determines the algebraic relation in the polynomial world corresponding to SSE-$\mathscr{R}$. It also gives another idea of how strong shift equivalence arises algebraically in a natural way. Recall for A and B invertible over $\mathscr{R}[t]$ we have

$$[A] = [B] \text{ in } K_1(\mathscr{R}[t]) \text{ if and only if } A \stackrel{\text{El-}\mathscr{R}[t]}{\sim} B.$$

In light of Theorem 4.7.8, if one tries to naively extend the K-theory of $\mathscr{R}[t]$ to not necessarily invertible matrices over $\mathscr{R}[t]$, then strong shift equivalence naturally appears.

Endomorphisms over $\mathscr{R}$	$\mathscr{R}[t]$-endomorphism relation	$\mathscr{R}[t]$-module relation
Similarity class of A_f	GL-$\mathscr{R}[t]$-conjugacy class of $t - A_f$	Isomorphism class of the $\mathscr{R}[t]$-module P
SSE-$\mathscr{R}$ class of A_f	El-$\mathscr{R}[t]$-equivalence class of $1 - tA_f$	??
SE-$\mathscr{R}$ class of A_f	Gl-$\mathscr{R}[t]$-equivalence class of $1 - tA_f$	Isomorphism class of the $\mathscr{R}[t,t^{-1}]$-module $P \otimes \mathscr{R}[t,t^{-1}]$

Table 4.1 *Endomorphisms and $\mathscr{R}[t]$-modules*

See (Remark 4.7.19) for a discussion of how $\det(1 - tf)$ fits into the above table in the case $\mathscr{R}$ is commutative.

As a nice corollary of Theorems 4.7.8 and 4.6.9, we have the following:

Corollary 4.7.9 *Let $\mathscr{R}$ be a ring, and let N be a nilpotent matrix over $\mathscr{R}$. Then*

$$0 = [N] \text{ in } \mathrm{Nil}_0(\mathscr{R}) \quad \text{if and only if} \quad [N]_{sse} = [0]_{sse}.$$

In other words, a nilpotent matrix is strong shift equivalent over $\mathscr{R}$ to the zero matrix if and only if its class in $\mathrm{Nil}_0(\mathscr{R})$ is trivial. One can use this together with Theorem 4.7.8 to show that the map defined in (4.5) is injective.

4.7.4 The refinement of shift equivalence over a ring by strong shift equivalence

Theorem 4.7.8 gives us a key tool to understand the refinement of shift equivalence by strong shift equivalence over $\mathscr{R}$, obtaining a description of the fibers of the map

$$\pi \colon SSE(\mathscr{R}) \to SE(\mathscr{R})$$

defined in (4.9). We do this as follows.

In light of Corollary 4.7.9 above, there is a well-defined action $\mathfrak{N}$ of the group $\mathrm{Nil}_0(\mathscr{R})$ on the set $SSE(\mathscr{R})$ by

$$\mathfrak{N}([N]) \colon [A]_{sse} \mapsto \left[\begin{pmatrix} A & 0 \\ 0 & N \end{pmatrix}\right]_{sse}, \quad [N] \in \mathrm{Nil}_0(\mathscr{R}), \quad [A] \in SSE(\mathscr{R}).$$

The following gives a description of the fibers of the map π.

Theorem 4.7.10 ([31, Theorem 6.6]) *Let $\mathscr{R}$ be a ring, and let A be a square matrix over $\mathscr{R}$. There is a bijection*

$$\pi^{-1}([A]_{se}) \overset{\cong}{\longrightarrow} \mathfrak{N}\text{-orbit of } [A]_{sse}.$$

In other words, there is a bijection between the set of strong shift equivalence classes of matrices which are shift equivalent to A, and the orbit of $[A]_{sse}$ under the action of $\mathrm{Nil}_0(\mathscr{R})$.

As a corollary, we get the following.

Corollary 4.7.11 *Let $\mathscr{R}$ be a ring. Then $NK_1(\mathscr{R}) = 0$ if and only if, for all square matrices A, B over $\mathscr{R}$,*

$$A \overset{se\text{-}\mathscr{R}}{\sim} B \text{ if and only if } A \overset{sse\text{-}\mathscr{R}}{\sim} B.$$

Proof First suppose $NK_1(\mathscr{R}) = 0$. By Theorem 4.6.9, this implies $\mathrm{Nil}_0(\mathscr{R}) = 0$, so the action $\mathfrak{N}$ is trivial. Thus if A is any square matrix over $\mathscr{R}$, by Theorem 4.7.10, the fiber $\pi^{-1}([A]_{se})$ is also trivial. It follows that if B is any square matrix over $\mathscr{R}$, then $A \overset{se\text{-}\mathscr{R}}{\sim} B \Leftrightarrow A \overset{sse\text{-}\mathscr{R}}{\sim} B$ as desired.

Now suppose matrices which are SE-$\mathscr{R}$ must be SSE-$\mathscr{R}$. Given N nilpotent over $\mathscr{R}$, N is clearly shift equivalent over $\mathscr{R}$ to the zero matrix, and hence by assumption, strong shift equivalent over $\mathscr{R}$ to the zero matrix. By Corollary 4.7.9, this implies $[N] = 0$ in $\mathrm{Nil}_0(\mathscr{R})$. Since N was a general nilpotent matrix, it follows that $\mathrm{Nil}_0(\mathscr{R}) = 0$. $\qquad\square$

So what is the behavior of the action $\mathfrak{N}$? In general, its orbit structure is far from trivial. Given A over $\mathscr{R}$, define the $\mathfrak{N}$-stabilizer of $[A]_{sse}$ to be

$$\mathrm{St}_{\mathfrak{N}}(A) = \{[N] \in \mathrm{Nil}_0(\mathscr{R}) \mid [A \oplus N]_{sse} = [A]_{sse}\}.$$

Note the $\mathfrak{N}$-stabilizer depends only on the SSE-$\mathscr{R}$ class of a matrix A, but to avoid cumbersome notation, we write simply $\mathrm{St}_{\mathfrak{N}}(A)$ instead of $\mathrm{St}_{\mathfrak{N}}([A]_{sse})$.

The notation used here differs from what is used in [31]; see (Remark 4.7.20).

There is a bijection between the $\mathfrak{N}$-orbit of $[A]_{sse}$ and $\mathrm{Nil}_0(\mathscr{R})/\mathrm{St}_{\mathfrak{N}}(A)$, the quotient given by mapping a coset of $[N]$ in $\mathrm{Nil}_0(\mathscr{R})/\mathrm{St}_{\mathfrak{N}}(A)$ to $[A \oplus N]_{sse}$.

For a commutative ring $\mathscr{R}$, we define

$$S\mathrm{Nil}_0(\mathscr{R}) = \{[N] \in \mathrm{Nil}_0(\mathscr{R}) \mid \det(I - tN) = 1\}.$$

It is straightforward to check that $S\mathrm{Nil}_0(\mathscr{R})$ is a subgroup of $\mathrm{Nil}_0(\mathscr{R})$, and that $S\mathrm{Nil}_0(\mathscr{R})$ is precisely the pullback via the isomorphism (4.5) of the subgroup $NK_1(\mathscr{R}) \cap SK_1(\mathscr{R}[t])$ in $NK_1(\mathscr{R})$.

Theorem 4.7.12 ([31, Theorems 4.7, 5.1]) *For any ring $\mathscr{R}$, both of the following hold.*

(1) *If A is nilpotent or invertible over $\mathscr{R}$, then $\mathrm{St}_{\mathfrak{N}}(A)$ is trivial.*

(2) *If $\mathscr{R}$ is commutative, then*

$$\bigcup_{[A]_{sse} \in SSE(\mathscr{R})} \mathrm{St}_{\mathfrak{N}}(A) = S\mathrm{Nil}_0(\mathscr{R}).$$

Exercise 4.7.13 (Remark 4.7.21) Prove that, if $\mathscr{R}$ is commutative and reduced (has no nontrivial nilpotent elements), then the groups $S\mathrm{Nil}_0(\mathscr{R})$ and $\mathrm{Nil}_0(\mathscr{R})$ coincide.

The nilpotent case of (1) is straightforward, and is Exercise 4.7.14 below. Both (2) and the invertible case of (1) are nontrivial to prove. Part (1) uses localization and K-theoretic techniques for localization; in the non-commutative case, this requires some deep K-theoretic results of Neeman and Ranicki [107] about non-commutative localization of rings. Part (2) uses work of Nenashev [108] on presentations for K_1 of exact categories. We will not go into more detail about the structure of these proofs, but instead refer the reader to [31].

Exercise 4.7.14 (Remark 4.7.22) Prove that, if A is a nilpotent matrix over $\mathscr{R}$, then $\mathrm{St}_{\mathfrak{N}}(A)$ vanishes.

There are rings $\mathscr{R}$ for which $S\mathrm{Nil}_0(\mathscr{R})$ does not vanish. For example, the ring $\mathbb{Q}[t^2, t^3, z, z^{-1}]$ is commutative and reduced, and has nontrivial NK_1 (see Example 4.6.4(1)).

By part (2) of the above theorem, it follows that the $\mathfrak{N}$-stabilizers can be nontrivial, and can change depending on the matrix. There is a conjectured analogous version of part (2) in the non-commutative case, which is more technical to state (see [31, Conjecture 5.20]). In general, we do not have a complete understanding of the groups $\mathrm{St}_{\mathfrak{N}}(A)$, and the following problem was posed in [31, Problem 5.21]:

Problem 4.7.15 Given a square matrix A over $\mathscr{R}$, give a satisfactory description of the elementary stabilizer $\mathrm{St}_{\mathfrak{N}}(A)$. In particular, determine when $\mathrm{St}_{\mathfrak{N}}(A)$ is trivial.

4.7.5 The SE and SSE relations in the context of endomorphisms

Using the results above, we can now give another view on what the relations SE-$\mathcal{R}$ and SSE-$\mathcal{R}$ mean in the context of endomorphisms, and how they fit in with the similarity relation over a ring. Given square matrices A, B over $\mathcal{R}$, we say

(1) B is a zero extension of A if there exists some matrix C over $\mathcal{R}$ such that $B = \left(\begin{smallmatrix} A & C \\ 0 & 0 \end{smallmatrix}\right)$ or $B = \left(\begin{smallmatrix} A & 0 \\ C & 0 \end{smallmatrix}\right)$.

(2) B is a nilpotent extension of A if there exists some matrix C over $\mathcal{R}$ and some nilpotent matrix N over $\mathcal{R}$ such that $B = \left(\begin{smallmatrix} A & C \\ 0 & N \end{smallmatrix}\right)$ or $B = \left(\begin{smallmatrix} A & 0 \\ C & N \end{smallmatrix}\right)$.

Zero and nilpotent extensions fit nicely into the context of the category **Nil**$\mathcal{R}$; see (Remark 4.7.23).

Theorem 4.7.16 *Let $\mathcal{R}$ be a ring.*

(1) *SSE-$\mathcal{R}$ is the equivalence relation on square matrices over $\mathcal{R}$ generated by similarity and zero extensions,*

(2) *SE-$\mathcal{R}$ is the equivalence relation on square matrices over $\mathcal{R}$ generated by similarity and nilpotent extensions.*

Proof Part (1) is Proposition 4.3.5 from Section 4.3.

For part (2), one direction is easy. If A and B are similar, they are certainly SE-$\mathcal{R}$. To see why A is shift equivalent to $\left(\begin{smallmatrix} A & B \\ 0 & N \end{smallmatrix}\right)$ for any B and N nilpotent, note there exists l such that

$$\begin{pmatrix} A & B \\ 0 & N \end{pmatrix}^l = \begin{pmatrix} A^l & C \\ 0 & 0 \end{pmatrix}$$

for some C. Then $\left(\begin{smallmatrix} A & B \\ 0 & N \end{smallmatrix}\right)$ and A are shift equivalent with lag l using

$$R = \begin{pmatrix} I \\ 0 \end{pmatrix}, \quad S = \begin{pmatrix} A^l & C \end{pmatrix}.$$

For the other direction, suppose A and B are SE-$\mathcal{R}$. Then the classes $[A]_{sse}, [B]_{sse}$ lie in the same fiber of the map π, so by Theorem 4.7.10, there exists a nilpotent matrix N over $\mathcal{R}$ such that $A \oplus N$ is SSE-$\mathcal{R}$ to B. Then $A \oplus N$ and B are connected by a chain of similarities and extensions by zero. Since A and $A \oplus N$ are related by an extension by a nilpotent, the result follows.

The proof that A is shift equivalent over $\mathcal{R}$ to $\left(\begin{smallmatrix} A & 0 \\ B & N \end{smallmatrix}\right)$ for any B and nilpotent N is analogous. $\qquad\square$

4.7.6 Notes

These notes contain some remarks and solutions for exercises for Section 4.7.

Remark 4.7.17 For a ring $\mathscr{R}$, shift equivalence and strong shift equivalence may be defined in the context of finitely generated projective $\mathscr{R}$-modules as follows. If $f\colon P \to P$ and $g\colon Q \to Q$ are endomorphisms of finitely generated projective $\mathscr{R}$-modules, then:

(1) f and g are strong shift equivalent (over **Proj**$\mathscr{R}$) if there exist module homomorphisms $r\colon P \to Q$ and $s\colon Q \to P$ such that $f = sr$ and $g = rs$;
(2) f and g are shift equivalent (over **Proj**$\mathscr{R}$) if there exist module homomorphisms $r\colon P \to Q$ and $s\colon Q \to P$ and an integer $l \geq 1$ such that $f^l = sr$ and $g^l = rs$.

Suppose now that $f\colon P \to P$ is an endomorphism of a finitely generated projective $\mathscr{R}$-module. Then there exists a finitely generated projective $\mathscr{R}$-module Q such that $P \oplus Q$ is free, and f is strong shift equivalent, over **Proj**$\mathscr{R}$, to the map $f \oplus 0\colon P \oplus Q \to P \oplus Q$ using $r\colon P \to P \oplus Q$ given by $r(x) = (x,0)$ and $s\colon P \oplus Q \to P$ given by $s(x,y) = f(x)$. It follows that, when considering strong shift equivalence and shift equivalence over **Proj**$\mathscr{R}$, we may without loss of generality work with free modules.

Remark 4.7.18 For an example of the relationship between endomorphisms of $\mathscr{R}$-modules and certain classes of modules over the polynomial ring $\mathscr{R}[t]$ worked out more formally, see Theorem 2 in [54] and the discussion on page 441 there.

Remark 4.7.19 Given a commutative ring $\mathscr{R}$ and an endomorphism $f\colon P \to P$ of a finitely generated projective $\mathscr{R}$-module where $\mathscr{R}$ is commutative, one may add the polynomial $\det(I - tf)$ as an additional entry to Table 4.1. The entry $\det(I - tf)$ corresponds to, on the endomorphism side, the class of $[f]$ in the endomorphism class group $\mathrm{End}_0(\mathscr{R})$ (see (Remark 4.6.21)).

Remark 4.7.20 The presentation here of the elementary stabilizers differs from the one given in [31]. Roughly speaking, here we use $\mathrm{Nil}_0(\mathscr{R})$ and the endomorphism side, whereas in [31] the notation and definitions are in terms of $NK_1(\mathscr{R})$ and the polynomial matrix side. More precisely, in [31] the elementary stabilizer of a polynomial matrix $I - tA$ is defined to be

$$E(A,\mathscr{R}) = \{U \in \mathrm{GL}(\mathscr{R}[t]) \mid U\,\mathrm{Orb}_{\mathrm{El}(\mathscr{R}[t])}(I - tA) \subset \mathrm{Orb}_{\mathrm{El}(\mathscr{R}[t])}(I - tA)\}$$

where $\mathrm{Orb}_{\mathrm{El}(\mathscr{R}[t])}(I - tA)$ denotes the set of matrices over $\mathscr{R}[t]$ which are elementary equivalent over $\mathscr{R}[t]$ to $I - tA$. There it is observed that $E(A,\mathscr{R})$ is a subgroup of $NK_1(\mathscr{R})$. Given A over $\mathscr{R}$, the map

$$E(A,\mathscr{R}) \to \mathrm{St}_{\mathfrak{N}}(A)$$
$$[I - tN] \mapsto [N]$$

(4.10)

defines a group isomorphism between $E(A,\mathscr{R})$ and $\mathrm{St}_{\mathfrak{N}}(A)$.

Remark 4.7.21 (Solution to Exercise 4.7.13)

This is essentially Exercise 4.6.15, just in the nilpotent endomorphism setting: if $\mathscr{R}$ is commutative and reduced, then $\det(I - tN) = 1$ for any nilpotent matrix N over $\mathscr{R}$.

Remark 4.7.22 (Solution to Exercise 4.7.14)

If A is a nilpotent matrix over $\mathscr{R}$ and $[N] \in \mathrm{St}_{\mathfrak{N}}(A)$, then $[A \oplus N]_{sse} = [A]_{sse}$. Since both A and N are nilpotent, $A \oplus N$ is nilpotent, and by Theorem 4.7.8 this implies $[A \oplus N] = [A]$ in the group $\mathrm{Nil}_0(\mathscr{R})$. Thus $[N] = 0$ in $\mathrm{Nil}_0(\mathscr{R})$.

Remark 4.7.23 The notion of zero and nilpotent extension can also be defined in terms of endomorphisms. Recall from (Remark 4.6.21) the category **End**$\mathscr{R}$ whose objects are pairs (P, f) where $f \colon P \to P$ is an endomorphism of a finitely generated projective $\mathscr{R}$-module and a morphism from (P, f) to (Q, g) is an $\mathscr{R}$-module endomorphism $h \colon P \to Q$ such that $hf = gh$. Given $(P, f), (Q, g)$ in **End**$\mathscr{R}$, we say

(1) (Q, g) is a zero extension of (P, f) if there exists some $\mathscr{R}$-module P_1 such that $Q = P \oplus P_1$ and either of the following happens:
   (i) there exists an $\mathscr{R}$-module homomorphism $h \colon P_1 \to P$ such that $g = \left(\begin{smallmatrix} f & h \\ 0 & 0 \end{smallmatrix} \right)$.
   (ii) there exists an $\mathscr{R}$-module homomorphism $h \colon P \to P_1$ such that $g = \left(\begin{smallmatrix} f & 0 \\ h & 0 \end{smallmatrix} \right)$.

(2) (Q, g) is a nilpotent extension of (P, f) if there exists some $\mathscr{R}$-module P_1 such that $Q = P \oplus P_1$ and either of the following happens:
   (i) there exists an $\mathscr{R}$-module homomorphism $h \colon P_1 \to P$ and a nilpotent endomorphism $j \colon P_1 \to P_1$ such that $g = \left(\begin{smallmatrix} f & h \\ 0 & j \end{smallmatrix} \right)$
   (ii) there exists an $\mathscr{R}$-module homomorphism $h \colon P \to P_1$ and a nilpotent endomorphism $j \colon P_1 \to P_1$ such that $g = \left(\begin{smallmatrix} f & 0 \\ h & j \end{smallmatrix} \right)$.

Recall that in **End**$\mathscr{R}$ we say that a sequence $(P_1, f_1) \to (P_2, f_2) \to (P_3, f_3)$ is exact if the corresponding sequence of $\mathscr{R}$-modules $P_1 \to P_2 \to P_3$ is exact. Zero extensions and nilpotent extensions have a nice interpretation in terms of certain exact sequences in the endomorphism category. Note that in **End**$\mathscr{R}$ the pair $(P, 0)$ means the zero endomorphism of the $\mathscr{R}$-module P, and $(0, 0)$ means the zero endomorphism of the zero $\mathscr{R}$-module. Since $(0, 0)$ serves as a zero object in **End$\mathscr{R}$**, we can consider short exact sequences in **End$\mathscr{R}$**, by which we mean an exact sequence of the form

$$(0,0) \to (P_1, f_1) \to (P_2, f_2) \to (P_3, f_3) \to (0,0).$$

Given this, the following shows that zero extensions and nilpotent extensions are given (up to isomorphism) by certain short exact sequences in **End$\mathscr{R}$**.

Proposition 4.7.24 *Let $\mathscr{R}$ be a ring, and suppose that*

$$(0,0) \to (P_1, f_1) \xrightarrow{\alpha_1} (P_2, f_2) \xrightarrow{\alpha_2} (P_3, f_3) \to (0,0)$$

*is a short exact sequence in **End$\mathscr{R}$**.*

(1) *If $f_1 = 0$ then (P_2, f_2) is isomorphic to a zero extension of (P_3, f_3).*

(2) *If $f_3 = 0$, then (P_2, f_2) is isomorphic to a zero extension of (P_1, f_1).*

(3) *If f_1 is nilpotent, then (P_2, f_2) is isomorphic to a nilpotent extension of (P_3, f_3).*

(4) *If f_3 is nilpotent, then (P_2, f_2) is isomorphic to a nilpotent extension of (P_1, f_1).*

Proof We will prove (4); the other parts are analogous. Since the sequence is exact there is a splitting map $\alpha_1': P_2 \to P_1$ such that $\alpha_1 \alpha_1' = id$ on P_1, and an $\mathscr{R}$-module isomorphism $\beta: P_2 \to P_1 \oplus P_3$ given by $\beta(x) = (\alpha_1'(x), \alpha_2(x))$ so that the following diagram commutes

$$
\begin{array}{ccccccccc}
0 & \longrightarrow & P_1 & \xrightarrow{\alpha_1} & P_2 & \xrightarrow{\alpha_2} & P_3 & \longrightarrow & 0 \\
& & \downarrow{\scriptstyle id} & & \downarrow{\scriptstyle \beta} & & \downarrow{\scriptstyle id} & & \\
0 & \longrightarrow & P_1 & \xrightarrow{i} & P_1 \oplus P_3 & \xrightarrow{q} & P_3 & \longrightarrow & 0
\end{array}
$$

where $i: P_1 \to P_1 \oplus P_3$ by $i(x) = (x, 0)$ and $q: P_1 \oplus P_3 \to P_3$ by $q(x, y) = y$. Define $g = \beta f_2 \beta^{-1}$, so $g: P_1 \oplus P_3 \to P_1 \oplus P_3$. We may write $g = \begin{pmatrix} g_1 & g_2 \\ g_2' & g_3 \end{pmatrix}$ where $(P_1, g_1), (P_3, g_3)$ are in **End$\mathscr{R}$**, and $g_2: P_3 \to P_1, g_2': P_1 \to P_3$. For any $x \in P_1$ we have

$$g(x, 0) = \beta f_2 \beta^{-1}(x, 0) = \beta f_2 \alpha_1(x) = \beta \alpha_1 f_1(x) = (f_1(x), 0)$$

so $g_2'(x) = 0$ and $g_1(x) = f_1(x)$. Since x was arbitrary, it follows that $g_2' = 0$ and $g_1 = f_1$. Likewise, one can check that $g_3 = f_3$. Altogether f_2 is isomorphic to $g = \begin{pmatrix} f_1 & g_2 \\ 0 & f_3 \end{pmatrix}$, and since f_3 is nilpotent, this is a nilpotent extension of (P_1, f_1). $\qquad\square$

4.8 Automorphisms of SFTs

We turn now to discussing automorphisms of shifts of finite type. In general, an automorphism of a dynamical system is simply a self-conjugacy of the given system. The collection of all automorphisms of a given system forms a group, the size of which can vary greatly depending on the system in question. It turns out that a nontrivial mixing shift of finite type possesses a very rich group of automorphisms.

It is maybe unsurprising that, even in the context of the classification problem for shifts of finite type (Problem 4.2.16 in Section 4.2), the study of automorphisms plays an important role. Partly, this role is indirect: various tools and ideas which were originally introduced to study automorphism groups of shifts of finite type (e.g. sign-gyration, introduced later in this section) in fact turned out to be important tools for the conjugacy problem. For example, the dimension representation plays a role in constructing counterexamples to Williams' conjecture in the reducible case (see [75]). Some of this we will discuss in Section 4.9.

The goal of this section is only to give a brief tour through some of main ideas in the study of automorphism groups for shifts of finite type. At the end of the section we mention some newer developments, as well as a small collection of problems and conjectures that have guided some of the direction for studying the automorphism groups.

We will continue to use the following notation. For a matrix A over $\mathbb{Z}_+$, we let (X_A, σ_A) denote the edge shift of finite type (as defined in Section 4.2.3) corresponding to the graph associated to A (i.e. the graph Γ_A as defined in Section 4.2). Since any shift of finite type is topologically conjugate to an edge shift (X_A, σ_A) for some $\mathbb{Z}_+$-matrix A (see e.g. [93, Theorem 2.3.2]), and automorphism groups of topologically conjugate systems are isomorphic, we will only consider edge shifts (X_A, σ_A). The fundamental case is when A is primitive with the topological entropy of the shift satisfying $h_{\text{top}}(\sigma_A) > 0$; with this in mind we make the following standing assumption.

Standing Assumption Throughout this section, unless otherwise noted, when considering an SFT (X_A, σ_A) we assume A is primitive with $\lambda_A > 1$, where λ_A denotes the Perron–Frobenius eigenvalue of A.

Since $h_{\text{top}}(\sigma_A) = \log \lambda_A$, where $h_{\text{top}}(\sigma_A)$ is the topological entropy of the shift σ_A, the assumption on λ_A is equivalent to the system (X_A, σ_A) having positive entropy.

Now let us say more precisely what we mean by an automorphism. We begin with a general definition, and specialize to shifts of finite type later. Recall by a topological dynamical system (X, f) we mean a self-homeomorphism f of a compact metric space X.

Definition 4.8.1 (Automorphism of dynamical system) Let (X, f) be a topological dynamical system. An automorphism of (X, f) is a homeomorphism $\alpha \colon X \to X$ such that $\alpha f = f\alpha$. The collection of automorphisms of (X, f) forms a group under composition, which we call the group of automorphisms of (X, f), and we denote this group by $\mathrm{Aut}(f)$. We define composition in $\mathrm{Aut}(f)$ left to right: given f, g in $\mathrm{Aut}(f)$ and an input x, the output $(fg)(x)$ is $g(f(x))$.[21]

In other words, an automorphism of (X, f) is simply a self-conjugacy of the system (X, f), and the automorphism group is the group of all self-conjugacies of (X, f).

It is straightforward to check that if two systems (X, f) and (Y, g) are topologically conjugate then their automorphism groups $\mathrm{Aut}(f)$ and $\mathrm{Aut}(g)$ are isomorphic.

Example 4.8.2 Let X be a Cantor set and $f \colon X \to X$ be the identity map, that is, $f(x) = x$ for all $x \in X$. Then $\mathrm{Aut}(f) = \mathrm{Homeo}(X)$ is the group of all homeomorphisms of the Cantor set.

Recall from Section 4.2.2 that a subshift is a system (X, σ) which is a subsystem of some full shift $(\mathscr{A}^{\mathbb{Z}}, \sigma)$.

Example 4.8.3 For a subshift (X, σ), the shift σ is itself always an automorphism of (X, σ), that is, $\sigma \in \mathrm{Aut}(\sigma)$. Whenever (X, σ) has an aperiodic point, σ clearly has infinite order in the group $\mathrm{Aut}(\sigma)$.

Example 4.8.4 Let (X_3, σ_3) denote the full shift on the symbol set $\{0, 1, 2\}$ and define an automorphism $\alpha \in \mathrm{Aut}(\sigma_3)$ using the block code

$$\alpha_0 \colon x \mapsto x + 1 \bmod 3, \ x \in \{0, 1, 2\}.$$

Thus, for example, α acts like the following:

$$\ldots 0102010201\overset{\bullet}{1}0202220102110\ldots$$

$$\downarrow \alpha$$

$$\ldots 1210121012\overset{\bullet}{2}1010001210221\ldots$$

This automorphism has order 3, so that $\alpha^3 = \mathrm{id}$.

---

[21] The choice of left-to right composition, at odds with our earlier convention, will imply that the dimension representation, defined later in this section, is a group homomorphism.

As we see later, automorphism groups of shifts of finite type contain a large supply of nontrivial automorphisms. Here is an interesting example of a subshift whose only automorphisms are powers of the shift.

Example 4.8.5 Let θ be an irrational, and consider the rotation map R_θ from $[0,1)$ to itself given by $R_\theta(x) = x + \theta \bmod 1$. Assume that $\theta \in (0,1)$. Consider the *indicator map* $I_\theta : [0,1) \to \{0,1\}$ given by

$$I_\theta(z) = \begin{cases} 0 & \text{if } z \in [0, 1-\theta), \\ 1 & \text{if } z \in [1-\theta, 1). \end{cases}$$

Now we can define a subshift $(X_\theta, \sigma_{X_\theta})$ of the full shift $(\{0,1\}^{\mathbb{Z}}, \sigma)$ on two symbols to be the orbit closure of locations of orbits of points under the map R_θ; that is, we let

$$X_\theta = \overline{\{I_\theta(R_\theta^k(z)) \mid k \in \mathbb{Z}, z \in [0,1)\}}.$$

The subshift $(X_\theta, \sigma_{X_\theta})$ is known as a Sturmian subshift, and it is a folklore result (see [111] or [40] for a proof) that $\mathrm{Aut}(\sigma_{X_\theta}) = \langle \sigma_{X_\theta} \rangle$, so as a group $\mathrm{Aut}(\sigma_{X_\theta})$ is isomorphic to the infinite cyclic group $\mathbb{Z}$.

The subshift in the last example has zero topological entropy, and the structure of its automorphism group is very easy to understand (as a group it is just $\mathbb{Z}$). In many cases, the automorphism groups of subshifts with such "low-complexity" dynamics (of which Example 4.8.5 is an example) have more constrained automorphism groups, in contrast to the automorphism groups of shifts of finite type (see (Remark 4.8.37) for a brief discussion of this, and for what we mean here by low-complexity).

By the Curtis–Hedlund–Lyndon theorem (Theorem 4.2.9), any automorphism of a subshift (X, σ) is induced by a block code. This leads immediately to the following observation:

Proposition 4.8.6 *If (X, σ) is a subshift, then $\mathrm{Aut}(\sigma)$ is a countable group.*

Thus for a shift of finite type (X_A, σ_A), $\mathrm{Aut}(\sigma_A)$ is always a countable group. Under our assumptions that (X_A, σ_A) is mixing with positive entropy, $\mathrm{Aut}(\sigma_A)$ is also always infinite.

It turns out that $\mathrm{Aut}(\sigma_A)$ possesses a rich algebraic structure. Example 4.8.4 above was induced by a block code of range 0, but for arbitrarily large $R \in \mathbb{N}$ there are automorphisms which can only be induced by block codes of range R or greater (see Definition 4.2.8). Indeed, given an SFT (X_A, σ_A) and a non-negative number R, there are only finitely many automorphisms in $\mathrm{Aut}(\sigma_A)$ having range $\leq R$. To give an indication that $\mathrm{Aut}(\sigma_A)$ is quite large, consider the following results regarding different types of subgroups that can occur in $\mathrm{Aut}(\sigma_A)$.

Theorem 4.8.7 *Let (X_A, σ_A) be a shift of finite type where A is a primitive matrix with $\lambda_A > 1$.*

(1) *(Boyle–Lind–Rudolph, in [27]) The group $\mathrm{Aut}(\sigma_A)$ contains isomorphic copies of each of the following groups:*

   (i) *any finite group;*

   (ii) $\bigoplus_{i=1}^{\infty} \mathbb{Z}$;

   (iii) *the free group on two generators, $\mathbb{F}_2$.*

(2) *(Kim–Roush) in [71]) For any $n \geq 2$, let (X_n, σ_n) denote the full shift on n symbols. Then $\mathrm{Aut}(\sigma_n)$ is isomorphic to a subgroup of $\mathrm{Aut}(\sigma_A)$.*

(3) *(Kim–Roush in [71]) Any countable, locally finite, residually finite group embeds into $\mathrm{Aut}(\sigma_A)$.*

In particular, by part (1), $\mathrm{Aut}(\sigma_A)$ is never amenable. By part (2), for full shifts, the isomorphism types of groups that can appear as subgroups of $\mathrm{Aut}(\sigma_n)$ is independent of n.

Recall a group G is residually finite if the intersection of all its subgroups of finite index is trivial. A finitely presented group G is said to have solvable word problem if there is an algorithm to determine whether a word made from generators is the identity in the group.

Exercise 4.8.8 (Remark 4.8.38) If (X, σ) is a subshift whose periodic points are dense in X, then $\mathrm{Aut}(\sigma)$ is residually finite.

Proposition 4.8.9 *Let (X_A, σ_A) be a shift of finite type where A is a primitive matrix with $\lambda_A > 1$. Then the group $\mathrm{Aut}(\sigma_A)$ is residually finite, and contains no finitely generated group with unsolvable word problem.*

Proof Such an SFT has a dense set of periodic points (see [93, Sec. 6.1]), so (1) follows from Exercise 4.8.8. For (2), see [27, Prop. 2.8]. □

Since a subgroup of a residually finite group must be residually finite, both parts of the previous proposition give some necessary conditions for a group to embed as a subgroup of $\mathrm{Aut}(\sigma_A)$. For example, it follows that the additive group of rationals $\mathbb{Q}$ cannot embed into $\mathrm{Aut}(\sigma_A)$, since $\mathbb{Q}$ under addition is not residually finite (however, the additive group of $\mathbb{Q}$ can embed into the automorphism group of a certain minimal subshift: see [27, Example 3.9]). Still, we do not have a good understanding of what types of countable groups can be isomorphic to a subgroup of $\mathrm{Aut}(\sigma_A)$.

An important tool for constructing automorphisms in $\mathrm{Aut}(\sigma_A)$ is the use of "markers". We forgo describing marker methods here, instead referring the reader to [27, Sec. 2]; but we note that, for example, all three parts of Theorem 4.8.7 make use of markers. We will see another perspective on marker automorphisms when discussing simple automorphisms below.

4.8.1 Simple automorphisms

In [102], Nasu introduced a class of automorphisms known as simple automorphisms, which we define shortly. The set of automorphisms built from compositions of these simple automorphisms encompasses the collection of automorphisms defined using marker methods (see [11] for a presentation of this), and gives rise to an important subgroup of $\mathrm{Aut}(\sigma_A)$ (Remark 4.8.39).

Let A be a square matrix over $\mathbb{Z}_+$, and let Γ_A be its associated directed graph. A *simple graph symmetry*[22] of Γ_A is a graph automorphism of Γ_A which fixes all vertices. A simple graph symmetry of Γ_A gives a 0-block code and hence a corresponding automorphism in $\mathrm{Aut}(\sigma_A)$. Given $\alpha \in \mathrm{Aut}(\sigma_A)$, we call α a *simple graph automorphism* if it is induced by a simple graph symmetry of Γ_A, and we call $\alpha \in \mathrm{Aut}(\sigma_A)$ a *simple automorphism* if it is of the form

$$\alpha = \Psi\gamma\Psi^{-1}$$

where $\Psi\colon (X_A, \sigma_A) \to (X_B, \sigma_B)$ is a conjugacy to some shift of finite type (X_B, σ_B) and $\gamma \in \mathrm{Aut}(\sigma_B)$ is a simple graph automorphism in $\mathrm{Aut}(\sigma_B)$.

Example 4.8.10 Let $A = \begin{pmatrix} 2 & 2 \\ 1 & 1 \end{pmatrix}$ and label the edges of Γ_A by $a, \cdots, f$. The graph automorphism of Γ_A defined by interchanging the edges c and d is a simple graph symmetry of Γ_A, and the corresponding simple graph automorphism in $\mathrm{Aut}(\sigma_A)$ is given by the block code of range 0 which swaps the letters c and d and leaves all other letters fixed (see Figure 4.1).

We define $\mathrm{Simp}(\sigma_A)$ to be the subgroup of $\mathrm{Aut}(\sigma_A)$ generated by simple automorphisms. It is immediate to check that $\mathrm{Simp}(\sigma_A)$ is a normal subgroup of $\mathrm{Aut}(\sigma_A)$.

Example 4.8.11 There is a conjugacy from the full 3-shift (X_3, σ_3) on symbols $\{0, 1, 2\}$ to the edge shift of finite type (X_A, σ_A) presented by the graph given in Figure 4.1 on symbol set $\{a, b, c, d, e, f\}$. Here the matrix A is given

---

[22] We use the term graph symmetry instead of graph automorphism to avoid confusion between automorphisms of graphs and automorphisms of subshifts.

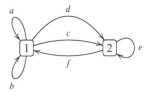

Figure 4.1 An example

by $A = \begin{pmatrix} 2 & 2 \\ 1 & 1 \end{pmatrix}$, and a conjugacy

$$\Psi \colon (X_3, \sigma_3) \to (X_A, \sigma_A)$$

is given by the block code:

$00 \mapsto a$	$10 \mapsto b$	$20 \mapsto f$
$01 \mapsto a$	$11 \mapsto b$	$21 \mapsto f$
$02 \mapsto d$	$12 \mapsto c$	$22 \mapsto e$

with inverse given by

$a \mapsto 0$	$d \mapsto 0$
$b \mapsto 1$	$c \mapsto 1$
$e \mapsto 2$	$f \mapsto 2$

Let γ denote the simple automorphism in $\mathrm{Aut}(\sigma_A)$ induced by the simple graph symmetry of Γ_A shown in Figure 4.1, which swaps the edges c and d, and let $\beta = \Psi\gamma\Psi^{-1}$. Then $\beta \in \mathrm{Simp}(\sigma_3)$, and acts for example like

$$\ldots 1120\overset{\bullet}{0}202120011\ldots$$

$$\Psi \downarrow$$

$$\ldots bcf\overset{\bullet}{a}dfdfcfaab\ldots$$

$$\gamma \downarrow$$

$$\ldots bdfac\overset{\bullet}{f}cfdfaab\ldots$$

$$\Psi^{-1} \downarrow$$

$$\ldots 1020\overset{\bullet}{1}21202001\ldots$$

Notice that β essentially scans a string of $0, 1, 2$'s, and swaps 12 with 02.

Simp(σ_A) is an important subgroup of Aut(σ_A), and we come back to it later.

4.8.2 The center of Aut(σ_A)

Understanding the structure of Aut(σ_A) as a group is not easy. One useful result is the following, proved by Ryan in 1972–1974.

Theorem 4.8.12 ([124, 125]) *If A is irreducible (in particular, if A is primitive) then the center of* Aut(σ_A) *is generated by σ_A.*

Ryan's theorem essentially says the center of Aut(σ_A) is as small as it could possibly be. In fact, for A irreducible, every normal amenable subgroup of Aut(σ_A) is contained in the subgroup generated by σ_A; see (Remark 4.8.40).

In [83], Kopra proved a finitary version of Ryan's theorem: namely, for any nontrivial irreducible shift of finite type, there exists a subgroup generated by two elements whose centralizer is generated by the shift map. In [84], Kopra extended this result to nontrivial transitive sofic shifts, and showed that it fails to hold for nonsofic S-gap shifts (see the cited paper for defninitons). Prior to Kopra's work, Salo in [126] had proved there is a finitely generated subgroup (needing more than two generators) of the automorphism group of the full shift on four symbols whose centralizer is generated by the shift map.

Ryan's theorem can be used to distinguish, up to isomorphism, automorphism groups of certain subshifts of finite type. The idea is to use Ryan's theorem in conjunction with the set of possible roots of the shift. For a subshift (X, σ), define the *root set* of σ to be

$$\text{root}(\sigma) = \{k \in \mathbb{N} \mid \text{there exists } \alpha \in \text{Aut}(\sigma) \text{ such that } \alpha^k = \sigma\}.$$

The following exercise demonstrates this technique.

Exercise 4.8.13 (Remark 4.8.41)

(1) Show that, if (X_A, σ_A) and (X_B, σ_B) are irreducible shifts of finite type such that Aut(σ_A) and Aut(σ_B) are isomorphic, then root(σ_A) = root(σ_B).
(2) Let $(X_2, \sigma_2), (X_4, \sigma_4)$ denote the full shift on 2 symbols and on 4 symbols, respectively. Show that root(σ_2) $\neq$ root(σ_4). Use part (1) to conclude that Aut(σ_2) and Aut(σ_4) are not isomorphic as groups.

The exercise above can be generalized, replacing 2 and 4 by some other values of m and n; one can find this written down in [60] (also see [27, Ex. 4.2] for an example where the method is used to distinguish automorphism groups in the non-full shift case). For a full shift (X_n, σ_n), it turns out that $k \in \mathrm{root}(\sigma_n)$ if and only if n has a kth root in $\mathbb{N}$ (see [91, Theorem 8]).

Currently, the technique of using Ryan's theorem in conjunction with the root set $\mathrm{root}(\sigma_A)$ is the only method known to us which can show two explicit nontrivial mixing shifts of finite type have non-isomorphic automorphism groups. We do not at the moment know how to distinguish automorphism groups with identical root sets; in particular, despite being introduced by Hedlund in the 60's, we still do not know whether $\mathrm{Aut}(\sigma_2)$ and $\mathrm{Aut}(\sigma_3)$ are isomorphic (see Problem 4.8.32 in Section 4.8.8).

4.8.3 Representations of $\mathrm{Aut}(\sigma_A)$

So how can we study $\mathrm{Aut}(\sigma_A)$? One way is to try to find good representations of it. There are two main classes of representations that we know of:

(1) periodic point representations, and representations derived from these;
(2) the dimension representation.

The first, the periodic point representations (and ones derived from them), are quite natural to consider. They also lead to the sign and gyration maps, which are also quite natural (once defined). The second, the dimension representation, is essentially a linear representation, and is based on the dimension group associated to the shift of finite type in question.

We start with the second one, the dimension representation.

4.8.4 Dimension representation

We briefly recall the definition, introduced in Section 4.3.6, of the dimension group associated to a $\mathbb{Z}_+$-matrix. Given an $r \times r$ matrix A over $\mathbb{Z}_+$, the eventual range subspace of A is $\mathrm{ER}(A) = \mathbb{Q}^r A^r$ (matrices will act on row vectors throughout), and the dimension group associated to A is

$$G_A = \{x \in \mathrm{ER}(A) \mid xA^k \in \mathbb{Z}^r \cap \mathrm{ER}(A) \text{ for some } k \geq 0\}.$$

Recall also the group G_A comes equipped with an automorphism (of abelian groups) $\delta_A : G_A \to G_A$ (the automorphism δ_A was denoted by $\hat{A}$ in Section 4.3, but we'll use the notation δ_A). The automorphism δ_A of G_A makes G_A into a $\mathbb{Z}[t,t^{-1}]$-module by having t act as δ_A^{-1}, but we will usually just refer to the pair (G_A, δ_A) to indicate we are considering both G_A and δ_A together. Then by an automorphism of (G_A, δ_A) we mean a group automorphism $\Psi : G_A \to G_A$ which satisfies $\Psi \delta_A = \delta_A \Psi$; in other words, an automorphism of the pair is equivalent to an automorphism of G_A as a $\mathbb{Z}[t,t^{-1}]$-module. Let $\mathrm{Aut}(G_A)$ denote the group of automorphisms of the pair (G_A, δ_A).

The group G_A is isomorphic, as an abelian group, to the direct limit group $\varinjlim\{\mathbb{Z}^r, x \mapsto xA\}$.

When A is over $\mathbb{Z}_+$ (which is the case for a matrix presenting an edge shift of finite type), G_A has a positive cone $G_A^+ = \{v \in G_A \mid vA^k \in \mathbb{Z}_+^r \text{ for some } k\}$ making G_A into an ordered abelian group. The automorphism δ_A maps G_A^+ into G_A^+, and when we want to keep track of the order structure we refer to the triple (G_A, G_A^+, δ_A). An automorphism of the triple (G_A, G_A^+, δ_A) then means an automorphism of (G_A, δ_A) which preserves G_A^+.

Exercise 4.8.14 (Remark 4.8.42) When $A = (n)$ (the case of the full-shift on n symbols), the triple (G_n, G_n^+, δ_n) is isomorphic to the triple $(\mathbb{Z}[\frac{1}{n}], \mathbb{Z}_+[\frac{1}{n}], m_n)$, where m_n is the automorphism of $\mathbb{Z}[\frac{1}{n}]$ defined by $m_n(x) = x \cdot n$.

The following exercise shows that for a mixing shift of finite type (X_A, σ_A), the group of automorphisms of (G_A, G_A^+, δ_A) is index two in $\mathrm{Aut}(G_A, \delta_A)$.

Exercise 4.8.15 (Remark 4.8.43) Let A be a primitive matrix and suppose Ψ is an automorphism of (G_A, δ_A). By considering G_A as a subgroup of $ER(A)$, show that Ψ extends to a linear automorphism $\tilde{\Psi} \colon ER(A) \to ER(A)$ which multiplies the Perron eigenvector of A by some quantity λ_Ψ. Show that Ψ is also an automorphism of the ordered abelian group (G_A, G_A^+, δ_A) if and only if λ_Ψ is positive.

Krieger [85] gave a definition of a triple (D_A, D_A^+, d_A) which is isomorphic to the triple (G_A, G_A^+, δ_A) using only topological/dynamical data intrinsic to the system (X_A, σ_A) (Remark 4.8.45).

A topological conjugacy between shifts of finite type $\Psi \colon (X_A, \sigma_A) \to (X_B, \sigma_B)$ induces an isomorphism $\Psi_* \colon (G_A, G_A^+, \delta_A) \xrightarrow{\cong} (G_B, G_B^+, \delta_B)$. This is easiest to see using Krieger's intrinsic definition of (G_A, G_A^+, δ_A) (see (Remark 4.8.45)). One can also see this in terms of the conjugacy/strong shift equivalence framework developed in Section 4.3, as follows. Given a conjugacy $\alpha \colon (X_A, \sigma_A) \to (X_B, \sigma_B)$, from Section 4.3 we know that corresponding to α is some strong shift equivalence from A to B

$$A = R_1 S_1, A_2 = S_1 R_1, \ldots, A_n = R_n S_n, B = S_n R_n.$$

Then we define an isomorphism $\pi(\alpha)$ from (G_A, G_A^+, δ_A) to (G_B, G_B^+, δ_B) by

$$\pi(\alpha) \colon v \mapsto vR_1 \cdots R_n .$$

A priori, it is not clear that $\pi(\alpha)$ is actually well-defined, since the strong shift equivalence we choose to associate to α may not be unique. However, it turns out that $\pi(\alpha)$ is indeed well-defined; this will be a consequence of material in Section 4.9.

Since an automorphism of (X_A, σ_A) is just a self-conjugacy of (X_A, σ_A), any $\alpha \in \text{Aut}(\sigma_A)$ induces an isomorphism $\alpha_* : (G_A, G_A^+, \delta_A) \xrightarrow{\cong} (G_A, G_A^+, \delta_A)$. Moreover, if for $i = 1, 2$ we have $\alpha_i : v \mapsto vR_i$, then $\alpha_1\alpha_2 : v \mapsto vR_1R_2$ (because composition in $\text{Aut}(\sigma_A)$ is defined left to right), hence $(\alpha_1)_*(\alpha_2)_* = (\alpha_1\alpha_2)_*$. Thus the rule $\alpha \mapsto \alpha_*$ defines a group homomorphism

$$\pi_A : \text{Aut}(\sigma_A) \to \text{Aut}(G_A, G_A^+, \delta_A) \tag{4.11}$$

which is known as the *dimension representation* of $\text{Aut}(\sigma_A)$[23].

Example 4.8.16 The automorphism $\sigma_A \in \text{Aut}(\sigma_A)$ corresponds to the strong shift equivalence

$$A = (A)(I), \ A = (I)(A).$$

In particular, we have for any shift of finite type (X_A, σ_A)

$$\pi_A(\sigma_A) = \delta_A \in \text{Aut}(G_A, G_A^+, \delta_A).$$

Example 4.8.17 When $A = (3)$, $\text{ER}(A) = \mathbb{Q}$, and as mentioned above, the dimension triple is isomorphic to $(\mathbb{Z}[\frac{1}{3}], \mathbb{Z}_+[\frac{1}{3}], m_3)$ where $m_3(x) = 3x$. Thus $\text{Aut}(G_3, G_3^+, \delta_3) \cong \mathbb{Z}$, where $\mathbb{Z}$ is generated by δ_3. The dimension representation then looks like

$$\pi_3 : \text{Aut}(\sigma_3) \to \text{Aut}(\mathbb{Z}[\tfrac{1}{3}], \mathbb{Z}_+[\tfrac{1}{3}], \delta_3) \cong \mathbb{Z} = \langle \delta_3 \rangle$$

$$\pi_3 : \sigma_3 \mapsto \delta_3.$$

More generally, the following proposition describes how the dimension representation behaves for full shifts. Given $n \in \mathbb{N}$, let $\omega(n)$ denote the number of distinct prime divisors of n.

Proposition 4.8.18 *Given $n \geq 2$, there is an isomorphism $\text{Aut}(G_n, G_n^+, \delta_n) \cong \mathbb{Z}^{\omega(n)}$ and the map $\pi_n : \text{Aut}(\sigma_n) \to \text{Aut}(G_n, G_n^+, \delta_n)$ is surjective.*

Proof From Exercise 4.8.14 we know $(G_n, G_n^+, \delta_n) \cong (\mathbb{Z}[\frac{1}{n}], \mathbb{Z}_+[\frac{1}{n}], \delta_n)$. The result follows since the group $\text{Aut}(\mathbb{Z}[\frac{1}{n}], \mathbb{Z}_+[\frac{1}{n}], \delta_n)$ is free abelian with basis given by the maps $\delta_{p_i} : x \mapsto x \cdot p_i$ where p_i is a prime dividing n. For the surjectivity part of π_n, see [27]. $\square$

---

[23] It can happen that $(\alpha_1)_*$ and $(\alpha_2)_*$ do not commute. In this case, the map π_A would be well defined, but would not be a group homomorphism.

In general, the dimension representation may not be surjective (see [78]), and the following question is still open:

Problem 4.8.19 Given a mixing shift of finite type (X_A, σ_A), what is the image of the dimension representation $\pi_A : \mathrm{Aut}(\sigma_A) \to \mathrm{Aut}(G_A, G_A^+, \delta_A)$?

Problem 4.8.19 is of relevance for the classification problem (see (Remark 4.8.47)).

In [27, Theorem 6.8] it is shown that if the nonzero eigenvalues of A are simple, and no ratio of distinct eigenvalues is a root of unity, then for all sufficiently large m the dimension representation $\pi_A^{(m)} : \mathrm{Aut}(\sigma_A^m) \to \mathrm{Aut}(G_{A^m}, G_{A^m}^+, \delta_{A^m})$ is onto. Long [95] showed the "elementary" construction method of [27, Theorem 6.8] is not in general sufficient to reveal the full image of the dimension representation.

An automorphism $\alpha \in \mathrm{Aut}(\sigma_A)$ is called *inert* if α lies in the kernel of π_A, and we denote the subgroup of inert automorphisms by

$$\mathrm{Inert}(\sigma_A) = \ker \pi_A.$$

The subgroup $\mathrm{Inert}(\sigma_A)$ is, roughly speaking, the heart of $\mathrm{Aut}(\sigma_A)$, and in general, we do not know how to distinguish the subgroup of inert automorphisms among different shifts of finite type. The following exercise shows that constructions using marker methods or simple automorphisms always lie in $\mathrm{Inert}(\sigma_A)$.

Exercise 4.8.20 (Remark 4.8.48) For any shift of finite type (X_A, σ_A), we have $\mathrm{Simp}(\sigma_A) \subset \mathrm{Inert}(\sigma_A)$. (Hint: Use (Remark 4.8.45))

Remark 4.8.21 As evidence that $\mathrm{Inert}(\sigma_A)$ contains much of the complicated algebraic structure of $\mathrm{Aut}(\sigma_A)$, consider the case of a full shift over a prime number of symbols, i.e. $A = (p)$ for some prime p. In this case, $\mathrm{Aut}(G_p, G_p^+, \delta_p) \cong \mathbb{Z}$ is generated by δ_p, and the map

$$\pi_p : \mathrm{Aut}(\sigma_p) \to \mathrm{Aut}(G_p, G_p^+, \delta_p)$$

is a split surjection, with a splitting map being given by $\delta_p \mapsto \sigma_p$. This shows $\mathrm{Aut}(\sigma_p)$ is isomorphic to a semi-direct product of $\mathrm{Inert}(\sigma_p)$ and $\mathbb{Z}$. Since σ_p lies in the center of $\mathrm{Aut}(\sigma_p)$, in fact this semi-direct product is isomorphic to a direct product, and we have

$$\mathrm{Aut}(\sigma_p) \cong \mathrm{Inert}(\sigma_p) \times \mathbb{Z}.$$

4.8.5 Periodic point representation

For an SFT (X_A, σ_A) and $k \in \mathbb{N}$ we let P_k denote the σ_A-periodic points of least period k, and Q_k the set of σ_A-orbits of length k (both P_k and Q_k depend on σ_A

of course; we suppress this in the notation since it's usually clear from context). For a shift of finite type, the set P_k is always finite, and we have

$$|P_k| = k|Q_k|.$$

Let $\alpha \in \mathrm{Aut}(\sigma_A)$ and let $k \in \mathbb{N}$. Since α is a bijection which commutes with σ_A, α maps P_k to itself and thus induces a permutation of P_k which we'll denote by $\rho_k(\alpha) \in \mathrm{Sym}(P_k)$, where $\mathrm{Sym}(P)$ denotes the group of permutations of the set P (we use the convention that if $P = \emptyset$ then $\mathrm{Sym}(P)$ is the group containing only one element).

It is straightforward to check that this assignment $\alpha \mapsto \rho_k(\alpha)$ defines a homomorphism

$$\rho_k \colon \mathrm{Aut}(\sigma_A) \to \mathrm{Sym}(P_k).$$

The automorphism α must also respect σ_A-orbits, and it follows that α induces a permutation of the set Q_k which we denote

$$\xi_k(\alpha) \in \mathrm{Sym}(Q_k).$$

Thus, we also get a homomorphism

$$\xi \colon \mathrm{Aut}(\sigma_A) \to \mathrm{Sym}(Q_k).$$

These homomorphisms assemble into homomorphisms

$$\rho \colon \mathrm{Aut}(\sigma_A) \to \prod_{k=1}^{\infty} \mathrm{Sym}(P_k)$$
$$\rho(\alpha) = (\rho_1(\alpha), \rho_2(\alpha), \ldots). \tag{4.12}$$

and

$$\xi \colon \mathrm{Aut}(\sigma_A) \to \prod_{k=1}^{\infty} \mathrm{Sym}(Q_k)$$
$$\xi(\alpha) = (\xi_1(\alpha), \xi_2(\alpha), \ldots). \tag{4.13}$$

The map ρ is called the *periodic point representation* of $\mathrm{Aut}(\sigma_A)$, and ξ is called the *periodic orbit representation*.

When A is irreducible, the map ρ is injective (this follows from the fact that for irreducible A, periodic points are dense in (X_A, σ_A): see [93, Sec. 6.1]). Clearly ξ cannot be injective since $\sigma_A \in \xi$. However, it turns out that σ_A generates the whole kernel of ξ, from a theorem of Boyle and Krieger [26].

Theorem 4.8.22 *If (X_A, σ_A) is an irreducible shift of finite type, then* $\ker \xi = \langle \sigma_A \rangle$.

Fix $k \in \mathbb{N}$ and $\alpha \in \mathrm{Aut}(\sigma_A)$. The periodic point representation $\rho_k(\alpha)$ is obtained by restricting α to the finite subsystem P_k of (X_A, σ_A), and $\rho_k(\alpha)$ lies in the automorphism group $\mathrm{Aut}(\sigma_A|_{P_k})$ of this finite system. It was observed in [26] that the automorphism group $\mathrm{Aut}(\sigma_A|_{P_k})$ is isomorphic to the semidirect product $(\mathbb{Z}/k\mathbb{Z})^{Q_k} \rtimes \mathrm{Sym}(Q_k)$ (Remark 4.8.49), and this leads to considering possible abelian factors of these automorphism groups $\mathrm{Aut}(\sigma_A|_{P_k})$. This motivates the following gyration maps, which were introduced by Boyle and Krieger in [26].

Definition 4.8.23 (Gyration map) Fix $k \in \mathbb{N}$. We define the kth gyration map $g_k \colon \mathrm{Aut}(\sigma_A) \to \mathbb{Z}/k\mathbb{Z}$ as follows. Let $\alpha \in \mathrm{Aut}(\sigma_A)$, let $Q_k = \{O_1, \ldots, O_{I(k)}\}$ denote the set of orbits in Q_k, and choose, for each $1 \leq i \leq I(k)$, some representative point $x_i \in O_i$. Then $\alpha(x_i) \in O_{\xi_k(\alpha)(i)}$, so there exists some $r(\alpha, i) \in \mathbb{Z}/k\mathbb{Z}$ such that $\alpha(x_i) = \sigma_n^{r(\alpha,i)}(x_{\xi_k(\alpha)(i)})$. Now define

$$g_k = \sum_{i=1}^{I(k)} r(\alpha, i) \in \mathbb{Z}/k\mathbb{Z}.$$

Boyle and Krieger showed this map is independent of the choices of x_i's, and is a homomorphism, so we get homomorphisms

$$g_k \colon \mathrm{Aut}(\sigma_n) \to \mathbb{Z}/k\mathbb{Z}.$$

Now we can define the *gyration representation* by

$$g \colon \mathrm{Aut}(\sigma_n) \to \prod_{k=1}^{\infty} \mathbb{Z}/k\mathbb{Z} \tag{4.14}$$
$$g(\alpha) = (g_1(\alpha), g_2(\alpha), \ldots).$$

Given k, consider $\mathrm{sign}\xi_k \colon \mathrm{Aut}(\sigma_A|_{P_k}) \to \mathbb{Z}/2\mathbb{Z}$, the map ξ_k composed with the sign map to $\mathbb{Z}/2\mathbb{Z}$. The gyration map g_k, together with $\mathrm{sign}\xi_k$, determines the abelianization of $\mathrm{Aut}(\sigma_A|_{P_k})$: any other map from $\mathrm{Aut}(\sigma_A|_{P_k})$ to an abelian group factors through the map

$$g_k \times \mathrm{sign}\xi_k \colon \mathrm{Aut}(\sigma_A|_{P_k}) \to \mathbb{Z}/k\mathbb{Z} \times \mathbb{Z}/2\mathbb{Z}$$

(see (Remark 4.8.50)

4.8.6 Inerts and the sign-gyration compatibility condition

A priori, it would seem that the dimension representation and the periodic point representation need not have any relationship. Remarkably, this turns out not to be the case, and there is in fact a connection between them: for inert automorphisms (recall inert automorphisms are precisely the kernel of the dimension representation), there are certain conditions which relate the periodic orbit representation and the periodic point representation of the automorphism. This is formalized in the following way.

Definition 4.8.24 (Sign-gyration compatibility condition) We say that $\alpha \in \text{Aut}(\sigma_A)$ *satisfies SGCC (sign-gyration compatibility condition)* if the following holds: for every positive odd integer m and every non-negative integer i, if $n = m2^i$, then

$$g_n(\alpha) = 0 \quad \text{if} \quad \prod_{j=0}^{i-1} \text{sign}\xi_{m2^j}(\alpha) = 1$$

$$g_n(\alpha) = \frac{n}{2} \quad \text{if} \quad \prod_{j=0}^{i-1} \text{sign}\xi_{m2^j}(\alpha) = -1 .$$

The empty product we take to have the value 1.

Thus for $\alpha \in \text{Aut}(\sigma_A)$ satisfying SGCC, $g(\alpha)$ and $\text{sign}\xi(\alpha)$ determine each other.

An important step is to rephrase the SGCC condition in terms of certain homomorphisms, which we describe now. Consider now the sign homomorphisms as taking values in the group $\mathbb{Z}/2\mathbb{Z}$ (so if τ is an odd permutation, $\text{sign}(\tau) = 1 \in \mathbb{Z}/2\mathbb{Z}$). Define for $n \geq 2$ the SGCC homomorphism

$$SGCC_n \colon \text{Aut}(\sigma_A) \to \mathbb{Z}/n\mathbb{Z}$$
$$SGCC_n(\alpha) = g_n(\alpha) + \left(\frac{n}{2}\right) \sum_{j>0} \text{sign}\xi_{n/2^j}(\alpha)$$

where we define $\text{sign}\xi_{n/2^j}(\alpha) = 0$ if $n/2^j$ is not an integer. The following is immediate to check, but very useful.

Proposition 4.8.25 *Let (X_A, σ_A) be a mixing shift of finite type, and let $\alpha \in \text{Aut}(\sigma_A)$. Then α satisfies SGCC if and only if, for all $n \geq 2$, $SGCC_n(\alpha) = 0$.*

So which automorphisms satisfy SGCC? Amazingly enough, every inert automorphism does. This fact was the culmination of results obtained over several years (see (Remark 4.8.51)), and was finally proved by Kim and Roush in [72], using an important cocycle lemma of Wagoner. A more complete picture was subsequently given by Kim–Roush–Wagoner in [78]; we'll describe this briefly here. The appropriate setting for a deeper understanding is Wagoner's CW complexes, which are the subject of the next section.

Suppose that $A = RS, B = SR$ is a strong shift equivalence over $\mathbb{Z}_+$, and let $\phi_{R,S} \colon (X_A, \sigma_A) \to (X_B, \sigma_B)$ be a conjugacy induced by this SSE. In [78],

Kim, Roush and Wagoner showed that, using certain lexicographical orderings on each set of periodic points, one can compute $SGCC_m$ values, with respect to this choice of ordering on periodic points, analogous to how the $SGCC_m$ homomorphisms are defined for automorphisms. Moreover, they showed that these values can be computed in terms of a (complicated) formula defined only using terms from the matrices R, S. In fact, this formula makes sense even if we start with a strong shift equivalence $A = RS, B = SR$ over $\mathbb{Z}$, and Kim–Roush–Wagoner showed that these formulas can be used to define homomorphisms $sgcc_m \colon \mathrm{Aut}(G_A, \delta_A) \to \mathbb{Z}/m\mathbb{Z}$. Note that the domain of this homomorphism is $\mathrm{Aut}(G_A, \delta_A)$, the set of automorphisms of the pair (G_A, δ_A) which don't necessarily preserve the positive cone G_A^+. Altogether, Kim–Roush–Wagoner proved the following.

Theorem 4.8.26 ([78]) *Let* (X_A, σ_A) *be a mixing shift of finite type. For every* $m \geq 2$ *there exists a homomorphism* $sgcc_m \colon \mathrm{Aut}(G_A, \delta_A) \to \mathbb{Z}/m\mathbb{Z}$ *such that the following diagram commutes:*

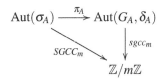

In particular, if $\alpha \in \mathrm{Inert}(\sigma_A)$, *then* $SGCC_m(\alpha) = 0$.

An explicit formula for $sgcc_2$ can be found in [78, Prop. 2.14], with a general formula for $sgcc_m$ described in [78, 2.31].

As shown in [72] and [78], the fact that SGCC vanishes on any inert automorphism can be used to rule out certain actions on finite subsystems of the shift system. For example, the following was shown in [78] (based on a suggestion by Ulf Fiebig). Consider an automorphism α of the period 6 points of the full 2 shift (X_2, σ_2) which acts by the shift on one of the orbits, and the identity on the remaining orbits. It is immediate to compute that $SGCC_6(\alpha) = 1 \in \mathbb{Z}/6\mathbb{Z}$. However $\mathrm{Aut}(G_2, \delta_2) \cong \mathbb{Z}$ is generated by δ_2, the image of the shift σ_2 under the dimension representation π_2, and $sgcc_6(\delta_2) = 3 \in \mathbb{Z}/6\mathbb{Z}$; by Theorem 4.8.26, this implies the image of $SGCC_6$ in $\mathbb{Z}/6\mathbb{Z}$ must be the subgroup $\{0, 3\} \subset \mathbb{Z}/6\mathbb{Z}$, which does not contain 1. Thus α cannot be the restriction of an automorphism in $\mathrm{Aut}(\sigma_2)$. This (along with an additional example given in [78]) resolved a long standing open problem about lifting automorphisms from finite subsystems (see Problem 4.8.29 in Section 4.8.8).

4.8.7 Actions on finite subsystems

The SGCC conditions give necessary conditions for the action of an inert automorphism on finite subsystems of the shift system. A natural question is whether one can determine precisely what possible actions can be realized: that is, what are sufficient conditions for an automorphism of a finite subsystem to be the restriction of an inert automorphism? (Remark 4.8.53) In [22], Boyle and Fiebig characterized the possible actions of finite-order inert automorphisms on finite subsystems of the shift. Then, in [79, 80, 81], Kim–Roush–Wagoner settled this question completely, by showing that the SGCC condition is also sufficient for lifting an automorphism of a finite subsystem to an automorphism of the shift. Together with the Boyle–Fiebig classification in [22], this is used in [80, 81] to resolve (in the negative) a long standing problem regarding finite-order generation of the inert subgroup $\mathrm{Inert}(\sigma_A)$; see Section 4.8.8.

4.8.8 Notable problems regarding $\mathrm{Aut}(\sigma_A)$

There have been a number of questions and conjectures that have been influential in the study of $\mathrm{Aut}(\sigma_A)$, and we describe a few of them here. This is by no means intended to be an exhaustive list; instead, we simply highlight some problems that have been important (both historically, and still), as well as some problems that demonstrate the state of our ignorance regarding the group $\mathrm{Aut}(\sigma_A)$. Some of these have been resolved in some cases, while some are open in all cases.

Given a group G, let $\mathrm{Fin}(G)$ denote the (normal) subgroup of G generated by elements of finite order.

Recall that, for any shift of finite type (X_A, σ_A), we have containments of subgroups $\mathrm{Simp}(\sigma_A) \subset \mathrm{Fin}(\mathrm{Inert}(\sigma_A)) \subset \mathrm{Inert}(\sigma_A)$. One general problem[24] is the following:

Problem 4.8.27 (Finite Order Generation (FOG) Problem)   When is it true that $\mathrm{Inert}(\sigma_A) = \mathrm{Fin}(\mathrm{Inert}(\sigma_A))$?

The FOG problem is an outgrowth of a conjecture, originally posed by F. Rhodes to Hedlund in a correspondence, asking whether $\mathrm{Aut}(\sigma_2)$ is generated by σ_2 and elements of finite order.

Kim, Roush and Wagoner in [80, 81] showed there exists a shift of finite type (X_B, σ_B) such that the containment $\mathrm{Fin}(\mathrm{Inert}(\sigma_B)) \subset \mathrm{Inert}(\sigma_B)$ is proper, showing the answer to FOG is 'not always' (see the discussion in Section 4.8.7).

[24] What we call the Finite Order Generation Problem here was historically posed as a conjecture. Here we opted instead for the word 'problem', since this conjecture is known to be false in general.

Prior to this, in [143] Wagoner considered a stronger form of FOG, asking whether it was always true that $\text{Simp}(\sigma_A) = \text{Inert}(\sigma_A)$; this was sometimes referred to as the Simple Finite Order Generation Conjecture (SFOG). Kim and Roush in [73] showed (prior to their example showing FOG does not always hold) that SFOG does not always hold, giving an example of a shift of finite type (X_A, σ_A) such that the containment $\text{Simp}(\sigma_A) \subset \text{Inert}(\sigma_A)$ is proper.

Expanding on FOG, we have the following more general problem:

Problem 4.8.28 (Index Problem)   Given a shift of finite type (X_A, σ_A), determine the index of the following subgroup containments:

(1) $\text{Simp}(\sigma_A) \subset \text{Inert}(\sigma_A)$.
(2) $\text{Fin}(\text{Inert}(\sigma_A)) \subset \text{Inert}(\sigma_A)$.

In particular, in each case, must the index be finite?

When $\text{Aut}(G_A)$ is torsion-free, every element of finite order in $\text{Aut}(\sigma_A)$ lies in $\text{Inert}(\sigma_A)$. In this case, the FOG problem is equivalent to determining whether the answer to Part (2) of the Index Problem is one.

In general, it is not known whether, for each part of the Index Problem, the index is finite or infinite. As noted earlier, in [80] an example is given of a mixing shift of finite type (X_A, σ_A) for which the index of $\text{Fin}(\text{Inert}(\sigma_A))$ in $\text{Inert}(\sigma_A)$ is strictly greater than one. This relies on being able to construct an inert automorphism in $\text{Aut}(\sigma_A)$ which cannot be a product of finite order automorphisms; this is carried out using the difficult constructions of Kim–Roush–Wagoner in [80, 81], in which the polynomial matrix methods (introduced in Section 4.4) play an invaluable role (we do not know how to do such constructions without the polynomial matrix framework).

However, whether FOG or even SFOG might hold in the case of a full shift (X_n, σ_n) is still unknown.

Finite order generation of the inert automorphisms for general mixing shifts of finite type is known to hold in the "eventual" setting; see (Remark 4.8.54).

Williams in [153] asked whether any involution of a pair of fixed points of a shift of finite type can be extended to an automorphism of the whole shift of finite type. More generally, this grew into the following problem (stated in [27, Question 7.1]) about lifting actions on a finite collection of periodic points of the shift:

Problem 4.8.29 (General lifting problem (LIFT))   Given a shift of finite type (X_A, σ_A) and an automorphism ϕ of a finite subsystem F of (X_A, σ_A), does there exist $\tilde{\phi} \in \text{Aut}(\sigma_A)$ such that $\tilde{\phi}|_F = \phi$?

The answer to LIFT is also 'not always': Kim and Roush showed in [72], based on an example of Fiebig, that there exists an automorphism of the set of periodic six points in the full 2-shift which does extend to an automorphism of the full 2-shift.

Roughly speaking, the LIFT problem involves two parts: determining the action of inert automorphisms on finite subsystems, and determining the range of the dimension representation. The first part has been resolved by Kim–Roush–Wagoner in [80, 81]; see Section 4.8.7. The second part, to determine the range of the dimension representation, is still open in general (this was also stated in Problem 4.8.19 in Section 4.8.4):

Problem 4.8.30   Given a mixing shift of finite type (X_A, σ_A), what is the image of the dimension representation $\pi_A \colon \mathrm{Aut}(\sigma_A) \to \mathrm{Aut}(G_A, G_A^+, \delta_A)$? Is the image always finitely generated?

In [78], Kim and Roush constructed a mixing shift of finite type for which the dimension representation is not surjective.

In [27, Example 6.9], an example is given of a primitive matrix A such that $\mathrm{Aut}(G_A, G_A^+, \delta_A)$ is not finitely generated. This does not resolve the second part of the Problem though, since the range of the dimension representation is not known.

Another question concerns the isomorphism type of the groups $\mathrm{Aut}(\sigma_A)$. It is straightforward to check that conjugate shifts of finite type have isomorphic automorphism groups, and that $\mathrm{Aut}(\sigma_A) = \mathrm{Aut}(\sigma_A^{-1})$ always holds (note that there exists shifts of finite type (X_A, σ_A) which are not conjugate to their inverse; see for example Proposition 4.3.50). In [27, Question 4.1] the following was asked:

Problem 4.8.31 (Aut-Isomorphism Problem)   If $\mathrm{Aut}(\sigma_A)$ and $\mathrm{Aut}(\sigma_B)$ are isomorphic, must (X_A, σ_A) be conjugate to either (X_B, σ_B) or (X_B, σ_B^{-1})?

A particular case of this which has been of interest is:

Problem 4.8.32 (Full Shift Aut-Isomorphism Problem)   For which m, n are the groups $\mathrm{Aut}(\sigma_m)$ and $\mathrm{Aut}(\sigma_n)$ isomorphic?

See Section 4.8.9 for some results related to Problem 4.8.32.

4.8.9 The stabilized automorphism group

Recently a new approach to the Aut-Isomorphism Problem, and the study of $\mathrm{Aut}(\sigma_A)$ in general, has been undertaken in [60]. The idea is to consider a certain stabilization of the automorphism group, using the observation that for all $k, m \geq 1$, $\mathrm{Aut}(\sigma_A^k)$ is naturally a subgroup of $\mathrm{Aut}(\sigma_A^{km})$. Define the *stabilized automorphism group* of (X_A, σ_A) to be

$$\mathrm{Aut}^{(\infty)}(\sigma_A) = \bigcup_{k=1}^{\infty} \mathrm{Aut}(\sigma_A^k)$$

where the union is taken in the group of all homeomorphisms of X_A. This is again a countable group. Similar to the definition of $\mathrm{Aut}^{(\infty)}(\sigma_A)$, one defines a stabilized group of automorphisms of the dimension group by

$$\mathrm{Aut}^{(\infty)}(G_A) = \bigcup_{k=1}^{\infty} \mathrm{Aut}(G_A, G_A^+, \delta_A^k).$$

The group $\mathrm{Aut}^{(\infty)}(G_A)$ is precisely the union of the centralizers of δ_A in the group $\mathrm{Aut}(G_A, G_A^+)$ of all order-preserving group automorphisms of G_A. Recall that, for a group G, we let G_{ab} denote the abelianization of G. In [60], the following was proved.

Theorem 4.8.33 *Let (X_A, σ_A) be a mixing shift of finite type. Then the dimension representation $\pi_A \colon \mathrm{Aut}(\sigma_A) \to \mathrm{Aut}(G_A)$ extends to a stabilized dimension representation*

$$\pi_A^{(\infty)} \colon \mathrm{Aut}^{(\infty)}(\sigma_A) \to \mathrm{Aut}^{(\infty)}(G_A)$$

and the composition

$$\mathrm{Aut}^{(\infty)}(\sigma_A) \xrightarrow{\pi_A^{(\infty)}} \mathrm{Aut}^{(\infty)}(G_A) \xrightarrow{\mathrm{ab}} \mathrm{Aut}^{(\infty)}(G_A)_{\mathrm{ab}}$$

has image isomorphic to the abelianization of the stabilized automorphism group $\mathrm{Aut}^{(\infty)}(\sigma_A)$. In particular, if $\mathrm{Aut}^{(\infty)}(G_A)$ is abelian, then the commutator subgroup of $\mathrm{Aut}^{(\infty)}(\sigma_A)$ coincides with the subgroup of stabilized inert automorphisms

$$\mathrm{Inert}^{(\infty)}(\sigma_A) = \ker \pi_A^{(\infty)} = \bigcup_{k=1}^{\infty} \mathrm{Inert}(\sigma_A^k).$$

For example, in the case of a full shift $A = (n)$, it follows from Theorem 4.8.33 that $\mathrm{Aut}^{(\infty)}(\sigma_n)_{\mathrm{ab}}$ is isomorphic to $\mathbb{Z}^{\omega(n)}$, where $\omega(n)$ denotes the number of distinct prime divisors of n. As a corollary of this, if $\omega(m) \neq \omega(n)$, then $\mathrm{Aut}^{(\infty)}(\sigma_m)$ and $\mathrm{Aut}^{(\infty)}(\sigma_n)$ are not isomorphic.

For a mixing shift of finite type, the classical automorphism group $\mathrm{Aut}(\sigma_A)$ is always residually finite. It turns out that in the stabilized case, $\mathrm{Aut}^{(\infty)}(\sigma_A)$ is never residually finite [60, Prop. 4.3]. In fact, in stark contrast, the following was proved in [60]:

Theorem 4.8.34 ([60])   *For any $n \geq 2$, the group of stabilized inert automorphisms* $\mathrm{Inert}^{(\infty)}(\sigma_n)$ *is simple.*

A significantly more general version of the above theorem was proved by Salo in [128]. A particular case, Corollary 1 of [128], shows that for any (nontrivial) mixing shift of finite type, the group of stabilized inert automorphisms is simple.

Subsequent to [60], a complete classification, up to isomorphism, of the stabilized automorphism groups of full shifts was given in [132]. Introduced there is a certain kind of entropy for groups[25] called local $\mathscr{P}$ entropy. Local $\mathscr{P}$ entropy is defined with respect to a chosen class $\mathscr{P}$ of finite groups which is closed under isomorphism. As a rough idea of what local $\mathscr{P}$ entropy measures, fix such a class $\mathscr{P}$, consider some group G with some distinguished element $g \in G$, and consider the conjugation map $C_g \colon G \to G$ given by $C_g(h) = g^{-1}hg$. One can try to measure the growth rate of the C_g-periodic point sets $\mathrm{Fix}(C_{g^n})$, which are precisely the centralizers of g^n in G; but these sets may be infinite. To proceed, instead one approximates these centralizer sets using groups belonging to the chosen class $\mathscr{P}$ (which are by definition finite), and then considers the doubly exponential[26] growth rate of such $\mathscr{P}$-approximations. This (when defined) leads to a nonnegative quantity $h_\mathscr{P}(G,g)$ called the local $\mathscr{P}$ entropy of the pair (G,g). A key thing proved in [132] is that the local $\mathscr{P}$ entropy of a pair (G,g) is an invariant of isomorphism of the pair: if there is an isomorphism of groups $G \overset{\cong}{\to} H$ taking $g \in G$ to $h \in H$, then assuming the local $\mathscr{P}$ entropies are defined, we have $h_\mathscr{P}(G,g) = h_\mathscr{P}(H,h)$.[27]

Using local $\mathscr{P}$ entropy, in [132] the following was proved.

Theorem 4.8.35 ([132])   *For a nontrivial mixing shift of finite type* (X_A, σ_A), *each of the following holds:*

(1) *There exists a class* $\mathscr{P}_A$ *of finite groups such that the local* $\mathscr{P}_A$ *entropy of the pair* $(\mathrm{Aut}^{(\infty)}(\sigma_A), \sigma_A)$ *is given by* $h_{\mathscr{P}_A}\left(\mathrm{Aut}^{(\infty)}(\sigma_A), \sigma_A\right) = h_{\mathrm{top}}(\sigma_A) = \log \lambda_A$.

(2) *If* (X_B, σ_B) *is any other shift of finite type such that the stabilized automorphism groups* $\mathrm{Aut}^{(\infty)}(\sigma_A)$ *and* $\mathrm{Aut}^{(\infty)}(\sigma_B)$ *are isomorphic, then* $\frac{\log \lambda_A}{\log \lambda_B}$ *is rational.*

---

[25] More precisely, it is defined for *leveled groups*, i.e. pairs (G,g) where g is a distinguished element in the group G.

[26] A related quantity is defined by considering just exponential growth; here we'll consider only the one using doubly exponential.

[27] It is also proved in the same paper that if there is an injective homomorphism $G \to H$ taking g to h, then $h_\mathscr{P}(G,g) \leq h_\mathscr{P}(H,h)$.

As a consequence this gives, as mentioned earlier, a complete classification of the stabilized automorphism groups of full shifts.

Corollary 4.8.36 ([132])   *Given natural numbers $m, n \geq 2$, the stabilized automorphism groups $\mathrm{Aut}^{(\infty)}(\sigma_m)$ and $\mathrm{Aut}^{(\infty)}(\sigma_n)$ are isomorphic if and only if there exists natural numbers k, j such that $m^k = n^j$.*

Finally, we make a few comments about the connection between the stabilized setting for automorphism groups described above and algebraic K-theory. In fact, the idea of the groups $\mathrm{Aut}^{(\infty)}(\sigma_A)$ is partly motivated by algebraic K-theory, where the technique of stabilization proves to be fundamental. Recall, as outlined in Section 4.6, as a starting point for algebraic K-theory, given a ring $\mathscr{R}$, one can consider the stabilized general linear group

$$\mathrm{GL}(\mathscr{R}) = \varinjlim \mathrm{GL}(n, \mathscr{R})$$

where $\mathrm{GL}(n, \mathscr{R}) \hookrightarrow \mathrm{GL}(n+1, \mathscr{R})$ via $A \mapsto \begin{pmatrix} A & 0 \\ 0 & 1 \end{pmatrix}$. Inside each $\mathrm{GL}(n, \mathscr{R})$ lies the subgroup $\mathrm{El}(n, \mathscr{R})$ generated by elementary matrices, and one likewise defines the stabilized group of elementary matrices by

$$\mathrm{El}(\mathscr{R}) = \varinjlim \mathrm{El}(n, \mathscr{R}).$$

Whitehead showed (see Section 4.6) that, upon stabilizing, the explicitly defined subgroup $\mathrm{El}(\mathscr{R})$ coincides with the commutator subgroup of $\mathrm{GL}(\mathscr{R})$. From this viewpoint, one may interpret Theorem 4.8.33 as a Whitehead-type result for shifts of finite type. In particular, in the case of a full shift (X_n, σ_n) (or more generally a shift of finite type (X_A, σ_A) where $\mathrm{Aut}^{(\infty)}(G_A)$ is abelian), after stabilizing, the commutator subgroup of $\mathrm{Aut}^{(\infty)}(\sigma_A)$ coincides with the subgroup $\mathrm{Inert}^{(\infty)}(\sigma_A)$.[28]

4.8.10 Mapping class groups of subshifts

Recall from Section 4.4.1 that two homeomorphisms are flow equivalent if there is a homeomorphism of their mapping tori which takes orbits to orbits and preserves the direction of the suspension flow. For a subshift (X, σ), an analog of the automorphism group in the setting of flow equivalence is given by the mapping class group $\mathscr{M}(\sigma)$, which is defined to be the group of isotopy classes of self-flow equivalences of the subshift (X, σ).

In [21] a study of the mapping class group for shifts of finite type was undertaken. There it was shown that, for a nontrivial irreducible shift of finite type

---

[28] In fact, something stronger is true: the commutator subgroup of $\mathrm{Aut}^{(\infty)}(\sigma_A)$ coincides with the stabilized group of simple automorphisms; see [60].

(X_A, σ_A), the mapping class group $\mathscr{M}(\sigma_A)$ is not residually finite. While the periodic point representations do not exist for $\mathscr{M}(\sigma_A)$, a vestige of the dimension representation survives in the form of the Bowen–Franks representation of $\mathscr{M}(\sigma_A)$. It was also shown that $\mathrm{Aut}(\sigma_A)/\langle\sigma_A\rangle$ embeds into $\mathscr{M}(\sigma_A)$, and there is an analog of block codes, known as flow codes. In [127], it was shown that Thompson's group V embeds into the mapping class group of a particular shift of finite type.

See also [131] for a study of the mapping class group in the context of minimal subshifts.

4.8.11 Notes

These notes contain some proofs, remarks, and solutions of various exercises through Section 4.8.

Remark 4.8.37 Recall from Section 4.2.4 that for a subshift (X, σ), we let $\mathscr{W}_n(X)$ denote the set of X-words of length n. We define the complexity function (of X) $P_X \colon \mathbb{N} \to \mathbb{N}$ by $P_X(n) = |\mathscr{W}_n(X)|$. Thus $P_X(n)$ simply counts the number of X-words of length n. For a shift of finite type (Y, σ) with positive entropy, the function $P_Y(n)$ grows exponentially in n; for example, for the full shift (X_m, σ_m) on m symbols, $P_{X_m}(n) = m^n$. For a subshift $(X_\alpha, \sigma_\alpha)$ of the form given in Example 4.8.5, the complexity satisfies $P_{X_\alpha}(n) = n + 1$ (such subshifts are called Sturmian subshifts). This is the slowest possible growth of complexity function for an infinite subshift: a theorem of Morse and Hedlund [101] from 1938 shows that for an infinite subshift (X, σ), we must have $P_X(n) \geq n + 1$.

There has been a great deal of interest in studying the automorphism groups of subshifts with slow-growing complexity functions. Numerous results show that such low complexity subshifts often have much more tame automorphism groups, in comparison to subshifts possessing complexity functions of exponential growth (e.g. shifts of finite type). We will not attempt to survey these results, but refer the reader to [40, 64, 35, 111, 129, 39, 37, 38].

Remark 4.8.38 (Solution to Exercise 4.8.8)

Given $n \in \mathbb{N}$, let $P_n(X)$ denote the set of points of least period n in X. Since X is a subshift, $|P_n(X)| < \infty$ for every n. If $\alpha \in \mathrm{Aut}(\sigma)$, then since α commutes with σ, for any n the set $P_n(X)$ is invariant under α. It follows there are homomorphisms

$$\rho_n \colon \mathrm{Aut}(\sigma) \to \mathrm{Sym}(P_n(X))$$
$$\rho_n \colon \alpha \mapsto \alpha|_{P_n(X)}$$

where $\text{Sym}(P_n(X))$ denotes the group of permutations of the set $P_n(X)$. Now suppose $\alpha \in \text{Aut}(\sigma)$ and $\rho_n(\alpha) = \text{id}$ for all n. Then α fixes every periodic point in X; since the periodic points are dense in X (by assumption) and α is a homeomorphism, α must be the identity. This shows that $\text{Aut}(\sigma)$ is residually finite.

Remark 4.8.39 Beyond introducing simple automorphisms, in his memoir [103] Nasu introduced the powerful machinery of "textile systems" for studying automorphisms and endomorphisms of shifts of finite type; he continued to apply and develop this theory in subsequent works (e.g. [104, 105, 106]). See [16, Appendices B,C] for a quick introduction to this theory.

Remark 4.8.40 Any discrete group G possesses a maximal normal amenable subgroup $\text{Rad}(G)$ known as the *amenable radical* of G. By Ryan's theorem, the center of $\text{Aut}(\sigma_A)$ is the subgroup generated by σ_A, and hence is contained in $\text{Rad}(\text{Aut}(\sigma_A))$. In [49] it was shown by Frisch, Schlank and Tamuz that, in the case of a full shift, $\text{Rad}(\text{Aut}(\sigma_n))$ is precisely the center of $\text{Aut}(\sigma_n)$, i.e. the subgroup generated by σ_n. In [155] Yang extended this result, proving that for any irreducible shift of finite type (X_A, σ_A), $\text{Rad}(\text{Aut}(\sigma_A))$ also coincides with the center of $\text{Aut}(\sigma_A)$ (in fact, Yang also proves the result for any irreducible sofic shift as well).

Remark 4.8.41 (Solution to Exercise 4.8.13)

For part (1), suppose that $\Psi\colon \text{Aut}(\sigma_A) \to \text{Aut}(\sigma_B)$ is an isomorphism and $k \in \text{root}(\sigma_A)$. By Ryan's theorem, $\Psi(\sigma_A) = \sigma_B$ or $\Psi(\sigma_A) = \sigma_B^{-1}$. Choose $\alpha \in \text{Aut}(\sigma_A)$ such that $\alpha^k = \sigma_A$. If $\Psi(\sigma_A) = \sigma_B$, then we have $(\Psi(\alpha))^k = \Psi(\alpha^k) = \Psi(\sigma_A) = \sigma_B$, so $k \in \text{root}(\sigma_B)$. If $\Psi(\sigma_A) = \sigma_B^{-1}$, then we have $(\Psi(\alpha^{-1}))^k = \Psi(\alpha^{-k}) = \Psi(\sigma_A^{-1}) = \sigma_B$ so again $k \in \text{root}(\sigma_B)$. Thus $\text{root}(\sigma_A) \subset \text{root}(\sigma_B)$. The proof that $\text{root}(\sigma_B) \subset \text{root}(\sigma_A)$ is analogous.

For part (2), choose a topological conjugacy $F\colon (X_4, \sigma_4) \to (X_2, \sigma_2^2)$. If we let $s = F^{-1}\sigma_2 F \in \text{Aut}(\sigma_4)$, then $s \in \text{Aut}(\sigma_4)$ and $s^2 = \sigma_4$, so $2 \in \text{root}(\sigma_4)$. We claim that $2 \notin \text{root}(\sigma_2)$. To see this, suppose toward a contradiction that $\beta \in \text{Aut}(\sigma_2)$ satisfies $\beta^2 = \sigma_2$. There are precisely two points x, y of least period 2 in (X_2, σ_2), so β^2 must act as the identity on the points x, y. But $\sigma_2(x) = y$, a contradiction.

Remark 4.8.42 (Solution to Exercise 4.8.14)

The eventual range of A is $\mathbb{Q}$. Given $\frac{p}{q} \in \mathbb{Q}$, $2^k \frac{p}{q} \in \mathbb{Z}_+$ if and only if $p \in \mathbb{Z}_+$ and q is a power of 2.

Remark 4.8.43 (Solution to Exercise 4.8.15)

That Ψ extends to a linear automorphism $\tilde{\Psi}$ of $ER(A)$ is immediate: given

$v \in ER(A)$, write $v = \frac{1}{q} w$ where w is integral, and define $\tilde{\Psi}(v) = \frac{1}{q} \Psi(w)$. The linear map Ψ commutes with δ_A on G_A, so $\tilde{\Psi}$ commutes with δ_A as a linear automorphism of $ER(A)$. Since A is primitive, a Perron eigenvector v_{λ_A} for λ_A spans a one-dimensional eigenspace for δ_A, which hence must be preserved by $\tilde{\Psi}$. Thus v_{λ_A} is also an eigenvector for $\tilde{\Psi}$, and has some corresponding eigenvalue λ_Ψ.

For the second part, we'll use the following proposition (a proof of which we include at the end).

Proposition 4.8.44 *Suppose that A is an $N \times N$ primitive matrix over $\mathbb{R}$. Let the spectral radius be λ and let v be a positive eigenvector, $vA = \lambda v$. Given x in $\mathbb{R}^N$, let c_x be the real number such that $x = c_x v + u_x$, with u_x a vector in the A-invariant subspace complementary to $\langle v \rangle$. Suppose that x is not the zero vector. Then xA^n is nonnegative for large n if and only if $c_x > 0$.*

To finish the exercise, suppose that $0 \neq w \in G_A^+$, and write $w = c_w v_{\lambda_A} + u_w$ as in the proposition. Since $w \in G_A^+$, $c_w > 0$. Then $\tilde{\Psi}(w) = c_w \lambda_\Psi v_{\lambda_A} + \tilde{\Psi}(u_w)$. Since $\lambda_\Psi > 0$, $c_w \lambda_\Psi > 0$, so the proposition implies $\Psi(w) \in G_A^+$ as desired.

Proof of Proposition 4.8.44. The Perron theorem tells us the positive eigenvector and complementary invariant subspace exist, with $\overline{\lim}_n \|u_x A^n\|^{1/n} < \lambda$. Consequently, for large n, xA^n is a positive vector if $c_x > 0$ and xA^n is a negative vector if $c_x < 0$. Given $c_x = 0$ and $x \neq 0$, no vector $w = u_x A^n$ can be nonnegative or nonpositive, because this would imply $\lim_n \|xA^n\|^{1/n} = \lim_n \|wA^n\|^{1/n} = \lambda$, a contradiction.

Remark 4.8.45 Consider an edge shift of finite type (X_A, σ_A). Here is an outline of Krieger's construction of an ordered abelian group which is isomorphic to (G_A, G_A^+, δ_A); our presentation follows the one given in [93, Sec. 7.5].

Recall we are assuming that A is a $k \times k$ irreducible matrix. By an *m-ray* we mean a subset of X_A given by

$$R(x, m) = \{ y \in X_A \mid y_{(-\infty, m]} = x_{(-\infty, m]} \}$$

for some $x \in X_A, m \in \mathbb{Z}$. An *$m$-beam* is a (possibly empty) finite union of m-rays. By a *ray* we mean an m-ray for some $m \in \mathbb{Z}$; likewise, by a *beam* we mean an m-beam for some m. It is easy to check that if U is an m-beam for some m, and $n \geq m$, then U is also an n-beam. Given an m-beam

$$U = \bigcup_{i=1}^{j} R(x^{(i)}, m),$$

define $v_{U,m} \in \mathbb{Z}^k$ to be the vector whose Jth component is given by

$$\#\{x^{(i)} \in U \mid \text{the edge corresponding to } x_m^{(i)} \text{ ends at state } J\}.$$

We define two beams U and V to be equivalent if there exists m such that $v_{U,m} = v_{V,m}$, and let $[U]$ denote the equivalence class of a beam U. We will make the collection of equivalence classes of beams into a semi-group as follows. Since A is an irreducible matrix and $0 < h_{\mathrm{top}}(\sigma_A) = \log \lambda_A$, given two beams U, V, we may find beams U', V' such that

$$[U] = [U'], \qquad [V] = [V'], \qquad U' \cap V' = \emptyset,$$

and we let D_A^+ denote the abelian monoid defined by the operation

$$[U] + [V] = [U' \cup V']$$

where the class of the empty set serves as the identity for D_A^+. Now let D_A denote the group completion of D_A^+; thus elements of D_A are formal differences $[U] - [V]$. Then D_A is an ordered abelian group with positive cone D_A^+. The map $d_A : D_A \to D_A$ induced by

$$d_A([U]) = [\sigma_A(U)]$$

is a group automorphism of D_A which preserves D_A^+, and the triple (D_A, D_A^+, d_A) is Krieger's dimension triple for the SFT (X_A, σ_A).

The connection between Krieger's triple (D_A, D_A^+, d_A) and the ordered abelian group triple (G_A, G_A^+, δ_A) is given by the following proposition.

Proposition 4.8.46 ([93], Theorem 7.5.3) *There is a semi-group homomorphism* $\theta : D_A^+ \to G_A^+$ *induced by the map*

$$\theta([U]) = \delta_A^{-k-n}(v_{U,n} A^k), \qquad U \text{ an } n\text{-beam}.$$

The map θ *satisfies* $\theta(D_A^+) = G_A^+$, *and induces an isomorphism* $\theta : D_A \to G_A$ *such that* $\theta \circ d_A = \delta_A \circ \theta$. *Thus* θ *induces an isomorphism of triples*

$$\theta : (D_A, D_A^+, d_A) \to (G_A, G_A^+, \delta_A).$$

Remark 4.8.47 In [74], Kim and Roush describe how the problem of classifying general (i.e. not necessarily irreducible) shifts of finite type up to topological conjugacy can be broken into two parts: classifying mixing shifts of finite type up to conjugacy, and determining the range of the dimension representation in the mixing shift of finite type case. That the dimension representation need not always be surjective was also instrumental in the Kim–Roush argument in [75] that shift equivalence over $\mathbb{Z}_+$ need not imply strong shift equivalence over $\mathbb{Z}_+$ in the reducible setting.

Remark 4.8.48 (Solution to Exercise 4.8.20)
This is most easily seen using Krieger's presentation (Remark 4.8.45). First suppose that $\alpha \in \mathrm{Simp}(\sigma_A)$ is induced by a simple graph symmetry of Γ_A. If

U is an m-beam in X_A, then $\alpha(U)$ is an m-beam, and $v_{\alpha(U),m} = v_{U,m}$. It follows that $[U] = [\alpha(U)]$, so α acts by the identity on the group D_A, and hence on G_A.

Now suppose $\beta = \Psi^{-1}\alpha\Psi$ where $\Psi\colon (X_A, \sigma_A) \to (X_B, \sigma_B)$ is a topological conjugacy and $\alpha \in \mathrm{Simp}(\sigma_B)$ is induced by a simple graph symmetry of Γ_B. If U is an m-beam in X_A, then by the previous part $\alpha\Psi([U]) = \alpha([\Psi(U)]) = [\Psi(U)] = \Psi([U])$, so

$$\beta([U]) = \Psi^{-1}\alpha\Psi([U]) = \Psi^{-1}\Psi([U]) = [U].$$

Thus β acts by the identity on G_A. Since $\mathrm{Simp}(\sigma_A)$ is generated by automorphisms in the form of β, this finishes the proof.

Remark 4.8.49   Let us write $\mathrm{Aut}(P_k, \sigma_A)$ for $\mathrm{Aut}(\sigma_A|_{P_k})$. For each orbit $q \in Q_k$ choose a point $x_q \in q$. There is a surjective homomorphism

$$\mathrm{Aut}(P_k, \sigma_A) \xrightarrow{\pi} \mathrm{Sym}(Q_k)$$

since any $\alpha \in \mathrm{Aut}(P_k, \sigma_A)$ must preserve σ_A-orbits, and the map π is split by the map $\rho\colon \mathrm{Sym}(Q_k) \to \mathrm{Aut}(P_k, \sigma_A)$ defined by, for $\tau \in \mathrm{Sym}(Q_k)$, setting

$$\rho(\tau)(\sigma_A^i(x_q)) = \sigma_A^i x_{\tau(q)}, \qquad 0 \le i \le k-1.$$

The kernel of π is isomorphic to $(\mathbb{Z}/k\mathbb{Z})^{Q_k}$ with an isomorphism given by

$$(\mathbb{Z}/k\mathbb{Z})^{Q_k} \to \ker \pi$$

$$g \mapsto \alpha_g, \qquad \alpha_g(\sigma_A^i x_q) = \sigma_A^{i+g(q)} x_q, \qquad 0 \le i \le k-1$$

and it follows $\mathrm{Aut}(P_k, \sigma_A)$ is isomorphic to the semidirect product $(\mathbb{Z}/k\mathbb{Z})^{Q_k} \rtimes \mathrm{Sym}(Q_k)$. The action of $\mathrm{Sym}(Q_k)$ on $(\mathbb{Z}/k\mathbb{Z})^{Q_k}$ is determined as follows. Let $g \in (\mathbb{Z}/k\mathbb{Z})^{Q_k}$, so $g\colon Q_k \to \mathbb{Z}/k\mathbb{Z}$. Then $\alpha_g \in \ker \pi$, and given some $\rho(\tau)$ for some $\tau \in \mathrm{Sym}(Q_k)$,

$$\rho(\tau)^{-1}\alpha_g\rho(\tau) = \alpha_{g\circ\tau}.$$

Remark 4.8.50   For a group G, let G_{ab} denote the abelianization. Using the notation from 4.8.49, we have an isomorphism $\Phi\colon \mathrm{Aut}(\sigma_A|_{P_k}) \to (\mathbb{Z}/k\mathbb{Z})^{Q_k} \rtimes \mathrm{Sym}(Q_k)$. The abelianization of $(\mathbb{Z}/k\mathbb{Z})^{Q_k} \rtimes \mathrm{Sym}(Q_k)$ is isomorphic to the group $\mathrm{Sym}(Q_k)_{\mathrm{ab}} \times ((\mathbb{Z}/k\mathbb{Z})^{Q_k})_{\mathrm{Sym}(Q_k)}$, where $((\mathbb{Z}/k\mathbb{Z})^{Q_k})_{\mathrm{Sym}(Q_k)}$ is the quotient of $(\mathbb{Z}/k\mathbb{Z})^{Q_k}$ by the subgroup generated by all elements of the form $\tau^{-1}g\tau - g$, for $\tau \in \mathrm{Sym}(Q_k), g \in (\mathbb{Z}/k\mathbb{Z})^{Q_k}_{\mathrm{ab}}$.

Now the abelianization of $\mathrm{Sym}(Q_k)$ is given by $\mathrm{sign}\colon \mathrm{Sym}(Q_k) \to \mathbb{Z}/2$, and the map

$$(\mathbb{Z}/k\mathbb{Z})^{Q_k} \to \mathbb{Z}/k\mathbb{Z}$$

$$g \mapsto \sum_{q \in Q_k} g(q)$$

maps elements of the form $\tau^{-1}g\tau - g$ to 0, and induces an isomorphism

$$((\mathbb{Z}/k\mathbb{Z})^{Q_k})_{\mathrm{Sym}(Q_k)} \xrightarrow{\cong} \mathbb{Z}/k\mathbb{Z}.$$

Remark 4.8.51   SGCC, and the question of which automorphisms satisfy SGCC, has a history spanning a number of years. The SGCC condition was introduced by Boyle and Krieger in [26], where it was also proved that, in the case of many SFTs, it holds for any inert automorphism which is a product of involutions. This was followed up by a number of more general results, summarized in the following theorem.

Theorem 4.8.52   *Let* (X_A, σ_A) *be a shift of finite type. An automorphism* $\alpha \in$ Aut(σ_A) *satisfies SGCC if any of the following hold:*

(1) *(Boyle–Krieger [26])* α *is inert and a product of involutions (not for all SFTs, but many, including the full shifts).*
(2) *(Nasu [102])* α *is a simple automorphism.*
(3) *(Fiebig [47])* α *is inert and has finite order.*
(4) *(Kim–Roush [72], with a key ingredient by Wagoner)* α *is inert.*

Remark 4.8.53   Williams first asked (around 1975) whether any permutation of fixed points of a shift of finite type could be lifted to an automorphism. Williams was motivated in part by the classification problem: he was studying an example of two shifts of finite type which were shift equivalent, one of which clearly had an involution of fixed points, while it was not obvious whether the other did. It is interesting to note that, many years later, the automorphism groups proved instrumental in addressing the classification problem.

Remark 4.8.54   In [143] Wagoner proved that the inert automorphisms are generated by simple automorphisms in the "eventual" setting: namely, given a primitive matrix A and inert automorphism $\alpha \in \mathrm{Inert}(\sigma_A)$, there exists some $m \geq 1$ such that, upon considering $\alpha \in \mathrm{Aut}(\sigma_A^m)$, α lies in $\mathrm{Simp}(\sigma_A^m)$. In [13] Boyle gave an alternative proof of this, and also gave a stronger form of the result.

4.9 Wagoner's strong shift equivalence complex, and applications

In the late 80's, Wagoner introduced certain CW complexes as a tool to study strong shift equivalence. These CW complexes provide an algebraic topological/combinatorial framework for studying strong shift equivalence, and have played a key role in a number of important results in the study of shifts of finite

type. Among these, one of the most significant was the construction of a counterexample to Williams' conjecture in the primitive case, which was found by
Kim and Roush in [77][29]. Wagoner independently developed another framework for finding counterexamples, and in [147] gave a different proof, using
matrices generated from Kim and Roush's method in [77], of the existence
of a counterexample to Williams' conjecture. Both the Kim and Roush strategy, and Wagoner's strategy, take place in the setting of Wagoner's strong shift
equivalence complexes.

The goal in this last section is to give a brief introduction to these complexes.
After defining and discussing them, we give a short introduction to how the
Kim–Roush and Wagoner strategies for producing counterexamples work. This
will be very much an overview, and we will not go into details.

In summary, our aim here is not to describe the construction of counterexamples to Williams' conjecture in any detail, but instead to give an overview
of how Wagoner's spaces are built, how the counterexample strategies make
use of them, and where they leave the state of the classification problem.

4.9.1 Wagoner's SSE complexes

Suppose that we have matrices A, B over $\mathbb{Z}_+$, and a strong shift equivalence
from A to B

$$A = A_0 \underset{\sim}{\overset{R_1, S_1}{}} A_1 \underset{\sim}{\overset{R_2, S_2}{}} \cdots \underset{\sim}{\overset{R_{n-1}, S_{n-1}}{}} A_{n-1} \underset{\sim}{\overset{R_n, S_n}{}} A_n = B,$$

where for each $i \geq 1$, $A_{i-1} \underset{\sim}{\overset{R_i, S_i}{}} A_i$ indicates an elementary strong shift equivalence

$$A_{i-1} = R_i S_i, \qquad A_i = S_i R_i.$$

We can visualize this as a path (at the moment we use the term path informally;
it will be made precise later).

where each arrow in this picture represents an elementary strong shift equivalence. From Williams' Theorem (Theorem 4.2.20), we have that there is a

---

Earlier counterexamples to Williams' conjecture in the reducible case were found by Kim and
Roush: see [75].

conjugacy $C: (X_A, \sigma_A) \to (X_B, \sigma_B)$ given by

$$C = \prod_{i=1}^{n} c(R_i, S_i)$$

where for each i, $c(R_i, S_i): (X_{A_{i-1}}, \sigma_{A_{i-1}}) \to (X_{A_i}, \sigma_{A_i})$ is a conjugacy induced by the ESSE $A_{i-1} \overset{R_i, S_i}{\sim} A_i$.

Now suppose, with the matrices A, B over $\mathbb{Z}_+$, we have two SSEs from A to B. We then have two paths of ESSEs from A to B:

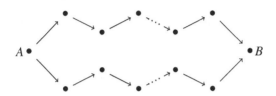

and a pair of conjugacies corresponding to each path

$$C_1: (X_A, \sigma_A) \to (X_B, \sigma_B)$$

$$C_2: (X_A, \sigma_A) \to (X_B, \sigma_B)$$

and one may ask: when do two such paths induce the same conjugacy? Can we determine this from the matrix entries in the paths themselves? Alternatively, is there a space in which we can actually consider these as paths, in which two paths are homotopic if and only if they give rise to the same conjugacy? Wagoner's complexes are a way to do this, and one of the key insights in Wagoner's complexes is determining the correct relations on matrices to accomplish this. These relations are known as the Triangle Identities. Since the Triangle Identities lead directly to the definition of Wagoner's Complexes (Remark 4.9.7), we'll define both simultaneously.

Definition 4.9.1 (Strong shift equivalence complex) Let $\mathscr{R}$ be a a subset of a semiring containing $\{0, 1\}$. We define a CW-complex $SSE(\mathscr{R})$ as follows:

(1) The 0-cells of $SSE(\mathscr{R})$ are square matrices over $\mathscr{R}$.
(2) An edge (R, S) from vertex A to vertex B corresponds to an elementary strong shift equivalence over $\mathscr{R}$ from A to B:

$$A \bullet \xrightarrow{\;(R,S)\;} \bullet B$$

where $A = RS$ and $B = SR$.

(3) 2-cells are given by triangles

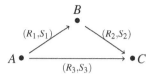

which satisfy the *Triangle Identities*:

$$R_1R_2 = R_3, \qquad R_2S_3 = S_1, \qquad S_3R_1 = S_2. \qquad (4.15)$$

For this section, we will consider the case where $\mathscr{R}$ may be one of:

$$ZO = \{0,1\} \subset \mathbb{Z}_+; \quad \mathbb{Z}_+; \quad \text{or} \quad \mathbb{Z}.$$

Note that ZO is not a semiring. Although our primary focus has been the case that $\mathscr{R}$ is a semiring, in Wagoner's theory the complex $SSE(ZO)$ plays a fundamental role. Wagoner also defines n-cells in $SSE(\mathscr{R})$ for $n \geq 3$ in [145], but we won't need these in this section.

Note that edges have orientations in $SSE(\mathscr{R})$. Recall also that, for an edge from A to B given by a SSE (R,S), we may choose an elementary conjugacy $c(R,S)$ (see Remarmk 4.2.21), and this choice of $c(R,S)$ does not depend only on R and S but also on some choice of simple automorphisms. By Williams' Decomposition Theorem (Theorem 4.2.20; see also (Remark 4.2.44)), if $C: (X_A, \sigma_A) \to (X_B, \sigma_B)$ is a topological conjugacy, then there is a strong shift equivalence

$$A = A_0 \overset{R_1, S_1}{\sim} A_1 \overset{R_2, S_2}{\sim} \dots \overset{R_{n-1}, S_{n-1}}{\sim} A_{n-1} \overset{R_n, S_n}{\sim} A_n = B$$

such that

$$C = \prod_{i=1}^{n} c(R_i, S_i)^{s(i)}$$

with each $c(R_i, S_i)$ an elementary conjugacy corresponding to the ESSE given by R_i, S_i, and

$$s(i) = \begin{cases} +1 & \text{if } A_{i-1} = R_i S_i \text{ and } A_i = S_i R_i, \\ -1 & \text{if } A_i = S_i R_i \text{ and } A_{i-1} = R_i S_i. \end{cases}$$

This presentation C of the conjugacy gives us a path in $SSE(\mathbb{Z}_+)$

Note that some arrows are drawn in reverse, as needed so that the conjugacy C matches the conjugacy given by following the path. Likewise, given a path γ in $SSE(\mathscr{R})$ between A and B

$$\gamma = \prod_{i=1}^{m} (R_i, S_i)^{s(i)}$$

there is a corresponding conjugacy

$$\tilde{\gamma} = \prod_{i=1}^{m} c(R_i, S_i)^{s(i)} : (X_A, \sigma_A) \to (X_B, \sigma_B).$$

In particular, vertices of $SSE(\mathbb{Z}_+)$ correspond to specific presentations of shifts of finite type (edge shift construction), and edges to specific conjugacies (elementary conjugacy coming from an elementary strong shift equivalence). Note that any path between two vertices in these complexes is homotopic to a path following a sequence of edges.

Recall from Section 4.2 that a matrix A is degenerate if it has a zero row or zero column; otherwise, it is nondegenerate. Following Wagoner, we only allow nondegenerate matrices as vertices. It is at times important to work with the larger space $SSE_{\deg}(\mathscr{R})$ which allows degenerate vertices; see for example [33]. It turns out that the inclusion $SSE(\mathbb{Z}_+) \to SSE_{\deg}(\mathbb{Z}_+)$ induces an isomorphism on π_0 [33] and also an isomorphism on π_1 [45] for each path-component.

4.9.2 Homotopy groups for Wagoner's complexes and $\mathrm{Aut}(\sigma_A)$

For a semiring $\mathscr{R}$ and square matrix A over $\mathscr{R}$, we let $SSE(\mathscr{R})_A$ denote the path-component of $SSE(\mathscr{R})$ containing the vertex A. From Williams' Theorem, the vertices A, B in $SSE(\mathbb{Z}_+)$ are in the same path-component if and only if the edge shifts (X_A, σ_A) and (X_B, σ_B) are topologically conjugate.

From the perspective of homotopy theory, the Triangle Identities dictate basic moves for paths in $SSE(\mathscr{R})$ to be homotopic. So why the Triangle Identities? The following result of Wagoner explains their importance. In the statement of the theorem, given A and B and two conjugacies $\phi_1, \phi_2 : (X_A, \sigma_A) \to (X_B, \sigma_B)$, we say $\phi_1 \sim_{\text{simp}} \phi_2$ if there exist simple automorphisms $\gamma_1 \in \mathrm{Simp}(\sigma_A)$, $\gamma_2 \in \mathrm{Simp}(\sigma_B)$ such that $\gamma_2 \phi_1 \gamma_1 = \phi_2$ Then $\sim_{\text{simp}}$ defines an equivalence relation on the set of conjugacies between (X_A, σ_A) and (X_B, σ_B)).

Theorem 4.9.2 ([142, 143, 144, 145]) (1) *Given vertices A, B in $SSE(ZO)$, two paths in $SSE(ZO)$ from A to B are homotopic in $SSE(ZO)$ if and only if they induce the same conjugacy from (X_A, σ_A) to (X_B, σ_B).*

(2) *Given vertices A,B in $SSE(\mathbb{Z}_+)$, two paths in $SSE(\mathbb{Z}_+)$ from A to B are homotopic in $SSE(\mathbb{Z}_+)$ if and only if they induce the same conjugacy from (X_A,σ_A) to (X_B,σ_B) modulo the relation $\sim_{\text{simp}}$.*

Item (2) in the above is perhaps expected; recall the construction given in Section 4.2.8 for associating conjugacies with SSEs over $\mathbb{Z}_+$ requires a choice of labels for certain edges. This choice is where ambiguity up to conjugating by simple automorphisms may arise.

Theorem 4.9.2 gives the first two parts of the following theorem of Wagoner. For a space X with point $x \in X$, let $\pi_k(X,x)$ denote the kth homotopy group based at x.

Theorem 4.9.3 ([142, 143, 144, 145]) *Let A be a square matrix over ZO. Then:*

(1) $\operatorname{Aut}(\sigma_A) \cong \pi_1(SSE(ZO),A)$.
(2) $\operatorname{Aut}(\sigma_A)/\operatorname{Simp}(\sigma_A) \cong \pi_1(SSE(\mathbb{Z}_+),A)$.
(3) $\operatorname{Aut}(G_A,\delta_A) \cong \pi_1(SSE(\mathbb{Z}),A)$.

It is immediate from the definition of the SSE spaces that the set $\pi_0(SSE(\mathbb{Z}_+))$ may be identified with the set of conjugacy classes of shifts of finite type. Moreover, $\pi_0(SSE(\mathbb{Z}))$ may be identified with the set of strong shift equivalence classes of matrices over $\mathbb{Z}$.

Upon using the identifications above, the composition map

$$\pi_1(SSE(ZO),A) \to \pi_1(SSE(\mathbb{Z}_+),A) \to \pi_1(SSE(\mathbb{Z}),A)$$

induced by the natural inclusions $SSE(ZO) \hookrightarrow SSE(\mathbb{Z}_+) \hookrightarrow SSE(\mathbb{Z})$ is isomorphic to the dimension representation factoring as

$$\operatorname{Aut}(\sigma_A) \to \operatorname{Aut}(\sigma_A)/\operatorname{Simp}(\sigma_A) \to \operatorname{Aut}(G_A),$$

i.e. the diagram

$$
\begin{array}{ccccc}
\pi_1(SSE(ZO),A) & \longrightarrow & \pi_1(SSE(\mathbb{Z}_+),A) & \longrightarrow & \pi_1(SSE(\mathbb{Z}),A) \\
\downarrow{\cong} & & \downarrow{\cong} & & \downarrow{\cong} \\
\operatorname{Aut}(\sigma_A) & \longrightarrow & \operatorname{Aut}(\sigma_A)/\operatorname{Simp}(\sigma_A) & \longrightarrow & \operatorname{Aut}(G_A,\delta_A)
\end{array}
$$

commutes.

Wagoner also proves that $\pi_k(SSE(ZO),A) = 0$ for $k \geq 2$. This implies that $SSE(ZO)_A$ is a model for the classifying space $B\operatorname{Aut}(\sigma_A)$ of $\operatorname{Aut}(\sigma_A)$, that is, $SSE(ZO)_A$ is homotopy equivalent to $B\operatorname{Aut}(\sigma_A)$ (Remark 4.9.8). Thus, for example, we have

$$\operatorname{Aut}(\sigma_A)_{\text{ab}} \cong H_1(\operatorname{Aut}(\sigma_A),\mathbb{Z}) \cong H_1(SSE(ZO)_A,\mathbb{Z}).$$

It is worth remarking that, at the moment, we do not know what the abelization Aut$(\sigma_A)_{ab}$ is for an arbitrary positive entropy shift of finite type (X_A, σ_A) (however it is at least known, from [27, Theorem 7.8], that Aut$(\sigma_A)_{ab}$ is not finitely generated).

Wagoner also introduced complexes $SE(\mathscr{R})$ defined analogously to $SSE(\mathscr{R})$ (see (Remark 4.9.9) for a definition). Since an ESSE over $\mathscr{R}$ also gives an SE over $\mathscr{R}$, there is a continuous inclusion map $i_{\mathscr{R}} \colon SSE(\mathscr{R}) \to SE(\mathscr{R})$. Wagoner proved in [144] that, in the case $\mathscr{R}$ is a principal ideal domain, this map $i_{\mathscr{R}}$ is a homotopy equivalence, and that $\pi_n(SSE(\mathscr{R}), A) = \pi_n(SE(\mathscr{R}), A) = 0$ for all $n \geq 2$ and any A. The map $i_{\mathscr{R}}$ cannot be a homotopy equivalence for a general ring $\mathscr{R}$ (Remark 4.9.10).

Wagoner's complexes, and the results of Theorem 4.9.3, have recently been generalized to a groupoid setting in [45]. This setting simplifies some of the proofs and extends Wagoner's construction to shifts of finite type carrying a free action by a finite group, as well as more general shifts of finite type over arbitrary finitely generated groups.

4.9.3 Counterexamples to Williams' conjecture

A counterexample to Williams' conjecture in the primitive case was given by Kim and Roush in [77]. In [147], Wagoner also verified the counterexamples using a different framework. Both methods for detecting the counterexamples take place in the setting of Wagoner's SSE complexes, and build on a great deal of work by many authors. We outline the techniques here; one may also see Wagoner's survey article [146] for an exposition regarding the counterexamples.

Since our goal is only to give a brief introduction to how these counterexamples arise, we won't actually list the explicit matrices involved; they can be found in [77] or [147]). Instead, we focus on the strategy used to prove that they in fact *are* counterexamples.

To start, both strategies roughly follow the same initial idea. As mentioned in Section 4.7.1, to find a counterexample to Williams' conjecture it is sufficient to find a pair of primitive matrices which are connected by a path in $SSE(\mathbb{Z})$, and show they cannot be connected by a path through $SSE(\mathbb{Z}_+)$. We can formalize this approach in terms of homotopy theory (this is an important viewpoint, although not necessary to understand the Kim–Roush counterexample, as we will see). Consider $SSE(\mathbb{Z}_+)$ as a subcomplex of $SSE(\mathbb{Z})$, and, upon fixing a base point A in $SSE(\mathbb{Z}_+)$, consider the long exact sequence in homotopy groups based at A for the pair $(SSE(\mathbb{Z}), SSE(\mathbb{Z}_+))$:

$$\cdots \to \pi_1(SSE(\mathbb{Z}),A) \to \pi_1(SSE(\mathbb{Z}), SSE(\mathbb{Z}_+),A) \to \pi_0(SSE(\mathbb{Z}_+),A)$$
$$\to \pi_0(SSE(\mathbb{Z}),A) \to \cdots$$

Here $\pi_1(SSE(\mathbb{Z}), SSE(\mathbb{Z}_+),A)$ denotes the set of homotopy classes of paths with base point in $SSE(\mathbb{Z}_+)$ and end point equal to A, and the map

$$\pi_1(SSE(\mathbb{Z}), SSE(\mathbb{Z}_+),A) \to \pi_0(SSE(\mathbb{Z}_+),A)$$

is defined by sending the homotopy class of a path γ to the component containing $\gamma(0)$ (details regarding this sequence can be found in [61, Ch. 4, Thm. 4.3]). The last three terms in this sequence are not actually groups, but just pointed sets. Still, exactness makes sense, by defining the kernel to be the pre-image of the base point. The base point in $\pi_1(SSE(\mathbb{Z}), SSE(\mathbb{Z}_+),A)$ is given by the homotopy class of a path which lies entirely in $SSE(\mathbb{Z}_+)$. In particular, the set $\pi_1(SSE(\mathbb{Z}), SSE(\mathbb{Z}_+),A)$ has only one element if and only if every path beginning in $SSE(\mathbb{Z}_+)$ and ending at A is homotopic to a path lying entirely in $SSE(\mathbb{Z}_+)$.

In this setup, the goal is to now find a function

$$F\colon \pi_1(SSE(\mathbb{Z}), SSE(\mathbb{Z}_+),A) \to G$$

to some group G; for computability, we would like G abelian. Then to find a counterexample, it would be enough to find matrices A and B and a path γ in $SSE(\mathbb{Z})$ from A to B such that $F(\gamma) \neq 0$, while $F(\beta) = 0$ for any $\beta \in \pi_1(SSE(\mathbb{Z}),A)$. From Theorem 4.9.3 we know $\pi_1(SSE(\mathbb{Z}),A) \cong \mathrm{Aut}(G_A, \delta_A)$, so in light of the long exact sequence in homotopy written above, being able to compute generators for $\mathrm{Aut}(G_A, \delta_A)$ plays an important role here.

Put another way, we want to find some abelian group G, a primitive matrix A, and some function F from edges in $SSE(\mathbb{Z})_A$ to G which satisfies all of the following, where $\alpha \star \beta$ denotes concatenation of paths:

$$F(\alpha \star \beta) = F(\alpha) + F(\beta) \qquad (4.16)$$

$$\text{If } \gamma_1 \text{ and } \gamma_2 \text{ are homotopic paths, then } F(\gamma_1) = F(\gamma_2) \qquad (4.17)$$

$$F(\gamma) = 0 \text{ if } \gamma \text{ lies in } SSE(\mathbb{Z}_+) \qquad (4.18)$$

$$F(\gamma_{A,B}) \neq 0 \text{ for some path } \gamma_{A,B} \text{ from } A \text{ to a primitive matrix } B \qquad (4.19)$$

Kim and Roush, and independently Wagoner, found functions F_m each satisfying (4.16), (4.17), (4.18) for $G = \mathbb{Z}/m$ for paths contained in any component of a matrix A satisfying $tr(A^k) = 0$ for all $1 \leq k \leq m$. Finally, for $m = 2$, Kim and Roush found a pair of matrices A, B and a path $\gamma_{A,B}$ satisfying (4.19).

4.9.4 Kim–Roush relative sign-gyration method

Let (X_A, σ_A) be a mixing shift of finite type, and recall from Section 4.8.6 the sign-gyration-compatability-condition homomorphisms

$$SGCC_m \colon \mathrm{Aut}(\sigma_A) \to \mathbb{Z}/m\mathbb{Z}$$

$$SGCC_m = g_m + \left(\frac{m}{2}\right) \sum_{j>0} \mathrm{sign}\xi_{m/2^j}.$$

Given $\alpha \in \mathrm{Aut}(\sigma_A)$, for any m, $SGCC_m(\alpha)$ is defined in terms of the action of α on the periodic points up to level m.

The idea behind the Kim and Roush technique is to define, for each m, a relative sign-gyration-compatibility-condition map

$$sgc_m \colon \pi_1\big(SSE(\mathbb{Z}), SSE(\mathbb{Z}_+), A\big) \to \mathbb{Z}/m\mathbb{Z}.$$

To start, suppose $A \xrightarrow{(R,S)} B$ is an edge in $SSE(\mathbb{Z}_+)$ given by a strong shift equivalence $A = RS, B = SR$ over $\mathbb{Z}_+$. Associated to this (by Theorem 4.2.20) is an elementary conjugacy

$$c(R,S) \colon (X_A, \sigma_A) \to (X_B, \sigma_B).$$

Recall this conjugacy $c(R,S)$ is not determined by (R,S), but is only defined up to composition with simple automorphisms in the domain and range. Given m, choose some orderings on the set of orbits whose lengths divide m, and a distinguished point in each such orbit, for each of (X_A, σ_A) and (X_B, σ_B); in [78], these choices are made using certain lexicographic rules on the set of periodic points. The conjugacy $c(R,S)$ induces a bijection between the respective periodic point sets for σ_A and σ_B, and we may define, with respect to the choices of orderings and distinguished points in each orbit, the sign and gyration maps, and hence define $SGCC_m(c(R,S)) \in \mathbb{Z}/m\mathbb{Z}$. If $RS \neq SR$, the value $SGCC_m(c(R,S))$ may depend on the choices of orderings and distinguished points.

In [78], Kim, Roush and Wagoner showed that for such a conjugacy $c(R,S)$, there is a formula $sgcc_m(R,S)$ for $SGCC_m(R,S)$ in terms of the entries from the matrices R, S. This was used to prove Theorem 4.8.26, that $SGCC$ factors through the dimension representation. We note that this formula for $sgcc_m(R,S)$ in general depends on the choice of orderings on the periodic points. Furthermore, the formulas defined in [78] are very complicated for large m. In [77], Kim and Roush defined sgc_m, a slightly different version (Remark 4.9.11) of $sgcc_m$, that also computes $SGCC_m$ in terms of entries from R and S; for $m = 2$, it takes the form

$$sgc_2(R,S) = \sum_{\substack{i<j \\ k>l}} R_{ik}S_{ki}R_{jl}S_{lj} + \sum_{\substack{i<j \\ k\geq l}} R_{ik}S_{kj}R_{jl}S_{li} + \sum_{i,j} \frac{1}{2}R_{ij}(R_{ij}-1)S_{ji}^2.$$

In other words, for an elementary strong shift equivalence $A = RS, B = SR$, we have $SGCC_m(R,S) = sgc_m(R,S)$. The formula given above for sgc_2 uses orderings on the fixed points and period two points defined by certain lexicographic rules given in [78]. For the counterexamples to Williams' conjecture, only sgc_2 is needed.

We can extend $SGCC_m$ from elementary conjugacies $c(R,S)$ to paths in $SSE(\mathbb{Z}_+)$: given a path

$$\gamma = \prod_{i=1}^{J}(R_i,S_i)^{s(i)}$$

define

$$SGCC_m(\gamma) = \sum_i^J s(i)SGCC_m(R_i,S_i).$$

Note from the above we also know that

$$SGCC_m(\gamma) = \sum_i^J s(i)sgc_m(R_i,S_i).$$

Now suppose we have a basic triangle in $SSE(\mathbb{Z}_+)$ with edges (R_1,S_1), (R_2,S_2) and (R_3,S_3). If $c(R_3,S_3) = c(R_1,S_1)c(R_2,S_2)$ then using the fact that $SGCC_m$ is defined in terms of dynamical data coming from the corresponding conjugacies, a calculation [78, Prop. 2.9] shows that

$$SGCC_m(R_1,S_1) + SGCC_m(R_2,S_2) = SGCC_m(R_3,S_3).$$

But by Theorem 4.9.2, up to conjugating by simple automorphisms, we do have $c(R_3,S_3) = c(R_1,S_1)c(R_2,S_2)$; since $SGCC_m$ vanishes on simple automorphisms (Theorem 4.8.26), this gives an addition formula for $SGCC_m$ over triangles in $SSE(\mathbb{Z}_+)$.

Now suppose we have an elementary strong shift equivalence $A = RS, B = SR$ over $\mathbb{Z}$ (so not necessarily in $\mathbb{Z}_+$). The sgc_m formulas still make sense, so we can define $sgc_m(\gamma)$ for any path γ in $SSE(\mathbb{Z})$. If sgc_m also satisfies an addition formula for triangles in $SSE(\mathbb{Z})$, then sgc_m will give us an extension of $SGCC_m$ to $SSE(\mathbb{Z})$. This turns out to be the case, and is a consequence of the following Cocycle Lemma.

Lemma 4.9.4 ([78, 77]) *If the edges* $(R_1, S_1), (R_2, S_2), (R_3, S_3)$ *form a basic triangle in* $SSE(\mathbb{Z})$, *then*

$$sgc_m(R_1, S_1) + sgc_m(R_2, S_2) = sgc_m(R_3, S_3).$$

The Cocycle Lemma was first proved in [78] in the case when the triangle contains a vertex which is strong shift equivalent over $\mathbb{Z}$ to a nonnegative primitive matrix. The version above, which does not require a primitivity assumption, was given in [77], with a much shorter proof suggested by Mike Boyle.

Putting all of the above together, for a matrix A, the map

$$sgc_m \colon \pi_1(SSE(\mathbb{Z}), SSE(\mathbb{Z}_+), A) \to \mathbb{Z}/m\mathbb{Z}$$

satisfies (4.16) and (4.17).

Now suppose that A satisfies $tr(A^k) = 0$ for all $1 \le k \le m$ and (R, S) is an edge in $SSE(\mathbb{Z}_+)$ from A to B. Then both (X_A, σ_A) and (X_B, σ_B) have no points of period k for any $1 \le k \le m$, and the dynamically defined $SGCC_m(R, S)$ must vanish; since $sgc_m = SGCC_m$ on edges in $SSE(\mathbb{Z}_+)$, this implies $sgc_m(R, S) = 0$. It follows that on path-components of matrices A with $tr(A^k) = 0$ for all $1 \le k \le m$, the map sgc_m also satisfies (4.18).

Finally, using $m = 2$, in [77] Kim and Roush found two primitive matrices A, B and a path γ in $SSE(\mathbb{Z})$ from A to B such that all of the following hold:

(1) $tr(A) = tr(A^2) = 0$.
(2) $sgc_2(\alpha) = 0$ for any $\alpha \in \pi_1(SSE(\mathbb{Z}), A)$.
(3) $sgc_2(\gamma) \ne 0$.

It follows these matrices A and B are strong shift equivalent over $\mathbb{Z}$, but not strong shift equivalent over $\mathbb{Z}_+$. The matrices A and B given in [77] are 7×7.

4.9.5 Wagoner's K_2-valued obstruction map

Wagoner, influenced by ideas from pseudo-isotopy theory, constructed a map F satisfying the three conditions 4.16 – 4.18 landing in the K-theory group $K_2(\mathbb{Z}[t]/(t^{m+1}))$. In [147] Wagoner then used this framework to detect counterexamples with matrices found using the technique given by Kim and Roush in [77]. The Kim–Roush relative-sign-gyration-compatability method of the previous section enjoys the fact that it is motivated by dynamical data relating directly to the shift systems, being based on ideas from sign-gyration. Wagoner's method is not as easily connected to the dynamics, but offers some alternative benefits, namely:

(1) Landing in K_2, it connects directly with algebraic K-theory.
(2) It operates within the polynomial matrix framework.
(3) It is perhaps suggestive of more general strategies for studying the refinement of strong shift equivalence over a ring by strong shift equivalence over the ordered part of a ring, i.e. part (3) in the picture in Section 4.7 describing Williams' problem.

So how does Wagoner's construction work? We recall two facts about the group $K_2(\mathscr{R})$ from Section 4.6.4:

(1) $K_2(\mathscr{R})$ is an abelian group.
(2) An expression of the form $\prod_{i=1}^{k} E_i = 1$, where E_i are elementary matrices over $\mathscr{R}$, can be used to construct an element of $K_2(\mathscr{R})$.

For $m \geq 1$, let $SSE_{2m}(\mathbb{Z}_+)$ denote the subcomplex of $SSE(\mathbb{Z}_+)$ consisting of path-components which have a vertex A such that $tr(A^k) = 0$ for all $1 \leq k \leq 2m$. Wagoner's construction proceeds as follows:

(1) Consider an edge in $SSE(\mathbb{Z})$ from A to B. As shown in Section 4.4, this gives matrices E_1, F_1 in $El(\mathbb{Z}[t])$ over $\mathbb{Z}[t]$ such that

$$E_1(I - tA)F_1 = I - tB.$$

(2) Suppose that the matrix A satisfies $tr(A^k) = 0$ for all $1 \leq k \leq m$. In [147, Prop. 4.9] it is shown there exist matrices E_2, F_2 in $El(\mathbb{Z}[t])$ and A' over $\mathbb{Z}[t]$ such that $E_2(I - tA)F_2 = I - t^{m+1}A'$. Doing the same for B yields matrices E_3, F_3 in $El(\mathbb{Z}[t])$ and some B' over $\mathbb{Z}[t]$ such that

$$E_2(I - tA)F_2 = I - t^{m+1}A'$$

$$E_3(I - tB)F_3 = I - t^{m+1}B'.$$

(3) Combining steps (1) and (2) we have matrices X, Y in $El(\mathbb{Z}[t])$ such that

$$X(I - t^{m+1}A')Y = I - t^{m+1}B'.$$

Passing to $\mathbb{Z}[t]/(t^{m+1})$, we get

$$XY = I.$$

We can now use this expression to produce an element of $K_2(\mathbb{Z}[t]/(t^{m+1}))$.

Wagoner shows this assignment defined above is additive with respect to concatenation of paths given by two subsequent edges, so one can extend it to arbitrary paths. Thus, given an edge γ in $SSE(\mathbb{Z})$, applying the above gives an element $F(\gamma) \in K_2(\mathbb{Z}[t]/(t^{m+1}))$. Then, given a path γ between two vertices A and B in $SSE_{2m}(\mathbb{Z})$, Wagoner shows:

(a) The element $F(\gamma)$ in $K_2(\mathbb{Z}[t]/(t^{m+1}))$ produced by the above construction is independent of the choices of elementary matrices made in the construction.

(b) If A, B are nonnegative and γ' is another path in $SSE(\mathbb{Z})$ from A to B such that γ and γ' are homotopic (with endpoints fixed), then $F(\gamma) = F(\gamma')$ in $K_2(\mathbb{Z}[t]/(t^{m+1}))$.

(c) If the path γ lies entirely in $SSE_{2m}(\mathbb{Z}_+)$, then the corresponding element $F(\gamma)$ in $K_2(\mathbb{Z}[t]/(t^{m+1}))$ vanishes.

Altogether this defines a function

$$\Phi_{2m} \colon \pi_1(SSE(\mathbb{Z}), SSE_{2m}(\mathbb{Z}_+), A) \to K_2(\mathbb{Z}[t]/(t^{m+1}))$$

satisfying the properties 4.16 – 4.18 for A in $SSE_{2m}(\mathbb{Z}_+)$.

Let $K_2(\mathbb{Z}[t]/(t^{m+1}), (t))$ denote the kernel of the split surjection map from $K_2(\mathbb{Z}[t]/(t^{m+1}))$ to $K_2(\mathbb{Z})$ induced by the ring map $\mathbb{Z}[t]/(t^{m+1}) \to \mathbb{Z}$ induced by $t \to 0$. Wagoner proved that the maps Φ_{2m} defined above actually land in $K_2(\mathbb{Z}[t]/(t^{m+1}), (t))$. This is a significant fact, since van der Kallen proved in [141] that $K_2(\mathbb{Z}[t]/(t^2), (t)) \cong \mathbb{Z}/2$. This calculation by van der Kallen was used by Wagoner to explicitly compute [147, Eq. 1.21] Φ_2, and to detect some explicit counterexamples in [147].

4.9.6 Some remarks and open problems

At the $m = 2$ level, each method outlined above gives a map

$$sgc_2 \colon \pi_1(SSE(\mathbb{Z}), SSE_2(\mathbb{Z}_+), A) \to \mathbb{Z}/2\mathbb{Z}$$

$$\Phi_2 \colon \pi_1(SSE(\mathbb{Z}), SSE_2(\mathbb{Z}_+), A) \to K_2(\mathbb{Z}[t]/(t^2), (t)) \cong \mathbb{Z}/2\mathbb{Z}.$$

While these were developed independently, remarkably, it was shown by Kim and Roush in the Appendix of [147] that $\Phi_2 = sgc_2$. Wagoner explicitly poses the problem in [146, Number 6] to determine, for larger m, the relationship between Φ_{2m} and sgc_m.

Finally, let us note that both the Kim–Roush method and Wagoner's method rely on the non-existence of periodic points at certain low levels. In Wagoner's case, without vanishing trace conditions, step (2) above cannot be carried out. Moreover, step (3) also relies on the vanishing trace conditions. As a result, Wagoner's construction is *only* defined in the case of shifts of finite type lacking periodic points of certain low order levels. For the Kim–Roush technique, the non-existence of low-order periodic points comes in when one wants to conclude that the assignment from edges to some element of $\mathbb{Z}/m$ vanishes along any path through $SSE(\mathbb{Z}_+)$: for an edge in $SSE(\mathbb{Z}_+)$, the assignment coincides with the relative sign-gyration numbers associated to a conjugacy, which, in the absence of any periodic points of the given levels, must vanish.

In light of this, neither method is able to produce more than a finite index refinement of the strong shift equivalence class of a given primitive matrix A over $\mathbb{Z}_+$, since (X_A, σ_A) will, above some level k depending on A, eventually contain periodic points at all levels larger than k.

To finish, we highlight two open problems (Problem 4.9.6 below was mentioned informally in the discussion following Conjecture 4.2.24 in Section 4.2):

Problem 4.9.5   If A is shift equivalent over $\mathbb{Z}_+$ to the 1×1 matrix (n), must A be strong shift equivalent over $\mathbb{Z}_+$ to (n)? In other words, does Williams' conjecture hold in the case of full shifts?

Problem 4.9.6   For a primitive matrix A, is the refinement of the SE-$\mathbb{Z}_+$-equivalence class of A by SSE-$\mathbb{Z}_+$ finite?

Finally, we think the complexes $SSE(ZO), SSE(\mathbb{Z}_+)$ and $SSE(\mathbb{Z})$ probably have much more to offer, and obtaining a deeper understanding of them would be valuable for studying both strong shift equivalence and the conjugacy problem for shifts of finite type.

4.9.7   Notes

These notes contain some proofs, remarks, and solutions of various exercises throughout Section 4.9.

Remark 4.9.7   Prior to considering the strong shift equivalence spaces $SSE(\mathscr{R})$, Wagoner also introduced a related 'space of Markov partitions' for a shift of finite type; we won't describe these here, and instead refer the reader to [145, 3, 45].

Remark 4.9.8   For a discrete group G, a classifying space is a path-connected space BG such that $\pi_1(BG) \cong G$ and $\pi_k(BG) = 0$ for all $k \geq 2$. The space BG has the property that $H_k(G, \mathbb{Z})$, the integral group homology of the group G, is isomorphic to $H_k(BG, \mathbb{Z})$, the integral singular homology of the space BG. See [149, 6.10.4] for details.

Remark 4.9.9   For a semiring $\mathscr{R}$, the shift equivalence space $SE(\mathscr{R})$ is the CW complex defined as follows.

(1) The 0-cells of $SE(\mathscr{R})$ are square matrices over $\mathscr{R}$.

(2) An edge from vertex A to vertex B corresponds to a shift equivalence over $\mathscr{R}$ from A to B, i.e. matrices R, S over $\mathscr{R}$ and $k \geq 1$ such that

$$A^k = RS, \qquad B^k = SR, \qquad AR = RB, \qquad SA = BS.$$

(3) 2-cells are given by triangles

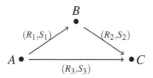

such that

$$R_1 R_2 = R_3. \qquad (4.20)$$

Higher cells are defined in the same way as for the SSE spaces. It is immediate from the definition that $\pi_0(SE(\mathscr{R}))$ is in bijective correspondence with the set of shift equivalence classes of matrices over $\mathscr{R}$.

Remark 4.9.10 We'll show here that $i_{\mathscr{R}}$ cannot in general be a homotopy equivalence. The map $i_{\mathscr{R}}$ induces a map $i_{\mathscr{R},*} \colon \pi_0(SSE(\mathscr{R})) \to \pi_0(SE(\mathscr{R}))$ of sets. We can identify $\pi_0(SSE(\mathscr{R}))$ with the set of SSE-classes of matrices over $\mathscr{R}$ and $\pi_0(SE(\mathscr{R}))$ with the set of SE-classes of matrices over $\mathscr{R}$, and upon making these identifications, the map $i_{\mathscr{R},*}$ agrees with the map π given in (4.9). Theorem 4.7.10 from Section 4.7 gives a description of the fibers of this map in terms of some K-theoretic data. In particular, from Corollary 4.7.11 we know that the map $i_{\mathscr{R},*} \colon \pi_0(SSE(\mathscr{R})) \to \pi_0(SE(\mathscr{R}))$ is not always an injection. Thus Wagoner's result that $i_{\mathscr{R}}$ is a homotopy equivalence when $\mathscr{R}$ is a principal ideal domain cannot hold in the case $NK_1(\mathscr{R}) \neq 0$; as we see, it need not even induce an injection on the level of π_0.

Remark 4.9.11 As pointed out in [77, Section 8], the maps sgc_m and $sgcc_m$ are not the same in general. However, they do yield the same value on path-components containing a primitive matrix whose trace is zero. The definition for $sgcc_m$ requires a component with a matrix which is shift equivalent to a primitive matrix, whereas the map sgc_m does not. See [77, Section 8] for more details regarding the difference between sgc_m and $sgcc_m$.

References

[1] Gert Almkvist, *K*-theory of endomorphisms, *J. Algebra*, **55** (1978), no. 2, 308–340.

[2] Gert Almkvist, Erratum: "*K*-theory of endomorphisms" [J. Algebra **55** (1978), no. 2, 308–340; MR 80i:18018], *J. Algebra*, **68** (1981), no. 2, 520–521.

[3] L. Badoian and J. B. Wagoner, Simple connectivity of the Markov partition space, *Pacific J. Math.*, **193** (2000), no. 1. 1–3.

[4] Kirby A. Baker, Strong shift equivalence of 2×2 matrices of nonnegative integers, *Ergodic Theory Dynam. Systems*, **3** (1983), no. 4, 501–508.

[5] Kirby A. Baker, Strong shift equivalence and shear adjacency of nonnegative square integer matrices, *Linear Algebra Appl.*, **93** (1987), 131–147.

[6] H. Bass, J. Milnor, and J.-P. Serre, Solution of the congruence subgroup problem for SL_n $(n \geq 3)$ and Sp_{2n} $(n \geq 2)$, *Inst. Hautes Études Sci. Publ. Math.*, **33** (1967), 59–137.

[7] Hyman Bass, *Algebraic K-theory*. W. A. Benjamin, Inc., New York-Amsterdam, 1968.

[8] Louis Block, John Guckenheimer, Michał Misiurewicz, and Lai Sang Young, Periodic points and topological entropy of one-dimensional maps, In *Global theory of dynamical systems (Proc. Internat. Conf., Northwestern Univ., Evanston, Ill., 1979)*, volume 819 of *Lecture Notes in Math.*, pages 18–34. Springer, Berlin, 1980.

[9] Rufus Bowen and John Franks, Homology for zero-dimensional nonwandering sets, *Ann. Math.* (2), **106** (1977), no. 1, 73–92.

[10] Mike Boyle, Shift equivalence and the Jordan form away from zero. *Ergodic Theory Dynam. Systems*, **4** (1984), no. 3, 367–379.

[11] M. Boyle, Nasu's simple automorphisms, In *Dynamical systems (College Park, MD, 1986–87)*, volume 1342 of *Lecture Notes in Math.*, pages 23–32. Springer, Berlin, 1988.

[12] Mike Boyle, Algebraic aspects of symbolic dynamics, In *Topics in symbolic dynamics and applications (Temuco, 1997)*, volume 279 of *London Math. Soc. Lecture Note Ser.*, pages 57–88, Cambridge Univ. Press, Cambridge, 2000.

[13] Mike Boyle, Eventual extensions of finite codes. *Proc. Amer. Math. Soc.*, **104** (1988), no. 3, 965–972.

[14] Mike Boyle, Symbolic dynamics and matrices. In *Combinatorial and graph-theoretical problems in linear algebra (Minneapolis, MN, 1991)*, volume 50 of *IMA Vol. Math. Appl.*, pages 1–38. Springer, New York, 1993.

[15] Mike Boyle, Positive K-theory and symbolic dynamics. In *Dynamics and randomness (Santiago, 2000)*, volume 7 of *Nonlinear Phenom. Complex Systems*, pages 31–52. Kluwer Acad. Publ., Dordrecht, 2002.

[16] Mike Boyle, Open problems in symbolic dynamics. In *Geometric and probabilistic structures in dynamics*, volume 469 of *Contemp. Math.*, pages 69–118. Amer. Math. Soc., Providence, RI, 2008.

[17] Mike Boyle, Notes on the Perron–Frobenius theory of nonnegative matrices. website www.math.umd.edu/~mboyle/papers/pfnotes.pdf, 2019. 7 pages.

[18] Mike Boyle, Toke Meier Carlsen, and Søren Eilers, Flow equivalence and isotopy for subshifts. *Dyn. Syst.*, **32** (2017), no. 3, 305–325.

[19] M. Boyle, T. M. Carlsen, and S. Eilers, Corrigendum: "Flow equivalence and isotopy for subshifts" [MR3669803]. *Dyn. Syst.*, **32** (2017), no. 3, ii.

[20] Mike Boyle, Toke Meier Carlsen, and Søren Eilers, Flow equivalence of G-SFTs. *Trans. Amer. Math. Soc.*, **373** (2020), no. 4, 2591–2657.

[21] Mike Boyle and Sompong Chuysurichay, The mapping class group of a shift of finite type. *J. Mod. Dyn.*, **13** (2018), 115–145.

[22] Mike Boyle and Ulf-Rainer Fiebig, The action of inert finite-order automorphisms on finite subsystems of the shift. *Ergodic Theory Dynam. Systems*, **11** (1991), no. 3, 413–425.

[23] Mike Boyle and David Handelman, The spectra of nonnegative matrices via symbolic dynamics (including Appendix 4 joint with Kim and Roush). *Ann. Math. (2)*, **133** (1991), no. 2, 249–316.

[24] Mike Boyle and David Handelman, Algebraic shift equivalence and primitive matrices. *Trans. Amer. Math. Soc.*, **336** (1993), no. 1, 121–149.

[25] Mike Boyle, K. H. Kim, and F. W. Roush, Path methods for strong shift equivalence of positive matrices. *Acta Appl. Math.*, **126** (2013), 65–115.

[26] Mike Boyle and Wolfgang Krieger, Periodic points and automorphisms of the shift. *Trans. Amer. Math. Soc.*, **302** (1987), no. 1, 125–149.

[27] Mike Boyle, Douglas Lind, and Daniel Rudolph, The automorphism group of a shift of finite type. *Trans. Amer. Math. Soc.*, **306** (1988), no. 1, 71–114.

[28] Mike Boyle, Brian Marcus, and Paul Trow, Resolving maps and the dimension group for shifts of finite type. *Mem. Amer. Math. Soc.*, **70** (1987), no. 377, vi+146pp.

[29] Mike Boyle and Scott Schmieding, Strong shift equivalence and the generalized spectral conjecture for nonnegative matrices. *Linear Algebra Appl.*, **498** (2016), 231–243.

[30] M. Boyle and S. Schmieding, Finite group extensions of shifts of finite type: K-theory, Parry and Livšic, *Ergodic Theory Dynam. Systems*, **37** (2017), no. 4, 1026–1059.

[31] Mike Boyle and Scott Schmieding, Strong shift equivalence and algebraic K-theory. *J. Reine Angew. Math.*, **752** (2019), 63–104.

[32] Mike Boyle and Michael C. Sullivan, Equivariant flow equivalence for shifts of finite type, by matrix equivalence over group rings. *Proc. London Math. Soc. (3)*, **91** (2005), no. 1, 184–214.

[33] M. Boyle and J. B. Wagoner, Positive algebraic *K*-theory and shifts of finite type. In *Modern dynamical systems and applications*, (ed. Michael Brin, Boris Hasselblatt and Yakov Pesin), pages 45–66. Cambridge Univ. Press, Cambridge, 2004.

[34] Alfredo Costa and Benjamin Steinberg, A categorical invariant of flow equivalence of shifts. *Ergodic Theory Dynam. Systems*, **36** (2016), No. 2. (2):470–513.

[35] Ethan M. Coven, Endomorphisms of substitution minimal sets. *Z. Wahrscheinlichkeitstheorie und Verw. Gebiete*, **20** (1971/2), 129–133.

[36] Joachim Cuntz and Wolfgang Krieger, Topological Markov chains with dicyclic dimension groups. *J. Reine Angew. Math.*, **320** (1980), 44–51.

[37] Van Cyr and Bryna Kra, The automorphism group of a shift of linear growth: beyond transitivity, *Forum Math. Sigma*, **3** (2015), e5 (27pp.)

[38] Van Cyr and Bryna Kra, The automorphism group of a minimal shift of stretched exponential growth. *J. Mod. Dyn.*, **10** (2016), 483–495.

[39] Van Cyr and Bryna Kra, The automorphism group of a shift of subquadratic growth. *Proc. Amer. Math. Soc.*, **144** (2016), no. 2, 613–621.

[40] Sebastián Donoso, Fabien Durand, Alejandro Maass, and Samuel Petite, On automorphism groups of low complexity subshifts. *Ergodic Theory Dynam. Systems*, **36** (2016), no. 1, 64–95.

[41] Fabien Durand and Dominique Perrin, *Dimension groups and dynamical systems—substitutions, Bratteli diagrams and Cantor systems*. Cambridge University Press, Cambridge, 2022.

[42] Edward G. Effros, *Dimensions and C*-algebras*, volume 46 of *CBMS Regional Conference Series in Mathematics*. Conference Board of the Mathematical Sciences, Washington, D.C., 1981.

[43] Edward G. Effros, David E. Handelman, and Chao-Liang Shen, Dimension groups and their affine representations. *Amer. J. Math.*, **102** (1980), no. 2, 385–407.

[44] Søren Eilers, Gunnar Restorff, Efren Ruiz, and Adam P.W. Sørenson, The complete classification of unital graph C*-algebras: geometric and strong. *Duke Math. J.* **170** (2021), no. 11, 2421–2517.

[45] Jeremias Epperlein, Wagoner's Complexes Revisited. arXiv:1911.06236, November 2019.

[46] F. T. Farrell, The nonfiniteness of Nil. *Proc. Amer. Math. Soc.*, **65** (1977), no. 2, 215–216.

[47] Ulf Fiebig, Dissertation (Ph.D.), University of Gottingen, Germany, 1987.

[48] John Franks, Flow equivalence of subshifts of finite type. *Ergodic Theory Dynam. Systems*, **4** (1984), no. 1, 53–66.

[49] Joshua Frisch, Tomer Schlank, and Omer Tamuz, Normal amenable subgroups of the automorphism group of the full shift. *Ergodic Theory Dynam. Systems*, **39** (2019), no. 5, 1290–1298.

[50] Patrick M. Gilmer, Topological quantum field theory and strong shift equivalence. *Canadian Math. Bulletin*, **42** (1999), no. 2, 190–197.

[51] Thierry Giordano, Hiroki Matui, Ian F. Putnam, and Christian F. Skau. Orbit equivalence for Cantor minimal $\mathbb{Z}^d$-systems. *Invent. Math.*, **179** (2010), no. 1, 119–158.

[52] Thierry Giordano, Ian F. Putnam, and Christian F. Skau, Topological orbit equivalence and C^*-crossed products. *J. Reine Angew. Math.*, **469** (1995), 51–111.

[53] K. R. Goodearl, *Partially ordered abelian groups with interpolation*, volume 20 of *Mathematical Surveys and Monographs*. American Mathematical Society, Providence, RI, 1986.

[54] Daniel R. Grayson, The K-theory of endomorphisms. *J. Algebra*, **48** (1977), no. 2, 439–446.

[55] Fritz J. Grunewald, Solution of the conjugacy problem in certain arithmetic groups. In *Word problems, II (Conf. on Decision Problems in Algebra, Oxford, 1976)*, volume 95 of *Stud. Logic Foundations Math.*, pages 101–139. North-Holland, Amsterdam-New York, 1980.

[56] Fritz Grunewald and Daniel Segal, Some general algorithms. I. Arithmetic groups. *Ann. Math.* (2), **112** (1980), no. 3, 531–583.

[57] David Handelman, Positive matrices and dimension groups affiliated to C^*-algebras and topological Markov chains. *J. Operator Theory*, **6** (1981), no. 1, 55–74.

[58] David Handelman, Eventually positive matrices with rational eigenvectors. *Ergodic Theory Dynam. Systems*, **7** (1987), no. 2, 193–196.

[59] Dennis R. Harmon, NK_1 of finite groups. *Proc. Amer. Math. Soc.*, **100** (1987), no. 2, 229–232.

[60] Yair Hartman, Bryna Kra, and Scott Schmieding. The stabilized automorphism group of a subshift. *Int. Math. Res. Not.*, **21** (2022), 17112–17186.

[61] Allen Hatcher, *Algebraic topology*. Cambridge University Press, Cambridge, 2002.

[62] G. A. Hedlund, Endomorphisms and automorphisms of the shift dynamical system. *Math. Systems Theory*, **3** (1969), :320–375.

[63] G. Higman, The units of group-rings, *Proc. London Math. Soc.* **46** (1940), 231–248.

[64] B. Host and F. Parreau, Homomorphismes entre systèmes dynamiques définis par substitutions. *Ergodic Theory Dynam. Systems*, **9** (1989), no. 3, 469–477.

[65] Emmanuel Jeandel, Strong shift equivalence as a category notion. arXiv:2107.10734, 2021.

[66] Charles R. Johnson, Row stochastic matrices similar to doubly stochastic matrices. *Linear and Multilinear Algebra*, **10** (1981), no. 2, 113–130.

[67] Charles R. Johnson, Thomas J. Laffey, and Raphael Loewy, The real and the symmetric nonnegative inverse eigenvalue problems are different. *Proc. Amer. Math. Soc.*, **124** (1996), no. 12, 3647–3651.

[68] Charles R. Johnson, Carlos Mariján, Pietro Paparella, and Miriam Pisonero. The NIEP. In *Operator theory, operator algebras, and matrix theory*, volume 267 of *Oper. Theory Adv. Appl.*, pages 199–220. Birkhäuser/Springer, Cham, 2018.

[69] Ki Hang Kim, Nicholas S. Ormes, and Fred W. Roush, The spectra of nonnegative integer matrices via formal power series. *J. Amer. Math. Soc.*, **13** (2000), no. 4, 773–806 (electronic).

[70] Ki Hang Kim and Fred W. Roush, Decidability of shift equivalence. In *Dynamical systems (College Park, MD, 1986–87)*, volume 1342 of *Lecture Notes in Math.*, pages 374–424. Springer, Berlin, 1988.

[71] K. H. Kim and F. W. Roush, On the automorphism groups of subshifts. *Pure Math. Appl. Ser. B*, **1** (1990), no. 4, 203–230.

[72] K. H. Kim and F. W. Roush, On the structure of inert automorphisms of subshifts. *Pure Math. Appl. Ser. B*, **2** (1991), no. 1, 3–22.

[73] K. H. Kim and F. W. Roush, Solution of two conjectures in symbolic dynamics. *Proc. Amer. Math. Soc.*, **112** (1991), no. 4, 1163–1168.

[74] K. H. Kim and F. W. Roush, Topological classification of reducible subshifts. *Pure Math. Appl. Ser. B*, **3** (1992), nos. 2–4, 87–102.

[75] K. H. Kim and F. W. Roush, Williams's conjecture is false for reducible subshifts. *J. Amer. Math. Soc.*, **5** (1992), no. 1, 213–215.

[76] K. H. Kim and F. W. Roush, The Williams conjecture is false for irreducible subshifts. *Electron. Res. Announc. Amer. Math. Soc.*, **3** (1997), 105–109 (electronic).

[77] K. H. Kim and F. W. Roush, The Williams conjecture is false for irreducible subshifts. *Ann. Math.*, (2) **149** (1999), no. 2, 545–558.

[78] K. H. Kim, F. W. Roush, and J. B. Wagoner, Automorphisms of the dimension group and gyration numbers. *J. Amer. Math. Soc.*, **5** (1992), no. 1, 191–212.

[79] K. H. Kim, F. W. Roush, and J. B. Wagoner, Inert actions on periodic points. *Electron. Res. Announc. Amer. Math. Soc.*, **3** (1997), 55–62 (electronic).

[80] K. H. Kim, F. W. Roush, and J. B. Wagoner, Characterization of inert actions on periodic points. I. *Forum Math.*, **12** (2000), no. 5, 565–602.

[81] K. H. Kim, F. W. Roush, and J. B. Wagoner, Characterization of inert actions on periodic points. II. *Forum Math.*, **12** (2000), no. 6, 671–712.

[82] Bruce P. Kitchens, *Symbolic Dynamics*. Springer-Verlag, Berlin, 1998.

[83] Johan Kopra, Glider automorphisms and a finitary Ryan's theorem for transitive subshifts of finite type. *Nat. Comput.*, **19** (2020), no. 4, 773–786.

[84] Johan Kopra, Glider automata on all transitive sofic shifts. *Ergodic Theory Dynam. Systems*, **42** (2022), no. 12, 3716–3744.

[85] Wolfgang Krieger, On dimension functions and topological Markov chains. *Invent. Math.*, **56** (1980), no. 3, 239–250.

[86] Thomas J. Laffey, A constructive version of the Boyle-Handelman theorem on the spectra of nonnegative matrices. *Linear Algebra Appl.*, **436** (2012), no. 6, 1701–1709.

[87] Thomas J. Laffey, Raphael Loewy, and Helena Šmigoc, Power series with positive coefficients arising from the characteristic polynomials of positive matrices. *Math. Ann.*, **364** (2016), nos. 1–2, 687–707.

[88] Thomas J. Laffey and Eleanor Meehan, A characterization of trace zero nonnegative 5×5 matrices, 1999. Special issue dedicated to Hans Schneider (Madison, WI, 1998).

[89] Claiborne G. Latimer and C. C. MacDuffee, A correspondence between classes of ideals and classes of matrices. *Ann. Math.*, **34** (1933), no. 2, 313–316.

[90] Chao-Hui Lin and Daniel Rudolph, Sections for semiflows and Kakutani shift equivalence. In *Modern dynamical systems and applications*, pages 145-161, Cambridge Univ. Press, Cambridge, 2004.

[91] D. A. Lind, The entropies of topological Markov shifts and a related class of algebraic integers. *Ergodic Theory Dynam. Systems*, **4** (1984), no. 2, 283–300.

[92] Douglas Lind and Brian Marcus, *An Introduction to Symbolic Dynamics and Coding*. Cambridge University Press, Cambridge, 1995.

[93] Douglas Lind and Brian Marcus, *An Introduction to Symbolic Dynamics and Coding*, second edition. Cambridge University Press, Cambridge, 2021.

[94] Raphael Loewy and David London, A note on an inverse problem for nonnegative matrices. *Linear and Multilinear Algebra*, **6** (1978/9), no. 1, 83–90.

[95] Nicholas Long, Mixing shifts of finite type with non-elementary surjective dimension representations. *Acta Appl. Math.*, **126** (2013), 277–295.

[96] M. Maller and M. Shub, The integral homology of Smale diffeomorphisms. *Topology*, **24** (1985), 153–164.

[97] Brian Marcus and Selim Tuncel, The weight-per-symbol polytope and scaffolds of invariants associated with Markov chains. *Ergodic Theory Dynam. Systems*, **11** (1991), no. 1, 129–180.

[98] Daniel A. Marcus, *Number fields*. Universitext, Springer-Verlag, New York–Heidelberg, 1977.

[99] John Milnor, *Introduction to algebraic K-theory*. Princeton University Press, Princeton, N.J.; University of Tokyo Press, Tokyo, 1971. Annals of Mathematics Studies, No. 72.

[100] Konstantin Mischaikow and Charles Weibel, Computing the Conley Index: a Cautionary Tale. arXiv:2303.06492, 2023.

[101] Marston Morse and Gustav A. Hedlund, Symbolic Dynamics. *Amer. J. Math.*, 60(4):815–866, 1938.

[102] Masakazu Nasu, Topological conjugacy for sofic systems and extensions of automorphisms of finite subsystems of topological Markov shifts. In *Dynamical systems (College Park, MD, 1986–87)*, volume 1342 of *Lecture Notes in Math.*, pages 564–607. Springer, Berlin, 1988.

[103] Masakazu Nasu, Textile systems for endomorphisms and automorphisms of the shift. *Mem. Amer. Math. Soc.*, **114** (1995), no. 546, viii+215pp.

[104] Masakazu Nasu, The dynamics of expansive invertible onesided cellular automata. *Trans. Amer. Math. Soc.*, **354** (2002), no. 10, 4067–4084.

[105] Masakazu Nasu, Nondegenerate q-biresolving textile systems and expansive automorphisms of onesided full shifts. *Trans. Amer. Math. Soc.*, **358** (2006), no. 2, 871–891.

[106] Masakazu Nasu, Textile systems and one-sided resolving automorphisms and endomorphisms of the shift. *Ergodic Theory Dynam. Systems*, **28** (2008), no. 1, 167-209.

[107] Amnon Neeman and Andrew Ranicki, Noncommutative localisation in algebraic K-theory. I. *Geom. Topol.*, **8** (2004), :1385–1425.

[108] A. Nenashev, K_1 by generators and relations. *J. Pure Appl. Algebra*, **131** (1998), no. 2, 195–212.

[109] Morris Newman, *Integral matrices*. Pure and Applied Mathematics, Vol. 45. Academic Press, New York–London, 1972.

[110] Robert Oliver, *Whitehead groups of finite groups*, volume 132 of *London Mathematical Society Lecture Note Series*. Cambridge University Press, Cambridge, 1988.

[111] Jeanette Olli, Endomorphisms of Sturmian systems and the discrete chair substitution tiling system. *Discrete Contin. Dyn. Syst.*, **33** (2013), no. 9, 4173–4186.

[112] William Parry, Intrinsic Markov chains. *Trans. Amer. Math. Soc.*, **112** (1964), 55–66.

[113] Bill Parry and Dennis Sullivan, A topological invariant of flows on 1-dimensional spaces. *Topology*, **14** (1975), no. 4, 297–299.

[114] William Parry and Selim Tuncel, *Classification problems in ergodic theory*, volume 67 of *London Mathematical Society Lecture Note Series*. Cambridge University Press, Cambridge, 1982.

[115] William Parry and Selim Tuncel, On the stochastic and topological structure of Markov chains. *Bull. London Math. Soc.*, **14** (1982), no. 1, 16–27.

[116] William Parry and R. F. Williams, Block coding and a zeta function for finite Markov chains. *Proc. London Math. Soc.* (3), **35** (1997), no. 3, 483–495.

[117] Dominique Perrin, On positive matrices. *Theoret. Comput. Sci.*, **94** (1992), no. 2, 357–366.

[118] I. Reiner, *Maximal orders*, volume 28 of *London Mathematical Society Monographs. New Series*. The Clarendon Press, Oxford University Press, Oxford, 2003. Corrected reprint of the 1975 original, With a foreword by M. J. Taylor.

[119] Gunnar Restorff, Classification of Cuntz-Krieger algebras up to stable isomorphism. *J. Reine Angew. Math.*, **598** (2006), 185–210.

[120] Norbert Riedel, An example on strong shift equivalence of positive integral matrices. *Monatsh. Math.*, **95** (1983), no. 1, 45–55.

[121] Leslie G. Roberts, K_2 of some truncated polynomial rings. In *Ring theory (Proc. Conf., Univ. Waterloo, Waterloo, 1978)*, volume 734 of *Lecture Notes in Math.*, pages 249–278. Springer, Berlin, 1979. With a section written jointly with S. Geller.

[122] Mikael Rørdam, Classification of Cuntz-Krieger algebras. *K-Theory*, **9** (1995), no. 1, 31–58.

[123] Jonathan Rosenberg, *Algebraic K-theory and its applications*, volume 147 of *Graduate Texts in Mathematics*. Springer-Verlag, New York, 1994.

[124] J. Patrick Ryan, The shift and commutativity. *Math. Systems Theory*, **6** (1972), 82–85.

[125] J. Patrick Ryan, The shift and commutativity. II. *Math. Systems Theory*, **8** (1974/5), no. 3, 249–250.

[126] Ville Salo, Transitive action on finite points of a full shift and a finitary Ryan's theorem. *Ergodic Theory Dynam. Systems*, **39** (2019), no. 6, 1637–1667.

[127] Ville Salo, Veelike actions and the MCG of a mixing SFT. arXiv:2103.15505, 2021.

[128] Ville Salo, Gate lattices and the stabilized automorphism group. *J. Mod. Dyn.*, **19** (2023), 717–749.

[129] Ville Salo and Ilkka Törmä, Block maps between primitive uniform and Pisot substitutions. *Ergodic Theory Dynam. Systems*, **35** (2015), no. 7, 2292–2310.

[130] Scott Schmieding, Explicit Examples in NK_1. arXiv:1506.07418, June 2015.

[131] Scott Schmieding and Kitty Yang, The mapping class group of a minimal subshift. *Colloq. Math.*, **163** (2021), no. 2, 233–265.

[132] Scott Schmieding, Local $\mathscr{P}$ entropy and stabilized automorphism groups of subshifts *Invent. Math.*, **227** (2022), no. 3, 963-995.

[133] Eugene Seneta, *Non-negative matrices and Markov chains*, Revised reprint of the second (1981) edition, Springer-Verlag, New York, 2006.

[134] Claude E. Shannon and Warren Weaver, *The Mathematical Theory of Communication*. The University of Illinois Press, Urbana, Ill., 1949.

[135] Desmond Sheiham, Whitehead groups of localizations and the endomorphism class group. *J. Algebra*, **270** (2003), no. 1, 261–280.

[136] Daniel S. Silver and Susan G. Williams, Knot invariants from symbolic dynamical systems. *Trans. Amer. Math. Soc.*, **351** (1999), no. 8, 3243-3265.

[137] John R. Silvester, *Introduction to algebraic K-theory*. Chapman & Hall, London-New York, 1981.

[138] H. R. Suleĭmanova, Stochastic matrices with real characteristic numbers. *Doklady Akad. Nauk SSSR (N.S.)*, **66** (1949), :343-345.

[139] Olga Taussky, On a theorem of Latimer and MacDuffee. *Canad. J. Math.*, **1** (1949), 300–302.

[140] Olga Taussky, On matrix classes corresponding to an ideal and its inverse. *Illinois J. Math.*, **1** (1957), 108–113.

[141] Wilberd van der Kallen, Le K_2 des nombres duaux. *C. R. Acad. Sci. Paris Sér. A-B*, **273** (1971), A1204–A1207.

[142] J. B. Wagoner, Markov partitions and K_2. *Inst. Hautes Études Sci. Publ. Math.*, **65** (1987), :91–129.

[143] J. B. Wagoner, Eventual finite order generation for the kernel of the dimension group representation. *Trans. Amer. Math. Soc.*, **317** (1990), no. 1, 331–350.

[144] J. B. Wagoner, Higher-dimensional shift equivalence and strong shift equivalence are the same over the integers. *Proc. Amer. Math. Soc.*, **109** (1990), no. 2, 527–536.

[145] J. B. Wagoner, Triangle identities and symmetries of a subshift of finite type. *Pacific J. Math.*, **144** (1990), no. 1, 181–205.

[146] J. B. Wagoner, Strong shift equivalence theory and the shift equivalence problem. *Bull. Amer. Math. Soc. (N.S.)*, **36** (1999), no. 3, 271–296.

[147] J. B. Wagoner, Strong shift equivalence and K_2 of the dual numbers. *J. Reine Angew. Math.*, **521** (2000), 119–160. With an appendix by K. H. Kim and F. W. Roush.

[148] Charles Weibel, NK_0 and NK_1 of the groups C_4 and D_4. Addendum to "Lower algebraic K-theory of hyperbolic 3-simplex reflection groups" by J.-F. Lafont and I. J. Ortiz [mr2495796]. *Comment. Math. Helv.*, **84** (2009), no. 2, 339–349.

[149] Charles A. Weibel, *An introduction to homological algebra*, volume 38 of *Cambridge Studies in Advanced Mathematics*. Cambridge University Press, Cambridge, 1994.

[150] Charles A. Weibel, *The K-book: an Introduction to Algebraic K-theory*, volume 145 of *Graduate Studies in Mathematics*. American Mathematical Society, Providence, RI, 2013.

[151] J. H. C. Whitehead, Simple homotopy types. *Amer. J. Math.*, **72** (1950), 1–57.

[152] R. F. Williams, Classification of one dimensional attractors. In *Global Analysis (Proc. Sympos. Pure Math., Vol. XIV, Berkeley, Calif., 1968)*, pages 341–361. Amer. Math. Soc., Providence, R.I., 1970.

[153] R. F. Williams, Classification of subshifts of finite type. *Ann. Math.* (2), **98** (1973), 120–153; erratum, ibid. **99** (1974), 380–381.

[154] R. F. Williams, Strong shift equivalence of matrices in $GL(2, \mathbf{Z})$. In *Symbolic dynamics and its applications (New Haven, CT, 1991)*, volume 135 of *Contemp. Math.*, pages 445–451. Amer. Math. Soc., Providence, RI, 1992.

[155] Kitty Yang, Normal amenable subgroups of the automorphism group of shifts of finite type. *Ergodic Theory Dynam. Systems*, **41** (2021), no. 4, 1250-1263.

[156] Inna Zakharevich, Attitudes of K-theory: topological, algebraic, combinatorial. *Notices Amer. Math. Soc*, **66** (2019), 1034–1044.

Author Index

Subject Index